AF323321

Topologically Ordered Zigzag Nanoribbon

e/2 Fractionally Charged Anyons and
Spin-Charge Separation

Topologically Ordered Zigzag Nanoribbon

e/2 Fractionally Charged Anyons and
Spin-Charge Separation

S-R Eric Yang

Korea University, South Korea

World Scientific

NEW JERSEY · LONDON · SINGAPORE · BEIJING · SHANGHAI · HONG KONG · TAIPEI · CHENNAI · TOKYO

Published by

World Scientific Publishing Co. Pte. Ltd.

5 Toh Tuck Link, Singapore 596224

USA office: 27 Warren Street, Suite 401-402, Hackensack, NJ 07601

UK office: 57 Shelton Street, Covent Garden, London WC2H 9HE

Library of Congress Control Number: 2022943355

British Library Cataloguing-in-Publication Data
A catalogue record for this book is available from the British Library.

TOPOLOGICALLY ORDERED ZIGZAG NANORIBBON
e/2 Fractionally Charged Anyons and Spin-Charge Separation

ISBN 978-981-126-189-3 (hardcover)
ISBN 978-981-126-190-9 (ebook for institutions)
ISBN 978-981-126-191-6 (ebook for individuals)

For any available supplementary material, please visit
https://www.worldscientific.com/worldscibooks/10.1142/13013#t=suppl

Desk Editor: Nur Syarfeena Binte Mohd Fauzi

Typeset by Stallion Press
Email: enquiries@stallionpress.com

To my wife, Joo Yeon Shin

Preface

This book is about $e^-/2$ fractionally charged anyons and spin–charge separation in topologically ordered zigzag nanoribbons. Zigzag graphene nanoribbons are not some artificial material imagined by theoretical physicists. They are a real material with great potential for industrial applications. Moreover, their Hamiltonian is simple and their properties can be accurately described by the well-known and straightforward Hartree–Fock approximation. I show in this book that an interacting disordered zigzag graphene nanoribbon has order beyond symmetry description and is a new quasi-one-dimensional topologically ordered insulator exhibiting $e^-/2$ fractional charges. I will try to explain how randomness can lead to fractional charge quantization. Odd-denominator fractional charges are now well established in fractional quantum systems. However, states with $e^-/2$ fractional charges that are robust against quantum fluctuations are rare in condensed matter physics (it is not enough that the probability density of an electron state splits in half; topological order is needed to stabilize $e^-/2$ fractional charges). This book explains how the subtle interplay between topology of the underlying lattice, electron correlation and disorder gives rise to $e^-/2$ fractional charges and spin–charge separation. Since this book covers several other topological topologically ordered insulators, I hope it may be used in an introductory physics graduate course. However, it is by no means meant to be a definitive textbook on topologically ordered insulators.

Interacting disordered zigzag ribbons display rich physics of soliton, instanton, spin–charge separation, fractional charge, anyon, localization–delocalization transition, Mott–Anderson insulator, and topological entanglement entropy. But their properties can all be derived from the simple Hubbard model using the well-known Hartree–Fock method, provided that one judiciously chooses the right Hartree–Fock ground state Ansatz. The prerequisites for this book are undergraduate condensed matter physics, quantum mechanics, and mathematical physics (acquaintance with the commutation relations for fermions and bosons may be helpful). It is especially important that students have a solid understanding of the basic band structure theory. In my experience, many problems a student has during a course often stem from the holes in his knowledge of the prerequisites. I avoid, as much as possible, the use of advanced math, technical terms, and jargons (if a result is not derived I give a reference where the derivation may be found). I believe that the book would best be read progressively from Chapters 1–21.

My experience in teaching students over many years has led to adopt the following approach in this book. I believe conceptual understanding is very important in learning physics and having a lucid picture is also a fundamental priority. I have used numerous figures to elucidate basic concepts and ideas. In many derivations, I have tried to give the "simplest" proof and have sacrificed mathematical rigor over heuristic reasoning and intuitive understanding. (But remember what Niels Bohr said: truth and clarity are complementary.) I do not claim any originality for these simplified derivations and cite the relevant references. Also I refer to many articles for non-specialists written by some of the key players in the field. Students are strongly encouraged to read them. They are gems. Also I tried to cite as many papers as I can that provide good insights and accessible reading. These references are marked with •. In addition, many problems are given with hints or answers. These exercises contain derivations that are left out in the main text and are an integral part of learning. Moreover, whenever possible, I have deliberately tried to write short chapters to keep students' attention focused. Conceptually difficult and mathematically involved sections are marked with $\star$ and may be skipped at first reading. I have also provided materials for further reading and studying them will deepen conceptual understanding of the basic ideas. (Some of them are for aficionados only.)

This does not mean that this book is easy. In teaching classes, I noticed that, although students could follow mathematical derivations, they often found concepts presented in this book difficult to assimilate. A lot of hard work and persistence are needed to overcome conceptual barriers (no pain, no gain). Students should heed the following quote on learning physics by Feynman: "I just hope that I haven't caused a serious trouble to you, and that you do not leave this exciting business. I hope that someone else can teach it to you in a way that doesn't give you indigestion, and that you will find someday that, after all, it isn't as horrible as it looks." (It is from *The Feynman Lectures on Physics*, Vol. 3.) I believe a good way to learn physics is to study several different treatments of the same subject by different teachers. Chances are that one of the explanations may click in you and that you suddenly understand the whole business. I don't believe that there is a universally good explanation for everybody. Each person has her own unique and irrational mental blocks.

The field of topological matter has become very wide. In this book, I focus on a narrow subfield of fractional charges and topologically ordered phases of zigzag nanoribbons. I select topics that have relevance to it and leave out other interesting and important subjects. But as background material, I explain several experimentally well-studied topologically ordered insulators and set the stage for study of topologically ordered phases of zigzag nanoribbons. (I do not cover several other important theoretical models of topologically ordered insulators, and hope that someone else will cover them in a more definitive and advanced textbook on topologically ordered insulators.) If the reader wishes to gain a broader perspective of topological matter, I recommend many excellent review papers and books that are available now. A good historical account of topological matter is given by F. D. M. Haldane in *Rev. Mod. Phys.* **89**, 040502 (2017). I also recommend especially several chapters in *Modern Condensed Matter Physics* by S. M. Girvin and K. Yang, written for graduate students.

The book has a remarkable breadth and depth, and explains several difficult subjects with conceptual clarity. For an advanced and comprehensible review, see "Colloquium: Zoo of quantum-topological phases of matter", *Rev. Mod. Phys.* **89**, 041004 (2017) by X.-G. Wen. I also recommend numerous articles listed in Chapter 1, written by some of the key players of the field for non-specialists

The content of this book is based on physics that I learned from many people over the years. I thank Professor Allan H. MacDonald for introducing me to quantum Hall physics and exact diagonalization. I am also grateful to my Ph.D. supervisor Professor Lu J. Sham for teaching me condensed matter physics in general. My deepest gratitude to Professor Henrik Smith for teaching me many-body physics and superconductivity and for introducing me to Professor Lu J. Sham. I also thank Professor Jens L. Peterson for introducing me to quantum field theory.

I am also grateful to Professor Hyun Chul Lee (Sogang University) and Professor Min Chul Cha (Hanyang University) for collaboration over many years. Professor Hyun Chul Lee read the draft of this book and made many valuable suggestions, especially regarding Luttinger liquids and field theoretical aspects. Professor Hyun Woo Lee (POSTECH) carefully read the chapter on Majorana zero modes and made numerous helpful suggestions. I also thank Professor Hyun-Yong Lee (Korea University) for helpful comments on entanglement spectrum and the matrix product states representation of density matrix renormalization group. His density matrix renormalization group calculation to check the results obtained from the Hartree–Fock approximation is deeply appreciated. In addition, he read the entire chapter on matrix product states and made many useful comments. I also thank hardworking and dedicated students who worked on zigzag graphene nanoribbons in my group over the years, especially Lê Hoàng Anh, Young Heon Kim, Hye Jeong Lee and In Hwan Lee. They also checked many derivations and drew numerous figures presented in this book. In addition, I thank Jinmo Yang, a former student of Korea University, for drawing many figures in the book.

The reader is welcome to send errors (including typographical) and points of confusion to eyang812@gmail.com. For updates and corrections, consult my webpage: https://cond-mat.tistory.com.

I would also like to thank Ms. Ania Levinson, who was an editor of World Scientific Publishing Co., for contacting me after my talk that I write a book about $e^-/2$ fractional charges. It was a pleasant surprise as the idea had never occurred to me. Also my sincere thanks go to my desk editor Ms. Nur Syarfeena. My heartwarming thanks to Ms. Yubing Zhai, VP/Executive Editor, for approving this project and guiding me from start to finish. I also thank the Basic Science Research Program through National Research Foundation of Korea for continuous funding of my graphene research over the past decade.

S.-R. Eric Yang

Contents

Topologically Ordered Phases   191

PART 1

Basics

Chapter 1

Introduction

"... we live in an emergent universe in which ceaseless unforeseeable creativity arises and surrounds us. And since we can neither prestate, let alone predict, all that will happen, reason alone is insufficient guide to living our lives forward."

Stuart Alan Kauffman

There are two different phases of topological insulators [1–3]: **symmetry protected** and **topologically ordered** phases.[a] The symmetry protected topological phase is characterized by a short-ranged entanglement, whereas the topologically ordered phase displays a long-range entanglement. Furthermore, the symmetry protected topological phase protects boundary gapless states, and it cannot be adiabatically connected to a trivial product state under perturbations preserving a certain symmetry. By contrast, in the topologically ordered phase, the global pattern of the entanglement causes the topological ground state degeneracy, depending on space topology, which is robust under the local perturbation regardless of its symmetry. A topologically ordered insulator has a non-local order and cannot be described by the usual Ginzburg–Landau theory of local order parameter $\Delta(\vec{r})$ [4]. (What local means in this context is usually that $\Delta(\vec{r})$ at point $\vec{r}$ can be constructed by looking at a small neighborhood around the point $\vec{r}$.) A non-local order is the topological entanglement entropy β defined by the entanglement entropy of a region of size L

$$S_D = \alpha L - \beta. \tag{1.1}$$

---

[a]I suggest that a student reader first skim through this chapter and come back for a second read after studying the rest of the book.

The first term is due to the short-range correlations and is non-universal. The second term represents the sub-dominant and universal topological entanglement entropy [5,6]. Topological order often causes topological excitations that carry a fractional quantum number. Entanglement spectrum may also display signs of topologically ordered insulators [7]. The term topological order was coined by X.-G. Wen in connection with chiral spin states [8].

Let us give some examples of symmetry protected and topologically ordered insulators. Examples of symmetry protected topological phases are quantum spin Hall states [9–14], polyacetylene [15–17], Haldane spin chains [18], and disorder-free zigzag graphene nanoribbons [19]. (Gapped one-dimensional systems are matrix product states that have no topological order [20].) Symmetry protected topological insulators can support robust edge states. Some of the earliest topologically ordered states are chiral spin liquids which break time reversal and parity symmetries [8, 21, 22]. States with time reversal and parity symmetries are given in [23–26]. A good but less appreciated example of a topologically ordered insulator is superconductors, which cannot be described by a simple local order parameter. The reason is subtle because it is not possible to find a gauge invariant local order parameter in the presence of the electromagnetic gauge field. Instead a **gauge invariant** non-local order parameter of a superconductor must be defined [3, 27]. Other topologically ordered phases include spin liquids [21, 28], anyon superconductivity [29, 30], resonating valence bound states for spin systems [31], the Kitaev chain [32], and quantum Hall states [2, 33–39]. We will see in this book that each topologically ordered phase has its own unique properties and often requires new concepts for its proper description. A historical account of topological matter can be found in Ref. [40].

It is not obvious how to find a new topologically ordered insulator since symmetry principles cannot be applied.[b] But recently a new class of topologically ordered system has been found, namely, interacting disordered graphene nanoribbons. It was shown in Ref. [41] that $e^-/2$ fractional charges exist in disordered interacting zigzag graphene nanoribbons. The topological implications of this charge fractionalization were theoretically

---

[b]As far as I know, there is no universal theory of topologically ordered phases and each phase has its own unique properties. Metaphorically one could say "... every unhappy family is unhappy in its own way" (Leo Tolstoy, Anna Karenina, 1878). There is no systematic method for finding a new topologically ordered state.

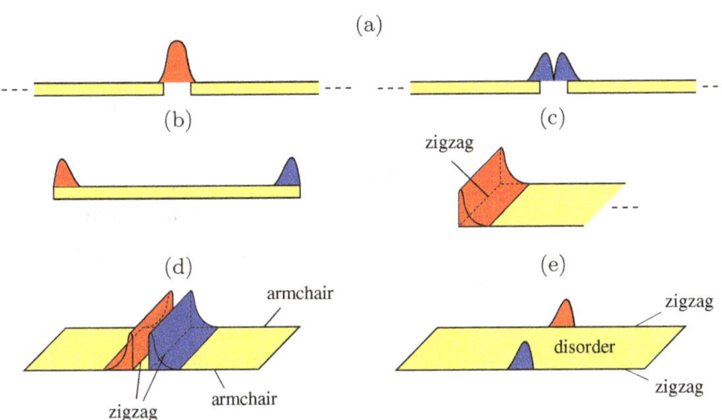

Fig. 1.1. The probability densities of a soliton on A and B carbon atoms are indicated by red and blue colors, respectively. (a) Chiral zero modes of polyacetylene: the domain-wall soliton (left) and antisoliton (right) have different chiralities. (b) End-state soliton of polyacetylene: this soliton can be represented by the bonding or antibonding linear combination of the chiral end states. (c) Chiral zero mode on zigzag edge of semi-infinite armchair graphene ribbon. (d) Domain-wall soliton of armchair graphene nanoribbon under local tensile strain. The probability densities on the A and B carbon atoms have a small overlap. (e) Charge fractionalization on zigzag edge of interacting disordered zigzag nanoribbon. The probability densities on the A and B carbon atoms are well separated and represent a non-local state. Reprinted with permission from Ref. [19].

investigated in Refs. [42,43]. In addition, the ground state of interacting disordered zigzag nanoribbons at **low doping** is expected to be a disordered anyon phase with unusual transport, magnetic, and inter-edge tunneling properties [44]. In the absence of disorder, a zigzag graphene nanoribbon is in a symmetry protected topological phase, but, in the presence of disorder, a zigzag ribbon is in a topologically ordered phase. Solitons appear in both of these systems (they are depicted in Fig. 1.1): In the symmetry protected phase of disorder-free zigzag graphene nanoribbons, they have an integer edge charge, but in a topologically ordered interacting disordered zigzag graphene nanoribbon well-defined soliton fractional charges appear. An excellent opportunity to observe these boundary charges has recently arisen, as rapid progress has been made in the fabrication of atomically precise graphene nanoribbons [45,46]. In the following, I give a brief overview of salient features of interacting disordered zigzag graphene nanoribbons and compare their properties with symmetry protected phases of polyacetylene and disorder-free zigzag nanoribbons.

1.1. Symmetry Protected Carbon-Based Systems

Polyacetylene and disorder-free zigzag graphene nanoribbons have particle–hole symmetry and both systems are in a symmetry protected topological phase. One can gain a deeper understanding of the symmetry protected topological phase of graphene nanoribbons by comparing it to that of polyacetylene. Polyacetylene and graphene nanoribbons have two inequivalent sublattices (here labeled A and B). Under the chiral operation Γ, the site annihilation operators transform according to $a_{iA} \to a_{iA}$ and $a_{iB} \to -a_{iB}$, where the index i denotes a unit cell containing one A and one B carbon atoms. The nearest-neighbor tight-binding Hamiltonians of polyacetylene and graphene nanoribbons satisfy the anticommutation relation $\{\Gamma, H\} = 0$.[c] Such a Hamiltonian can yield a massless Dirac equation, which describes well low-energy excitations. Single-particle eigenstates with non-zero eigenenergies E and $-E$ are related by the chiral operation $\Gamma|\psi_E\rangle = |\psi_{-E}\rangle$ and their probability densities are identical. Thus, chiral symmetry implies particle–hole symmetry. Of special interest are the single-particle eigenstates of the chiral operator, which represent chiral topological edge states (examples are shown in Figs. 1.1(b) and 1.1(c)).

- Zak Phase and Edge Charge

One can show that chiral symmetry leads to a boundary charge. It may be related to a bulk topological number called the Zak phase. The Zak phase of a one-dimensional band structure is a topological Berry phase acquired by an electron as it moves adiabatically through the first Brillouin zone [47]

$$Z_{1D} = \sum_{l \in occ} i \oint_{B.Z.} \langle u_{lk} | \nabla_k u_{lk} \rangle \, dk, \qquad (1.2)$$

where $|u_{lk}\rangle$ is the periodic part of the Bloch wave function of the lth band and the sum is over the occupied bands. Note that Z_{1D} is a **bulk** property determined by the band structure of a periodic system. Even when the chiral symmetry is broken, a one-dimensional periodic insulator with inversion/reflection symmetry has a quantized Zak phase: Z_{1D} equal to either 0 or π mod 2π [47] (the modular 2π is a consequence of the gauge invariance).

According to the modern theory of polarization, Z_{1D} is related to the bulk polarization P of a periodic system [48, 49]

$$P = \frac{e}{2\pi} Z_{1D}. \qquad (1.3)$$

---

[c]This symmetry is not exact when next the nearest-neighbor hopping is included.

(Note that, for one-dimensional polarization, P is defined as the dipole moment per length.) Consider an **insulating** edge of the finite-length system generated by cutting a periodic system. When quantum fluctuations are small, the Zak phase of the periodic system can be related to the magnitude of the edge charge Q of such a finite-length system [31]

$$Q = \vec{P} \cdot \hat{n} = \frac{e}{2\pi} Z_{1D}. \tag{1.4}$$

(When the edge charge is non-integer this result may not be applied; this will be discussed in Sec. 17.3.) This is an example of a bulk-edge correspondence. Note that Q is located at the edge and the direction of $\hat{n}$ is perpendicular to the edge. As Z_{1D} is multi-valued, Eq. (1.4) gives possible values of Q only. The actual value of Q must be computed with consideration of the coupling between the edge and bulk. In some cases, a boundary edge can affect the bulk property: In a zigzag graphene nanoribbon a disorder or staggered potential induces spin-splitting, see Ref. [50]. This does not happen in armchair ribbons. (Note that in symmetry protected topological insulators quantum fluctuations between quasi-degenerate states may be significant and the result of Eq. (1.4) may not be applicable. Polyacetylene is an example, as we will argue below.)

- Polyacetylene

An object of one-dimensional insulators qualifies as a soliton if it derives half its fractional spectral weight from each of the conduction and valence bands. Jackiw and Rebbi [16] showed that a twist in the scalar mass potential in the Dirac equation, connecting two degenerate ground states of polyacetylene, can produce a zero-energy soliton state. Furthermore, the independent work of Su, Schrieffer, and Heeger [15, 17] showed the presence of a soliton (kink) in polyacetylene. This kink connects two different dimerized phases and is called a domain-wall soliton (see Fig. 1.1(a)). A domain wall supports either a soliton or an antisoliton, but not both.

Polyacetylene with the solitonic state unoccupied by an electron has an unusual charge q and spin s relation: $q = e$ and $s = 0$ (here, the contribution from the positive ion background is included; $e > 0$ is the elementary charge). On the other hand, when the solitonic state is occupied, the following values are obtained: $q = 0$ and $s = 1/2$. A soliton may exist with fractional **boundary** charges at the two endpoints of a finite-length polyacetylene (one for each end), as shown in Figs. 1.1(b) and 1.2. The Zak phase of polyacetylene Z_{1D} is equal to 0 or π mod 2π, depending on the choice of the unit cell. From the relation between the

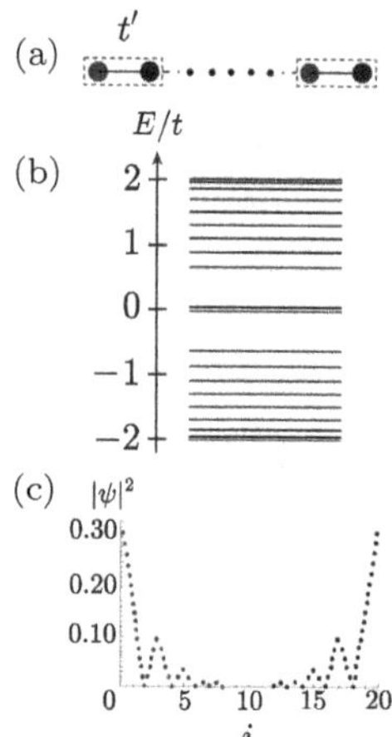

Fig. 1.2. (a) Finite-length dimer chain with unit cell containing two carbon atoms connected by single bond. The intra cell hopping t' is smaller than the inter cell hopping t. (b) Tight-binding energy spectrum. Two nearly degenerate gap states exist. (c) Probability density of a gap state as function of site index i. Peaks are present at the left and right end sites. The probability densities of the bonding and antibonding states are almost identical. Reprinted with permission from Ref. [42].

Zak phase and edge charge, Eq. (1.4), one finds an integer or a half-integer edge charge. However, an accurate charge fractionalization has not been observed in polyacetylene. The splitting of the probability density of an electronic state does not guarantee a robust charge fractionalization. This is because quantum charge fluctuations between quasi-degenerate states may be significant. Polyacetylene displays non-local domain wall excitations. Despite this, polyacetylene does not display topological order.

- Disorder-Free Graphene Nanoribbon

In carbon-based systems, the formation of solitons is usually based on the presence of different dimerization phases. However, Jeong *et al.* showed that a soliton can form in the absence of dimerization when zigzag edges are present [41]. A soliton in a zigzag graphene nanoribbon connects an A-chiality zigzag edge to a B-chiality zigzag edge. (We will call such a state a mixed chiral state. It is analogous to a soliton of polyacetylene that connects two different dimerized phases). A domain-wall soliton can also exist in a graphene nanoribbon [41] (see Fig. 1.1(d)). A single domain wall in a graphene nanoribbon can support both a soliton and antisoliton, in contrast to polyacetylene. Consequently, no unusual spin and charge relation holds.

The possible values of the zigzag edge charge of a **rectangular** armchair graphene nanoribbon[d] may also be computed using the Z_{1D} of the periodic armchair graphene nanoribbon: one finds $Z_{1D} = 2\pi N$, where N is an integer [50, 51]. According to Eq. (1.4), the possible values of the edge charge is an **integer** with $Q = eN$, which agrees with the result obtained from **antiferromagnetic** coupling between the opposite zigzag edges [52].

1.2. Topologically Ordered Phase of Zigzag Nanoribbon

The Hamiltonian of interacting disordered zigzag ribbons is given by the simple Hubbard model with the nearest-neighbor hopping term and with on-site repulsion. However, as I mentioned in the preface, it contains rich physics of soliton, spin–charge separation, fractional charge, anyon, localization–delocalization transition, Mott–Anderson insulator, and topological entanglement entropy. These are all non-perturbative effects. One does not need sophisticated numerical methods or abstruse mathematical approaches to understand them. We will see in this book that the simple Hartree–Fock approximation can account for these diverse phenomena. Below we briefly explain the properties of topologically ordered of zigzag nanoribbons.

- Edge states of a topological insulator are usually not signficantly affected by a disorder potential, but the symmetry protected topological phase of a zigzag graphene nanoribbon is profoundly changed by it. A disorder potential behaves similarly to a **singular** perturbation on zigzag edge electronic states. This is analogous to the singular coupling between the two wells of a quantum double well, which generates an instanton. In zigzag ribbons, it connects A- and B-sublattices, analogous to a soliton in polyacetylene that connects two different dimerized phases. As a result, disorder partly mitigates the effect of antiferromagnetic coupling between the opposite zigzag edges and generates numerous gap states (instantons) with $e^-/2$ **fractional boundary charges** on the opposite zigzag edges, i.e., one for each edge [41], as shown in Figs. 1.1(e) and 1.3 (zigzag edges are located along the ribbon direction). They are a mixed chiral state and represent a non-local state. Actually, a random potential generates a continuous charge values between 0 and e^-. The states near the midgap energy display $e^-/2$ fractional anyon charges. Note that

---

[d]A rectangular armchair graphene nanoribbon has two long armchair edges and two short zigzag edges.

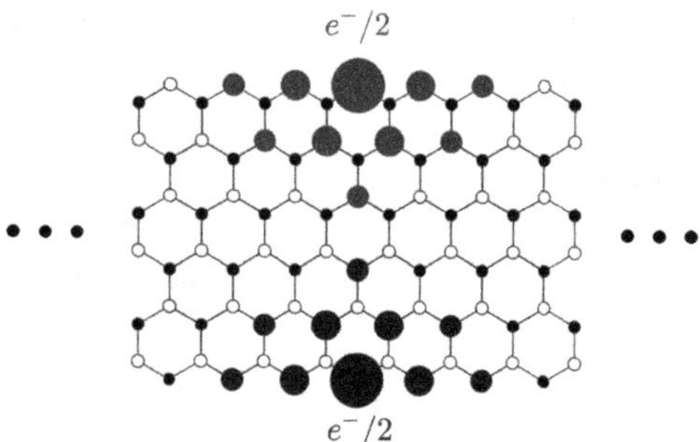

Fig. 1.3. Two non-local $e^-/2$ charges residing on the zigzag edges of a interacting disordered zigzag graphene nanoribbon.

numerous fractional charges are present in the ground state. The proliferation of fractional charges is related to the fact that the number of gap-edge states is proportional to the ribbon length. This is in contrast to the ground state of an odd-denominator fractional quantum Hall state which does not contain fractional charges (only the quasiparticle excitations have fractional charges). The gapless edge modes of fractional quantum Hall states are a bulk property. However, as mentioned in Sec. 1.1, edge properties of a disordered interacting zigzag ribbon are not necessarily so because zigzag edges induce a bulk spin-splitting, as opposed to the armchair edges.

- Zigzag edges profoundly affect the properties of the ribbon. Interacting disordered zigzag graphene nanoribbons represent a Mott–Anderson insulator with **spin-split** gap-edge states and localized **magnetic moments** residing on the zigzag edges. Spin-splitting is a non-perturbative effect and the Hartree–Fock Ansatz wave function must incorporate this feature.

- In the presence of disorder, the Mott–Hubbard gap Δ is transformed into an exponentially decaying soft gap of size Δ_s [42], see Fig. 1.5. The electronic band structure of zigzag graphene nanoribbons with a gap is displayed in Figs. 1.4 (b) and 1.5 in the presence/absence of disorder and on-site repulsion. The new gap will protect the fractional charges from quantum fluctuations.

- The zigzag edge states decay exponentially perpendicular and parallel to the zigzag edges. The gap states are localized whereas the states outside the gap are delocalized, and the usual localization theory does not apply.

- Also spin–charge separation can take place in the presence of disorder [42]. The origin of this phenomenon is rather similar to that of spin–charge separation in polyacetylene. The presence of fractional edge

(a)

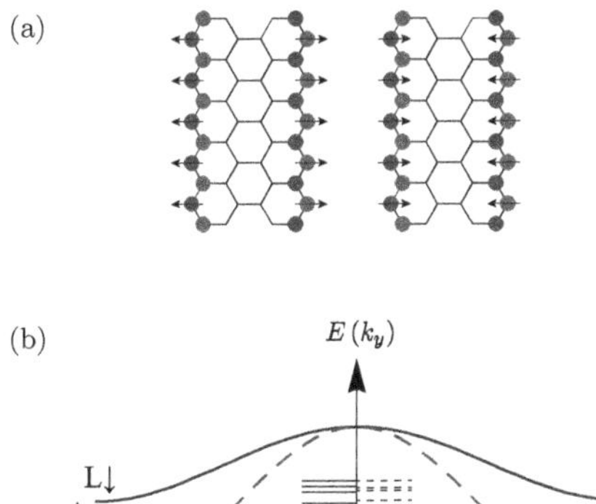

(b)

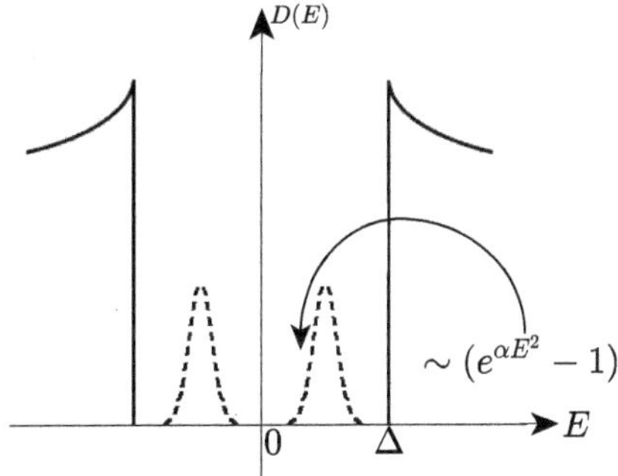

Fig. 1.4. (a) Zigzag edge antiferromagnetism of an interacting zigzag graphene nanoribbon without disorder showing the two degenerate ground states. (b) Schematic of interacting (solid curves) and non-interacting (dashed curves) zigzag graphene nanoribbon band structures. The unoccupied and occupied states near the wave vectors $k = \pm\pi/a_0$ are shown: R and L represent states confined on the right and left zigzag edges, respectively ($a_0 = 1.73a$ is the unit cell length of the zigzag graphene nanoribbon and $a = 1.42\text{Å}$ is the C–C distance). The small arrows indicate spins. Disorder induces spin-split gap states, shown as short horizontal lines, solid lines (dashed lines) represent the spin-up (-down) states. Reprinted with permission from Ref. [42].

Fig. 1.5. Dashed line: the exponentially small *soft* gap with $\alpha \sim 1/\sqrt{\Delta_s}$ near the Fermi energy of the tunneling density of states of a half-filled interacting disordered zigzag graphene nanoribbon in the weak disorder regime. Solid line: the *hard* gap, Δ, of the tunneling density of states of an interacting zigzag graphene nanoribbon in the absence of disorder. Reprinted with permission from Ref. [43]

charges and spin–charge separation leads to a magnetic reconstruction of boundary zigzag edges.

- Interacting disordered zigzag graphene nanoribbons exhibit two-fold ground state degeneracy [42], as depicted in Fig. 1.4(a).

- The matrix product state representation of density matrix renormalization group shows that no topological order exists in gapped one-dimensional systems [20]. One might thus think that the presence of a gap may be compatible with zero topological entanglement entropy. However, this is not necessarily so in the presence of an exponentially decaying soft gap. The topological entanglement entropy is found to be finite and universal [43], i.e., disorder induces a phase change from a symmetry protected topological phase to a **topologically ordered** phase (this is a topological phase transition). The ground state of an interacting disordered zigzag nanoribbon can be treated the matrix product state representation of density matrix renormalization group [44]. The results of this approach show that the Hartree–Fock approximation provides an excellent description of disordered zigzag nanoribbons.

- There is a significant difference in the entanglement spectrum [7] between symmetry protected and topologically ordered zigzag graphene nanoribbons: degeneracy near zero eigenvalues of the reduced density matrix is split in the topologically ordered phase and the shape of the entanglement spectrum is similar to the DOS of the edge states [44].

- The ground state of a doped disorder-free ribbon displays an edge spin density wave, in contrast to the uniform spin density of undoped ribbons. When extra electrons are inserted into an interacting disordered zigzag nanoribbon, those electrons divide into anyons. As doping level increases, the soft gap of the tunneling density of states in the undoped case is replaced by a sharp peak at the midgap energy with two accompanying peaks. The $e^-/2$ fractional charges that reside on the boundary of the zigzag edges are responsible for these peaks. The midgap peak disappears as the doping level increases. The detection of these peaks will provide strong evidence for the presence of $e^-/2$ fractional charges. Doped zigzag ribbons may display unusual transport, magnetic, and inter-edge tunneling properties. The ground state of interacting disordered zigzag nanoribbons at **low doping** represents a new disordered anyon phase [44].

It is instructive to compare the key features of topologically ordered of zigzag nanoribbons and fractional quantum Hall states, see Table 1.1.

Table 1.1. Comparison between properties of topologically ordered of zigzag nanoribbons and the Laughlin $1/m$ fractional quantum Hall states (m is an odd integer). The edge states of topologically ordered of zigzag nanoribbons are non-propagating, unlike those of fractional quantum Hall states. The DOS of a ribbon develops a soft gap that is exponentially suppressed. Moreover, ribbons do not display the quantized Hall effect. In doped systems fractional quasiparticles are present in both systems and their correlations may lead to new phases.

	Disordered zigzag ribbon	$\frac{1}{m}$ Fractional quantum Hall state
ground-state degeneracy	2-fold degeneracy	m-fold degeneracy
fractional charge	$\frac{e^-}{2}$ fractional charge	$\frac{e^-}{m}$ fractional charge
correlation between quasiparticles	distorted edge spin density wave	hierachial quantum Hall states
edge excitation	exponentially suppressed excitations	gapless chiral excitations
gap	exponentially suppressed soft gap	hard gap

1.3. Particle Fractionalization in Other Systems

Kitaev's chain [32], polyacetylene, and interacting disordered zigzag graphene nanoribbons all have end states. There are similarities and differences between them [19]. It is both interesting and instructive to take note of them.

Chiral symmetry guarantees existence of edge states in zigzag graphene nanoribbons and polyacetylene. However, in the presence of disorder there is an important difference between zigzag graphene nanoribbons and polyacetylene. Let us consider the Su–Schrieffer–Heeger effect in detail [15]. Consider finite-length polyacetylene in one of the dimerized phases, see Fig. 1.2. The electron density is uniform with occupation number $n_i = 1$ at all sites i. Here two nearly degenerate soliton end states exist (bonding and antibonding states); one will be occupied and the other unoccupied. Disorder will split these states because their wave functions are somewhat different. Then two possibilities are present. (a) The boundary fractional charges will suffer quantum charge fluctuations because the energy splitting is small (only a large energy splitting will suppress quantum charge fluctuations, see an insightful discussion by Girvin [53]). (b) Disorder will most likely not couple the left and right ends equally. Hence the boundary charge will not be exactly $e^-/2$. There is as yet no conclusive experimental evidence for fractional charges in polyacetylene (but spin–charge separation was observed).

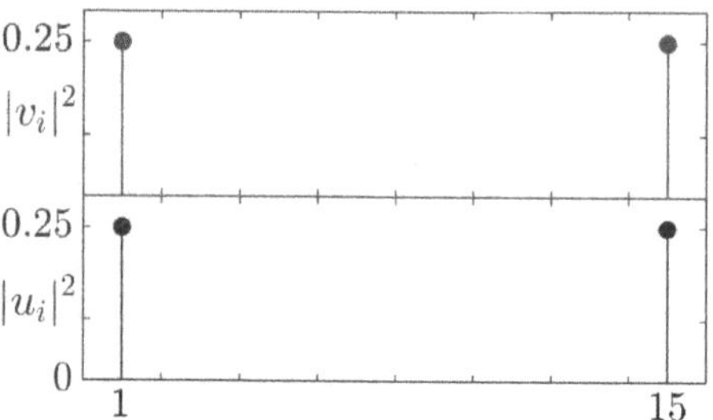

Fig. 1.6. Kitaev's chain has two degenerate zero energy states. They can be combined to give one bonding and one antibonding states. For each of these states, the probability to find an electron (hole) at site $i = 1, \ldots 15$ is $|u_i|^2$ ($|v_i|^2$). We have set the hopping parameter equal to the value of the gap $t = \Delta$ and the chemical potential $\mu = 0$. A Majorana zero mode (half a real fermion mode) at each end of the chain is displayed. Reprinted with permission from Ref. [42].

Consider Kitaev's toy model of one-dimensional p-wave superconductivity, which has relevance to topological superconductors [54]. Particle–hole symmetry plus bulk edge correspondence guarantees the presence of zero energy modes. It exhibits a charge **neutral** particle that is divided between the two ends of the chain. These Majorana zero modes are expected to display non-Abelian statistics. Figure 1.6 displays such zero modes of a finite length chain. (This is a well-known result and we show it here just for comparison with polyacetylene and zigzag graphene nanoribbons). However, it should be noted that these end states do not constitute a new particle and they are not a Majorana fermion. (One-dimensional p-wave superconductors are an example of an inverted topologically ordered insulator, like the integer quantum Hall state.)

Now let us discuss edge states of a zigzag graphene nanoribbon. The probability density of the midgap states is fractionalized equally between the left and right zigzag edges [41]. It is similar to fractionalization occurring at the endpoints of polyacetylene [15] and Kitaev's chain [32]. However, in interacting disordered zigzag graphene nanoribbons fractional charges reside on the zigzag edges that form the **side** boundary of the ribbon, see Fig. 1.1. In addition, the presence of a gap Δ_s protects the fractional charges against quantum charge fluctuations. These fractional charges are Abelian anyons.

1.4. Organization of the Book

This book is organized as follows. In Part 1, consisting of Chapters 2–7, I will go through basic concepts and techniques that are needed to study topological insulators. These are prerequisites and students should study all of them carefully. In Chapters 2 and 3, I will first briefly review geometric Berry phases and then explain basic topological properties of simple electron systems. They will introduce basic concepts such as the Berry curvature, the Chern number, non-Abelian matrix Berry phase, Aharonov–Bohm phase, and Abelian anyons. Then in Chapter 4, a short account of polyacetylene exhibiting solitons is presented. In Chapter 5, the anomalous velocity of a band structure is discussed. Here the connection between the Berry phase of an electronic band structure (Zak phase) and polarization/boundary charge is explained. Also we will discuss the connection between the anomalous velocity and the off-diagonal conductivity given by the Chern number. These chapters provide background materials for understanding symmetry protected topological disorder-free zigzag graphene nanoribbons. (However, these chapters are not meant to be a comprehensive introduction to symmetry protected topological insulators. Only minimum material needed to understand zigzag ribbons is included since symmetry protected topological insulators are not the main subject of this book.) Before we move on to topologically ordered phases, I give a short introduction to many-body physics in Chapters 6–7.

Chapters 8–13 constitute Part 2, where I will review basic properties of several topologically ordered insulators. First I explain two phases that have been thoroughly explored experimentally, namely quantum Hall states and superconductivity. Then, I will discuss a spin liquid system called the toric code which can be solved analytically. These chapters provide some background materials for understanding topologically ordered phase of disordered interacting zigzag graphene nanoribbons. They are given to provide a broader perspective. (In my search for fractional charges in interacting disordered zigzag graphene nanoribbons, I was especially inspired by the physics of quantum Hall effects and chiral Luttinger liquid.)

In Part 3, consisting of Chapters 14–21, I will argue that graphene nanoribbons are a topologically ordered insulator. In Chapter 14, I explain some basic properties of graphene. The quantum anomalous Hall effect is described in Chapter 15, using a model Hamiltonian of graphene introduced by Haldane. In Chapters 16–17, I will describe disorder-free zigzag nanoribbons from the perspective of symmetry protected topological insulator. We need to understand symmetry protected topological zigzag nanoribbons because they will act as a springboard for studying topologically ordered interacting disordered zigzag nanoribbons. In the rest of the book, Chapters 18–21, I discuss topologically ordered phase of disordered

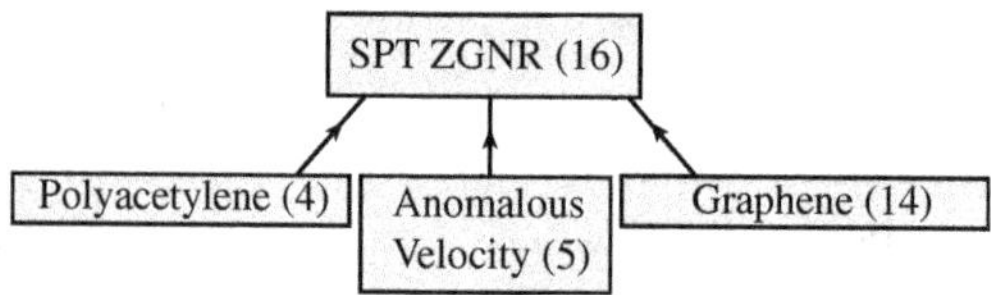

Fig. 1.7. In this book, to give a deeper understanding, some background materials are provided before I explain symmetry protected topological zigzag graphene nanoribbons. Numbers in the boxes indicate the relevant chapters in the book.

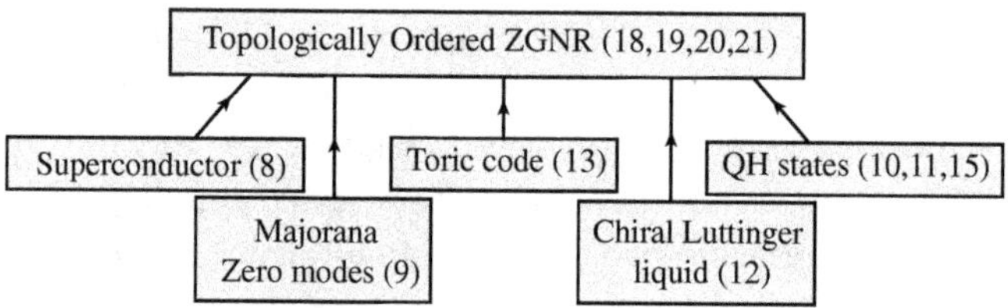

Fig. 1.8. The reader is suggested to read some background materials before studying topologically ordered interacting disordered zigzag graphene nanoribbons.

interacting zigzag graphene nanoribbons. The organization of this book is presented graphically in Figs. 1.7 and 1.8.

Bibliography

[1] X.-G. Wen, Colloquium: Zoo of quantum-topological phases of matter, *Rev. Mod. Phys.* **89**, 041004 (2017).

[2] X.-G. Wen, Topological order: From long-range entangled quantum matter to a unified origin of light and electrons, *ISRN Conden. Matter Phys.* **2013**, 198710 (2013). •

[3] S. M. Girvin and K. Yang, *Modern Condensed Matter Physics* (Cambridge University Press, Cambridge. 2019). •

[4] N. Goldenfeld, *Lectures on Phase Transitions and the Renormalization Group* (CRC Press, Boca Raton, 1992). •

[5] A. Y. Kitaev and J. Preskill, Topological entanglement entropy, *Phys. Rev. Lett.* **96**, 110404 (2006).

[6] M. Levin and X.-G. Wen, Detecting topological order in a ground state wave function, *Phys. Rev. Lett.* **96**, 110405 (2006).

[7] H. Li and F. D. M. Haldane, Entanglement spectrum as a generalization of entanglement entropy: identification of topological order in non-abelian fractional quantum Hall effect states, *Phys. Rev. Lett.* **101**, 010504 (2008).

[8] X.-G. Wen, Vacuum degeneracy of chiral spin state in compactified spaces, *Phys. Rev. B* **40**, 7387 (1989); X.-G. Wen, Topological order in rigid states, *Int. J. Mod. Phys. B* **4**, 239 (1990).

[9] C. L. Kane and E. J. Mele, Quantum spin Hall effect in graphene, *Phys. Rev. Lett.* **95**, 226801 (2005).

[10] C. L. Kane and J. Moore, Topological insulators, *Phys. World* **24**, 2, 32 (2011). •

[11] B. A. Bernevig and S.-C. Zhang, Quantum spin Hall effect, *Phys. Rev. Lett.* **96**, 106802 (2006).

[12] B. A. Bernevig, T. L. Hughes and S.-C. Zhang, Quantum spin Hall effect and topological phase transition in HgTe quantum wells, *Science* **314**, 1757 (2006).

[13] M. König, S. Wiedmann, C. Brüne, A. Roth, H. Buhmann, L. W. Molenkamp, X.-L. Qi, and S.-C. Zhang, Quantum spin Hall insulator state in HgTe quantum wells, *Science* **318**, 766 (2007).

[14] X.-L. Qi and S.-C. Zhang, The Quantum spin Hall effect and topological insulators, *Phys. Today* **63**, 1, 33 (2010). •

[15] W. P. Su, J. R. Schrieffer, and A. J. Heeger, Solitons in polyacetylene, *Phys. Rev. Lett.* **42**, 1698 (1979). •

[16] R. Jackiw and C. Rebbi, Solitons with fermion number 1/2, *Phys. Rev. D* **13**, 3398 (1976).

[17] A. J. Heeger, S. Kivelson, J. R. Schrieffer, and W. P. Su, Solitons in conducting polymers, *Rev. Mod. Phys.* **60**, 781 (1988).

[18] F. D. M. Haldane, O(3) Nonlinear σ model and the topological distinction between integer–and half–integer-spin antiferromagnets in two dimensions, *Phys. Rev. Lett.* **61**, 1029 (1988).

[19] S.-R. Eric Yang, Soliton fractional charges in graphene nanoribbon and polyacetylene: Similarities and differences, *Nanomaterials* **9**, 885 (2019). •

[20] M. B. Hastings, An area law for one-dimensional quantum systems, *J. Stat. Mech.: Theory Exp.* P08024 (2007).

[21] V. Kalmeyer and R. B. Laughlin, Equivalence of the resonating–valence–bond and fractional quantum Hall states, *Phys. Rev. Lett.* **59**, 2095 (1987).

[22] X.-G. Wen, F. Wilczek, and A. Zee, Chiral spin states and superconductivity, *Phys. Rev. B* **39**, 11413 (1989).

[23] G. Misguich, C. Lhuillier, B. Bernu, and C. Waldtmann, Spin-liquid phase of the multiple-spin exchange Hamiltonian on the triangular lattice, *Phys. Rev. B* **60**, 1064 (1999).

[24] R. Moessner and S. L. Sondhi, Resonating valence bond phase in the triangular lattice quantum dimer model, *Phys. Rev. Lett.* **86**, 1881 (2001).

[25] N. Read and Subir Sachdev, Large-N expansion for frustrated quantum antiferromagnets, *Phys. Rev. Lett.* **66**, 1773 (1991).

[26] X.-G. Wen, Mean-field theory of spin-liquid states with finite energy gap and topological orders, *Phys. Rev. B* **44**, 2664 (1991).

[27] T. H. Hansson, V. Organesyan, and S. L. Sondhi, Superconductors are topologically ordered, *Ann. Phys.* **313**, 497 (2004).

[28] J. Knolle and R. Moessner, A field guide to spin liquids, *Annu. Rev. Condens. Matter Phys.* **10**, 451 (2019).

[29] Y. H. Chen, F. Wilczek, E. Witten, and B. I. Halperin, On anyon superconductivity, *Int. J. Mod. Phys. B* **3**, 1001 (1989).

[30] A. Fetter, C. Hanna, and R. B. Laughlin, Random-phase approximation in the fractional-statistics gas, *Phys. Rev. B* **39**, 9679 (1989).

[31] S. A. Kivelson, D. S. Rokhsar, and J. P. Sethna, Topology of the resonating valence-bond state: Solitons and high-T_c superconductivity, *Phys. Rev. B* **35**, 8865 (1987).

[32] A. Y. Kitaev, Unpaired Majorana fermions in quantum wires, *Phys.–Usp* **44**, 131 (2001).

[33] K. v. Klitzing, G. Dorda, and M. Pepper, New method for high-accuracy determination of the fine-structure constant based on quantized Hall resistance, *Phys. Rev. Lett.* **45**, 494 (1980).

[34] B. I. Halperin, Quantized Hall conductance, current-carrying edge states, and the existence of extended states in a two-dimensional disordered potential, *Phys. Rev. B* **25**, 2185 (1992). •

[35] D. Thouless, M. Kohmoto, M. Nightingale, and M. den Nijs, Quantized Hall conductance in a two-dimensional periodic potential, *Phys. Rev. Lett.* **49**, 405 (1982).

[36] R. B. Laughlin, Anomalous quantum Hall effect: an incompressible quantum fluid with fractionally charged excitations, *Phys. Rev. Lett.* **50**, 1395 (1983).

[37] G. Moore and N. Read, Nonabelions in the fractional quantum Hall effect, *Nucl. Phys. B* **360**, 362 (1991).

[38] N. Read and D. Green, Paired states of fermions in two dimensions with breaking of parity and time-reversal symmetries and the fractional quantum Hall effect, *Phys. Rev. B* **61**, 10267 (2000).

[39] N. Read, Topological phases and quasiparticle braiding, *Phys. Today* **65**, 7, 38 (2012). •

[40] F. D. M. Haldane, Nobel lecture: topological quantum matter, *Rev. Mod. Phys.* **89**, 040502 (2017). •

[41] Y. H. Jeong, S.-R. Eric Yang, and M. C. Cha, Soliton fractional charge of disordered graphene nanoribbon, *J. Phys.: Condens. Matter* **31**, 265601 (2019).

[42] S.-R. Eric Yang, M. C. Cha, H. J. Lee, and Y. H. Kim, Topologically ordered zigzag nanoribbon: $e/2$ fractional edge charge, spin–charge separation, and ground–state degeneracy, *Phys. Rev. Res.* **2**, 033109 (2020).

[43] Y. H. Kim, H. J. Lee and S.-R. Eric Yang, Topological entanglement entropy of interacting disordered zigzag graphene ribbons, *Phys. Rev. B* **103**, 115151 (2021).

[44] Y. H. Kim, H. J. Lee, Hyun-Yong Lee, and S.-R. Eric Yang, New disordered anyon phase of doped graphene zigzag nanoribbon, *Sci. Rep.* **12**, 14551 (2022).

[45] P. Ruffieux, S. Wang, B. Yang, C. Sanchez-Sanchez, J. Liu, T. Dienel, L. Talirz, P. Shinde, C. A. Pignedoli, and D. Passerone, On-surface synthesis of graphene nanoribbons with zigzag edge topology, *Nature* **531**, 489 (2016).

[46] M. Kolmer, A.-K. Steiner, I. Izydorczyk, W. Ko, M. Engelund, M. Szymonski, A.-P. Li, and K. Amsharov, Rational synthesis of atomically precise graphene nanoribbons directly on metal oxide surfaces, *Science* **369**, 571 (2020).

[47] J. Zak, Berry's Phase for energy bands in solids, *Phys. Rev. Lett.* **62**, 2747 (1989).

[48] R. Resta, Polarization as a Berry phase, *Europhys. News* **28**, 18 (1997). •

[49] D. Vanderbilt, *Berry Phases in Electronic Structure Theory* (Cambridge University Press, Cambridge, 2018). •

[50] Y. H. Jeong and S.-R. Eric Yang, Topological end states and Zak phase of rectangular armchair ribbon, *Ann. Phys.* **385**, 688 (2017).

[51] Y. H. Jeong, S. C. Kim, and S.-R. Eric Yang, Topological gap states of semiconducting armchair graphene ribbons, *Phys. Rev. B* **91**, 205441 (2015).

[52] M. Fujita, K. Wakabayashi, K. Nakada, and K. Kusakabe, Peculiar localized state at zigzag graphite edge, *J. Phys. Soc. Jpn.* **65**, 1920 (1996).

[53] S. M. Girvin, The quantum Hall effect: Novel excitations and broken Symmetries, in *Les Houches Lecture Notes: Topological Aspects of Low Dimensional Systems*, edited by A. Comtet, T. Joliceur, S. Ouvry, and F. David (Springer Verlag, Berlin and Les Editions de Physique, Les Ulis, 2000). •

[54] C. W. J. Beenakker, Search for Majorana fermions in superconductors, *Annu. Rev. Condens. Matter Phys.* **4**, 113 (2013). •

Chapter 2

Geometric Berry Phase and Chern Number

"Analogy as the fuel and fire of thinking."

Douglas Hofstadter and Emmanuel Sander [1]

"A mathematician is a person who can find analogies between theorems; a better mathematician is one who can see analogies between proofs, and the best mathematician can notice analogies between theories. One can imagine that the ultimate mathematician is one who can see analogies between analogies."

Stefan Banach

"... most of our ordinary conceptual system is metaphorical in nature.[a]"

George Lakoff and Mark Johnson

A geometric phase is related to the curvature of the relevant manifold of the physical system of our interest. Often it is not a manifold in real space but in the parameter space of the system. Geometric phases can be explained using the language of gauge theory, and there is a useful analogy between electromagnetism and geometric Berry phase, see Table 2.1. Two gauge theories are relevant in this connection: Abelian and non-Abelian theories which give rise to complex phase factors and matrix phases, respectively. They may be applied to both symmetry protected topological phases and topologically ordered phases.

---

[a] A metaphor is often poetically saying something is something else. An analogy is saying something is like something else to make some sort of an explanatory point (Written by MasterClass).

Table 2.1. Well-known analogy between electromagnetism and geometric Berry phase.

Electromagnetism	Geometric Berry phase
Vector potential	Berry connection
Magnetic field	Berry curvature
Magnetic monopole	Degenerate point
Magnetic flux	Berry phase

How does a gauge theory for geometric phases arise? Suppose the Hamiltonian depends on some physical parameters $\vec{R} = (R_1, \ldots, R_N)$, and suppose that one of its quantum state changes adiabatically as $\vec{R}$ follows a closed path C in the parameter space. The phase change after a cycle is

$$\Delta\gamma = \oint_C \vec{A}(\vec{R}) \cdot d\vec{R}. \tag{2.1}$$

It is determined by some abstract vector potential $\vec{A}(\vec{R})$ in the parameter space. This phase is called the Berry phase. Although the phase change is unique the vector potential is **not unique**: any vector potential

$$\vec{A}(\vec{R}) \to \vec{A}(\vec{R}) - \nabla_{\vec{R}}\theta(\vec{R}) \tag{2.2}$$

can be used as long as $\theta(\vec{R})$ is single-valued. So gauge degree of freedom appears! For an introduction to the Berry phase for layperson, see [2] written by Berry.

2.1. Classical Geometric Phase and Gauss-Bonnet Theorem

The Foucault pendulum was introduced in 1851. It was the first experiment to give simple, direct evidence of the earth's rotation. Let us see how it does that. Using Newton's laws in the Earth-bound coordinate system of Fig. 2.1, we can describe the motion of the pendulum bob as follows [3]

$$\frac{d^2x}{dt^2} = -\omega^2 x + 2\Omega\frac{dy}{dt}\sin\theta,$$

$$\frac{d^2y}{dt^2} = -\omega^2 y - 2\Omega\frac{dx}{dt}\sin\theta. \tag{2.3}$$

The origin of the coordinate system is at the supporting point of the pendulum and its position is fixed in the Earth-bound coordinate system. The terms containing the frequency of the bob ω originate from the restoring

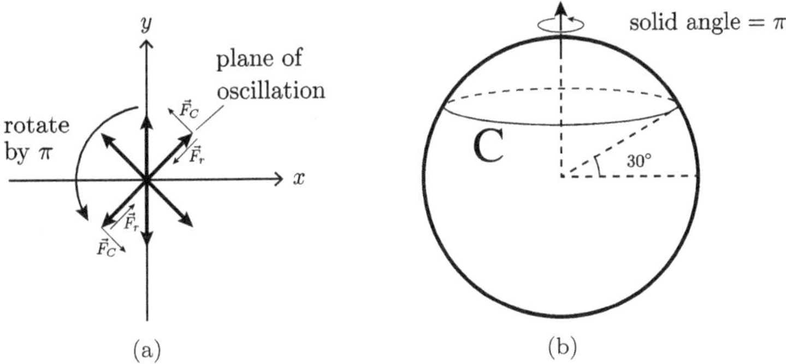

Fig. 2.1. (a) The vectors indicate the plane of oscillation of the Foucault pendulum located at the origin of the coordinate system. The coordinate system rotates with earth and the y-axis always points north. Note that the coordinate system is in the tangent plane of the point where the pendulum is located. At the latitude $\theta = 30°$ the plane of oscillation is rotated by $180°$ in one day. The Coriolis force $\vec{F}_C$ and restoring force $\vec{F}_r$ are shown. (b) The circle represents the pendulum position at $\theta = 30°$ seen from an observer out in space. The solid angle subtended by this circle is also $180°$.

force $\vec{F}_r$ (see Fig. 2.1(a); it is a combined effect of the gravitational force and string tension acting on the bob of the pendulum). The terms containing the rotational frequency of Earth Ω represent the Coriolis force $\vec{F}_C = -2m\vec{\Omega} \times \vec{v}$, where $\vec{v}$ is the velocity of the bob, see Fig. 2.1(a). The latitude of the supporting point of the pendulum is θ. (Here small oscillations are assumed and only the z-component $\Omega_z = \Omega \sin \theta$ is important. Other components may be ignored since they give rise to a Coriolis force that is small compared to the gravitational force.)

Changing to complex coordinates $z = x + iy$ and assuming $\Omega/\omega \ll 1$ one finds

$$\frac{d^2 z}{dt^2} + 2i\Omega \frac{dz}{dt} \sin \theta + \omega^2 z = 0,$$

$$z = e^{-i\Omega t \sin \theta}(c_1 e^{i\omega t} + c_2 e^{-i\omega t}). \tag{2.4}$$

The plane of oscillation rotates by an angle of $\delta = -2\pi \sin \theta$ in one day (in our units, one day is $t = 1$, implying the rotational frequency $\Omega = 2\pi$). This angle is precisely the solid angle subtended by the closed path C of the bob, see Fig. 2.1(b). Let us mention two examples of the pendulum motion. As Earth rotates, a Foucault pendulum at the North Pole ($\theta = 90°$) swings in the same plane ($\delta = 2\pi$). At $\theta = 30°$, the pendulum returns to the original position, following a circle C shown in Fig. 2.1(b). But the **plane of oscillation is rotated** by $180°$, as shown in Fig. 2.1(a). The solid angle subtended by the circle representing the fixed position of the pendulum at $\theta = 30°$ is also $180°$, see Fig. 2.1(b).

What is remarkable is that the rotation angle of the plane of oscillation can be expressed as the twisted angle of a vector under **parallel transport**. Let us explain this concept. First, we explain parallel transport of a vector $\vec{v}$ in the tangent plane T_P at point P. (Note that tangent vectors all lie in the tangent plane; some tangent planes are shown in Fig. 2.2(a)). This will provide the general idea behind parallel transport. One parallel transports $\vec{v} \in T_P$ from P to another point Q on a curve C in three-dimensional space (in the usual way in Euclidian space). Then the transported vector is projected on the tangent plane T_Q, as shown in Fig. 2.2(b). The resulting vector is the parallel transported vector of $\vec{v}$. Note that the projection of a vector on a plane is its orthogonal projection on that plane. There is a different way to explain the concept of parallel transport. Consider the

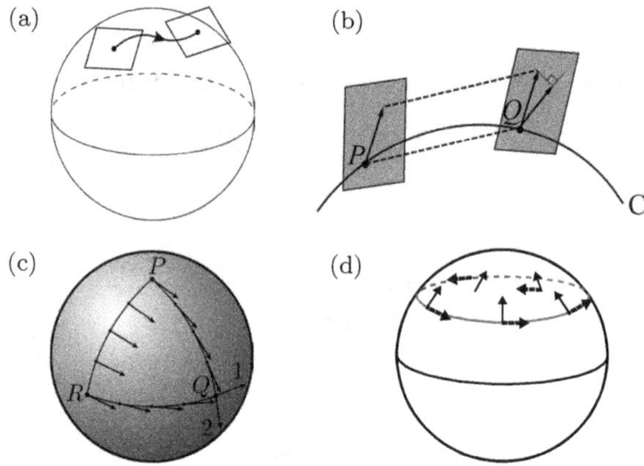

Fig. 2.2. (a) Two tangent planes T_P of a manifold are shown. (b) Parallel transport of a tangent vector $\vec{v} \in T_P$ and projection of the transported vector on the tangent plane T_Q are illustrated, see the explanation in the text. (c) Parallel transport of a vector 1 following a closed path $Q \to R \to P \to Q$ (the path consists of great circles) on a sphere gives in a different vector 2, see the two thick vectors located at Q. However, following a path on a cylinder returns the original vector [4]. (d) Parallel transport on a small circle is not easy to visualize. Suppose that a vector is perpendicular to the tangent vector of the small circle and that it is in the tangent plane. The drawing shown here may look like a parallel transport but it is really **not** a parallel transport because the initial vector is returned to itself. The true situation is analogous to the example of driving a motorcycle on the globe. If you're driving on a great circle of the globe you never have to turn the front wheel from the straight position. However, on the latitudinal drive you need to turn the wheel.

parallel transport of an **arbitrary** vector along a curve on a manifold. It is defined through the preservation of its angle with tangent vectors on the curve as it is transported. In other words, a change $d\vec{v}$ must be perpendicular to $\vec{v}$, i.e., the following condition must hold:

$$\frac{d\vec{v}}{d\lambda} \cdot \vec{v} = 0, \tag{2.5}$$

where λ parameterizes C. The velocity vector of a circular motion satisfies this condition.

The rotation angle of the plane oscillations of the Foucault pendulum can be calculated by cutting the top cone, see Fig. 2.3(a), and unfolding and flattening it [5]. The advantage of this trick is that parallel transported vectors can be found in the usual way in Euclidian space, see Fig. 2.3(c). Repatch the flattened piece of paper to a cone, as shown in Figs. 2.4(a) and 2.4(b). Then the angle between the end vectors $\vec{v}_1$ and $\vec{v}_2$ is the angle of rotation, as shown in Fig. 2.4(b). The solid angle subtended by the circle of the Foucault pendulum can be computed differently using the concept of curvature. It is related to the curvature by

$$\delta = \int_S \vec{B} \cdot d\vec{\sigma}, \tag{2.6}$$

where the curvature at each point on C is denoted by

$$\vec{B} = K\hat{n} \tag{2.7}$$

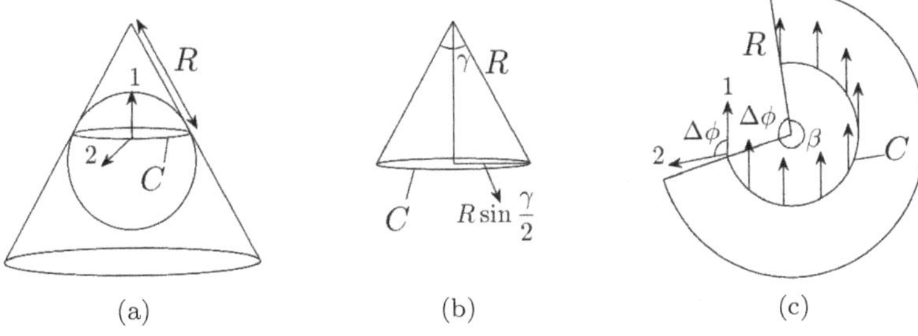

(a)   (b)   (c)

Fig. 2.3. (a) The circle in Fig. 2.1(b) is shown again. Note that the relevant tangent planes all lie in the surface of the cone. The two vectors 1 and 2 show the directions of the plane oscillations after one day, see Fig. 2.1(a). (b) The enlarged view of the top cone in (a). (c) Unfolding of the cone. We choose the cut position to coincide with the initial position of the pendulum. Parallel transported vectors on C are shown as vertical arrows. The rotation angle of the plane oscillations after one day can be calculated from $R\beta = 2\pi R \sin \gamma/2$, where $2\pi - \Delta\phi = \beta$.

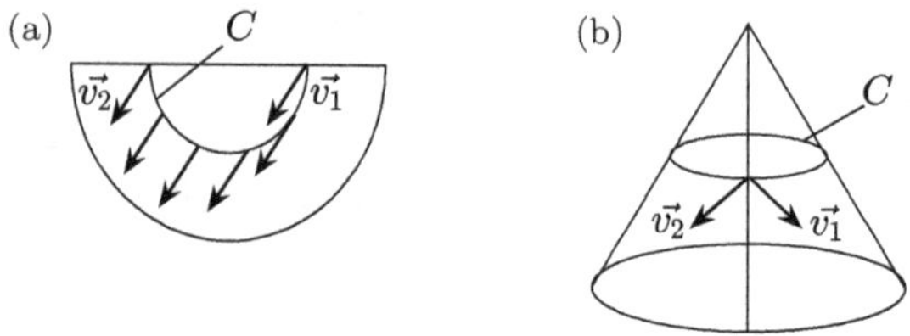

Fig. 2.4. Parallel transport of a vector on a circle on the sphere (see Fig. 2.3(a)) may be understood using the usual parallel transport on a plane. (a) Cut the cone, unroll it and parallel-transport a vector along the arc. Note that **folding** a piece of paper does not change its intrinsic geometry, only the extrinsic geometry changes. (An example is folding a piece of paper into a cylinder.) (b) Repatch the flattened cone to find that $\vec{v_2}$ does not point in the direction of $\vec{v_1}$ (b).

with the Gaussian curvature of a sphere $K = \frac{1}{R^2}$ (R is the radius of the earth). Here S is the area bounded by C and $\hat{n}$ is the normal vector. This angle is precisely equal to the **solid angle** subtended by path C. The curvature can be determined as follows. Consider a point O on a smooth surface. Suppose the origin of an xyz coordinate system coincides with O. The change of the normal vector at a point infinitesimally close to O is related to the curvature at point O:

$$\frac{d\vec{n}}{dx} \times \frac{d\vec{n}}{dy} = K\hat{z}. \tag{2.8}$$

Notice the appearance of the cross product. (A similar cross product will appear in connection with the Berry curvature, see below.) A cylinder and a cone have no intrinsic curvature but a sphere does. This can be seen by parallel transporting a vector. However, except for parallel transport on a great circle of a sphere, it is not easy to visualize a parallel transport on a sphere.[b]

Geometry and topology are intimately related. The Gauss–Bonnet theorem says the integral of the local curvature of a closed surface is an integer

$$\frac{1}{2\pi} \int_S K\,dA = 2(1 - g) = \text{integer}, \tag{2.9}$$

where g is number of holes of the manifold (A torus has one hole and a sphere has zero hole). Note that this integer is necessarily even. Quantum

---

[b]Consult any book on general relativity on how the curvature and parallel transport may be computed from the change of basis vectors at each point on a manifold. (Note that the basis vectors can be different at different points on C (e.g., think about the basis vectors $\hat{e}_r$, $\hat{e}_\theta$ and $\hat{e}_\varphi$ of a spherical coordinate system. This leads to the connection coefficients [4]).

generalization of the Gauss–Bonnet theorem leads to the Chern number which is the integral of the Berry curvature over a surface (see Sec. 2.4).

2.2. Berry Connection

Let us study the quantum version of a geometric phase, namely the Berry phase. There are several excellent introductory papers on the Berry phase [6–8]. Here I give only a short introduction. Let us derive the Berry phase from the time-dependent Schrödinger equation

$$i\hbar\frac{\partial\psi_n(t)}{\partial t} = H(\vec{R}(t))\psi_n(t).\qquad(2.10)$$

The Hamiltonian $H(\vec{R}(t))$ depends on physical parameters $\vec{R} = (R_1,\ldots,R_N)$, which are varied **adiabatically**. They can be a magnetic field, an electric field, etc., that can be **controlled externally**. When $\vec{R}$ follows a closed loop, the electron wave function can acquire a non-trivial phase. This result looks rather peculiar. How can an isolated system acquire an extra phase when it is returned to its original state? In this approach, the system is treated as if it were isolated, but the coupling with the rest of the universe is hidden in the parametric dependence. During the adiabatic process the system is interacting with the environment. The Berry phase is thus a subtle consequence of the coupling to the outside world, see [9].

Consider the nth eigenstate of a quantum system. Assume that the nth quantum state may acquire an additional Berry phase γ during an adiabatic process [7]

$$|\psi_n(t)\rangle = e^{i\gamma(t)}e^{-\frac{i}{\hbar}\int_0^t dt'\, E_n(t')}|n(\vec{R}(t))\rangle,\qquad(2.11)$$

where $|n(\vec{R}(t))\rangle$ satisfies the instantaneous Schrödinger equation

$$H(t)|n(\vec{R}(t))\rangle = E_n(t)|n(\vec{R}(t))\rangle.\qquad(2.12)$$

Plugging the Ansatz, Eq. (2.11), into the time-dependent Schrödinger equation, Eq. (2.10), we find

$$i\hbar\frac{\partial}{\partial t}|\psi_n(t)\rangle = -\hbar\dot{\gamma}(t)e^{i\gamma(t)}e^{-\frac{i}{\hbar}\int_0^t dt'\, E_n(t')}|n(t)\rangle + E_n(t)|\psi_n(t)\rangle$$

$$+ e^{-\frac{i}{\hbar}\int_0^t dt'\, E_n(t')}e^{i\gamma(t)}i\hbar\dot{\vec{R}}(t)\cdot\nabla_{\vec{R}}|n(\vec{R}(t))\rangle$$

$$= E_n(t)|\psi_n(t)\rangle.\qquad(2.13)$$

Here the gradient[c] is defined as

$$\nabla_{\vec{R}} = \left(\frac{\partial}{\partial R_1}, \frac{\partial}{\partial R_2}, \dots \right) \tag{2.14}$$

and the product $\dot{\vec{R}}(t){\cdot}\nabla_{\vec{R}} = \frac{d}{dt}$ is used. Multiplying both sides with $\langle \psi_n(t)|$, we find that the phase satisfies

$$\dot{\gamma}(t) = i\langle n(\vec{R}(t))|\nabla_{\vec{R}}|n(\vec{R}(t))\rangle \cdot \dot{\vec{R}}(t). \tag{2.15}$$

From this we find

$$\gamma = i \int_C dt\langle n(\vec{R})|\nabla_{\vec{R}}|n(\vec{R})\rangle \dot{\vec{R}}(t) = \int_C \vec{A}(\vec{R}) \cdot d\vec{R}, \tag{2.16}$$

where **the Berry connection** [10] is

$$\vec{A}(\vec{R}) = i\langle n(\vec{R}) \left| \frac{\partial}{\partial \vec{R}} \right| n(\vec{R})\rangle. \tag{2.17}$$

The Berry phase γ can be also represented in the language of a fiber bundle, see Fig. 2.5. Non-zero imaginary part of $\langle n(\vec{R})|\frac{\partial}{\partial \vec{R}}|n(\vec{R})\rangle$ is responsible for the Berry phase.

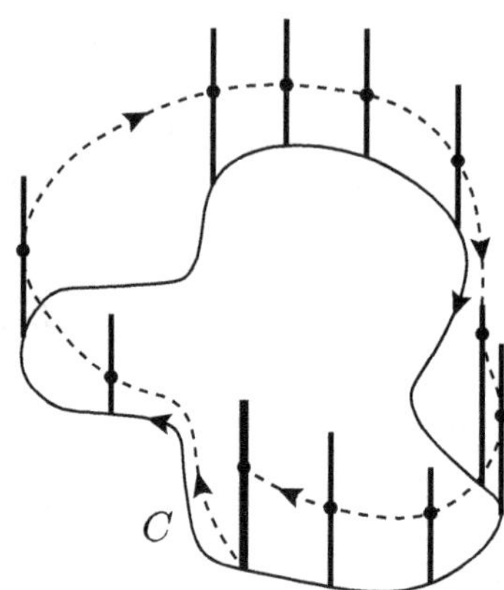

Fig. 2.5.   Phase changes can be represented as position of a dot on vertical bars of curve C. Position of each dot denotes the value of the phase change $\gamma(t) \in [0, 2\pi]$. Note that the dot does not return to the original position. These bars constitute a part of a fiber bundle (see [11] for an easy introduction).

---

[c]It should not be confused with the gradient of a curvilinear coordinate system.

> **Exercise 2.1.** A gauge theory associates a group element to each path. For an infinitesimal path labeled with index m, the group element is $1 + i\frac{q}{\hbar c}\vec{A}_m \cdot d\vec{R}_m$. What is the group element for a finite path parameterized between $t = 0$ and T? Answer: In a group the following properties must be present: associativity, identity element, and inverse element. For infinitesimal paths, the group element is close to the identity. For longer paths, the group element is the successive product of the infinitesimal group elements
>
> $$\prod_m \left(1 + i\frac{q}{\hbar c}\vec{A}_m \cdot \frac{d\vec{R}_m}{dt} dt\right) = e^{i\frac{q}{\hbar c}\int_{\vec{R}(0)}^{\vec{R}(T)} \vec{A}\cdot d\vec{R}}. \tag{2.18}$$
>
> Associativity is clearly satisfied for these transformations. An inverse path corresponds to an inverse element. The group is $U(1)$ gauge group of electromagnetism, consisting of the unit circle in the complex plane. The group operation corresponds to multiplication of these complex numbers.

For an open loop in the parameter space, the Berry phase given in Eq. (2.16) is gauge-dependent. Consider a **gauge transformation** of multiplying the wave function with an arbitrary phase[d]

$$|\tilde{\psi}\rangle = e^{i\theta(\vec{R})}|\psi\rangle. \tag{2.19}$$

This leads to the following change in the Berry connection:

$$\tilde{\vec{A}}(\vec{R}) = \vec{A}(\vec{R}) - \frac{\partial\theta(\vec{R})}{\partial\vec{R}} \tag{2.20}$$

and in the Berry phase

$$\tilde{\gamma} = \gamma + \theta(\vec{R}(0)) - \theta(\vec{R}(t)). \tag{2.21}$$

Thus the phase depends on the choice of the gauge. But for a closed loop the condition of single-valued wave function, $\theta(\vec{R}(0)) = \theta(\vec{R}(T))$, must hold

---

[d]Note that this is **not** the local gauge transformation used in electromagnetism. Here the phase does not depend on the spatial coordinates $\vec{r}$. It is a gauge transformation in the parameter space and not in real space. In this book the term "gauge transformation" will be only used in connection with the Berry phase. In other cases we will use the term "local gauge transformation".

(T is the time it takes to go around the loop). The resulting Berry phase

$$\gamma = \oint_C \vec{A}(\vec{R}) \cdot d\vec{R} \tag{2.22}$$

is **gauge-independent** modulo 2π. This is very reminiscent of a magnetic field not being affected by a gauge transformation of the vector potential.

It is instructive to give an alternative derivation of the Berry phase. The change in the phase factor is

$$e^{-i\Delta\gamma} = \frac{\langle \psi(\vec{R})|\psi(\vec{R}+\Delta\vec{R})\rangle}{|\langle \psi(\vec{R})|\psi(\vec{R}+\Delta\vec{R})\rangle|} \approx 1 + \langle \psi(\vec{R})|\nabla_{\vec{R}}\psi(\vec{R})\rangle \cdot \Delta\vec{R}. \tag{2.23}$$

Since the phase change $\Delta\gamma$ is very small, it follows that

$$-i\Delta\gamma = \langle \psi(\vec{R})|\nabla_{\vec{R}}\psi(\vec{R})\rangle \cdot \Delta\vec{R}. \tag{2.24}$$

When the closed loop C is divided into N intervals, the total accumulated phase can be written as

$$\gamma = \sum_{m=1}^{N} \Delta\gamma_n \approx i \oint_C d\vec{R} \cdot \langle \psi(\vec{R})|\nabla_{\vec{R}}\psi(\chi)\rangle = \int_C \vec{A}(\vec{R}) \cdot d\vec{R}, \tag{2.25}$$

in agreement with our previous result.

Let us now discuss a discrete formulation of the Berry phase which is ideally suited for numerical computing of the Berry phase. We divide the path C into N subintervals. For a subinterval, we have

$$e^{-i\Delta\gamma_m} \approx \frac{\langle \psi_{m-1}|\psi_m\rangle}{|\langle \psi_{m-1}|\psi_m\rangle|}. \tag{2.26}$$

The total Berry phase is given by

$$e^{-i\gamma} = \frac{\langle \psi_0|\psi_1\rangle}{|\langle \psi_0|\psi_1\rangle|} \cdots \frac{\langle \psi_{N-1}|\psi_0\rangle}{|\langle \psi_{N-1}|\psi_0\rangle|}, \tag{2.27}$$

which can be written as

$$\gamma = -\text{Im}\,\ln[\langle \psi_0|\psi_1\rangle\langle \psi_1|\psi_2\rangle \cdots \langle \psi_{m-1}|\psi_m\rangle\langle \psi_m|\psi_{m+1}\rangle \cdots \langle \psi_{N-1}|\psi_0\rangle], \tag{2.28}$$

where the periodic boundary condition $|\psi_0\rangle = |\psi_N\rangle$ is used. In numerical words, an arbitrary phase $e^{i\alpha}$ may appear in $|\psi_m\rangle$, but since its complex conjugate $e^{-i\alpha}$ appear in $\langle \psi_m|$, the phases cancel each other (note that they appear in the product $|\psi_m\rangle\langle \psi_m|$). Thus the final result is gauge invariant. This gives a numerically stable method to evaluate the Berry phase.

The Berry phase can be thought of as a twisted angle of a vector parallel transported along a loop on a manifold, see Fig. 2.2(c). This is explained in Exercise 2.2.

Exercise 2.2. How is the adiabatic transport related to parallel transport? Consider each interval i on a loop C. Let us make a gauge transformation such that each $|\psi_i\rangle$ is transformed to $|\tilde{\psi}_i\rangle$ with the following conditions:

$$|\tilde{\psi}_0\rangle = |\psi_0\rangle,$$

$$|\tilde{\psi}_0\rangle = |\tilde{\psi}_N\rangle\langle\tilde{\psi}_N|\tilde{\psi}_0\rangle. \tag{2.29}$$

Before the gauge transformation, $|\psi_0\rangle = |\psi_N\rangle$ but after the gauge transformation, $|\tilde{\psi}_0\rangle$ and $|\tilde{\psi}_N\rangle$ differ by a phase factor. Furthermore, we assume that the following condition holds under the gauge transformation

$$\mathrm{Im}\,\ln\langle\tilde{\psi}_i|\tilde{\psi}_{i+1}\rangle = 0 \text{ for } i = 0,\dots N-2. \tag{2.30}$$

It is equivalent to choosing the phase such that $\langle\tilde{\psi}_i|\tilde{\psi}_{i+1}\rangle$ is real and positive, which is also equivalent to $\vec{\tilde{A}}(\vec{R}) = 0$ or $\langle\tilde{\psi}|\frac{d\tilde{\psi}}{dR_\alpha}\rangle = 0$. This means that the state is being parallel transported along the curve: the change of the state along the curve, $|d\tilde{\psi}\rangle$, is perpendicular to $|\tilde{\psi}\rangle$ (it is analogous to the parallel transport on a manifold, see Eq. (2.5)). The new gauge is called a **parallel-transport gauge**. Show that the Berry phase, given by the phase mismatch at the end of the loop, is

$$\gamma = -\mathrm{Im}\,\ln\langle\tilde{\psi}_N|\tilde{\psi}_0\rangle = -\mathrm{Im}\,\ln\langle\tilde{\psi}_{\lambda=1}|\tilde{\psi}_{\lambda=0}\rangle, \tag{2.31}$$

where λ is an adiabatic parameter. A good example of $\tilde{\psi}(x)$ is the periodic part of the Bloch wave function $u_k(x)(k = \lambda)$. See Sec. 4.1. Hint: Plug $|\tilde{\psi}_0\rangle = |\tilde{\psi}_N\rangle\langle\tilde{\psi}_N|\tilde{\psi}_0\rangle$ into $\langle\tilde{\psi}_{N-1}|\tilde{\psi}_0\rangle$, which appears in the transformed version of Eq. (2.28). Consult also Ref. [12].

2.3. Berry Curvature

Using Stokes' theorem, one may write **the Berry phase** as

$$\gamma = \oint_C \vec{A}(\vec{R}) \cdot d\vec{R} = \int_S \nabla_{\vec{R}} \times \vec{A}(\vec{R}) \cdot d\vec{S} = \int_S \vec{b} \cdot d\vec{S}, \tag{2.32}$$

where S is any surface bounded by a loop C. The Berry phase is analogous to a magnetic flux, and the quantity

$$\vec{b} = \vec{\nabla} \times \vec{A}(\vec{R}) \tag{2.33}$$

is called **the Berry curvature**, which is analogous to a magnetic field. A good way of thinking about these quantities is to imagine a fictitious magnetic monopole sitting inside a closed surface. (More about it in the next section.) One can rewrite the Berry curvature of the nth level as (do Exercise 2.3 below)

$$\vec{b} = \nabla_{\vec{R}} \times \vec{A}(\vec{R}) = i\nabla_{\vec{R}} \times \langle \psi_n | \nabla_{\vec{R}} | \psi_n \rangle$$
$$= i \langle \nabla_{\vec{R}} \psi_n | \times | \nabla_{\vec{R}} \psi_n \rangle. \tag{2.34}$$

Notice the similarity of this expression to the definition of a curvature of a point on a surface, see Eq. (2.8). Another way to derive Eq. (2.34) is as follows. It is easy to show from the definition of the Berry connection (Eq. (2.17)) that the z-component of the Berry curvature of the nth quantum level can be expressed as[e]

$$b_z = \frac{\partial}{\partial R_x} A_y(\vec{R}) - \frac{\partial}{\partial R_y} A_x(\vec{R})$$
$$= i \left\langle \frac{\partial \psi_n}{\partial R_x} \Big| \frac{\partial \psi_n}{\partial R_y} \right\rangle - i \left\langle \frac{\partial \psi_n}{\partial R_y} \Big| \frac{\partial \psi_n}{\partial R_x} \right\rangle. \tag{2.35}$$

The Berry curvature can be written in a gauge invariant form. Since

$$\vec{b} = i \sum_{m \neq n} \langle \nabla_{\vec{R}} \psi_n | \psi_m \rangle \times \langle \psi_m | \nabla_{\vec{R}} \psi_n \rangle, \tag{2.36}$$

one can write it as (see Exercise 2.5)

$$\vec{b} = i \sum_{m \neq n} \frac{\langle \psi_n | \nabla_{\vec{R}} H(\vec{R}) | \psi_m \rangle \times \langle \psi_m | \nabla_{\vec{R}} H(\vec{R}) | \psi_n \rangle}{(E_m(\vec{R}) - E_n(\vec{R}))^2}. \tag{2.37}$$

This expression is **gauge invariant**: the phase factors cancel out since each wave function appears with its complex conjugate. Note that the sum of all $m \neq n$ is telling us that the phase of the nth quantum state depends on all other states of the system. Why is this so? A non-trivial Berry phase only occurs in a non-isolated quantum system, and originates from the interaction with the environment, see Ref. [9]. Note also that if the contributions

---

[e]It should not be confused as a component of the curl in a curvilinear coordinate system.

from all the levels are added, the resulting Berry curvature is zero (this can be seen by observing $\vec{a} \times \vec{b} = -\vec{b} \times \vec{a}$ for any two vectors).

Exercise 2.3. Equation (2.34) is derived using

$$\nabla_{\vec{R}} \times [f(\vec{R})\nabla_{\vec{R}}g(\vec{R})] = \nabla_{\vec{R}}f(\vec{R}) \times \nabla_{\vec{R}}g(\vec{R}). \qquad (2.38)$$

Show this.

Exercise 2.4. In Eq. (2.36) the term with $m = n$ is omitted. This is because the Berry curvature is real while the omitted term is imaginary. (a) Show that the Berry curvature is real, i.e., that the Berry connection $i\langle\psi_n|\frac{d}{d\vec{R}}|\psi_n\rangle$ is real. (b) Show that the omitted term is purely imaginary. Hint: use

$$\langle\psi_n|\psi_n\rangle = 1. \qquad (2.39)$$

Exercise 2.5. Demonstrate Eq. (2.37) by showing

$$\langle\psi_m|\nabla_{\vec{R}}|\psi_n\rangle = \frac{\langle\psi_m|\nabla_{\vec{R}}H(\vec{R})|\psi_n\rangle}{E_n(\vec{R}) - E_m(\vec{R})} \; for \; m \neq n. \qquad (2.40)$$

Hint: Operate $\nabla_{\vec{R}}$ on

$$H(\vec{R})|\psi_n\rangle = E_n(\vec{R})|\psi_n\rangle. \qquad (2.41)$$

2.4. Magnetic Monopole and Chern Number

• Two-Level Hamiltonian

Not all systems display a Berry phase. The parameter space must be "curved". Is there some way to find out whether a system will display a Berry phase? One usually looks for a crossing of 2 energy levels in the parameter space [8]. According to Eq. (2.37) the Berry connection $\vec{b}$ is singular when

$$E_n(\vec{R}) = E_m(\vec{R}). \qquad (2.42)$$

Any 2×2 Hermitian matrix $\mathbf{A}$ can be written as

$$\mathbf{A} = \beta\mathbf{I} + \frac{\vec{R}}{2} \cdot \vec{\sigma}, \qquad (2.43)$$

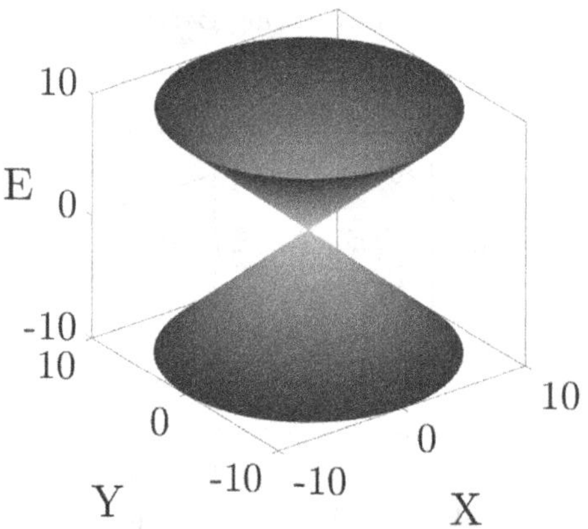

Fig. 2.6. Energy of the complex Hermitian Hamiltonian, Eq. (2.46), as a function of X and Y (with $Z = 0$). The energy surface has the shape of a double cone (diabolo). $E = 0$ is the degenerate point.

where β is a real number, $\vec{\sigma} = (\sigma_x, \sigma_y, \sigma_z)$ are the Pauli spin matrices, and **I** is a unit matrix. Without loss of generality, one can shift the origin of the parameter space to the point of double degeneracy. This is equivalent to setting $\beta = 0$. Then the resulting 2×2 complex Hermitian Hamiltonian near the degenerate point $\vec{R} = 0$ is[f]

$$H = \vec{R} \cdot \vec{\sigma}/2 = \frac{1}{2} \begin{pmatrix} Z & X - iY \\ X + iY & -Z \end{pmatrix}, \qquad (2.46)$$

where $\vec{R} = (X, Y, Z)$. (This Hamiltonian is identical to the **spin 1/2** Hamiltonian: $\vec{\sigma}/2$ and $\vec{R}$ correspond, respectively, to spin operator $\vec{S}$ and magnetic field $\vec{d}$.) This Hamiltonian is **ubiquitous** in topological systems, such as graphene, polyactylene, and the Kitaev model.

The eigenvalues of H are shown as a function of $(X, Y, 0)$ in Fig. 2.6. The energy surface is a diabolo[g] with the diabolical degeneracy point at $\vec{R} = 0$ [13]. In graphene, this point is called the Dirac point.

---

[f] Note that

$$A\mathbf{I} + X\sigma_x + Y\sigma_y + Z\sigma_z = \begin{pmatrix} A + Z & X - iY \\ X + iY & A - Z \end{pmatrix}, \qquad (2.44)$$

where the Pauli spin matrices are

$$\sigma_x = \begin{pmatrix} 0 & 1 \\ 1 & 0 \end{pmatrix}, \quad \sigma_y = \begin{pmatrix} 0 & -i \\ i & 0 \end{pmatrix}, \quad \sigma_z = \begin{pmatrix} 1 & 0 \\ 0 & -1 \end{pmatrix}. \qquad (2.45)$$

[g] It is an Italian toy of similar shape shown in Fig. 2.6.

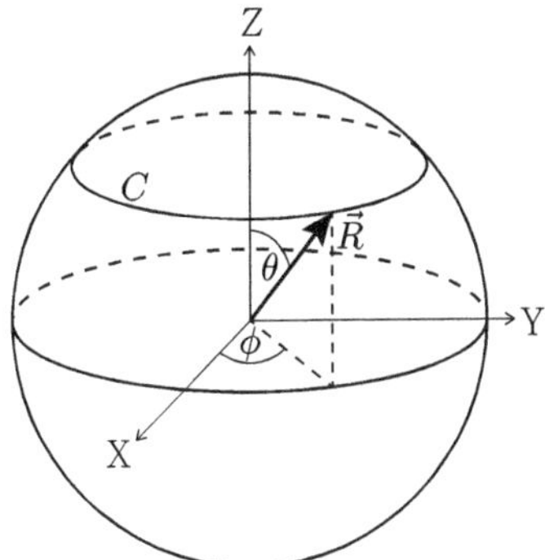

Fig. 2.7. An adiabatic path traces a circle on a sphere in the parameter space.

• Berry Connection and Phase

The point $\vec{R} = 0$ has interesting topological properties. Let us compute the Berry phase of an electron spin in a magnetic field $\vec{d}$. Consider a closed adiabatic path on a sphere, as shown in Fig. 2.7. The position on the path is determined by the spherical coordinates θ and ϕ. We will use (R, θ, ϕ) as our new **adiabatic parameters**.[h] The eigenstates and eigenenergies of

$$H = \vec{R} \cdot \vec{\sigma}/2 = \vec{d} \cdot \vec{\sigma} \tag{2.47}$$

are

$$|-\rangle = \begin{pmatrix} \sin(\theta/2)e^{-i\varphi} \\ -\cos(\theta/2) \end{pmatrix} \quad \text{and} \quad |+\rangle = \begin{pmatrix} \cos(\theta/2) \\ \sin(\theta/2)e^{i\varphi} \end{pmatrix} \tag{2.48}$$

respectively, and

$$E_{\pm} = \pm R/2. \tag{2.49}$$

Note that the spin expectation value is given by

$$\langle \pm | \vec{\sigma} | \pm \rangle = \pm \vec{d}/|\vec{d}|. \tag{2.50}$$

In other words, the expectation value is either parallel or anti-parallel to the "magnetic field" $\vec{d} = \vec{R}/2$.

---

[h]The new adiabatic parameters are $\vec{R} = (R_1, R_2, R_3) = (R, \theta, \phi)$ and not the coordinates (X, Y, Z). Note that the gradient appearing in the Berry connection Eq. (2.17) is thus defined as $\nabla_{\vec{R}} = (\frac{\partial}{\partial R_1}, \frac{\partial}{\partial R_2}, \frac{\partial}{\partial R_3}) = (\frac{\partial}{\partial R}, \frac{\partial}{\partial \theta}, \frac{\partial}{\partial \varphi})$, see Eq. (2.14). It should not be confused with the gradient in the spherical coordinate system.

The Berry connections are

$$A_\theta(\theta,\varphi) = i\left\langle -\left|\frac{\partial}{\partial\theta}\right|-\right\rangle = 0,$$

$$A_\varphi(\theta,\varphi) = i\left\langle -\left|\frac{\partial}{\partial\varphi}\right|-\right\rangle = \frac{1-\cos\theta}{2}. \tag{2.51}$$

We use this result and the contour integration given in Eq. (2.16) to calculate the Berry phase of an electron with energy E_- moving adiabatically on a loop C (see Fig. 2.7)

$$\gamma_- = \int_0^{2\pi} d\varphi A_\varphi = \pi(1-\cos\theta). \tag{2.52}$$

One can also calculate the Berry phase from the Berry curvature

$$b(\theta,\varphi) = \frac{\partial A_\varphi}{\partial\theta} - \frac{\partial A_\theta}{\partial\varphi} = \frac{1}{2}\sin\theta. \tag{2.53}$$

From this, we compute the Berry phase

$$\gamma_- = \int d\varphi d\theta b(\theta,\varphi) = \int d\phi d\theta \frac{1}{2}\sin\theta = \pi(1-\cos\theta) = \frac{1}{2}\Omega, \tag{2.54}$$

where Ω is the solid angle subtended by C. This result may be verified by noticing $\int d\Omega = \int \sin\theta d\theta d\varphi$. Note that, in contrast to the case of pendulum bob of Sec. 2.1, it is **one-half** of the solid angle subtended by the trajectory.

- **Berry Curvature and Magnetic Monopole**

We will now show that the effect of a degenerate point is equivalent to that of a fictitious magnetic monopole with charge $g = s$ sitting at $\vec{R} = 0$, see Fig. 2.8(a). (Here we use the analogy given in Table 1 and derive the result given in Eq. (2.54).) This monopole will generate a Berry curvature, i.e., a magnetic field. We switch to the Cartesian coordinates $\vec{R} = (X, Y, Z)$ and use them as our adiabatic parameters. The Berry curvature of the lower energy state $|-\rangle$ is

$$\vec{b} = s\frac{\vec{R}}{R^3}, \tag{2.55}$$

where the spin value is $s = 1/2$ (this expression for $\vec{b}$ holds for other values of s; for the higher energy state $|+\rangle$ the Berry curvature has opposite sign, see Eq. (2.50)). Note that this magnetic field exists in the **parameter space** and not in a real space! (Later we will study Dirac

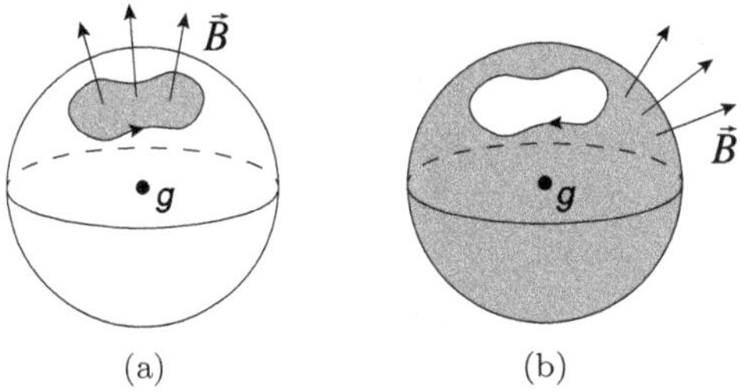

Fig. 2.8. Consider a loop C on a sphere. The enclosed area defined by C depends on how the loop is encircled: (a) for counterclockwise rotation it is S and (b) for clockwise rotation it is $\overline{S}$ (b). If someone circles C with their head pointing in the direction of the unit normal $\vec{n}$ then both S and $\overline{S}$ are on the left-hand side. (The unit normal vector $\vec{n}$ is defined to point outside of the manifold.) The Berry curvature is analogous to a magnetic field emanating from a fictitious magnetic monopole g. Only magnetic fields piercing S or $\overline{S}$ are shown.

magnetic monopoles that exist in real space, see Sec. 3.4). Its Berry phase is

$$\gamma_- = R^2 \int_S b\,d\Omega = g\Omega. \tag{2.56}$$

Identifying the magnetic charge g with spin value s we see that this result is identical to the result given in Eq. (2.54). **Thus, if a magnetic monopole is present, the Berry phase is finite.**

2.5. Quantized Winding Number

Consider a closed surface and the integral over it

$$C = \frac{1}{2\pi} \oint_M \vec{b} \cdot d\vec{S} = \gamma/2\pi = \text{integer}. \tag{2.57}$$

Its value is an integer. This is the quantum analog of the Gauss–Bonnet theorem, Eq. (2.9). Let us show how this quantity is related to the Berry phase.

The Berry phase is

$$\gamma_1 = \int_S \vec{b} \cdot d\vec{S}. \tag{2.58}$$

(Here S is the shaded area shown in Fig. 2.8(a). The area vector is pointing outward away from the center of sphere). When the particle completes the loop, it acquires a phase factor $e^{i\gamma_1}$. Let us adiabatically transport the eigenstate in the **opposite** direction around the loop (the enclosing area $\overline{S}$

of the new curve is the complement of S, see Fig. 2.8(b)). The new phase factor should be $e^{-i\gamma_2}$ where γ_2 is defined as

$$\gamma_2 = \int_{\overline{S}} \vec{b} \cdot d\vec{S} > 0. \tag{2.59}$$

(The area vector $d\vec{S}$ points again away from the center of the sphere.) But, going around the loop clockwise should result in the same phase factor as going around the loop counterclockwise

$$e^{i\gamma_1} = e^{-i\gamma_2}. \tag{2.60}$$

After completing the loop clockwise and then counterclockwise, no phase factor should appear

$$e^{i(\gamma_1+\gamma_2)} = 1. \tag{2.61}$$

This total phase is equal to

$$\gamma = \gamma_1 + \gamma_2 = \int_{S+\overline{S}} \vec{b} \cdot d\vec{S} = 0 \quad \mathrm{mod} \ \ 2\pi \tag{2.62}$$

from which we find the desired result given in Eq. (2.57). Using the result for the Berry phase, one can show that the monopole charge inside the surface is quantized. From the result

$$\gamma = \oint_M g \frac{\vec{R}}{R^3} d\vec{S} = 4\pi g, \tag{2.63}$$

we find the quantization of the magnetic monopole charge/spin quantization

$$g = s = \text{integer} \times 1/2. \tag{2.64}$$

We have seen that any traceless 2×2 Hermitian matrix can be written in terms of the Pauli spin matrices $\vec{\sigma} = (\sigma_x, \sigma_y, \sigma_z)$ (see Eq. (2.46))

$$H = \vec{d}(\vec{k}) \cdot \vec{\sigma}, \tag{2.65}$$

where $\vec{k}$ denotes points in the parameter space. The unit vector $\hat{d}(\vec{k}) = \frac{\vec{d}}{|\vec{d}|}$ represents a point on a unit sphere, see Fig. 2.9. One can define the winding number as follows

$$w = \frac{1}{2\pi} \int dk_x dk_y \hat{d} \cdot \frac{1}{2} \frac{\partial \hat{d}}{\partial k_x} \times \frac{\partial \hat{d}}{\partial k_y}. \tag{2.66}$$

It counts how many times $\hat{d}$ passes through each point on the sphere as $\vec{k}$ goes through each point in the parameter space. The quantity

$$\vec{b} = \frac{1}{2} \frac{\partial \hat{d}}{\partial k_x} \times \frac{\partial \hat{d}}{\partial k_x} \tag{2.67}$$

corresponds to the **curvature** of a point on a manifold, see Eq. (2.8). The winding number thus represents a Chern number. This result is similar to the Gauss–Bonnet theorem, see Eq. (2.9).

Now consider a one-dimensional manifold. When chiral symmetry is present, $d_z = 0$. In this case, $\vec{d}(\vec{k})$ maps a circle into another circle. But a circle is a one-dimensional manifold and the definition given in Eq. (2.66)

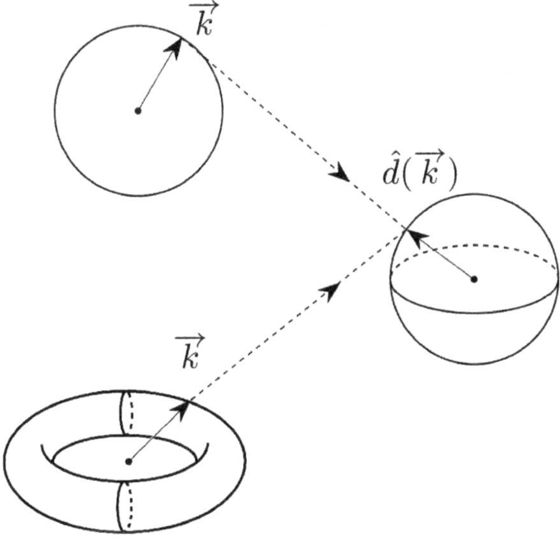

Fig. 2.9. $\vec{d}(k)$ maps a manifold into a manifold. In one example it maps a circle into a sphere and in another example from a torus to a sphere. Note that a torus is equivalent to a rectangle under periodic boundary conditions; an example is a two–dimensional first Brillouin zone.

cannot be used to compute the winding number. Instead, the winding number is defined **without** the factor $1/2$

$$w = \frac{1}{2\pi} \oint dk\, \hat{z}\hat{d}(k) \times \frac{d}{dk}\hat{d}(k), \tag{2.68}$$

where $\hat{z}$ is the unit vector along the z-axis,

$$(\hat{d}_x, \hat{d}_y) = (d_x, d_y)/|\vec{d}|, \tag{2.69}$$

and

$$\hat{d}(k) \times \frac{d}{dk}\hat{d}(k) = \left(\hat{d}_x \frac{d}{dk}\hat{d}_y - \hat{d}_y \frac{d}{dk}\hat{d}_x \right) \hat{z}. \tag{2.70}$$

Let us apply these results to one-dimensional band structures. How many times does $\vec{d}(\vec{k})$ rotate on the circle as k varies from $-\pi/a$ to π/a in a one-dimensional first Brillouin zone? This number is defined as the winding number w (this number is **not** called a Chern number, which is reserved for surfaces). One can show [14] that

$$w = -\frac{1}{2\pi i} \oint \frac{d}{dk}\log(f(k))dk = -\frac{1}{2\pi i} \oint \frac{\dot{f}(k)}{f(k)}dk, \tag{2.71}$$

where

$$f(k) = d_x(k) - id_y(k) \tag{2.72}$$

and the dot stands for differentiation with respect to k. Let us make a change of variable: $z = e^{ika}$ ($ka \in [-\pi, \pi]$) with $F(z) = f(k)$. According to Cauchy's argument principle, the winding number is

$$w = -\frac{1}{2\pi i} \oint \frac{\dot{F}(z)}{F(z)}dz = N_z - N_p, \tag{2.73}$$

where N_z and N_p are, respectively, number of roots and poles of $F(z)$ counting multiplicities.[i] Alternatively, one can simply use Mathematica to investigate how many time $\hat{d}(k)$ rotates and compute the winding number.

---

[i]The multiplicity k of a zero at a_z can be found by writing the function as $f(z) = (z - a_z)^k g(z)$ with $g(a_z) \neq 0$. The multiplicity p of a pole at z_p can be found by writing the function as $f(z) = (z - a_p)^{-p} g(z)$ with $g(a_p) \neq 0$.

2.6. Non-Abelian Berry Phase of Semiconductor Quantum Dot

Single electron control in semiconductor quantum dots[j] would be valuable for spintronics, quantum information, and spin qubits. Adiabatic time evolution of **degenerate** eigenstates of a quantum system may provide a means for controlling individual quantum states. Degenerate states can exhibit non-Abelian gauge structures and often give rise to finite non-Abelian Berry phases (they are also called **matrix Berry** or holonomic phases). These phases depend on the geometry of the path traversed in the parameter space of the Hamiltonian. Nuclei, superconducting nanocircuits, optical systems, and atomic systems have suitable degenerate quantum states that may exhibit non-Abelian phases. However, the presence of degenerate states is a necessary but not a sufficient condition as there is no general way to know in advance whether such a system displays a finite non-Abelian phase.

N-type semiconductor quantum dots with the Rashba and Dresselhaus spin–orbit interactions[k] may be used for single electron manipulation through adiabatic transformations between degenerate states [16]. The Hamiltonian of the semiconductor dot can be written in the envelope wave function approach (see Appendix A)

$$H = -\frac{\hbar^2 \nabla^2}{2m^*} + U(x, y) + V(z) + H_R, \tag{2.74}$$

where m^* is the effective mass of electron. The confining potentials $U(x, y)$ and $V(z)$ are shown in Fig. 2.10. They can be changed **electrically**. The Rashba spin–orbit term is

$$H_R = c_R \left[\sigma_x k_y - \sigma_y k_x \right], \tag{2.75}$$

where $\hbar k_{x,y}$ are the momentum operators (here we will ignore the Dresselhaus term). The constant c_R can be tuned by an external electric field E, which also defines the confining potential $V(z)$. All the energy levels are discrete in this quantum dot, and the energy levels possess a **double degeneracy** due to time reversal symmetry in the presence of the Rashba and/or Dresselhaus spin–orbit coupling terms (this is the well-known Kramers double degeneracy in the presence of spin–orbit interaction [17]). The presence

---

[j]A quantum dot is an artificial atom. It is the smallest transistor made by men with dimension $\lesssim 100\,\text{Å}$ [15].

[k]The Dresselhaus spin–orbit interaction arises from the lack of inversion symmetry in the material. Inversion symmetry may also be broken by an electric field, giving rise to the Rashba spin–orbit interaction.

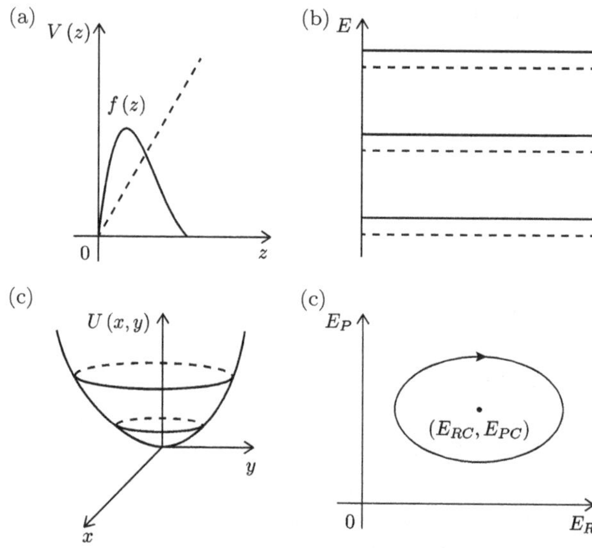

Fig. 2.10. (a) An electric field along the z-axis quantizes the electronic motion in a triangular potential along the axis. In our work, the triangular potential is sufficiently strong and only the lowest-energy subband is included. The structural inversion asymmetry in $V(z)$ leads to the Rashba interaction. (b) Each electron energy level is doubly degenerate (an artificial energy splitting is shown in the figure). It is crucial that the degeneracy remains intact as adiabatic parameters change; the degeneracy is symmetry protected. (c) The electron motion in the xy-plane is confined by an electric potential. Inversion symmetry is broken. (d) An adiabatic change implemented by changing the energy parameters E_R and E_p that characterize the strength of the Rashba term and the lateral distortion potential $U(x, y)$. Reprinted with permission from Ref. [16].

of double degeneracy does not always give rise to a finite non-Abelian matrix Berry phase. However, in this particular quantum dot, non-Abelian Berry phases are present. Note that in this system, the matrix Berry phase is controlled **electrically**.

Below we explain the concepts of matrix Berry connection, matrix Berry phase, and matrix Berry curvature. These concepts are needed to understand topological quantum computing using Majorana zero modes.

• Matrix Berry Connections

Consider a Hamiltonian that depends adiabatically on some parameters $\vec{R}(t) = (R_1(t), R_2(t), \ldots, R_N(t))$. We label these components by $n = 1, 2, \ldots$. Suppose that the system has a set of degenerate states with wave functions ψ_i and a degenerate energy E_d. An initial state is given as a linear combination of degenerate states. Suppose the degeneracy does not break under the adiabatic transformation. Then time-dependent wave function can be written as

$$\psi(t) = \sum_i c_i(t)\psi_i(t). \tag{2.76}$$

For convenience, the wave functions ψ_i are denoted by i or k (these are a degenerate subspace index) while points in the parameter space are denoted $a, b, c \ldots$. We will show that the wave function may acquire a matrix Berry phase. (Here we partly follow Zee's derivation [18].) From the time-dependent Schrödinger equation, one can show that the expansion coefficients satisfy

$$i\frac{dc_i}{dt} = -\sum_j A_{ij}c_j, \tag{2.77}$$

where, for simplicity, the degenerate energy E_d is set to zero. The Berry connections are defined as

$$A_{ij} = i\left\langle \psi_i \left| \frac{d\psi_j}{dt} \right. \right\rangle. \tag{2.78}$$

One can then rewrite the Berry connections as

$$A_{ij} = \sum_n (A^n)_{ij}\frac{dR_n}{dt}, \tag{2.79}$$

where the **matrix** Berry connections are

$$(A^n)_{ij} = i\left\langle \psi_i(t) \left| \frac{\partial\psi_j}{\partial R_n} \right. \right\rangle. \tag{2.80}$$

(Here $\hbar = 1$.) The matrix Berry connections can be written as a vector

$$\vec{\mathbf{A}} = (\mathbf{A}^1, \ldots, \mathbf{A}^N), \tag{2.81}$$

where each component is a matrix. Note that the matrix dimension is determined by the number of degenerate states and the number of the vector components is determined by the number of the adiabatic parameters.

• Matrix Berry Phase

The Berry connections are related to the matrix Berry phase, which relates the initial and final states. The final and initial expansion coefficients are connected to each other:

$$c_i(T) = \sum_j \Gamma_{ij}(0 \to T)c_j(0). \tag{2.82}$$

If the degeneracy is two, the operator $\boldsymbol{\Gamma}(t_{p-1} \to t_p)$ in each incremental interval can be represented by the following 2×2 matrix equation

$$\begin{bmatrix} c_1(t_p) \\ c_2(t_p) \end{bmatrix} = \begin{pmatrix} \Gamma_{11} & \Gamma_{12} \\ \Gamma_{21} & \Gamma_{22} \end{pmatrix} \begin{bmatrix} c_1(t_{p-1}) \\ c_2(t_{p-1}) \end{bmatrix}. \tag{2.83}$$

For an arbitrary closed loop in the adiabatic parameter space, the rotation of the initial state can be computed from the matrix Berry phase

$$\boldsymbol{\Gamma}(a \to a) = \boldsymbol{\Gamma}(0 \to T) = \boldsymbol{\Gamma}(t_{N-1} \to T) \ldots \boldsymbol{\Gamma}(t_1 \to t_2)\boldsymbol{\Gamma}(0 \to t_1). \tag{2.84}$$

The interpretation of this result is as follows: At each incremental path, a unitary transformation is performed and successive transformations produce the final state. An example is displayed in Fig. 2.11.

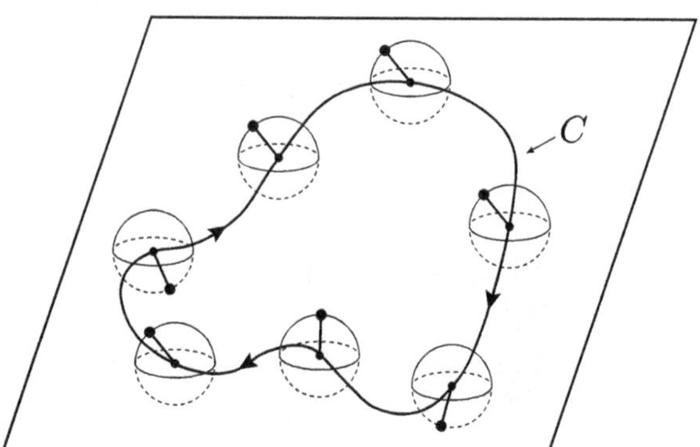

Fig. 2.11. Non-Abelian gauge transformation gives a multi-component state a matrix phase at each point on a loop C in the adiabatic parameter space. An example is given in the figure. In the case of two degenerate states, $|0\rangle$ and $|1\rangle$, an internal state, $|\psi\rangle = \cos\frac{\theta}{2}|0\rangle + e^{i\varphi}\sin\frac{\theta}{2}|1\rangle$, can be represented by a dot on the Bloch sphere, i.e., by the spherical angles θ and φ. Note that, as the parameters change, the state is not transformed in the parameter space but in the internal space. At each point on C, a sphere is attached; this structure is a fiber bundle, see Refs. [11, 19]. Note that the **degeneracy must be preserved** during the adiabatic changes. If a matrix Berry phase is present, the state does not return to the original state when the full loop is traversed. It is rotated in the degenerate Hilbert space.

• Matrix Berry Curvature

How are the Berry connections related to the matrix Berry phase? Suppose the system moves from a point and back to the same point via the path shown in Fig. 2.12. Then the acquired matrix Berry phase is

$$\mathbf{\Gamma}(a \to a) = P e^{i \oint_C \vec{\mathbf{A}} \cdot d\vec{s}}, \tag{2.85}$$

where the operator P stands for a contour-ordering. Note the similarity of Γ to the time evolution operator in the interaction picture $U_I(t) = T e^{-i \int_0^t dt' V_I(t')}$, which solves $\frac{dU_I}{dt} = V_I(t) U_I$. Here T is the time ordering operator. The solution of the coupled differential equation Eq. (2.77) has the similar form given by Eq. (2.85). This loop is also called a **Wilson loop**.[1] For the contour shown in Fig. 2.12, we have

$$P e^{i \oint_C \vec{\mathbf{A}} \cdot d\vec{s}} = \mathbf{\Gamma}(d \to a) \mathbf{\Gamma}(c \to d) \mathbf{\Gamma}(b \to c) \mathbf{\Gamma}(a \to b)$$

$$= e^{i \int_d^a \vec{\mathbf{A}} \cdot d\vec{s}} e^{i \int_c^d \vec{\mathbf{A}} \cdot d\vec{s}} e^{i \int_b^c \vec{\mathbf{A}} \cdot d\vec{s}} e^{i \int_a^b \vec{\mathbf{A}} \cdot d\vec{s}}. \tag{2.86}$$

Here $\mathbf{\Gamma}(a \to b)$ takes a state from point "a" to point "b" in the parameter space.

An **infinitesimally** small loop C is used to define the value of the **matrix Berry curvature $\vec{\mathbf{B}}$**.

$$P e^{i \oint_C \vec{\mathbf{A}} \cdot d\vec{s}} = e^{i \vec{\mathbf{B}} \cdot d\vec{S}}, \tag{2.87}$$

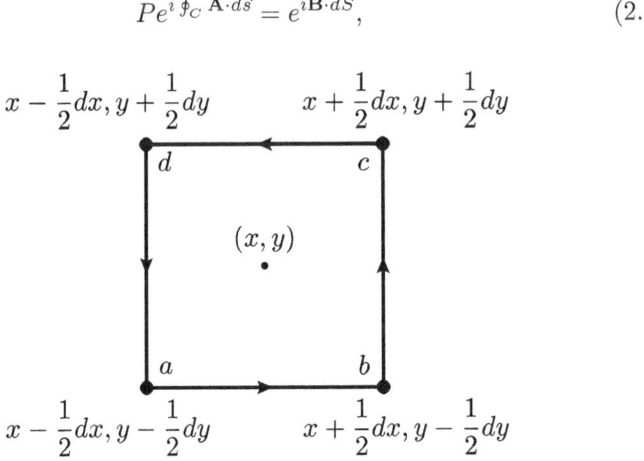

Fig. 2.12. The strength of the Berry curvature $\vec{B}$ is evaluated using an infinitesimally small square.

---

[1] We will encounter it again in connection with topological entanglement entropy in Sec. 20.3.

where each component of the magnetic field $\vec{\mathbf{B}}$ is a matrix and $d\vec{S} = \vec{n}dxdy$ is the normal areal vector (do Exercise 2.6). From this, we find that a component l of the generalized magnetic field is related to the vector potential

$$\mathbf{B}^l = \partial_m \mathbf{A}^n - \partial_n \mathbf{A}^m + i[\mathbf{A}^m, \mathbf{A}^n]. \tag{2.88}$$

Or in a vector form

$$\vec{\mathbf{B}} = \nabla \times \vec{\mathbf{A}} + i\vec{\mathbf{A}} \times \vec{\mathbf{A}}. \tag{2.89}$$

These results given here are valid when the dimension of the adiabatic parameter space is three. In other dimensions, the Berry curvature is

$$\mathbf{F}_{ij}^{mn} = (\partial_m \mathbf{A}^n)_{ij} - (\partial_n \mathbf{A}^m)_{ij} + i[\mathbf{A}^m, \mathbf{A}^n]_{ij}. \tag{2.90}$$

One can also define non-Abelian particles using the statistical gauge field $\vec{\mathbf{A}}$. The statistical phase factor upon exchange of two particles is given by

$$\psi_i(\vec{r}_\alpha, \vec{r}_\beta) \rightarrow e^{i\sum_a \theta_a T_{ij}^a} \psi_j(\vec{r}_\alpha, \vec{r}_\beta), \tag{2.91}$$

where the parameters θ_a and matrices $\mathbf{T}^a$ may be obtained from the matrix $\mathbf{\Gamma}$. These phases θ_a can be thought of as the Aharonov–Bohm phase due to $\vec{\mathbf{A}}$.

Exercise 2.6. Compute Eq. (2.87). Answer: Represent points in Fig. 2.12 as

$$a \rightarrow \vec{s},$$

$$b \rightarrow \vec{s} + d\vec{x},$$

$$c \rightarrow \vec{s} + d\vec{x} + d\vec{y}, \tag{2.92}$$

$$d \rightarrow \vec{s} + d\vec{y}.$$

Then perform the line integrals shown in Fig. 2.12. Using Eq. (2.96) one can show the following results for the matrix Berry phase $\mathbf{\Gamma}$

$$\mathbf{\Gamma}(b \rightarrow c)\mathbf{\Gamma}(a \rightarrow b) = e^{i\mathbf{A}_y(\vec{s}+d\vec{x})dy}e^{i\mathbf{A}_x(\vec{s})dx} = e^{i[\mathbf{A}_y+\partial_x\mathbf{A}_y dx]dy}e^{i\mathbf{A}_x dx}$$

$$\approx e^{i[\mathbf{A}_x dx + \mathbf{A}_y dy + \partial_x\mathbf{A}_y dxdy] - \frac{1}{2}[\mathbf{A}_y, \mathbf{A}_x]dxdy} \tag{2.93}$$

and

$$\Gamma(d \to a)\Gamma(c \to d) = \Gamma^{-1}(a \to d)\Gamma^{-1}(d \to c)$$
$$= e^{-i\mathbf{A}_y(\vec{s})dy} e^{-i\mathbf{A}_x(\vec{s}+d\vec{y})dx}$$
$$\approx e^{-i[\mathbf{A}_x dx + \mathbf{A}_y dy + \partial_y \mathbf{A}_x dxdy] - \frac{1}{2}[\mathbf{A}_y, \mathbf{A}_x]dxdy}.$$
$$(2.94)$$

Combining these results and using Eq. (2.96), we find

$$\Gamma(d \to a)\Gamma(c \to d)\Gamma(b \to c)\Gamma(a \to b) = e^{i\vec{\mathbf{B}}\cdot d\vec{S}}. \qquad (2.95)$$

(Small higher order terms are neglected). Hint: Use the following well-known approximation

$$e^{\epsilon \mathbf{C}} e^{\epsilon \mathbf{D}} \approx e^{\epsilon(\mathbf{C}+\mathbf{D}) + \frac{\epsilon^2}{2}[\mathbf{C},\mathbf{D}]}. \qquad (2.96)$$

- **Gauge Structure**

Under a **unitary** transformation to a different basis, the wave function transforms as

$$\psi_i' = \sum_j U_{ij}^* \psi_j. \qquad (2.97)$$

(Note the complex conjugation in U). At each point $\vec{R}$, the Berry connection transforms as

$$(A'^n)_{ij} = \sum_{kq} U_{ik}(A^n)_{kq}U_{qj}^+ + i\sum_k U_{ik}\frac{\partial U_{kj}^+}{\partial R_n}, \qquad (2.98)$$

or in a matrix form

$$\mathbf{A}'^n = \mathbf{U}\mathbf{A}^n\mathbf{U}^+ + i\mathbf{U}\frac{\partial \mathbf{U}^+}{\partial R_n}. \qquad (2.99)$$

(See Exercise 2.7.) These unitary transformations form a group, i.e., a gauge group. For a closed curve, the matrix phase Γ is no longer gauge invariant since the matrices do not commute. However, its trace is invariant

$$\boldsymbol{\Gamma}_{\mathrm{inv}} = \mathrm{Tr}\,\boldsymbol{\Gamma}. \tag{2.100}$$

When the degenerate states are Slater determinant many-body states, one can show that antisymmetry of the Slater wave functions does not contribute to the matrix Berry phase [20].

Exercise 2.7. Derive Eq.(2.98) from $A_{ij}^{n'} = i\langle\psi_i'|\frac{d\psi_j'}{dR_n}\rangle$. Hint: Differentiate $\psi_j' = \sum_k U_{jk}^*\psi_k$ with respect to R_n. Contract the result thus obtained with $i\psi_i'^* = i\sum_p U_{ip}\psi_p^*$.

Further reading. Read about the difference between Abelian and non-Abelian theories [18, 21]. Quantum electrodynamics is an Abelian theory with the action

$$S = \int d^4x \left[-\frac{1}{4e^2} F_{\mu\nu} F^{\mu\nu} + \bar{\Psi}(i\slashed{D} - m)\Psi \right], \tag{2.101}$$

where the second term gives the Dirac equation and $F_{\mu\nu}$ is the electromagnetic field tensor (repeated upper and lower indices are summed over). Quantum chromodynamics is a non-Abelian theory with the action

$$S = \int d^4x \left[-\frac{1}{4g^2} \mathrm{Tr}\{\mathbf{F}_{\mu\nu}\mathbf{F}^{\mu\nu}\} + \sum_j \bar{\Psi}_j(i\slashed{D} - m)\Psi_j \right], \tag{2.102}$$

where g is the coupling constant. (See Ref. [22] for how the factor $\frac{1}{4g^2}$ appears.) Note the similarity between these two actions. The covariant derivative is

$$i\slashed{D} = i\gamma^\mu \mathbf{D}_\mu, \tag{2.103}$$

where

$$\mathbf{D}_\mu = \partial_\mu - i\mathbf{A}_\mu \tag{2.104}$$

and γ^μ are the gamma matrices with $\mu = 0, 1, 2, 3$ denoting the Minkowski indices for time and space coordinates (in contrast to this the matrix Berry connections, Eq. (2.80), do not depend on time and space coordinates).

Note that each connection is given by the following linear combination:

$$\mathbf{A}_\mu = \lambda_f A_\mu^f, \tag{2.105}$$

where λ_f are 3×3 matrices, unlike the Abelian case. The reason λ_f are 3×3 matrices is because quarks represent the fundamental representation of SU(3) gauge group. There are 8 of them, $f = 1, \ldots, 8$. (There are $3^2 - 1$ gauge fields. The space of 3×3 traceless Hermitian matrices is eight dimensional). Note that λ_a represent rotations in the internal space of three quark charges: blue, green, and red; in other words, the eight fields A_μ^n represent the eight gluon fields mediating interactions between quarks.[m] Each field strength $\mathbf{F}_{\mu\nu} = \partial_\mu \mathbf{A}_\nu - \partial_\nu \mathbf{A}_\mu + \mathbf{A}_\mu \times \mathbf{A}_\nu$ is also a matrix (the cross product here is really a wedge product).

Bibliography

[1] D. Hofstadter and E. Sander, *Surfaces and Essences: Analogy as the Fuel and Fire of Thinking* (Basic Books, New York, 2013).

[2] M. V. Berry, The geometric phase, *Sci. American* **259**, 6, 46 (1988). •

[3] G. R. Fowles and G. L. Cassiday, *Analytical Mechanics* (Cengage Learning, Boston, 2004).

[4] R. J. A. Lambourne, *Relativity, Gravitation and Cosmology* (Cambridge University Press, Cambridge, 2010).

[5] V. I. Arnold, *Mathematical Methods of Classical Mechanics* (Springer, New York, 1989).

[6] M. V. Berry, Quantal phase factors accompanying adiabatic changes, *Proc. R. Soc. Lond.* A **392**, 45 (1984).

[7] J. J. Sakurai, *Advanced Quantum Mechanics* (Addison-Wesley, Boston, 1967). •

[8] A. Garg, Berry phases near degeneracies: Beyond the simplest case, *Am. J. Phys.* **78**, 661 (2010). •

[9] R. Resta, Polarization as a Berry phase, *Europhys. News* **28**, 18 (1997). •

[10] B. Simon, Holonomy, the Quantum Adiabatic Theorem, and Berry's Phase, *Phys. Rev. Lett.* **51**, 2167 (1983).

[11] H. J. Bernstein and A. V. Phillips, Fiber bundles and quantum theory, *Sci. American* **245**, 1, 122 (1981). •

[12] D. Vanderbilt, *Berry Phases in Electronic Structure Theory* (Cambridge University Press, Cambridge, 2018). •

---

[m]There are 9 possible gluon fields mediating interaction between quarks: red anti-red, red anti-blue, red anti-green, blue anti-red, blue anti-blue, blue anti-green, green anti-red, green anti-blue, and green anti-green. But since baryons are colorless and do not interact with each other only eight of them are physical.

[13] M. V. Berry and M. Wilkinson, Diabolical points in the spectra of triangles, *Proc. R. Soc. Lond.* A **392**, 15 (1984).

[14] J. K. Asbóth, L. Oroszlány, and A. Pályi, *A Short Course on Topological Insulators* (Springer, New York, 2016).

[15] M. Kastner, Artificial atoms, *Phys. Today* **46**, 1, 24 (1993). •

[16] S.-R. Eric Yang and N. Y. Hwang, Single electron control in n-type semiconductor quantum dots using non-Abelian holonomies generated by spin–orbit coupling, *Phys. Rev.* B **73**, 125330 (2006).

[17] S. M. Girvin and K. Yang, *Modern Condensed Matter Physics* (Cambridge University Press, Cambridge, 2019).

[18] A. Zee, *Quantum Field Theory in a Nutshell* (Princeton University Press, Princeton, 2010).

[19] F. Wilczek, *Fantastic Realities* (World Scientific, Singapore, 2006). •

[20] S.-R. Eric Yang, Control of many electron states in semiconductor quantum dots by non-Abelian vector potentials, *Phys. Rev.* B **75**, 245328 (2007).

[21] T. Lancaster and S. J. Blundell, *Quantum Field Theory for the Gifted Amateur* (Oxford University Press, Oxford, 2014).

[22] D. Tong, *Lectures on Gauge Theory*, Unpublished Notes.

Chapter 3

Basic Topological Properties

"Elegance may produce the feeling of the unforeseen by the unexpected meeting of objects we are not accustomed to bring together; there again it is fruitful, since it thus unveils for us kinships before unrecognized. It is fruitful even when it results only from the contrast between the simplicity of the means and the complexity of the problem set; it makes us then think of the reason for this contrast and very often makes us see that chance is not the reason; that it is to be found in some unexpected law."

Henri Poincaré

Here we study several simple models that exhibit topological properties. Many important basic concepts are introduced, such as the Aharonov–Bohm phase, twisted boundary conditions, topological action, topological quantization, instantons, and Abelian anyons. The reader should try to understand them thoroughly.

3.1. Aharonov–Bohm Phase as a Berry Phase

A charge that is far away from a magnetic field can still be affected by it. When a charge q encircles a flux, it will acquire the Aharonov–Bohm phase [1]

$$\gamma = 2\pi \frac{\Phi}{\Phi_q} = 2\pi \frac{q}{e} \frac{\Phi}{\Phi_0}, \tag{3.1}$$

where $\Phi_q = \frac{hc}{q}$ and $\Phi_0 = \frac{hc}{e}$ with $e > 0$ is the quantum unit of flux. The Aharonov–Bohm phase can be thought of as a Berry phase [2,3]. Consider a particle with a charge q in a small quantum box and in the absence of a

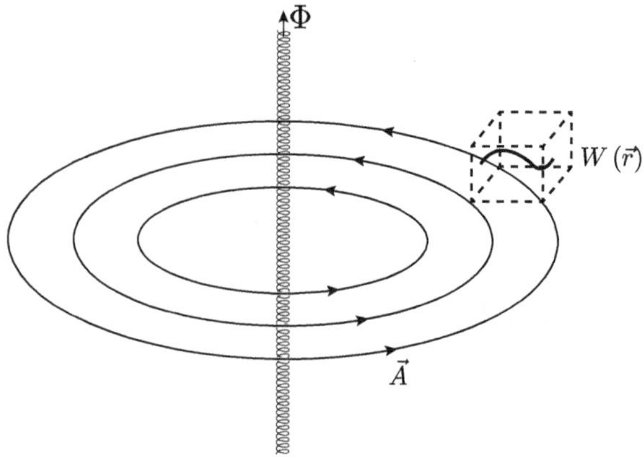

Fig. 3.1. An electron in a vector potential is transported in a box around a solenoid. Note that the wave function of a confined electron is real.

magnetic field, see Fig. 3.1. Its wave function is **real** and is determined by the Schrödinger equation

$$\left[\frac{\vec{p}^{\,2}}{2m} + V(\vec{r})\right] W(\vec{r}) = E W(\vec{r}), \tag{3.2}$$

where $\vec{p} = \frac{\hbar}{i}\nabla$ and m is the particle mass. Here the potential $V(\vec{r})$ confines the particle in a quantum box. Suppose we place a solenoid far away from the box and assume that the box encircles the solenoid.

The particle now satisfies the Schrödinger equation

$$\left[\frac{\vec{\Pi}^{2}}{2m} + V(\vec{r} - \vec{R})\right] \psi(\vec{r}) = E\psi(\vec{r}), \tag{3.3}$$

where $\vec{\Pi} = \frac{\hbar}{i}\nabla - \frac{q}{c}\vec{A}$, $\vec{A}$ is the electromagnetic vector potential due to the solenoid, and $\vec{R}$ is the position of the box. The particle eigenfunction is

$$\psi(\vec{r}) = e^{i\gamma(\vec{r})} W(\vec{r} - \vec{R}), \tag{3.4}$$

where the phase is

$$\gamma(\vec{r}) = \frac{2\pi}{\Phi_q} \int_{\vec{R}}^{\vec{r}} \vec{A}(\vec{r}\,') \cdot d\vec{r}\,'. \tag{3.5}$$

This follows from local gauge transformation. When the vector and scalar potentials transform as follows

$$\vec{A} \to \vec{A} + \nabla\lambda(\vec{r}, t),$$

$$\varphi \to \varphi - \frac{1}{c}\frac{\partial\lambda(\vec{r}, t)}{\partial t}, \tag{3.6}$$

the electron wave function transforms as

$$\psi(x, y) \to \psi(x, y)e^{i\frac{q}{\hbar c}\lambda(x,y)}. \tag{3.7}$$

When $\vec{R}$ traces a circle C round the solenoid (see Fig. 3.1), the particle in the quantum box acquires the Berry connection

$$i\langle\psi(\vec{R})|\nabla_{\vec{R}}\psi(\vec{R})\rangle = i\langle W(\vec{R})|\nabla_{\vec{R}}W(\vec{R})\rangle + \frac{2\pi}{\Phi_q}\vec{A}. \tag{3.8}$$

The first term is zero since W is real. So

$$i\langle\psi(\vec{R})|\nabla_{\vec{R}}\psi(\vec{R})\rangle = \frac{2\pi}{\Phi_q}\vec{A}. \tag{3.9}$$

The Berry phase is thus exactly equal to the Aharonov–Bohm phase

$$i\oint_C \langle\psi(\vec{R})|\nabla_{\vec{R}}\psi(\vec{R})\rangle \cdot d\vec{R} = \frac{2\pi}{\Phi_q}\int_{C_1}\vec{A}\cdot d\vec{R} - \frac{2\pi}{\Phi_q}\int_{C_2}\vec{A}\cdot d\vec{R}$$

$$= \frac{2\pi}{\Phi_q}\oint_C \vec{A}\cdot d\vec{R} = 2\pi\frac{\Phi}{\Phi_q}, \tag{3.10}$$

where the loop is divided into two half circles C_1 and C_2 and $C = C_1 - C_2$. In this problem, the geometric Berry connection is the real magnetic vector potential!

> **Exercise 3.1.** Show that $\langle W(\vec{R})|\nabla_{\vec{R}}W(\vec{R})\rangle = 0$ in Eq. (3.8). Hint: The gradient in the parameter space is not a gradient in real space.

Initially, there was a lot of skepticism about this Aharonov–Bohm effect [1]. Tonomura *et al.* [4] measured the Aharonov–Bohm effect of an electron using a superconducting ring. The experimental set up is shown

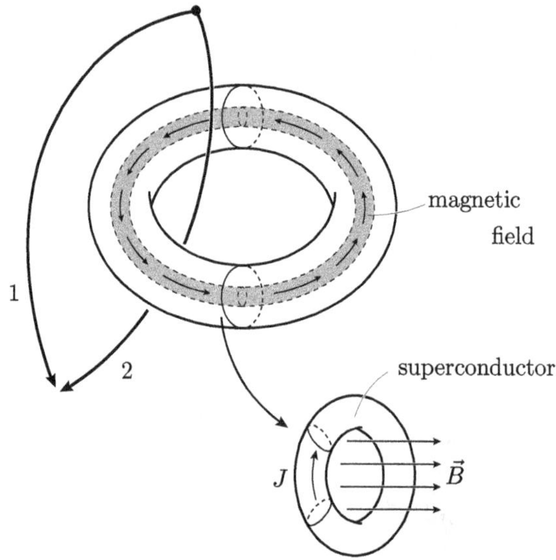

Fig. 3.2. Beautiful experiment of Tonomura *et al.* to measure the Aharonov–Bohm effect. A solenoid is placed inside a superconducting ring. The magnetic field outside the solenoid is zero because the superconductor expels magnetic fields (this is called the Meissner effect). An electron can follow either path 1 or path 2 and form interference patterns due to the vector potential (outside the torus there is no magnetic field).

schematically in Fig. 3.2. Since the flux is quantized as[a]

$$\Phi = \frac{N}{2}\Phi_0, \tag{3.11}$$

in a superconductor ring, it follows from Eq. (3.1) that the Aharonov–Bohm phase of an electron encircling the flux is

$$2\pi\frac{\Phi}{\Phi_0} = \pi N, \tag{3.12}$$

where N is an integer.

One can show that all physical properties of an annulus enclosing a magnetic flux Φ through the opening are periodic in the flux with period $\Phi_0 = hc/e$. This is the Byers–Yang theorem [5].

---

[a]See Sec. 8.9 for an explanation. Note also that a Cooper pair of a superconductor has charge $2e^-$.

3.2. Topology of Multiply-Connected Domain, Singular Gauge Transformation, Twisted Boundary Conditions

Topology of a multiply-connected domain profoundly affects the energy and wave function of an electron. Suppose a ring is threaded by a flux. Using a singular gauge transformation, we will show that its effect is equivalent to imposing a twisted boundary condition.

• Flux and Periodic Boundary Conditions

Suppose a flux Φ from a solenoid with a small diameter r_0 threads a ring with radius R, see Fig. 3.3. The vector potential outside solenoid, $r > r_0$, is

$$\vec{A}_\phi(r) = \frac{\Phi}{2\pi r}\hat{\phi}, \tag{3.13}$$

where $\hat{\phi}$ is the azimuthal unit vector. Consider a particle with a mass m confined to move on a ring. Its Hamiltonian in cylindrical coordinates is

$$H = \frac{\hbar^2}{2mR^2}\left(\frac{1}{i}\frac{\partial}{\partial\varphi} - f\right)^2, \tag{3.14}$$

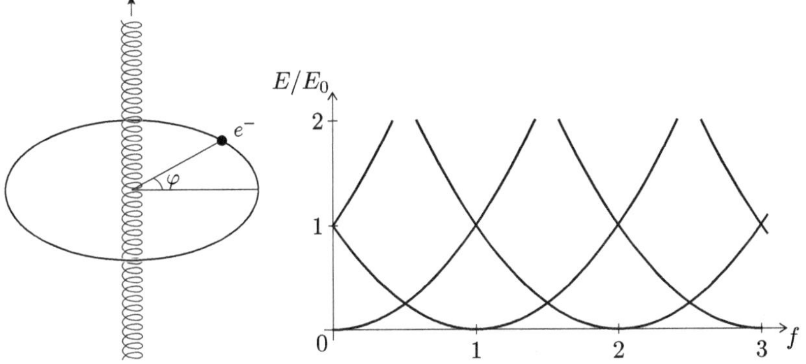

Fig. 3.3. **Left**: A flux threads a ring and an electron moves on the ring. **Right**: The single electron energy of an ideal one-dimensional ribbon threaded by a magnetic flux Φ is $E_n(f) = E_0(n-f)^2$, where $E_0 = \hbar^2/2mR^2$ with the radius of the ring R. Each energy curve is spin degenerate and parabolic. It takes zero value at integer values $f = n$. Note that at half-integer values of flux the ground state is **degenerate**.

where $f = \frac{\Phi}{\Phi_0}$ is the flux. Under periodic boundary conditions, the eigenstates and eigenenergies are, respectively,

$$\psi_b(\varphi) = e^{in\varphi} \text{ and } E = \frac{\hbar^2}{2mR^2}(n - f)^2, \tag{3.15}$$

where $n = 0, 1, 2, \ldots$ is proportional to the z-component of the electron angular momentum. We see that the electron energy does depend on the flux, even though the particle never goes near the region where the magnetic field is finite.

> **Exercise 3.2.** Derive the Hamiltonian Equation (3.14). Hint: Use a cylindrical coordinate system with $\vec{A} = A_\rho \hat{\rho} + A_\phi \hat{\phi} + A_z \hat{z}$ and $\int A_\varphi R d\varphi = \Phi$.

> **Exercise 3.3.** Suppose that an electron is initially in the state $n = 0$. If the flux Φ is **adiabatically** changed the electron energy increases. This phenomenon is called "**spectral flow**". Is this in contradiction with local gauge transformation? Hint: Not any local gauge transformation is possible in a ring. The only legitimate local gauge transformations are the large gauge transformations, i.e., discrete insertion of an integer multiple of the elementary flux, see Sec. 3.3. In this case, the energy spectrum is invariant.

- **Twisted Boundary Conditions Replace Flux**

Let us first explain what twisted boundary conditions are. It is a generalization of the more standard periodic boundary conditions. Consider a free particle moving on a circle with length L.[b] Its Hamiltonian in cylindrical coordinates is

$$H = -\frac{\hbar^2}{2mR^2}\frac{\partial^2}{\partial\varphi^2}. \tag{3.16}$$

The most general boundary condition that the eigenstates satisfy is a **twisted** boundary condition

$$\psi(\varphi + 2\pi) = e^{i\theta}\psi(\varphi), \tag{3.17}$$

---

[b]A ring is not simply connected, because any loop that encloses the hole cannot be contracted to a point without exiting the region.

where $\varphi \in [0, 2\pi]$. Note that the absolute values of the wave functions are identical. (Bulk properties are **not** expected to depend on the type of the boundary conditions used.) When $\theta = 0$, the eigenstates are periodic and are given by

$$\psi(\varphi) = e^{ikR\varphi} \tag{3.18}$$

and their eigenenergies are

$$E = \frac{\hbar^2 k^2}{2m}, \tag{3.19}$$

where $k = \frac{2\pi n}{L}$ (n is an integer). When $\theta \neq 0$, the solutions are aperiodic and are given by

$$\psi(\varphi) = e^{i\frac{2\pi n+\theta}{L}R\varphi}, \tag{3.20}$$

and they satisfy a twisted boundary condition $\psi(0)e^{i\theta} = \psi(2\pi)$. Their energies are

$$E_n = \frac{\hbar^2}{2mL^2}(2\pi n + \theta)^2. \tag{3.21}$$

Note that if you adiabatically change $\theta = 0 \to 2\pi$, the electron ends up in a new state $\psi_{n+1}(\varphi)$ and not in the original state $\psi_n(\varphi)$.

Exercise 3.4. Show that θ in the boundary condition (3.17) may be regarded as the wave vector k of a Bloch wave function

$$k = \theta/2\pi R. \tag{3.22}$$

Now reconsider a particle moving on a ring threaded by a magnetic flux. Since the magnetic field is zero in the region outside the solenoid, $r > r_0$, the vector potential may be **removed** via a local gauge transformation given in Eq. (3.7): $\vec{A}' = \vec{A} + \nabla\lambda$ for $r > r_0$, where the scalar function is $\lambda = -\Phi\frac{\varphi}{2\pi}$. The transformed vector potential is zero

$$A'(r) = \vec{A} + \frac{1}{r}\frac{\partial\lambda}{\partial\varphi} = \vec{A} - \frac{\Phi}{2\pi r} = 0. \tag{3.23}$$

(Note that the scalar function λ is singular, i.e., it is multi-valued: its values at ϕ and $\phi + 2\pi n$ are different.) Since the new vector potential vanishes in the region $r > r_0$, the Hamiltonian becomes the free Hamiltonian:

$$H = -\frac{\hbar^2}{2mR^2}\frac{\partial^2}{\partial\varphi^2}.$$
(3.24)

The vector potential does not appear in the Hamiltonian, but is, instead, shifted to the boundary conditions imposed on the wave function: When the above local gauge transformation is performed the initial wave function $\psi_b(\varphi)$ changes into

$$\psi_a(\varphi) = e^{-i\frac{\Phi}{\Phi_0}\varphi}\psi_b(\varphi).$$
(3.25)

Note that

$$\psi_a(\varphi) \neq \psi_a(\varphi + 2\pi),$$
(3.26)

i.e., it satisfies a twisted boundary condition (see Fig. 3.4)

$$\psi_a(2\pi) = e^{-i\frac{\Phi}{\Phi_0}2\pi}\psi_a(0).$$
(3.27)

It is interesting to note the following: When $\Phi = n\Phi_0$ (n is an integer) the wave function is periodic. There is no trace of the solenoid flux even in the wave function. Thus when the flux is an **integer multiple** of the elementary quantum flux, its effect can be **removed** by performing a singular gauge transformation. (This will be discussed more when we study fractional quantum Hall effect in Sec. 11.6.)

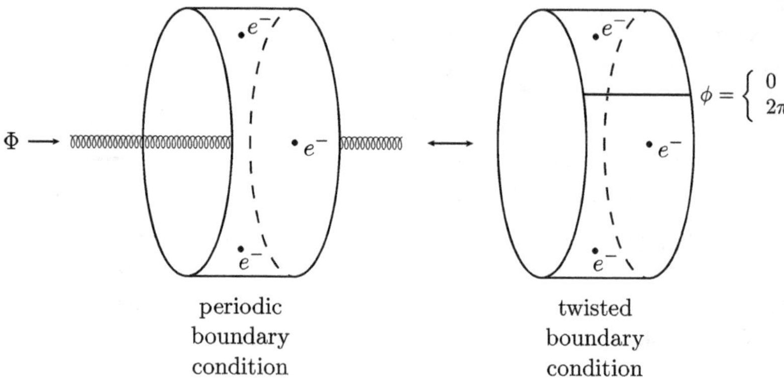

Fig. 3.4.   Consider electrons living on a ribbon threaded by a magnetic flux. By performing a local gauge transformation, the vector potential that the electrons experience can be replaced by a twisted boundary condition, which means that the values of the wave function at $\varphi = 0$ and 2π differ by a phase factor. This **singular** gauge transformation of course does not affect the magnetic field inside the solenoid.

> **Exercise 3.5.** Derive the wave function given in Eq. (3.25). **Hint:** It cannot be derived from the free Hamiltonian. Instead try to find out how the wave function changes under the local gauge transformation.

Interesting physics can be extracted by computing how much the eigenvalues change when a periodic boundary condition is changed to a twisted boundary condition. The Thouless number is a measure for how much the eigenenergies change when the boundary conditions are altered. Investigation of the Thouless number is useful because it is expected to be related to the conductivity, see Sec. 10.11.

3.3. Degeneracy and Combined Operation of Time-Reversal Operation and Large Gauge Transformation

Consider narrow semiconductor rings in the presence of the Rashba and Dresselhaus spin–orbit terms. The electron Hamiltonian has the following symmetry: it is invariant under a time-reversal operation[c] followed by a large gauge transformation, which is defined in Eq. (3.36) below.

One can show [6] that these wave functions of a degenerate pair are related to each other by the combined symmetry operation mentioned above. The eigenstates are doubly degenerate when an integer or half-integer quantum flux threads the quantum ring, see Fig. 3.5. These results are valid even in the presence of a disorder potential. When the Zeeman term is present, only some of these degenerate levels anticross. Below we explain in detail how these results are obtained.

Let us now consider an electron with the effective mass m^* in a semiconductor ring in the presence of the spin-orbit terms and under the influence of a vector potential due to a solenoid at the center of the ring, see Fig. 3.3. The total effective mass Hamiltonian (see Appendix A) is $H = H_0 + H_R$. The first part is

$$H_0 = \hbar^2 \Pi^2 / 2m^* + U(\vec{\rho}) + V(z), \tag{3.28}$$

---

[c]Note that one does not change the directions of the **external** magnetic field under time-reversal operation. Only variables of the system under consideration are changed.

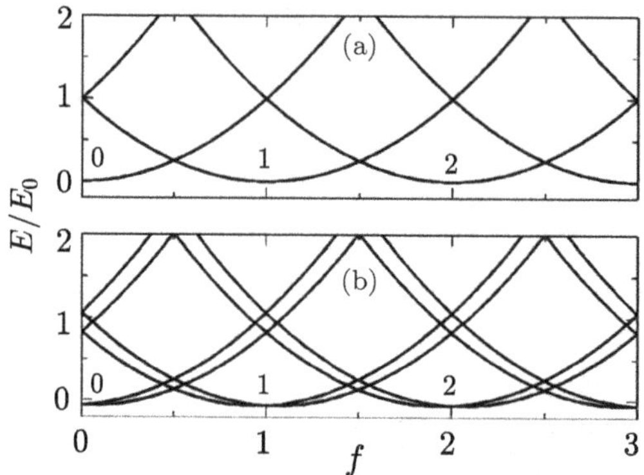

Fig. 3.5. Energy levels of a narrow ring as a function dimensionless flux f in the presence of the Rashba spin–orbit term, which lifts spin degeneracy. The energy scale associated with the Rashba constant is set to E_0. The numbers near the curves label the energy curves; they are no longer z components of angular momentum since the rotational symmetry is broken. At integer and half-integer values of f the energy levels are doubly degenerate.

where $\vec{\Pi}$ is the canonical momentum, $U(\vec{\rho})$ is the radial potential energy that confines an electron to a ring, $V(z)$ is the Rashba confinement potential (see Fig. 2.10(a)), and $\vec{\rho} = (x, y)$. The Rashba spin–orbit term is

$$H_R = c_R(\vec{\sigma} \times \vec{\Pi}) \cdot \hat{z} = c_R \left[\sigma_x \left(k_y + \frac{e}{\hbar c} A_y \right) - \sigma_y \left(k_x + \frac{e}{\hbar c} A_x \right) \right],$$

$$(3.29)$$

where $\hbar k_{x,y}$ are the kinematical momentum operators. The vector potential of the solenoid is $\vec{A}$. The constant c_R depends on the external electric field E applied along the z-axis. The constant represents breaking of inversion symmetry by the crystal in zinc blende structures. The confinement potentials along the z-axis and radial direction are assumed to be sufficiently strong that only the lowest-energy subbands are relevant. (There is another term called the Dresselhaus spin–orbit term but its expectation value with respect to the first subband wave function vanishes.)

The total effective mass electron wave function in a narrow semiconductor ring can be written as

$$\Psi(\vec{r}) = f(z)R(\rho)G(\varphi)$$

$$(3.30)$$

The lowest subband wave functions along the radial and z-axis directions are $R(\rho)$ and $f(z)$, respectively. We will assume that the wave functions

$R(\rho)$ and $f(z)$ are known except for $G(\varphi)$. The effective Hamiltonian for $R(\rho)$ and $G(\varphi)$ can be written as

$$H = -\frac{\hbar^2}{2m^*}\left(\frac{\partial^2}{\partial\rho^2} + \frac{1}{\rho}\frac{\partial}{\partial\rho} - \frac{1}{\rho^2}\left(\frac{1}{i}\frac{\partial}{\partial\varphi} - f\right)^2\right) + \zeta\sigma_z$$

$$-\frac{c_R}{\rho}(\sigma_x\cos\varphi + \sigma_y\sin\varphi)\left(\frac{1}{i}\frac{\partial}{\partial\varphi} - f\right)$$

$$-ic_R(\sigma_y\cos\varphi - \sigma_x\sin\varphi)\frac{\partial}{\partial\rho}, \tag{3.31}$$

where the Zeeman term $\zeta\sigma_z$ is added (the z-component of the wave function $f(z)$ is ignored here because it just contributes a constant, which may be absorbed in the eigenenergy). If the ring is infinitely thin one may set $\rho = R$ in the Hamiltonian. However, in the presence of the Rashba term, one cannot just set $\rho = R$ because there is a term involving the derivative $\frac{\partial}{\partial\rho}$. One needs to take into account properly the confinement of the wave function in the radial direction and investigate the limit $\rho \to R$. The resulting effective one-dimensional Hamiltonian is

$$H = H_1 + H_2,$$

$$H_1 = \frac{\hbar^2}{2m^*R^2}\left(\frac{1}{i}\frac{\partial}{\partial\varphi} - f\right)^2 + \zeta\sigma_z - \frac{c_R}{R}(\sigma_x\cos\varphi + \sigma_y\sin\varphi)\left(\frac{1}{i}\frac{\partial}{\partial\varphi} - f\right),$$

$$H_2 = -i\frac{c_R}{2R}(\sigma_y\cos\varphi - \sigma_x\sin\varphi). \tag{3.32}$$

Here H_1 is obtained from Eq. (3.31) by setting $r = R$ and ignoring the derivative terms. A correction term H_2 is added [7].

The azimuthal wave function $G(\varphi)$ is an eigenstate of the following Hamiltonian matrix:

$$H_{\mathrm{eff}} = \begin{pmatrix} E_0(\frac{1}{i}\frac{\partial}{\partial\varphi} - f)^2 + \zeta & e^{-i\varphi}E_R(\frac{1}{i}\frac{\partial}{\partial\varphi} - f - 1/2) \\ e^{i\varphi}E_R(\frac{1}{i}\frac{\partial}{\partial\varphi} - f + 1/2) & E_0(\frac{1}{i}\frac{\partial}{\partial\varphi} - f)^2 - \zeta \end{pmatrix}, \tag{3.33}$$

where $E_0 = \hbar^2/2m^*R^2$, $E_R = c_R/R$, and the Zeeman term is $\zeta = \tilde{g}\mu_0 B/2$ (B, $\tilde{g}$, and μ_0 are, respectively, the strength of magnetic field, the electron g–factor of the semiconductor, and the Bohr magneton). Since the magnetic field is confined to the solenoid, the Zeeman term ζ can be ignored. Note that the off–diagonal terms contain $1/2$ and $-1/2$ that originate from H_2. They make the Hamiltonian matrix **Hermitian**: The eigenstates have the following structure:

$$G(\varphi) = \begin{bmatrix} a_m e^{im\varphi} \\ b_{m+1} e^{i(m+1)\varphi} \end{bmatrix}, \tag{3.34}$$

and a_m and b_{m+1} satisfy the Hermitian matrix equation

$$\begin{pmatrix} E_0(m-f)^2 & E_R(m-f+1/2) \\ E_R(m-f+1/2) & E_0(m-f+1)^2 \end{pmatrix} \begin{bmatrix} a_m \\ b_{m+1} \end{bmatrix} = E \begin{bmatrix} a_m \\ b_{m+1} \end{bmatrix}. \tag{3.35}$$

> **Exercise 3.6.** Derive the Hamiltonian H_{eff} given in Eq. (3.35) from Eq. (3.32).

The energy spectrum is plotted as a function of flux in Fig. 3.5. There are degenerate pairs of eigenstates at integer or half-integer fluxes $g\Phi_0$ ($g = 1/2$ or 1). Let us try to understand these degeneracies using a symmetry argument [6]. Consider a **discrete** transformation called a **large gauge transformation** which changes the flux by an integer multiple of Φ_0. When the flux change is $\Delta\Phi_0 = -2\Phi_0$,

$$g\Phi_0 \to g\Phi_0 - 2g\Phi_0. \tag{3.36}$$

It flips the direction of the vector potential $(\vec{A} \to -\vec{A})$. We will combine this transformation with time-reversal operation.[d] Under time-reversal operation, spins transform as $\vec{s} \to -\vec{s}$ ($\vec{s} = \vec{\sigma}/2$). Under the combined transformation of these two operations, the resulting electron Hamiltonian is invariant but the wave function transforms as follows:

$$\begin{bmatrix} F_\uparrow(\varphi) \\ F_\downarrow(\varphi) \end{bmatrix} \to e^{i2g\varphi} \begin{bmatrix} F_\uparrow(\varphi) \\ F_\downarrow(\varphi) \end{bmatrix} \to \begin{bmatrix} -e^{-i2g\varphi} F_\downarrow^*(\varphi) \\ e^{i2g\varphi} F_\uparrow(\varphi) \end{bmatrix}. \tag{3.38}$$

---

[d]Under time reversal operation, a spinor transforms as (see [8])

$$\begin{bmatrix} a \\ b \end{bmatrix} \to \begin{bmatrix} -b^* \\ a \end{bmatrix}. \tag{3.37}$$

Note that when a flux $\Delta\Phi = -2g\Phi_0$ is added to the solenoid, the electron wave functions $F_\sigma(\varphi)$ acquires a phase factor

$$e^{-i\frac{\Delta\Phi}{\Phi_0}\varphi} = e^{i2g\varphi}. \tag{3.39}$$

This procedure is equivalent to applying time-reversal operation first and then adding the flux

$$\begin{bmatrix} F_\uparrow(\varphi) \\ F_\downarrow(\varphi) \end{bmatrix} \rightarrow \begin{bmatrix} -F_\downarrow^*(\varphi) \\ F_\uparrow^*(\varphi) \end{bmatrix} \rightarrow \begin{bmatrix} -e^{-i2g\varphi}F_\downarrow^*(\varphi) \\ e^{i2g\varphi}F_\uparrow^*(\varphi) \end{bmatrix}. \tag{3.40}$$

Note that when a flux $-2g\Phi_0$ is added, the complex conjugate of the wave function $F_\downarrow^*(\varphi) \rightarrow e^{-i2g\varphi}F_\downarrow^*(\varphi)$. From these results, we infer that the presence of the pairs of degenerate states at integer and half-integer values of $f = \Phi/\Phi_0$ is consistent with the combined symmetry operation of the Hamiltonian, namely a large gauge transformation followed by time-reversal operation. This result is valid even in the presence of a disorder potential and in noncircular rings.

3.4. Topological Quantization and Dirac Monopole

In Sec. 2.4, we learned about a magnetic monopole in the parameter space. Here we discuss a magnetic monopole in the **real** space. We will argue that its magnetic charge is quantized.

Suppose a magnetic monopole is at the origin of a coordinate system. Its magnetic field $\vec{b}$ behaves as g/r^2 and is directed in the radial direction (g is called a **magnetic charge**). What is the vector potential that gives rise to the magnetic field of a monopole? Suppose $\vec{A}$ is such a vector potential. But then its magnetic flux is zero

$$\int_S \nabla \times \vec{A} \cdot dS = \int \nabla \cdot (\nabla \times \vec{A})dV = 0 \tag{3.41}$$

because the divergence of a curl is always zero. (Here S is a closed surface enclosing the magnetic monopole.) Something is not right: It turns out that our assumption $\vec{b} = \nabla \times \vec{A}$ is incorrect — one cannot find a non-singular $\vec{A}$ that gives the correct $\vec{b}$ everywhere.

One can resolve this problem by introducing an infinitely thin solenoid (a Dirac string) carrying an **integer multiple** n of the quantum unit flux Φ_0, see Fig. 3.6. A Dirac string links a monopole and antimonopole of opposite magnetic charges. In this toy model of a magnetic monopole the vector

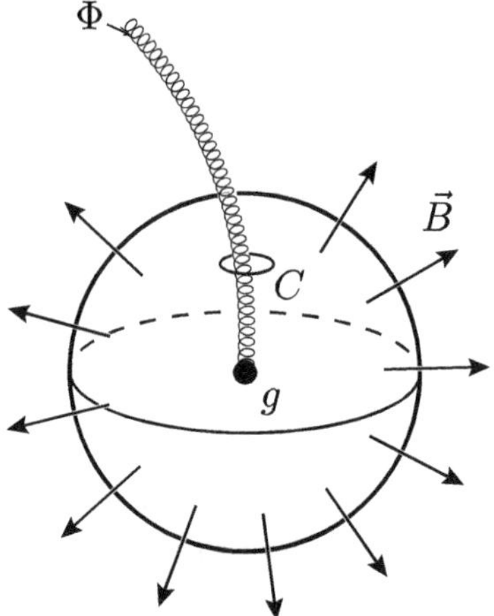

Fig. 3.6. Toy model of a monopole. Its magnetic charge is g. Its flux is an integer multiple of Φ_0.

potential is singular on the z-axis, either for $z > 0$ or $z < 0$. The contribution from the solenoid string may be neglected because it produces an Aharonov–Bohm phase of $\pm 2\pi n$. The Dirac monopole behaves like a point magnetic charge with no string attached. However, the vector potential is singular where the string is.

Using this toy model of a magnetic monopole, let us derive the quantization rule for the magnetic charge g. Assume that the solenoid goes through a small hole C near "the north pole". There are now two magnetic fields $\vec{b}$ and $\vec{B}$. The first one exists only inside the solenoid. The second exists only outside the solenoid (it exists because the solenoid is abruptly ended). Now we demand

$$\nabla \times \vec{A} = \vec{b} + \vec{B}. \tag{3.42}$$

The total flux of the solenoid is equal to the flux that threads the sphere

$$\Phi = 4\pi R^2 \times \frac{g}{R^2} = 4\pi g, \tag{3.43}$$

where R is the radius of the sphere. A particle with electric charge q that orbits a small circle C "feels" the magnetic field $\vec{b}$ confined inside the solenoid. (On the other hand, the flux of $\vec{B}$ through the area defined by C is negligible since the area is infinitely small.) The flux of the solenoid induces the Aharonov–Bohm phase

$$\gamma = 2\pi\Phi/\Phi_q = 2\pi(4\pi g)/\Phi_q = 8\pi^2 gq/(hc). \tag{3.44}$$

We see that it is proportional to the electric charge q of the particle, as well as to the magnetic charge g of the source. Because the flux of the solenoid is equal to an integer multiple of the quantum unit flux, the phase factor $e^{i\gamma}$ must be 1. Hence the phase γ must be an integer multiple of 2π: $\gamma = 2\pi n$. Thus we find that the quantization of the magnetic charge g is given by

$$\frac{2gq}{\hbar c} = n. \tag{3.45}$$

One can also use a method of differential geometry[e] to derive the quantization of the magnetic charge [9–11]. The method is called a coordinate patch. The introduction of an artificial Dirac string can be avoided by defining **two** singular vector potentials and patch them together. We define a function for the vector potential on the "northern hemisphere" (the half-space $z > 0$ above the particle), and another function for the southern hemisphere. Away from the southern pole the vector potential is

$$\vec{A}_N = \frac{g}{r}\frac{(1-\cos\theta)}{\sin\theta}\hat{e}_\varphi, \tag{3.46}$$

where $\theta < \pi$ (here cgs units are used). Away from the northern hemisphere the vector potential is

$$\vec{A}_S = -\frac{g}{r}\frac{(1+\cos\theta)}{\sin\theta}\hat{e}_\varphi, \tag{3.47}$$

where $\theta > 0$. The difference between them is

$$\vec{A}_N - \vec{A}_S = 2\frac{g}{r}\frac{1}{\sin\theta}\hat{e}_\varphi = \vec{\nabla}(2g\varphi). \tag{3.48}$$

Thus $\vec{A}_N$ and $\vec{A}_S$ are related by a local gauge transformation $\vec{A} \to \vec{A} + \vec{\nabla}\lambda(\varphi)$, where $\lambda = 2g\varphi$.

If the magnetic monopole has an **electric charge** q, the wave functions of a particle in the vector potentials $\vec{A}_S$ and $\vec{A}_N$ are related by the local

---

[e]See Ref. [12]. This book is the author's choice of a good introduction to differential geometry. For a short and gentle introduction, see [13].

gauge transformation

$$\psi_S(\vec{r}) = \psi_N(\vec{r})e^{i\frac{q}{\hbar c}2g\varphi}. \tag{3.49}$$

Since the wave function must be single-valued at $\varphi = 0$ and 2π, we have

$$\frac{q}{\hbar c}4\pi g = 2\pi n, \tag{3.50}$$

where n is an integer. This implies the following quantization rule given in Eq. (3.45). It follows from this result that if electric charges are quantized then magnetic charges are also quantized.

Exercise 3.7. Find magnetic field of a magnetic monopole whose vector potential is given Eqs. (3.46) and (3.47). Hint: Use

$$\nabla \times \vec{A} = \frac{1}{r\sin\theta}\frac{\partial}{\partial\theta}(A_\varphi \sin\theta)\hat{e}_r - \frac{1}{r}\frac{\partial}{\partial r}(rA_\varphi)\hat{e}_\theta. \tag{3.51}$$

Exercise 3.8. Find out how the quantization equation, Eq. (3.45), changes in SI units. Note

$$\nabla\vec{E} = \rho(\vec{r})/\epsilon_0 \tag{3.52}$$

in SI units. In contrast,

$$\nabla\vec{E} = 4\pi\rho(\vec{r}) \tag{3.53}$$

in cgs units. Answer:

$$\frac{gq}{2\pi\hbar} = n. \tag{3.54}$$

Further reading. Imagine two magnetic monopoles of opposite charges in a superconductor. The magnetic fields between them will be confined into a region that has the shape of a sausage. And in this region superconductivity is destroyed. This object is called the Nielson–Olesen vortex [14]. It was a toy model for quark confinement.[f] We have studied the

---

[f] For a popular account, see "In search of the ultimate building blocks" by G. 't Hooft [15].

Dirac magnetic monopole in an Abelian gauge theory. There is also a magnetic monopole in a non-Abelian gauge theory. It is called the 't Hooft–Polyakov monopole [11]. It does not have a mathematical singularity as the Dirac monopole does.

3.5. Instanton and Singular Coupling in Double Well Potential

Singular perturbation is abound in condensed matter physics [3]. An example is an instanton. When it is applied to a double quantum well, one finds that the energy splitting between the bonding and antibonding states is $\Delta E \propto e^{-1/\eta}$, which is singular in the coupling constant η. An instanton is a non-perturbative effect and originates from the topology of a double well. (In field theory an instanton is used to connect two classical vacua.)

Below we demonstrate the singular aspect of the energy splitting in a double quantum well (Disorder will also couple the left and right zigzag edges in a singular manner, see Sec. 18.1.) An instanton can be thought of as a soliton moving in the imaginary time domain. Consider the following double well potential with minima at $x = \pm a$

$$V(x) = \lambda(x^2 - a^2)^2. \tag{3.55}$$

According to the path integral formulation of quantum mechanics, the amplitude for going from $-a$ to a is given by

$$\langle a|e^{-iHT/\hbar}|-a\rangle = \int \mathcal{D}x(t)e^{iS/\hbar}, \tag{3.56}$$

where S is the classical action and the symbol $\mathcal{D}x(t)$ means integration over all possible paths from $-a$ to a, see [16]. The time to go from the position $-a$ to a is T. We now switch to imaginary time $t \to -i\tau$. (This operation of a multiplication with the imaginary unit can be interpreted as a $\pi/2$-rotation in the complex plane. It turns Minkowski space into Euclidean space and is called a Wick's rotation.) The new amplitude is

$$\langle a|e^{-HT/\hbar}|-a\rangle = \int \mathcal{D}x(\tau)e^{-S/\hbar}, \tag{3.57}$$

where the action is

$$S = \int_{-T/2}^{T/2} d\tau \left[\frac{1}{2}m^* \left(\frac{dx}{d\tau}\right)^2 + V(x)\right]. \tag{3.58}$$

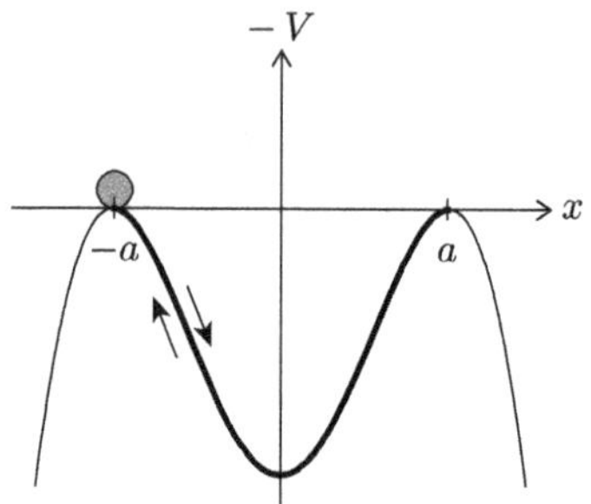

Fig. 3.7. The classical solution of the equation of motion that connects the two potential maxima $(-a \to a)$ is called an instanton solution while a solution traversing the same path but in the opposite direction $(a \to -a)$ is called an anti–instanton.

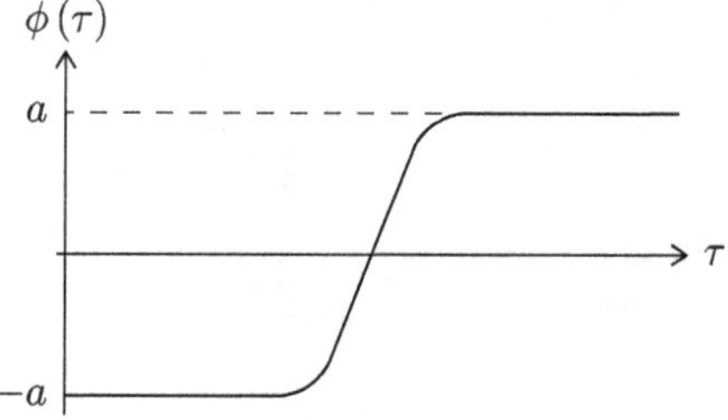

Fig. 3.8. Trajectory of a single instanton.

(Here $T \to -i\mathcal{T}$.) This action describes a particle moving in the **inverted** potential $-V(x)$, see Fig. 3.7. An extremum path called an instanton is shown in Figs. 3.7 and 3.8. Its probability amplitude is

$$A = K e^{-S/\hbar}. \tag{3.59}$$

(The constant K is not of interest to us and we do not compute it here.)

There are other extremum paths with odd number of instantons (3, 5, etc.), see Fig. 3.9. Each number characterizes a different **topological sector** (see Sec. 3.6). Suppose a multiple instanton consists of n successive instanton transitions at τ_1, τ_2, etc. We classify each multiple instanton solution by the centers $(\tau_1, \tau_2, \ldots, \tau_n)$ where instant changes of particle positions take place. The n instanton bounces contributing to the probability amplitude A can take place at arbitrary times $\tau_i \in [-\mathcal{T}/2, \mathcal{T}/2]$, $i = 1, \ldots, n$, and all these possibilities have to be added (i.e. integrated). We assume that each instanton in a multiple instanton is independent of each other (dilute instanton gas approximation). Instantons thus give the

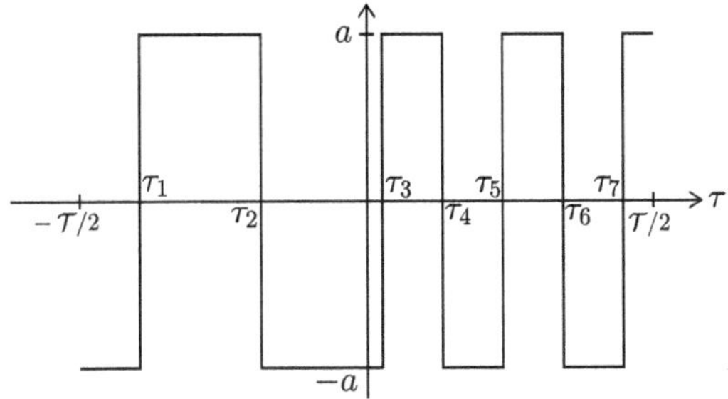

Fig. 3.9. Multiple instantons.

following contribution to the path integral

$$\int_{-\mathcal{T}/2}^{\mathcal{T}/2} dt_n e^{-S_{\tau_n}/\hbar} \int_{-\mathcal{T}/2}^{\tau_n} d\tau_{n-1} e^{-S_{\tau_{n-1}}/\hbar} \cdots \int_{-\mathcal{T}/2}^{\tau_2} d\tau_1 e^{-S_{\tau_1}/\hbar}. \quad (3.60)$$

Here the action of a multiple instanton is split into the sum of single instanton actions. So n instantons for a given time set $(\tau_1, \tau_2, \ldots, \tau_n)$ contribute A^n to the amplitude. To find the total amplitude all possible time values $(\tau_1, \tau_2, \ldots, \tau_n)$ must be summed:

$$\int_{-\mathcal{T}/2}^{\mathcal{T}/2} dt_n \int_{-\mathcal{T}/2}^{\tau_n} d\tau_{n-1} \cdots \int_{-\mathcal{T}/2}^{\tau_2} dt_1 = \frac{\mathcal{T}^n}{n!}. \quad (3.61)$$

The total amplitude for all possible values of n is

$$\sum_{n \in \mathrm{odd}} (\mathcal{T}A)^n/n!. \quad (3.62)$$

We thus obtain

$$\langle a|e^{-H\mathcal{T}/\hbar}|-a\rangle = \left(\frac{m^*\omega}{\hbar}\right)^{1/2} e^{-\omega\mathcal{T}/2} \sinh(\mathcal{T}Ke^{-S/\hbar}), \quad (3.63)$$

where

$$S = \int_{-a}^{a} dx \sqrt{2m^*V(x)} = \frac{4}{3}a^3\sqrt{2m^*\lambda} = \frac{m^{*2}}{12}\frac{\omega^3}{\lambda}. \quad (3.64)$$

Here ω is the frequency of a harmonic oscillator at the bottom of each potential well

$$\omega = \sqrt{\frac{8\lambda a^2}{m^*}}. \quad (3.65)$$

Below we will take m^*, λ, and ω as independent physical parameters instead of m^*, λ, and a. For the transition $-a \to -a$ we find the following amplitude

$$\langle -a | e^{-H\mathcal{T}/\hbar} | -a \rangle = \left(\frac{m^*\omega}{\hbar} \right)^{1/2} e^{-\omega\mathcal{T}/2} \cosh(\mathcal{T}Ke^{-S/\hbar}). \qquad (3.66)$$

The amplitude can be resolved using a complete set of eigenstates of the double well $|n\rangle$

$$\langle -a | e^{-H\mathcal{T}/\hbar} | -a \rangle = \sum_n \langle -a | n \rangle \langle n | -a \rangle e^{-\mathcal{T}E_n/\hbar}$$
$$\approx \langle -a | + \rangle \langle + | a \rangle e^{-\mathcal{T}E_+/\hbar}$$
$$+ \langle -a | - \rangle \langle - | a \rangle e^{-\mathcal{T}E_-/\hbar}. \qquad (3.67)$$

Here only two dominant terms are kept: In the limit $\mathcal{T} \to \infty$ the dominant terms are the nearly degenerate lowest-energy states, namely symmetric and antisymmetric states $|+\rangle$ and $|-\rangle$. The energies of the symmetric and antisymmetric states are (see Exercise 3.9)

$$E_\pm = \frac{\hbar\omega}{2} \mp K\hbar e^{-S/\hbar}, \qquad (3.68)$$

where the value of S is given in Eq. (3.64). The energy splitting is

$$\Delta E \propto e^{-1/\eta}, \quad \eta = 12\frac{\hbar\lambda}{m^*\omega^3}. \qquad (3.69)$$

As the strength of the potential λ increases the dimensionless coupling constant η increases and the splitting grows. Also as the coupling between the two wells increases, i.e., as the distance a between them decreases, the splitting also grows (as a decreases ω decreases and η increases). Note that ΔE has an essential singularity at $\eta = 0$.[g]

> **Exercise 3.9.** Show the result given in Eq. (3.68). Hint: Write $e^{-\frac{\omega\mathcal{T}}{2}} \cosh(\mathcal{T}Ke^{-S/\hbar}) \propto e^{-E_+\mathcal{T}} + e^{-E_-\mathcal{T}}$.

---

[g] The function

$$e^{1/z} = 1 + \frac{1}{z} + \frac{1}{2!z^2} + \cdots \qquad (3.70)$$

has an essential singularity at $z = 0$.

3.6. Topological Actions

There are physical systems whose Lagrangians have non-perturbative topological terms. Here we give two simple examples.

• Topological Term of Ring

The following example shows what a topological action is. Consider again a particle moving on a ring threaded by a solenoid, see Fig. 3.3. Its position may be specified by the azimuthal angle φ. The Lagrangian is

$$L = \frac{m^* R^2}{2} \dot{\varphi}^2 + \hbar f \dot{\varphi}, \tag{3.71}$$

and the action is

$$S = \frac{m^* R^2}{2} \int dt \dot{\varphi}^2 + \hbar f \int dt \dot{\varphi}, \tag{3.72}$$

where $f = \Phi/\Phi_0$ is the flux ratio. This action leads to the Hamiltonian

$$H = \frac{\hbar^2}{2m^* R^2} \left(\frac{1}{i} \frac{\partial}{\partial \varphi} - f \right)^2. \tag{3.73}$$

The resulting equation of motion is

$$\frac{d^2 \varphi}{dt^2} = 0, \tag{3.74}$$

whose solution is

$$\varphi(t) = 2\pi N t/T, \tag{3.75}$$

where the winding number N is an integer and the period is T. This solution satisfies the periodic boundary condition $\varphi(T) = \varphi(0) \bmod 2\pi T$. The second term in the action depends on the winding number N

$$S_{\text{top}} = 2\pi N \hbar f. \tag{3.76}$$

This is called a topological term in the action. The probability amplitude, $e^{\frac{i}{\hbar} S}$, for the path going from point $(\varphi, t) = (0, 0)$ to point $(0, T)$ is

$$\langle 0, T | e^{-iHT/\hbar} | 0, 0 \rangle = \sum_N e^{i 2\pi N f} \int_N D\varphi e^{\frac{i}{\hbar} \frac{m R^2}{2} \int dt \dot{\varphi}^2}, \tag{3.77}$$

where the functional integral is over paths with different winding numbers N. Different winding numbers correspond to different **topological sectors**. We saw that a similar phenomenon appears in the case of instantons (see Sec. 3.5).

> **Exercise 3.10.** Derive the Hamiltonian equation (3.73) from the action given in Eq. (3.72). Hint: Use the canonical momentum $p_\varphi = \frac{\partial L}{\partial \dot{\varphi}}$ and the Hamiltonian $H = \dot{\varphi} p_\varphi - L$.

• Topological Action and Spin Coherent States

Let us give another example of a topological action. Consider a free spin. Since it has no kinetic and potential energies, one may expect that its Lagrangian is zero. However, the presence of a topological term is manifest in the Lagrangian formulation. It was not obvious how to formulate path integral formulation of spins. People realized that it can be done using spin coherent wave functions (for a good introduction to spin coherent path integral, see [17]). Consider a spin $s = 1/2$ in a magnetic field $\vec{B}$. Its Hamiltonian is

$$H = \vec{B} \cdot \vec{s}, \tag{3.78}$$

where $\vec{s}$ is the spin operator. We define a spin coherent state as the state obtained by rotating fully polarized state $|s, s_z\rangle = |s, s\rangle$ by an angle θ around the y-axis and then by ϕ around the z-axis[h]

$$|z\rangle = e^{-i\varphi} \cos\frac{\theta}{2} \left|\frac{1}{2}, \frac{1}{2}\right\rangle + \sin\frac{\theta}{2} \left|\frac{1}{2}, -\frac{1}{2}\right\rangle, \tag{3.80}$$

which can be written as

$$|z\rangle = \begin{bmatrix} z_1 \\ z_2 \end{bmatrix} = \begin{bmatrix} e^{-i\varphi} \cos\frac{\theta}{2} \\ \sin\frac{\theta}{2} \end{bmatrix} = |\vec{n}\rangle, \tag{3.81}$$

------

[h]The rotated state is

$$|z\rangle = \begin{bmatrix} e^{-i\varphi/2} \cos\frac{\theta}{2} \\ e^{i\varphi/2} \sin\frac{\theta}{2} \end{bmatrix}. \tag{3.79}$$

Multiplication of a phase factor $e^{-i\varphi/2}$ leads to Eq.(3.80).

where $\vec{n}$ is a unit vector. This state is a **spin coherent state** [18] because it satisfies

$$\vec{n} \cdot \vec{s} \, |\vec{n}\rangle = s|\vec{n}\rangle. \tag{3.82}$$

(The same definition is used to define spin coherent states for $s \neq 1/2$, see Eq. (3.97) below.) Note the normalization condition

$$\langle z|z \rangle = 1. \tag{3.83}$$

The expectation value of the spin direction is given by $\vec{n}$

$$\vec{n} = \langle z|\vec{\sigma}|z \rangle. \tag{3.84}$$

Suppose the magnetic field changes adiabatically. The time evolution operator $U(t,0)$ can be evaluated as follows. The time interval $[0,T]$ is divided into small intervals at $t_i = iT/N$ $(i = 0, \ldots, N)$. We resolve the amplitude for the time evolution operator as follows ($\hbar = 1$ and $\epsilon = T/N$):

$$\langle \vec{n}_2|U(T,0)|\vec{n}_1 \rangle = \int \left[\prod_i \frac{d\vec{n}(t_i)}{2\pi} \right]$$

$$\times \langle \vec{n}(T)|e^{-i\epsilon H}|\vec{n}(t_{N-1})\rangle \ldots \langle \vec{n}(t_1)|e^{-i\epsilon H}|\vec{n}(0)\rangle. \tag{3.85}$$

Here we have used the amplitude $\langle \vec{n}(t_{i+1})|e^{-i\epsilon H}|\vec{n}(t_i)\rangle$ in the time interval $[t_i, t_{i+1}]$. In addition, the identity

$$\int \frac{d^2\vec{n}}{2\pi}|\vec{n}\rangle\langle \vec{n}| = I \tag{3.86}$$

is repeatedly applied (I is a 2×2 unit matrix). Using the useful identity $1 - \langle z_i|z_i \rangle = 0$ we rewrite

$$\begin{aligned}
\langle \vec{n}(t_{i+1})|e^{-i\epsilon H}|\vec{n}(t_i)\rangle &= \langle z_{i+1}|e^{-i\epsilon H}|z_i\rangle \approx \langle z_{i+1}|z_i\rangle - i\epsilon\langle z_{i+1}|\vec{B}\cdot\vec{s}|z_i\rangle \\
&= 1 - \langle z_i|z_i\rangle + \langle z_{i+1}|z_i\rangle - i\epsilon\langle z_{i+1}|\vec{B}\cdot\vec{s}|z_i\rangle \\
&= e^{\langle z_{i+1}|z_i\rangle - \langle z_i|z_i\rangle - i\epsilon\langle z_{i+1}|\vec{B}\cdot\vec{s}|z_i\rangle}.
\end{aligned} \tag{3.87}$$

Note the spin coherent states are not orthogonal, i.e., $\langle z_{i+1}|z_i\rangle \neq 0$. Using these results, we find the path integral

$$U = \int \frac{dz_1 \ldots dz_N}{(2\pi)^N} e^{-i\epsilon \sum_i \langle z_{i+1}|\vec{B}\cdot\vec{s}|z_i\rangle + \epsilon \sum_i \frac{\langle z_{i+1}|z_i\rangle - \langle z_i|z_i\rangle}{\epsilon}}$$

$$= \int Dz\, e^{-i\int_0^T dt(i\langle z|\partial_t z\rangle + \langle z|\vec{B}\cdot\vec{s}|z\rangle)}, \tag{3.88}$$

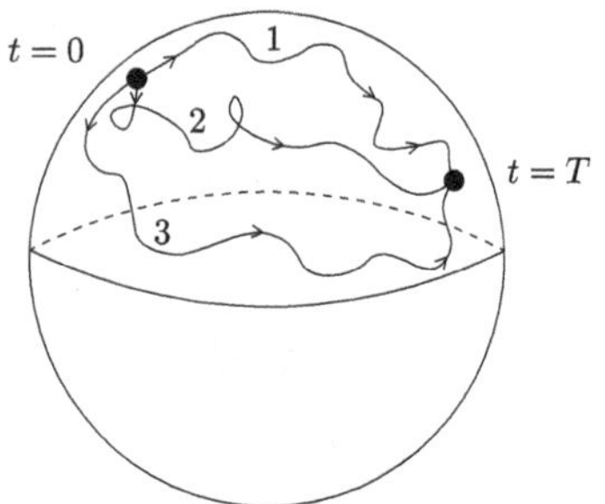

Fig. 3.10. Three examples of spin paths $\vec{n}(t)$ on the Bloch sphere are shown.

where $Dz = dz_1 \ldots dz_N/(2\pi)^N$ implies the sum over all possible paths $z(t)$ (see Fig. 3.10). Here we have used

$$\langle z|\partial_t z\rangle = -\langle \partial_t z|z\rangle. \tag{3.89}$$

Note that this quantity has the same form as the Berry connection

$$\langle z|\partial_t z\rangle, \tag{3.90}$$

and is called the **spin Berry connection**.

The resulting **action** in e^{-iS} is

$$S[n(t)] = \int dt[2siz^\dagger \dot{z} + \vec{B}\cdot\vec{n}s]. \tag{3.91}$$

The first term is the topological Berry term. Suppose the magnetic field changes slowly and the electron spin moves on a closed loop. Then, after time T we get (see Exercise 3.11)

$$\langle \vec{n}|e^{-i\int_0^T dt\vec{B}\cdot\vec{s}}|\vec{n}\rangle = e^{-iE_0 T}e^{iA}, \tag{3.92}$$

where E_0 is the ground state energy and the Berry connection A is

$$A = i\int_0^T dt\langle \vec{n}|\frac{d}{dt}|\vec{n}\rangle. \tag{3.93}$$

It is instructive to note that this result is identical to the previous result of the Berry connections for a two–level system.

Exercise 3.11. Derive Eq. (3.92). Answer: By splitting time integration into small intervals we can write

$$\langle \vec{n}|e^{-i\int_0^T dt H(t)}|\vec{n}\rangle = \langle \vec{n}|\prod_i [1 - i dt H(\tau_i)]|\vec{n}\rangle$$

$$= \prod_i \int \frac{d\vec{n}_i}{2\pi} \langle \vec{n}_i|\vec{n}_{i+1}\rangle [1 - i dt \tilde{H}(t)]$$

$$= \prod_i \int \frac{dz_i}{2\pi} e^{-dt z_i^\dagger \frac{dz_i}{dt}} e^{-i dt \tilde{H}(t)}$$

$$= \int D[z] e^{-\int_0^T dt z^\dagger \frac{dz}{dt} - i \int dt \tilde{H}(t)}$$

$$= \int D[z] e^{-\int_0^T dt z^\dagger \frac{dz}{dt} - i E_0 T}$$

$$= e^{-i E_0 T} e^{-\int_0^T dt \langle \vec{n}|\frac{d}{dt}|\vec{n}\rangle}, \tag{3.94}$$

where

$$\tilde{H}(t) = \frac{\langle \vec{n}_i|H|\vec{n}_{i+1}\rangle}{\langle \vec{n}_i|\vec{n}_{i+1}\rangle} = E_0. \tag{3.95}$$

Going from the first line to the second line we have used Eq. (3.86). From the second to the third line, we have used

$$\langle \vec{n}_i|\vec{n}_{i-1}\rangle = z_i^\dagger z_{i-1} = 1 - z_i^\dagger [z_i - z_{i-1}] = e^{-dt z^\dagger \frac{dz}{dt}}. \tag{3.96}$$

So far we have considered $s = 1/2$. For spin $s = 1$, the coherent states can be formed by considering the direct product of 2 fully polarized states

$$|s, s_z\rangle = |s, s\rangle = \left|\frac{1}{2}, \frac{1}{2}\right\rangle \otimes \left|\frac{1}{2}, \frac{1}{2}\right\rangle. \tag{3.97}$$

We will get a tensor product of rotated states $|z_1\rangle \otimes |z_2\rangle$, or, equivalently, coherent states of the following form:

$$|Z\rangle = \prod_{i=1}^{2} |z_i\rangle \tag{3.98}$$

with the obvious generalization when more than two spins are present. It is more convenient to write these coherent states as

$$|Z\rangle = (z_1, z_2).$$ (3.99)

(We can do this because spins are independent.) Then note that

$$\langle Z|\partial_t Z\rangle = \sum_i \langle z_i|\partial_t z_i\rangle = 2\langle z|\partial_t z\rangle.$$ (3.100)

(Here we have replaced $z_i = z$.) As in the case of $s = 1/2$, the action of the spin Hamiltonian has a topological term

$$S_{\text{top}} = i2s \int_0^T dt\langle z|\partial_t z\rangle.$$ (3.101)

Thus we see that different spins $s = 1/2, 1, \ldots$ have **different topological terms**, which has an important effect in one-dimensional spin chains where the site spins interact with each other. When $s = 1/2$, no excitation gap exists but when $s = 1$ a gap exists, see [19]. This is analogous to the result of an electron moving on a ring threaded by a flux. At half-integer values of the flux, the ground state is gapless while at integer values it is gapful, see Fig. 3.3. At half-integer values of the flux, the Aharonov–Bohm phase is present and is equal to π, see Eq. (3.27). Note that first-order time derivatives appear in both actions given by Eqs. (3.72) and (3.93).

Further reading. For an introduction to topological field theory, see [8, 11]. The fractional quantum Hall effect can be described by a topological (Chern–Simons) field theory. Quantum Hall effects are explained in Chapters 10 and 11.

3.7. Toy Model of Abelian Anyons

An anyon is neither a fermion nor a boson. An anyon is defined as follows: When two anyons are exchanged, a statistical angle is generated, see Fig. 3.11(a). The **anticlockwise**[i] exchange of one anyon around another gives rise to a complex phase factor

$$|\psi_1\psi_2\rangle = e^{i\theta}|\psi_2\psi_1\rangle,$$ (3.102)

---

[i]A clockwise exchange produces $e^{-i\theta}$.

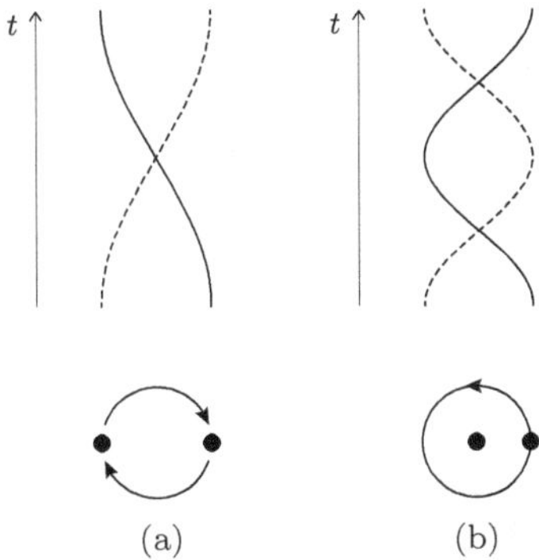

(a)   (b)

Fig. 3.11. (a) Worldlines of a particle exchange. It represents a braiding of worldlines. (b) Worldlines of a particle encircling another particle. Such an encircling is equivalent to two exchanges.

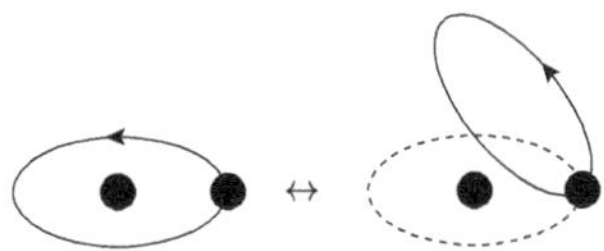

Fig. 3.12. A particle encircles another particle in two dimensions. Anyons can thus exist in two dimensions. However, in three dimensions the encircling path can be lifted out of the two–dimensional plane so that the path no longer encircles the other particle. An anyon thus cannot be defined in three dimensions.

which can be different from ± 1 ($\theta = 2\pi$ for bosons and $\theta = \pi$ for fermions). Anyons do not exist in three dimensions but they can exist in two dimensions. This is because of different **topological** properties of three and two dimensional spaces, see Fig. 3.12. A good example of an anyon is $\frac{e^-}{3}$ fractional quasi-particles of the 1/3 Laughlin state, as we will see in Sec. 11.6.

In this section, we give an artificial toy model of an anyon [20–23], which helps us to think about anyons. A toy model of an anyon consists of a fictitious charge q and flux Φ, see Fig. 3.13(a). The statistical phases can be understood using this model of anyons. Suppose one anyon is encircling the other, as shown in Fig. 3.13(b). We consider this process as consisting of two separate processes put together, as shown in Fig. 3.14. In the first process, the flux of anyon 1 **encircles** the charge q of anyon 2. In the second process, the charge q of anyon 1 encircles the flux of anyon 2 and acquires the Aharonov–Bohm phase $e^{i2\pi\Phi/\Phi_q}$ with $\Phi_q = \frac{hc}{q}$. So the total phase change is the sum of these two phases: $e^{i4\pi\Phi/\Phi_q}$. The encircling operation can be thought of as two successive exchanges. Thus the phase change after

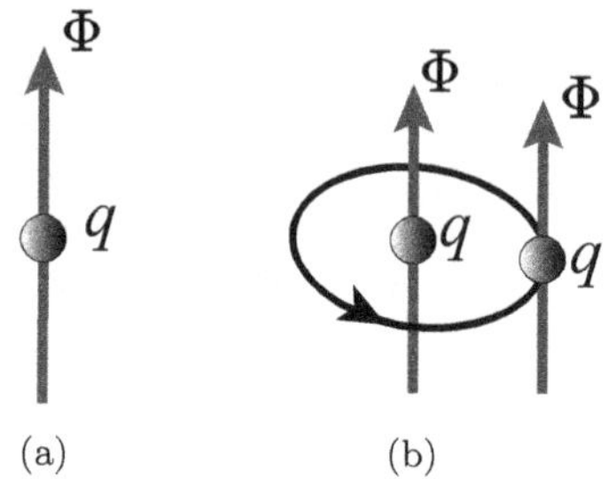

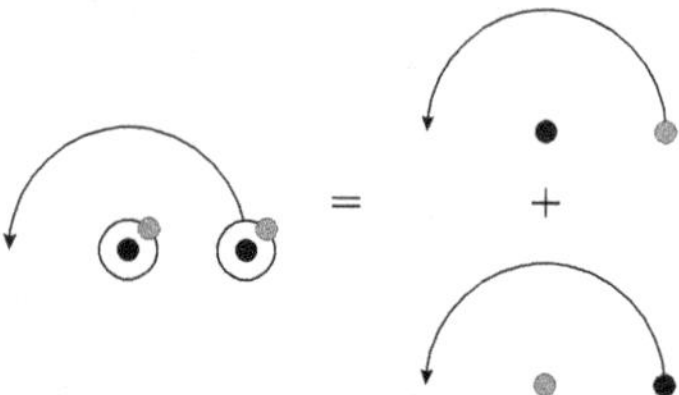

Fig. 3.13.   (a) An anyon is a charge q with a flux Φ attached to it. Its charge encircles its own flux. (b) An anyon encircles another anyon.

Fig. 3.14.   Exchange of two anyons. This is equivalent to the sum of the following two processes: the charge (gray circle) q of the right anyon rotates π around the flux (black circle) of the left anyon and the flux of the right anyon rotates π around the charge q of the left anyon.

one exchange is half of it

$$\theta = 2\pi\Phi/\Phi_q = 2\pi\frac{q}{e}\frac{\Phi}{\Phi_0}. \tag{3.103}$$

If this phase is not π or 2π, the particle is an anyon.

One can formally assign a spin value to an anyon by attributing its spin to the circular motion of its charge around the attached flux.[j] When the anyon charge rotates around its own flux, it acquires the Aharonov–Bohm phase $e^{i2\pi\Phi/\Phi_q}$. When a particle with spin s moves counterclockwisely on a circle, it acquires a phase $e^{i2\pi s}$, see Fig. 3.15. Setting this and the Aharonov–Bohm phases equal to each other we get

$$e^{i2\pi s} = e^{i2\pi\Phi/\Phi_q}. \tag{3.104}$$

---

[j]The spin statistics theorem of $3+1$ relativistic quantum mechanics dictates that a fermion has a half integer spin and a boson an integer spin. Anyons live in two dimensions. Wilczek described their spin statistics as "interpolating continuously between the usual boson and fermion cases" [21].

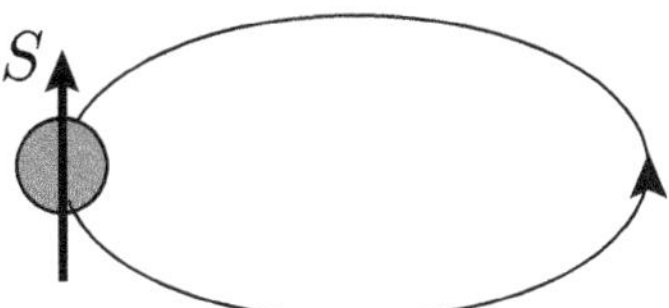

Fig. 3.15. When a particle with spin s rotates 2π counterclockwise it acquires a phase $e^{i2\pi s}$. When an anyon with charge q encircles counterclockwise its **own** flux tube it acquires a phase factor $e^{i2\pi\Phi/\Phi_0}$. This effect may be interpreted as the particle possessing a spin with the acquired phase $e^{i2\pi s}$.

From this, we find the spin of an anyon

$$s = \frac{\Phi}{\Phi_q} = \frac{q}{e}\frac{\Phi}{\Phi_0}. \tag{3.105}$$

Further reading. Anyon statistics can be formulated using a local quantum field theory called Chern–Simons theory. A concise introduction is given in Refs. [8] and [24].

3.8. Fractional Charge and Quantum Fluctuations

What is a fractional charge? A fractional charge must satisfy the following conditions. (1) A fractional charge profile decays fast so that it does not overlap with other fractional charges. (2) It is important that a fractional charge is quantized to a high precision. This means that quantum charge fluctuations must be small. (3) Correlations between fractionalized charges are negligible, i.e., fractionalized charges behave like real independent particles.[k]

The second condition about quantum fluctuations is subtle and needs to be explained in some detail. Consider two nearly degenerate bonding and antibonding end states of a double quantum well, as shown in Figs. 3.16 and 3.17. Each of these states appears to have a bound end fractional charge $e^-/2$. Suppose one applies an external potential $V_L(x)$ to measure whether the electron is in the left well. Is such a fractional charge robust against quantum charge fluctuations generated by the measurement process? When the charge is measured, an electron will fluctuate between the bonding and antibonding states [25, 26], because these states overlap significantly with

---

[k]Fractional charges contribute independently to the density of states. Non-local entanglement between fractional charges are expected to be immaterial (see Sec. 19.2).

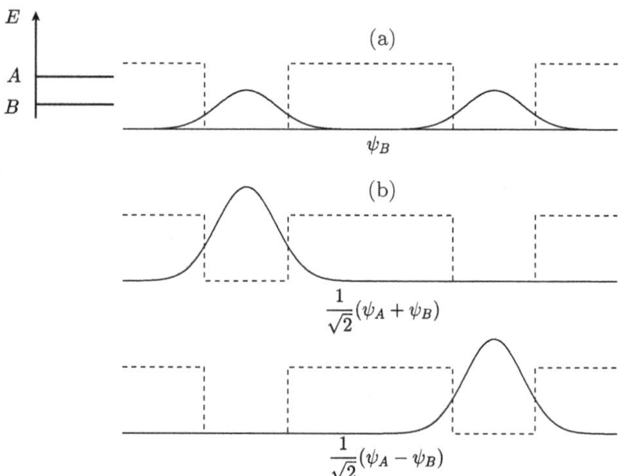

Fig. 3.16. (a) Energy diagram of the bonding and antibonding states of a double quantum well is shown with the probability density of the bonding state. Note the energy splitting between the bonding and antibonding states. (b) Linear combinations of bonding and antibonding states that are localized either in the left or right well.

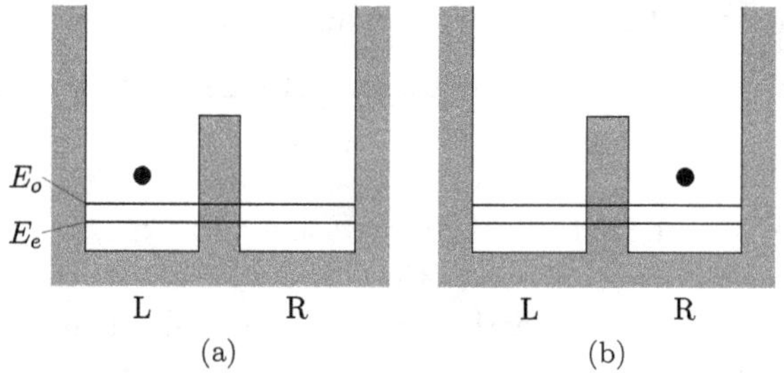

Fig. 3.17. An electron in a double well. Nearly degenerate energy levels E_e and E_0 of the even and odd states are shown. A measurement finds the electron in the left well (a) or in the right well (b). The probability to find the electron in the left (right) well is 1/2.

each other. This means that the electron can be in a **virtual** state for short times.[1] What is the consequence of such fluctuations? The time-dependent wave function of the virtual state is given by the linear combination of the

---

[1] This is a consequence of the time-energy uncertainty relation, see Sec. 5.7.3 of Ref. [10].

bonding and antibonding states, ψ_A and ψ_B,[m]

$$\psi(t) \approx A(t)\psi_A + B(t)\psi_B. \tag{3.106}$$

When $A(t) = B(t) = 1/\sqrt{2}$ or $A(t) = -B(t) = 1/\sqrt{2}$ the electron is localized either in the left or right well, see Fig. 3.16(b) and 3.16(c). The expansion coefficients $A(t)$ and $B(t)$ will vary in a complicated manner over time. This implies that the charge in each well fluctuates, Fig. 3.16(b) and 3.16(c).

The following argument is equivalent to the reasoning given above. Suppose one quickly measures the charge of the right well N_T times: half of the time the measured value is 1 and the other half it is 0. The averaged measured values are $\langle q \rangle = \frac{1}{2}$ and $\langle q^2 \rangle = \frac{N_T}{2} \times 1^2 \times \frac{1}{N_T} = \frac{1}{2}$. Thus the charge variance is significant

$$\langle q^2 \rangle - \langle q \rangle^2 = \frac{1}{4}. \tag{3.107}$$

(Do also Exercise 3.12.)

Exercise 3.12. Consider a double quantum well with bonding and antibonding states $|b\rangle$ and $|a\rangle$. Assume that no other relevant states exist, i.e., other states are well separated in energy. This means that the identity operator may be taken as $I = |a\rangle\langle a| + |b\rangle\langle b|$. Let the operator q_R measure the electron charge in the right well. Show

$$\langle b|q_R^2|b\rangle = \langle b|q_R|b\rangle\langle b|q_R|b\rangle + \langle b|q_R|a\rangle\langle a|q_R|b\rangle = \frac{1}{4} + \frac{1}{4} = \frac{1}{2}. \tag{3.108}$$

This is gives a large charge variance $(\Delta q_R)^2 = \langle q_R^2 \rangle - \langle q_R \rangle^2 = 1/4$, in agreement with Eq. (3.107). Hint: Use $\langle b|q_R|b\rangle = \langle b|q_R|a\rangle = 1/2$.

How can one reduce these fluctuations? Suppose that there is a small overlap between the bonding and antibonding gap states and that the fractional charges of the bonding state ψ_A are well-separated between the two wells, see the middle panel in Fig. 3.16(a). (Such a state is realized in interacting disordered zigzag nanoribbons. Its wave function is **nonlocal**.) In

---

[m]The effective Hilbert space consists of only bonding and antibonding states ψ_A and ψ_B and the matrix element between them $\langle \psi_A \,|\, V_L(x) \,|\, \psi_B$ is non-zero.

this case, fractional charges will be well-defined, i.e., the charge variance will be zero. Suppose there is a many-body gap between occupied and unoccupied gap-edge states. In this case, functional changes are also well-defined. However, if one tries to take such a fractional charge out of the system and isolate it, the intervention will destroy the many-body gap Δ. Fractional charges are thus well-defined only inside the system. This is somewhat analogous to quarks existing only in hadrons. Anderson localization may also reduce charge fluctuations between quasi-degenerate states. This is because quasi-degenerate states no longer overlap with each other when Anderson localization occurs; they tend to be separated spatially by large distances (this is a non-trivial effect of Anderson localization [27]). This should be contrasted with the overlapping bonding and antibonding states of a double well. Electron localization effect stabilizes fractional charges when electrons are added to the interior of a fractional quantum state (for a detailed explanation, see Sec. 1.8 of Ref. [25]; see also Sec. 19.4).

3.9. Crystalline Line Defects

So far we have considered electronic topological properties. Let us close this chapter with a brief discussion of crystalline line defects. Reference [28] gives a good introduction to topological line defects, and we follow the discussion given in it.

Suppose a line of atoms are taken out in a square lattice of period d. Atoms find new equilibrium positions forming a line defect, as shown in see Fig. 3.18. Consider a loop surrounding the defect, as shown in Fig. 3.19. Displacement vectors $\vec{u}$ from the equilibrium positions along the loop are shown. (The equilibrium positions near the starting point of the line defect are displayed in Fig. 3.20.) The order parameter is the local displacement

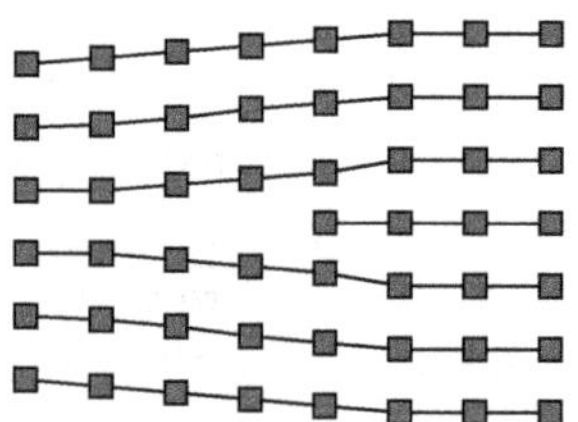

Fig. 3.18. A line of atoms is taken out. Atoms find new equilibrium positions.

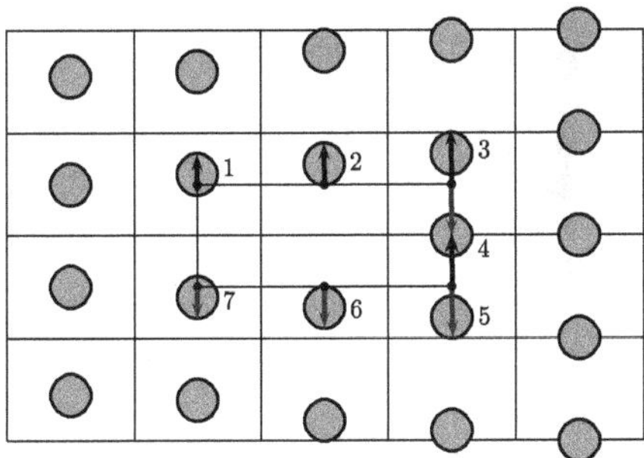

Fig. 3.19. Small dots indicate undistorted atomic positions (see Fig. 3.20) while larger dots show distorted atomic positions. Displacement vectors $\vec{u}$ are shown. They are all parallel or antiparallel to each other. The atom labeled 4 may thought of as originating from the nearest neighbor site below such that $u = d/2$ (see Eq. (3.109)).

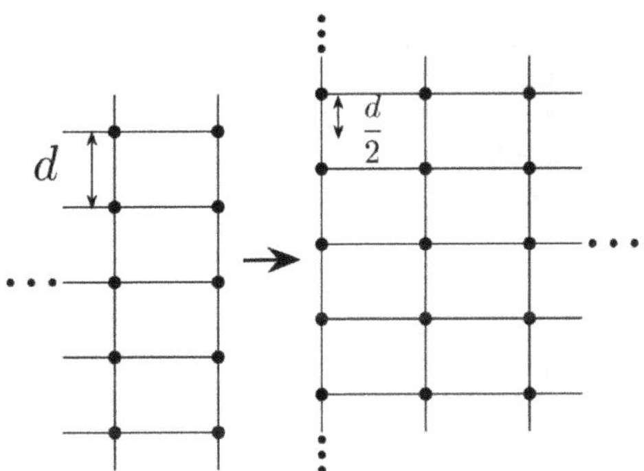

Fig. 3.20. Two lattices are joined together. One is shifted vertically by amount $d/2$.

vector of an atom $\vec{u}$. There is an ambiguity in the definition of $\vec{u}$ since it may be defined within Bravais lattice vectors $\vec{R}$

$$\vec{u} \to \vec{u} + \vec{R}. \tag{3.109}$$

This means independent values of $\vec{u}$ form a square lattice with periodic boundary conditions, i.e., a torus, see Fig. 3.21. Note that the length of the square is the lattice period d. In Fig. 3.21, these vectors are shown in the Brillouin zone or equivalently in a torus. The winding number of the order parameter vector is one.

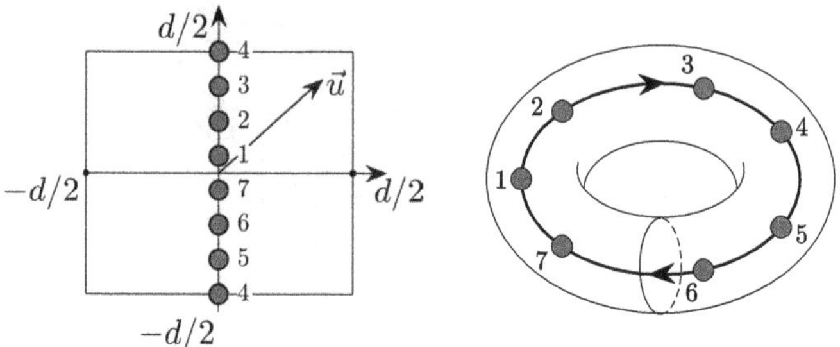

Fig. 3.21. **Upper**: Displacement vectors at sites $1, 2, 3, 4, 5, 6, 7$. Dots denote the end points of $\vec{u}$. **Lower**: Displacement vectors shown in torus. See Ref. [28].

Bibliography

[1] Y. Aharonov and D. Bohm, Significance of electromagnetic potentials in the quantum theory, *Phys. Rev.* **115**, 485 (1959).

[2] M. V. Berry, Quantal phase factors accompanying adiabatic changes, *Proc. R. Soc. Lond. A* **392**, 45 (1984).

[3] M. V. Berry, Singular limits, *Phys. Today* **55**, 5, 10 (2002). •

[4] A. Tonomura, S. Yano, N. Osakabe, T. Matsuda, H. Tamada, T. Kawasaki, and J. Endo, Evidence for Aharonov–Bohm effect with magnetic field completely shielded from electron wave, *Phys. Rev. Lett.* **56**, 792 (1986).

[5] N. Byers and C. N. Yang, Theoretical considerations concerning quantized magnetic flux in superconducting cylinders, *Phys. Rev. Lett.* **7**, 46 (1961).

[6] S.-R. Eric Yang, Degenerate states of narrow semiconductor rings in the presence of spin–orbit coupling: Role of time-reversal and large gauge transformations, *Phys. Rev. B* **74**, 075315 (2006).

[7] F. E. Meijer, A. F. Morpurgo, and T. M. Klapwijk, One-dimensional ring in the presence of Rashba spin–orbit interaction: Derivation of the correct Hamiltonian, *Phys. Rev. B* **66**, 033107 (2002).

[8] A. Zee, *Quantum Field Theory in a Nutshell* (Princeton University Press, Princeton, 2010). •

[9] T. T. Wu and C. N. Yang, Dirac monopole without strings: Monopole harmonics, *Nucl. Phys. B* **107**, 365 (1976).

[10] J. J. Sakurai and J. Napolitano, *Modern Quantum Mechanics* (Cambridge University Press, Cambridge, 2021). •

[11] T. Lancaster and S. J. Blundell, *Quantum Field Theory for the Gifted Amateur* (Oxford University Press, Oxford, 2014).•

[12] J. P. Fortney, *A Visual Introduction to Differential Forms and Calculus on Manifolds* (Springer, New York, 2018). •

[13] P. Collier, *A Beginner's Guide to Differential Forms* (Incomprehensible Books, Harlow, 2021). •

[14] H. B. Nielson and P. Olesen, Vortex-line models for dual strings, *Nucl. Phys. B* **61**, 45 (1973).

[15] G.'t Hooft, *In Search of the Ultimate Building Blocks* (Cambridge University Press, Cambridge, 1996). •

[16] R. P. Feynman and A. R. Hibbs, *Quantum Mechanics and Path Integrals: Emended Edition* (Dover Publications, Mineola, 2010). •

[17] R. Shankar, *Principles of Quantum Mechanics* (Springer, New York, 1994). •

[18] Y. V. Nazarov and J. Danon, *Advanced Quantum Mechanics* (Cambridge University Press, Cambridge, 2013).

[19] F. D. M. Haldane, O(3) Nonlinear σ model and the topological distinction between integer-and half–integer-spin antiferromagnets in two dimensions, *Phys. Rev. Lett.* **61**, 1029 (1988).

[20] J. M. Leinaas and J. Myrheim, On the theory of identical particles, *Nuovo Cimento B* **37**, 1 (1977).

[21] F. Wilczek, Quantum mechanics of fractional-spin particles, *Phys. Rev. Lett.* **49**, 957 (1982). •

[22] F. Wilczek, Anyons for anyone, *Phys. World* **4**, 1, 40 (1991). •

[23] F. Wilczek, Anyons, *Sci. American* **264**, 5, 58 (1991). •

[24] A. Altland and B. Simons, *Condensed Matter Field Theory* (Cambridge University Press, Cambridge, 2006).

[25] S. M. Girvin, "The quantum Hall effect: Novel excitations and broken symmetries," in *Les Houches Lecture Notes: Topological Aspects of Low Dimensional Systems*, edited by A. Comtet, T. Joliceur, S. Ouvry, and F. David (Springer-Verlag, Berlin and Les Editions de Physique, Les Ulis, 2000). •

[26] R. Jackiw, A. K. Kerman, I. Klebanov, and G. Semenoff, Fluctuations of fractional charge in soliton anti-soliton systems, *Nucl. Phys. B* **225**, 233 (1983).

[27] B. Altshuler, "Inductory Anderson Localization", in *Advanced Workshop on Anderson Localization, Nonlinearity and Turbulence: a Cross-Fertilization.*, (International Centre for Theoretical Physics, Italy, 2010).

[28] J. P. Sethna, *Entropy, Order Parameters, and Complexity* (Oxford University Press, Oxford, 2006).•

Chapter 4

Polyacetylene

"He asked the students to play a very easy warm-up exercise that any beginning trumpet player might be given. They played the handful of simple notes, which sounded childish compared to the dramatically fast, high notes from the earlier, more sophisticated pieces. After they played the simple phrase, Tony, for the first time during the lesson, picked up the trumpet. He played that same phrase, but when he played it, it was not childish. It was exquisite. Each note was a rich, delightful sound. Tony explained that mastering an efficient, nuanced performance of simple pieces allows one to play spectacularly difficult pieces with greater control and artistry." [1]

Teaching of Anthony Plog[a]

"Understand simple things deeply."

Edward B. Burger and Michael Starbird

In this chapter, we will study some basic properties of polyacetylene.[b] The interesting physics of polyacetylene is the formation of a domain-wall soliton, which connects two degenerate ground states, i.e., two different dimerized phases. Its essence is that half of its spectral weight originates from

---

[a] An internationally acclaimed trumpet virtuoso, composer, and teacher.

[b] The Nobel Prize in Chemistry 2000 was awarded jointly to Alan J. Heeger, Alan G. MacDiarmid and Hideki Shirakawa "for the discovery and development of conductive polymers".

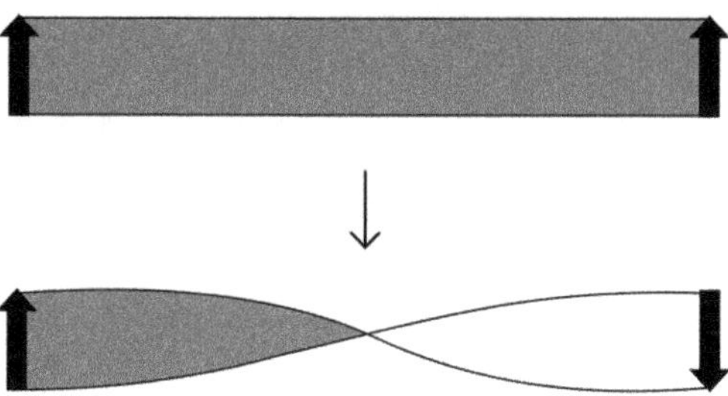

Fig. 4.1. Twist of a ribbon. The directions of the vectors indicate two types of dimerized phases. Note that this is a non-local topological excitation in the lattice configuration.

the conduction band and the other half from the valence band. This effect is what is usually meant by fractionalization in polyactylene. Such a soliton is formed from a ribbon twist, see Fig. 4.1. A domain-wall soliton is a stable configuration with a **local** energy minimum [2,3]. It is energetically favorable for a tunneling electron to enter the zero energy soliton state rather than go to the top of the conduction band.

A soliton gives rise to spin–charge separation, which is another form of electron fractionalization. Many topologically ordered phases, such as the integer quantum Hall effect, superconductivity, and disordered interacting zigzag nanoribbons, exhibit spin–charge separation. Polyacetylene serves as a good introduction to spin–charge separation.

Solitons also give rise to gap states that are localized at the ends of a finite-length polyacetylene. Both graphene and polyacetylene are in a **symmetry protected topological phase** (their similarities and differences are reviewed in Ref. [4]). Polyacetylene and graphene nanoribbons have two inequivalent sublattices (here labeled A and B). Under the chiral operation $\mathbf{C}$, the site annihilation operators transform according to $a_m \to a_m$ and $b_m \to -b_m$, where the index m denotes a unit cell containing one A and one B carbon atom.[c] Chiral symmetry **guarantees** the existence of boundary (edge) states [5], which is an example of the bulk-edge correspondence.

Only some salient features of solitons in polyacetylene are reviewed here (see Ref. [2] for a comprehensive review; Refs. [3,6] also give a nice overview of solitons.)

---

[c]This symmetry is not exact when next nearest-neighbor hopping is included.

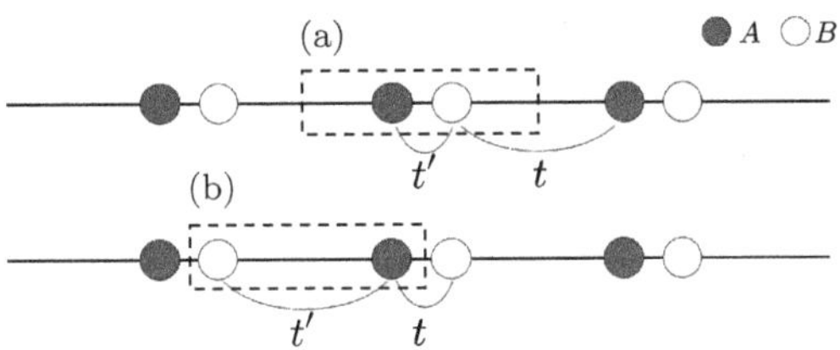

Fig. 4.2. Two primitive cells of a dimer chain are displayed, (a) and (b). **Intra** site tunneling is t' and **inter** site tunneling is t. In (a) $t' > t$ while $t' < t$ in (b). A primitive cell consists of A and B atoms.

4.1. Band Structure of Polyacetylene

Let us derive the band structure of polyacetylene.[d] We consider a half-filled dimer with one electron per site on average. The tight-binding Hamiltonian of the dimer chain is

$$H = - \sum_{m=1} [t' b_m^\dagger a_m + t a_{m+1}^\dagger b_m] + \text{h.c.}, \tag{4.1}$$

where $a_m^\dagger$ is the creation (annihilation) operator of an A-electron at mth site and is $b_m^\dagger$ the creation (annihilation) operator of a B-electron at mth site. The intra-site tunneling is t' and inter-site tunneling is t, see Fig. 4.2.

● **Chiral Symmetry**

The nearest-neighbor tight-binding Hamiltonian is

$$H = - \begin{pmatrix} \mathbf{O} & \mathbf{T} \\ \mathbf{T}^\dagger & \mathbf{O} \end{pmatrix}. \tag{4.2}$$

Here the matrix

$$\mathbf{T} = \begin{pmatrix} t' & 0 & 0 & 0 & \cdots \\ t & t' & 0 & 0 & \cdots \\ 0 & t & t' & 0 & \cdots \\ 0 & 0 & t & t' & \cdots \\ \vdots & \vdots & \vdots & \vdots & \end{pmatrix}, \tag{4.3}$$

---

[d]Students who need to refresh their understanding of basic band structure theory should read this section carefully; this method will be used repeatedly in later chapters. They should do all the exercises in this section.

and $\mathbf{O}$ is a zero matrix. In writing this matrix, the basis states are ordered as $(a_1, a_2, \ldots, b_1, b_2, \ldots)$. The chiral operator transforms the fermion operators according to $a_m \to a_m$ and $b_m \to -b_m$. Its matrix representation is

$$\mathbf{C} = \begin{pmatrix} \mathbf{I} & \mathbf{O} \\ \mathbf{O} & -\mathbf{I} \end{pmatrix}. \tag{4.4}$$

The nearest-neighbor tight-binding Hamiltonians of polyacetylene and graphene nanoribbons satisfy the anticommutation relation

$$\{\mathbf{C}, H\} = 0. \tag{4.5}$$

From this, it follows that single-particle eigenstates with non-zero eigenenergies E and $-E$ are related by the chiral operation

$$\mathbf{C}|\psi_E\rangle = |\psi_{-E}\rangle. \tag{4.6}$$

Thus, chiral symmetry implies that the energy spectrum has particle-hole symmetry. The probability densities of $|\psi_E\rangle$ and $\mathbf{C}|\psi_E\rangle$ are identical. Of special interest are the single-particle eigenstates of the chiral operator, which represent chiral topological states.

- **Eigenstates and Eigenenergy**

The Bloch wave function of a periodic system with a unit cell consisting of two basis atoms is

$$\psi_k(x) = \sum_m e^{ikX_m}[A_k\phi(x - X_m) + B_k\phi(x - d - X_m)]$$

$$= A_k\psi_{Ak}(x) + B_k\psi_{Bk}(x), \tag{4.7}$$

where

$$\psi_{Ak}(x) = \sum_m e^{ikX_m}\phi(x - X_m), \quad \psi_{Bk}(x) = \sum_m e^{ikX_m}\phi(x - d - X_m).$$

$$\tag{4.8}$$

Here $\phi(x - X_m)$ is the atomic orbital wave function, X_m is the position of mth unit cell, and d is the distance between the two carbon atoms in a unit cell. The wave function with wave vector k can also be written as

$$\psi_k(x) = e^{ikx}u_k(x), \tag{4.9}$$

where $u_k(x)$ is a cell periodic function: $u_k(x + X_n) = u_k(x)$ (this can be easily shown from Eq. (4.7)). Note that $\psi_{k+K}(x) = \psi_k(x)$ but $u_{k+K}(r) = e^{-iKx}u_k(x)$, where $K = \frac{2\pi}{a}m$ (a is the lattice constant).

The amplitudes $\begin{bmatrix} A_k \\ B_k \end{bmatrix}$ of the Bloch wavefunction, Eq. (4.7), may be determined as follows. Let us Fourier transform the site operators

$$a_m^\dagger = \frac{1}{\sqrt{M}} \sum_k a_k^\dagger e^{-ikam},$$

$$b_m^\dagger = \frac{1}{\sqrt{M}} \sum_k b_k^\dagger e^{-ikam}, \tag{4.10}$$

where M is the number of unit cells. Since the system is **periodic** with the period a, the wave vectors are restricted to the first Brillouin zone (see Exercise 4.2)

$$-\pi/a < k < \pi/a. \tag{4.11}$$

The Hamiltonian, Eq. (4.1), is transformed into

$$H = -\sum_k [t' b_k^\dagger a_k + te^{-ika} a_k^\dagger b_k] + \text{h.c.} \tag{4.12}$$

Adding the Hermitian conjugate part to the first term this can be written as

$$H = \sum_k (a_k^\dagger, b_k^\dagger) H(k) \begin{bmatrix} a_k \\ b_k \end{bmatrix}. \tag{4.13}$$

Here the k-space Hamiltonian $H(k)$ has the following matrix representation

$$H(k) = -t|\rho(k)| \begin{bmatrix} 0 & e^{-i\phi(k)} \\ e^{i\phi(k)} & 0, \end{bmatrix}, \tag{4.14}$$

where

$$\rho(k) = t'/t + e^{-ika} = |\rho(k)|e^{-i\phi(k)} \tag{4.15}$$

and

$$\tan\phi(k) = \frac{\sin ka}{t'/t + \cos ka}. \tag{4.16}$$

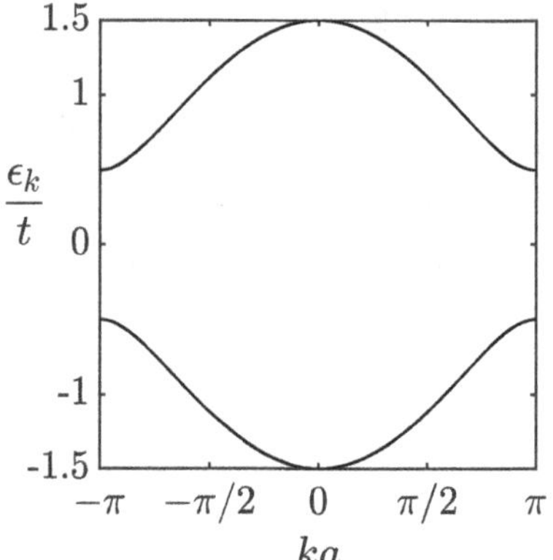

Fig. 4.3. Band structure for $t' = t/2$ is plotted as a function k. Since the unit cell has two carbon atoms there are two bands.

The eigenstates of $H(k)$ are the amplitudes of the Bloch wave function of Eq. (4.7)

$$\begin{bmatrix} A_k \\ B_k \end{bmatrix} = \frac{1}{\sqrt{2}} \begin{bmatrix} e^{-i\phi(k)} \\ \pm 1 \end{bmatrix}. \tag{4.17}$$

The eigenenergy is

$$\epsilon_{k,\pm} = \pm t |\rho(k)|, \tag{4.18}$$

where minus (plus) sign is for the valence (conduction) band. These eigenenergies are shown in Fig. 4.3.

Exercise 4.1. Suppose the length of polyacetylene is L. The Bloch wave functions satisfy the following boundary condition:

$$\psi_k(x + L) = \psi_k(x). \tag{4.19}$$

Show that this condition restricts the allowed values of k

$$k = \frac{2\pi n}{L}, \tag{4.20}$$

where $L = Na$ and $n = -N/2, \ldots, 0, \ldots, +N/2$ (n and N are integers).

Exercise 4.2. Explain why the allowed values of k may be restricted to the first Brillouin zone

$$-\pi/a < k < \pi/a. \tag{4.21}$$

Hint: Other states with k outside this zone may be represented by a Bloch state inside the first Brillouin zone. Show $\psi_{k+K_m}(x) = \psi_k(x)$, where the reciprocal vector is $K_m = 2\pi m/a$.

Exercise 4.3. Derive Eqs. (4.12), (4.13), and (4.14) from Eq. (4.1). Show also Eqs. (4.17) and (4.18).

- ## Winding Number

Let us compute a topological number associated with the band structure. This number is related to chiral symmetry. Consider an infinitely long polyacetylene specimen. The tight-binding Hamiltonian can be written as Eq. (4.14)

$$H = -t\vec{g}(k) \cdot \vec{\sigma}, \tag{4.22}$$

where

$$\vec{g}(k) = |\rho(k)|(\cos\phi(k), \sin\phi(k)) = \left(\frac{t'}{t} + \cos ka, \sin ka\right). \tag{4.23}$$

This Hamiltonian has the standard form given in Eq. (2.46). (The vector $\vec{g}(k)$ represents the complex number $\rho(k)$.) The chiral operator is $\Gamma = \sigma_z$ and chiral symmetry means $\{\sigma_z, H\} = 0$. This anticommutation relation holds for the tight-binding Hamiltonian Eq. (4.22) since σ_z is **absent** from H, which is a **consequence of chiral symmetry**.

Let us compute the one-dimensional winding number defined previously in Eq. (2.68). Note that the vector $\vec{g}(k)$ does not have a z-component [6]; this feature leads to **non-trivial** values of the winding number and the Zak phase. We plot $\vec{g}(k)$ as a function of k (note that $t\vec{g}(k)$ is equivalent to $\vec{R}/2$ in Eq. (14.3)). When will $\vec{g}(k)$ encircle a magnetic monopole located at the origin? In other words when is polyacetylene topologically non-trivial? One finds it is non-trivial when $t'/t < 1$, see Fig. 4.4. A topological phase transition occurs at $t = t'$.

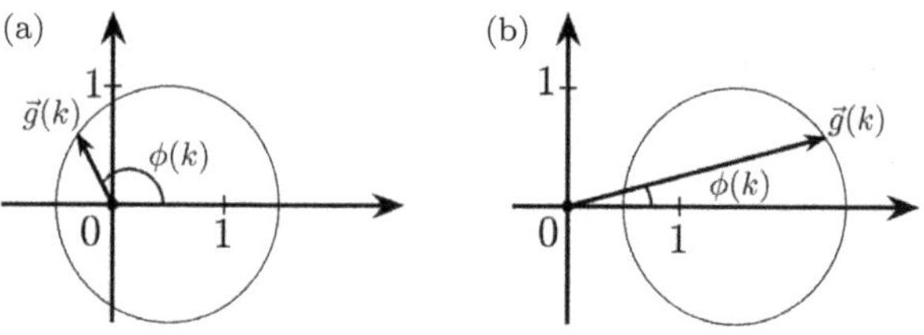

Fig. 4.4. Plot of $\vec{g}(k)$ for (a) $t'/t < 1$ and (b) $t'/t > 1$.

4.2. End Fractional Charges?

Let us investigate the end charges. Consider **finite-length** polyacetylene in one of the dimerized phases (no domain wall exists as only one type of dimerized phase is present). The electron density is uniform with the occupation number $n_i = 1$ at all sites i. There are two types of **finite** length polyacetylene, one with long and the other with short end bonds, as shown in Figs. 4.5 and 4.6, respectively.

In the case of the short-bond unit cell, no end state exists, i.e., no gap state exists, as shown in Fig. 4.5. Tight-binding calculations show that, for the long-bond unit cell shown in Fig. 4.6, two nearly degenerate bonding ϕ_B and antibonding ϕ_A **gap** states exist with almost zero-energy (the energy splitting vanishes when the system length becomes infinitely large). Since $E_F = 0$ one state is occupied while the other is empty. One half of the spectral weight of each of these gap states is derived from the conduction band, while the other half is from the valence band. The probability density of such a state **splits** into two parts, located near the left and right end points. If an electron is added into a solitonic state, the resulting electron density

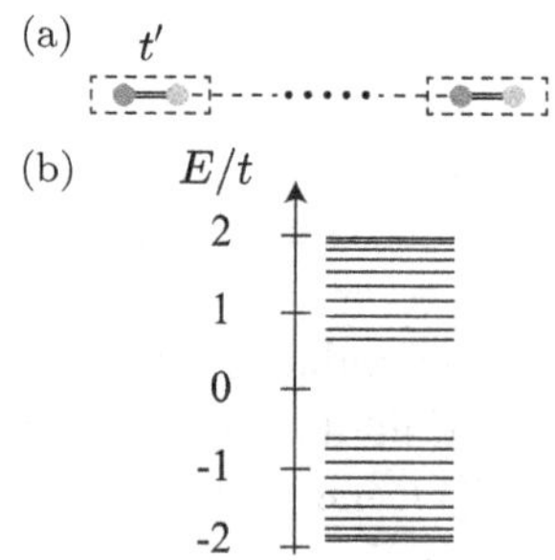

Fig. 4.5. (a) Finite-length dimer chain with unit cell containing two carbon atoms connected by double bond. The intra cell hopping t' is larger than the inter cell hopping t. Note that the unit cell is different from that shown in Fig. 1.2. (b) Energy spectrum. No gap states exist. Reprinted with permission from Ref. [4].

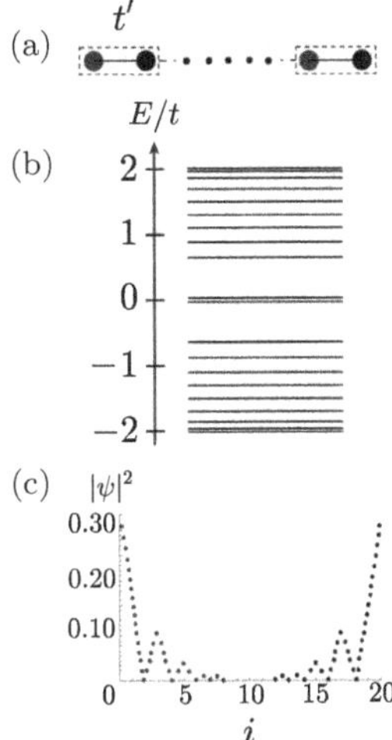

Fig. 4.6. (a) Finite-length dimer chain with unit cell containing two carbon atoms connected by single bond. The intra cell hopping t' is smaller than the inter cell hopping t. (b) Tight-binding energy spectrum. Two nearly degenerate gap states exist. (c) Probability density of a gap state is plotted as function of site index i. Peaks are present at the left and right end sites. The probability densities of the bonding and antibonding states are almost identical. Reprinted with permission from Ref. [9].

$\rho(x)$ has $e^-/2$ fractional charges near the **two ends** of the polyacetylene. Note that these solitons have mixed chirality with different chirality at the opposite ends. However, their linear combinations $\phi_B \pm \phi_A$ are **chiral** and are located near either the left or right end points.

Let us consider the robustness of these fractional charges. Two effects must be considered. (a) The boundary fractional charges will suffer quantum charge fluctuations because the energy splitting is small (only a large energy splitting will suppress fluctuations, see an insightful discussion by Girvin, Ref. [8]; see also Sec. 3.8). (b) Disorder may affect these states. Most likely, the left and right end states will be affected differently. Hence the boundary charge will not be exactly $e/2$. There is as yet no conclusive experimental evidence for fractional charges in polyacetylene (but spin–charge separation was observed).

4.3. Domain-Wall Soliton

A domain wall of the polyacetylene connects the two dimerized degenerate ground states labeled m and $-m$, as shown in Fig. 4.7. The total system (lattice plus electrons) with a domain wall between two dimerized phases

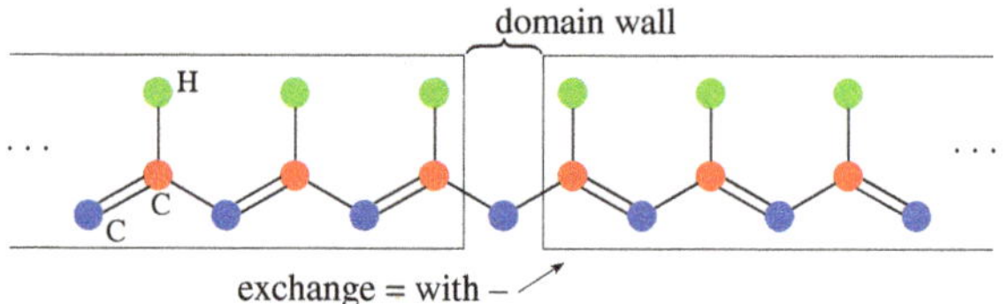

Fig. 4.7. Two semi-infinite polyacetylene chains with a domain wall, which connects two different dimerized phases labeled $-m$ and m. A double bond is shorter than a long bond and a Peierls lattice distortion is present. The directions of the double bonds vary in these two phases. Periodic boundary conditions cannot be imposed on this system as there is only one domain wall. Either a soliton or antisoliton is possible, but not both. Reprinted with permission from Ref. [4].

m and $-m$ has a **higher** energy than that of the ground state in one of the dimerized phases with m or $-m$. How does one go about creating solitons in an experiment? One way is to add extra electrons. There is a structural relaxation of the polymer chain around charges added to the chains. These charges become localized solitons [10].

This lattice distortion supports an electronic state called a soliton. The size of this object ξ is large compared to the lattice constant and varies slowly over several lattice constants. Its property can be found by numerically solving the tight-binding Hamiltonian. Instead, here we use the Dirac equation, which is a long wavelength (a continuum) version of the tight-binding equation. (The Dirac equation may be derived from the tight-binding Hamiltonian equation (4.1), see also Exercise 4.4.) **The Dirac equation** [6] has the form

$$H = -iv_F\hbar\partial_x\sigma_x + m(x)\sigma_y, \qquad (4.24)$$

where $\{\sigma_x, \sigma_y\}$ are the Pauli spin matrices and $v_F = ta$ is the characteristic velocity. The second term is the twisted scalar mass potential, see Fig. 4.8. The twist consists of $-m$ for $x < 0$ and m for $x > 0$, where $m = |t' - t|$ is a constant (this system is not periodic as the $\pm m$ terms represent two different dimerized phases). In the continuum description of this system, a fermion mass potential with no twist in the Dirac equation produces an excitation gap. A twist in this scalar potential produces a **zero**-energy soliton and fermion fractionalization [11].

> **Exercise 4.4.** Derive the Dirac equation, Eq. (4.24) when $m(x) = m = $ constant. Hint: Expand the tight-binding Hamiltonian equation (4.22) near the momentum $k = \frac{\pi}{a}$ (set $q = k - \frac{\pi}{a}$) and replace $q \to \frac{1}{i}\frac{d}{dx}$ (for justification of this replacement, see Exercise 14.5). (This approach is called the envelope wave function or effective mass approximation, see Appendix A.) After expanding momentum near $k = \frac{\pi}{a}$, perform a clockwise "spin" rotation (internal space rotation) of $-\pi/2$ about the z-axis with the operator $U = e^{i\theta\sigma_z/2}$. Use $U\sigma_x U^{-1} = -\sigma_y$ and $U\sigma_y U^{-1} = \sigma_x$.

The Dirac equation has **one** zero-energy soliton (kink) mode $\psi_0(x)$, which is bound to the domain wall and decays **exponentially**:

$$\psi_0(x) \sim e^{\int_0^x -\frac{m(x')}{\hbar v_F}dx'} \begin{bmatrix} 1 \\ 0 \end{bmatrix}. \tag{4.25}$$

Its first and second components give the probability amplitudes of finding the electron on A and B carbons, respectively. Note that only the A-component of the wave function is non-zero. It corresponds to a soliton that is very well localized near the A-carbon atom at $x = 0$, as the tight-binding solution of Fig. 4.9 shows. A soliton **is a chiral mode** and is topologically robust because it originates from a twist in the variation of the dimerization $m(x)$.

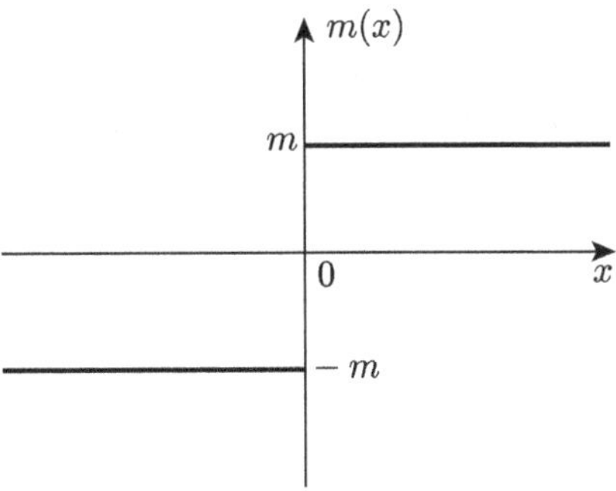

Fig. 4.8.   An abrupt change in the dimerization is plotted.

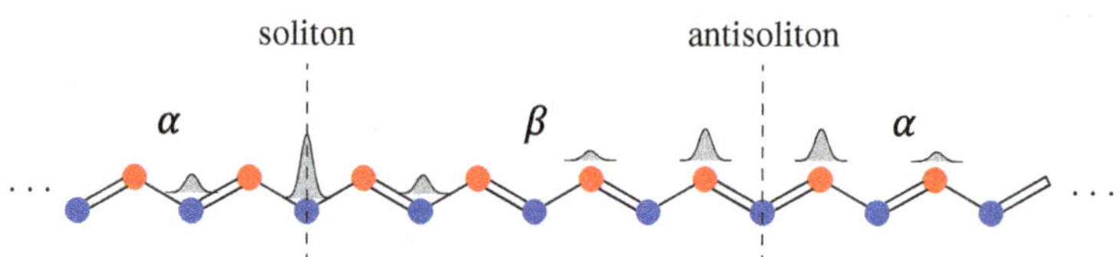

Fig. 4.9. In a periodic system with two domain walls, a soliton–antisoliton pair must exist. Their site probability densities computed from the tight-binding Hamiltonian are shown. For a soliton (antisoliton), the probability density is finite on A (B) carbon atoms only. Reprinted with permission from Ref. [4].

Exercise 4.5. Show that the soliton wave function Eq. (4.25) is an eigenstate of the Hamiltonian equation (4.24). Suppose $m(x)$ is given by a step function (see Fig. 4.8). Sketch the shape of $\psi(x)$ as a function of x. Answer: it is localized and is centered around $x = 0$. First plot $\psi(x)$ for $x < 0$ and then for $x > 0$.

Exercise 4.6. The Dirac Hamiltonian has the form

$$H = c \sum_i \alpha_i p_i + \beta m c^2. \tag{4.26}$$

with $\beta^2 = 1$, $\{\alpha_i, \beta\} = 0$, and $\{\alpha_i, \alpha_j\} = 2\delta_{ij}$ (c is the velocity of light and m is the electron mass). (a) Show that

$$H^2 = c^2 p^2 + m^2 c^4. \tag{4.27}$$

(It is reminiscent of $E^2 = c^2 p^2 + m^2 c^4$ in Einstein's special theory of relativity.) Hint: Look up what α_i and β are in one dimension, two dimensions, and three dimensions. (b) Think about how α_i and β are related to the electron angular momentum. Hint: Angular momentum is quantized in three dimensions. In two dimensions only the z-component of angular momentum is well-defined. In one dimension it is meaningless.

Next, consider a **periodic** polyacetylene specimen having two domain walls, see Fig. 4.9. It is instructive to consider tight-binding solutions. There are a soliton and an antisoliton solutions, as shown in Fig. 4.9. For infinitely long polyacetylene, the energy difference between a soliton and an antisoliton vanishes. When both a soliton and an antisoliton are present, they must be located at **different** positions along a periodic ring: a soliton connects two dimerized phases, $-m \to m$, whereas an antisoliton connects $m \to -m$, as shown in Fig. 4.9. A soliton and an antisoliton are chiral modes with different chirality: a soliton has only the A-component wave function and an antisoliton has only the B-component wavefunction, as apparent from Fig. 4.9. An antisoliton is also an eigenstate of the chiral operator. The energy spectrum varies strongly in the vicinity of the kink/antikink. The total number of electrons or states in the filled valence band in the vicinity of the kink/antikink decreases by precisely $1/2$ per spin (see the derivation below).

• Spectral Weight of a Soliton

The following discussion demonstrates that the conduction and valence bands each contribute a fractional spectral weight of $1/2$ to a soliton. For simplicity, **spinless** electrons in disorder-free and half-filled polyacetylene are considered first. The conduction and valence bands obey chiral symmetry so that the wave functions of states with opposite energies E and $-E$ are identical: $\psi_E(x) = \psi_{-E}(x)$ (particle–hole symmetry follows from chiral symmetry; the Fermi energy is $E_F = 0$).

A spectral analysis is performed using the local density of states $\rho^0(E, x)$ and $\rho^{\mathrm{kink}}(E, x)$ of **translationally invariant** and non-invariant systems. As the total weight is conserved before and after the translational symmetry is broken, we have

$$\sum_E \rho^{\mathrm{kink}}(E, x) = \sum_E \rho^0(E, x). \tag{4.28}$$

Before a soliton is introduced there is no zero-energy state; thus, $\sum_E \rho^0(E, x) = \sum_{E \neq 0} \rho^0(E, x)$. After a kink is introduced, the total density of states (DOS) at x is $\sum_E \rho^{\mathrm{kink}}(E, x) = |\psi_0(x)|^2 + \sum_{E \neq 0} \rho^{\mathrm{kink}}(E, x)$. From this and that the total integrated density of states is unchanged when a kink is introduced[e] and we find

$$\sum_{E \neq 0} \rho^0(E, x) = |\psi_0(x)|^2 + \sum_{E \neq 0} \rho^{\mathrm{kink}}(E, x). \tag{4.30}$$

---

[e] This follows from the closure property of the eigenstates $\psi_E(x)$

$$\sum_E \rho(E, x) = \sum_E \psi_E^*(x)\psi_E(x) = 1. \tag{4.29}$$

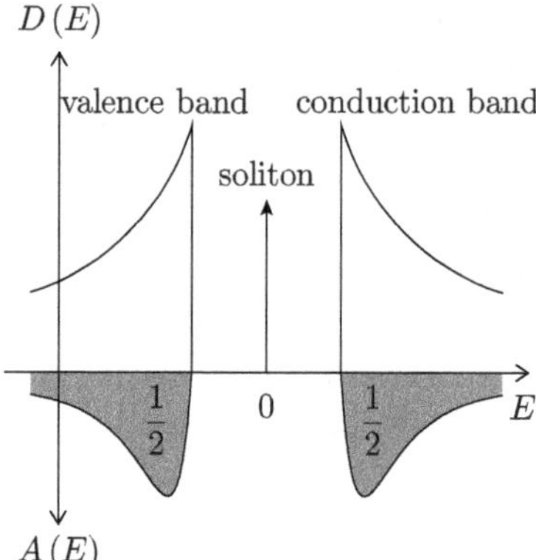

Fig. 4.10. Midgap state represents a soliton state. Its spectral weight stems equally from the conduction and valence bands.

The induced DOS excluding the $E = 0$ state is

$$\sum_{E\neq 0} \delta\rho(E,x) = \sum_{E\neq 0}(\rho^{\text{kink}}(E,x) - \rho^0(E,x)) = -|\psi_0(x)|^2. \qquad (4.31)$$

As the conduction and valence bands are symmetric, the contribution from the **occupied** valence band states is $1/2$ (see Fig. 4.10)

$$\sum_{E<0} \delta\rho(E,x) = -\frac{1}{2}|\psi_0(x)|^2. \qquad (4.32)$$

Thus, the occupied valence band contributes a **fraction** $1/2$ to the total spectral contribution of the solitonic state, while the unoccupied conduction band contributes another half (one half of a state is missing from the valence band and the corresponding charge is assumed to be in the vicinity of the soliton [2]). What all this means is as follows: suppose there were 1000 states in the conduction band and 1000 states in the valence band before a kink is introduced. When a kink is introduced there will be 999.5 states both in the conduction and valence bands, and there will be one state in the kink. But only half of a kink is occupied

4.4. Spin–Charge Separation

The charge of a soliton state is the **total** charge which includes the depletion or surplus of the electron charge density of the many-body state, and

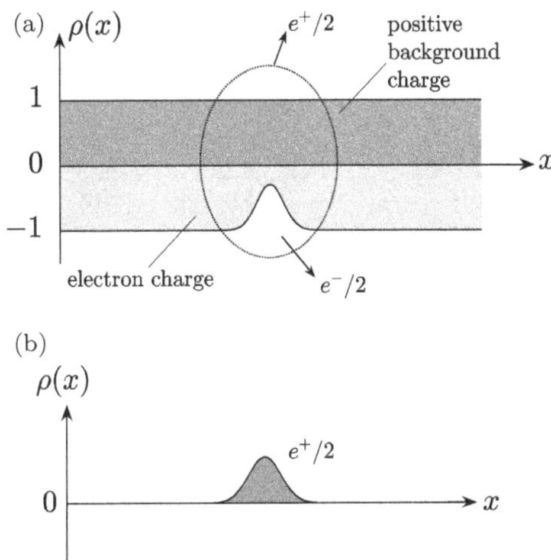

Fig. 4.11. (a) When a soliton gap state is formed, $e^-/2$ is removed from each of the uniform densities for spin-up and -down electrons. The effective charge $e^+/2$ represents the solitonic charge in each spin-channel (it includes the contribution from the positive background change). The electron density is depleted where the soliton is located. (b) Each spin state of a soliton has an effective charge $e^+/2$, as shown in the figure.

in addition to the positive background charge, see Fig. 4.11. Each spin-up and -down soliton gap state takes half the spectral weight from the occupied valence band. If the spin-up and -down soliton states are both empty, effectively the charge $e^-/2 + e^-/2 = e^-$ is removed from the total electron charge of the ground state, i.e., a hole with charge e^+ is present. (Here e^+ is the contribution from the positive background charge at the location where the soliton is.) Thus, if a soliton state is **empty**, its effective charge is $q = e^+$ and its spin is $s = 0$, as shown in Fig. 4.12 [2, 12]. Spin and charge thus separate. There is an experimental evidence for this effect: the conductivity and magnetic susceptibility of polyacetylene, as a function of doping density, display independent behavior from each other [13]. When the spin-up soliton state is occupied while the spin-down state is empty, the resulting values are $q = e^- + e^+ = 0$ and $s = 1/2$. When both spin-up and -down states are both occupied, the soliton state has $q = e^-$ and $s = 0$. Therefore, the solitonic states have unusual charge and spin relations [2]. Note that, independent of the number of occupied electrons, the net charge of a domain-wall soliton state is never fractional.

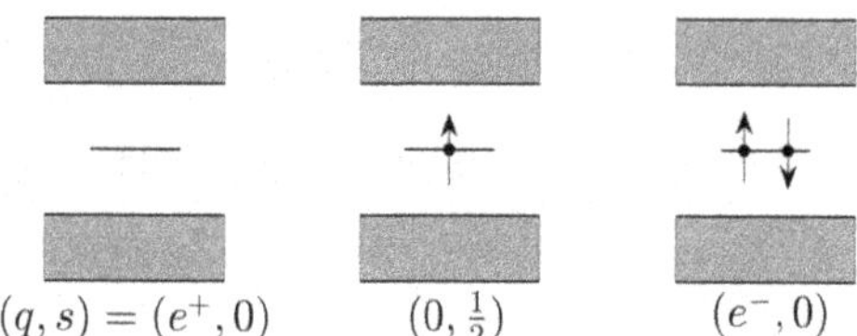

Fig. 4.12. Gap states, conduction, and valence bands are shown. Only the valence band is occupied. A soliton midgap state can be empty, singly occupied, or doubly occupied. The charge a and spin q values of these states are different from the usual spin–charge relation for fermions.

4.5. Schrieffer Counting Argument and Dimerized Polyacetylene

Let us give a different but an equivalent argument for fractional charges [14]. It is called Schrieffer's counting argument [15]. Consider a dimerized chain with a possible ground state with a unit cell 12 (α) or 21 (β), see Fig. 4.13. Here a double bond is represented by 2 and a single bond by 1. Note that a **single bond** represents a **pair of electrons** with **no** net spin (a double bond has four electrons). Removing a bond, i.e., removing two electrons with opposite spins, creates two defects (domain walls) with zero spin. Such a defect, denoted by 11, has charge one but no spin, see Fig. 4.14(a). (If a single bond consisted of a single electron then defect would have charge $\frac{e^-}{2}$. Its spin would be 1/4 because spin 1/2 is divided into two.) Hence, spin and charge **separate** from each other. These defects connect the degenerate ground states $\alpha = \ldots 12$ and $\beta = 21 \ldots$. Note that a bond decomposes into two **non-local** defects. The probability density of a soliton gap state is divided into two locations, similar to the case of a double quantum-well system, see Fig. 4.14(a). These two objects are non-local because they are separated by a region that is in the opposite phase of β in comparison to the phase before a soliton is formed. One cannot thus create these objects locally. Figure 4.14(b) displays the energy spectrum. There are two nearly degenerate gap states representing defects. In such a case, **quantum**

$$\alpha \cdots \text{(1 2)(1 2)(1 2)} \cdots$$

$$\beta \cdots \text{(2 1)(2 1)(2 1)} \cdots$$

Fig. 4.13. Two degenerate phases of polyacetylene.

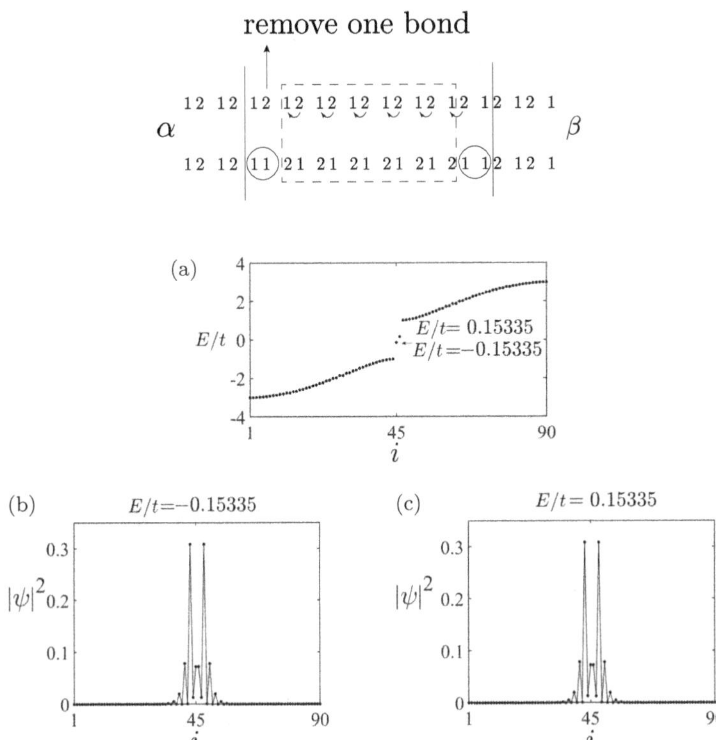

Fig. 4.14. **Upper:** A removal of a bond from the α phase is depicted. As a result of this removal two domain walls are formed. Note that the region in the dashed box between these two domain walls is in the opposite phase β, i.e., these domain walls connect α and β degenerate ground states, as shown in Fig. 4.7. Such a domain wall has charge e^+. **Lower:** (a) Tight-binding energy spectrum of a system with two defects is plotted with increasing magnitude for $t'/t = 2$ (the hopping parameter of a double (single) bond is $t'(t)$). (b) and (c) display probability densities of nearly degenerate gap states indicated in (a).

charge fluctuations [8] are significant and a fractional charge may not be precisely quantized, as we discussed in Sec. 3.8.

4.6. Schrieffer Counting Argument and Trimerized Polyacetylene

Now consider a trimerized polyacetylene chain [15] with a unit cell containing two single bonds and one double bond. There are three degenerate ground states: each unit cell in each ground state contains $\alpha = 112$, $\beta = 211$, and $\gamma = 121$, respectively (see Fig. 4.15).

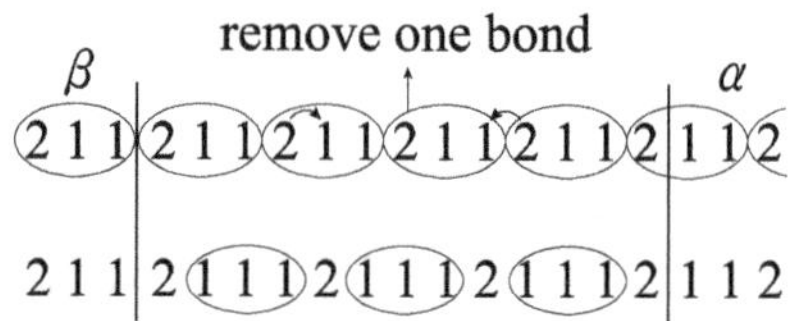

Fig. 4.15. Three degenerate ground states α, β, and γ. Each oval represents a unit cell.

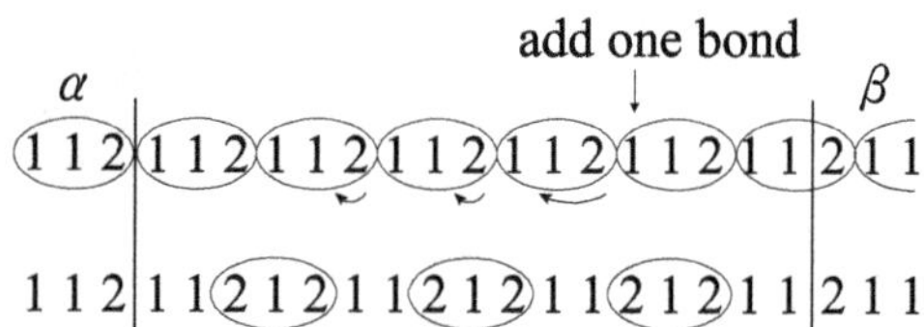

Fig. 4.16. **Upper:** First drawing depicts a removal of a bond, followed by moving two bonds to neighboring sites. As a result, three domain wall defects are created (a defect is denoted by 111). Each domain wall has a charge $\frac{2}{3}e^+$. These domain walls connect α (right) and β (left) degenerate ground states. **Lower:** By adding a bond three domain walls can be created.

In Fig. 4.16 we show that removing a bond, i.e., two electrons from the ground state type 112 can lead to 3 defects, generating fractional charges of $\frac{2}{3}e^+$. This is because $2e^+$ is divided into 3 defects. Note that a bond decomposes again into three **non-local** defects. When an extra bond is added (see Fig. 4.16) fractional charges of $\frac{2}{3}e^-$ may also be formed. Thus fractional charges may occur as a consequence of ground state degeneracy. These defects are antiparticles of $\frac{2}{3}e^+$.

The energy spectrum of this system with the total charge $2e^+$ can be investigated using the tight-binding Hamiltonian, see Fig. 4.17. Charge and spin values of gap states are shown. When the lower energy gap states with spin-up and -down (see the left panel in Fig. 4.17) are unoccupied the effective charge of the domain wall is $q = \frac{e^+}{3} + \frac{e^+}{3} = \frac{2e^+}{3}$ and its spin is $s = 0$. If it is occupied by one electron, $(q, s) = (\frac{e^-}{3}, \frac{1}{2})$. If two electrons occupy it we have $(q, s) = (\frac{4e^-}{3}, 0)$.

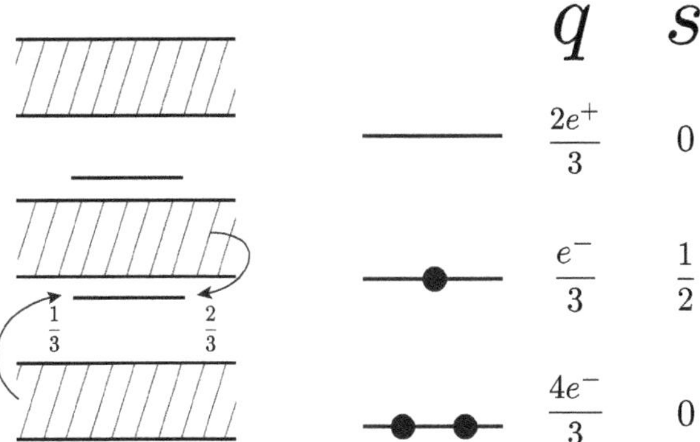

Fig. 4.17. **Left**: A gap state when a bond is removed: 1/3 of its spectral weight originates from the bottom band and 2/3 from the upper bands [15] (see Exercise 4.7). **Right**: Charge q and spin s of the lower energy gap state depend on the number of occupied electrons: the empty state has $(q,s) = (\frac{2}{3}e^+, 0)$, the filled state with one electron has $(q,s) = (e^-/3, 1/2)$, and the filled state with two electrons has $(q,s) = (\frac{4}{3}e^-, 0)$.

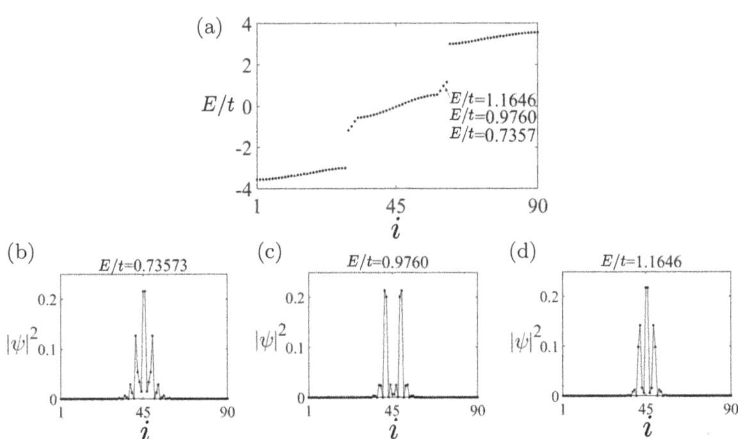

Fig. 4.18. When a bond is removed polyacetylene can have three defects, see Fig. 4.16. (a) Tight-binding energy spectrum of such a system with $t'/t = 3$ is plotted. Here index i labels eigenenergies (b), (c), and (d) display probability densities of nearly degenerate gap states.

There are three nearly degenerate gap states representing defects. Therefore, again, quantum charge fluctuations between them are significant and a fractional charge is not expected to be precisely quantized. The probability density of a gap state is approximately similar to the case of a triple quantum-well system, see Fig. 4.18.

> **Exercise 4.7.** How can you show that 1/3 of the spectral weight of a gap state originates from the bottom band and 2/3 from the upper bands? (Reread the caption of Fig. 4.17.) Hint: One can first compute the electronic states of the trimerized polyacetylene without removing a bond. This will give three bands. Then the perturbation of removing a bond is treated in the degenerate perturbation theory with the basis consisting of the unperturbed electronic states. After this, the spectral weight of the eigenstate corresponding to the gap state can be analyzed.

Bibliography

[1] E. B. Burger and M. Starbird, *The 5 Elements of Effective Thinking* (Princeton University Press, Princeton, 2012).

[2] A. J. Heeger, S. Kivelson, J. R. Schrieffer, and W. P. Su, Solitons in conducting polymers, *Rev. Mod. Phys.* **60**, 781 (1988).

[3] S. M. Girvin and K. Yang, *Modern Condensed Matter Physics* (Cambridge University Press, Cambridge, 2019). •

[4] S.-R. Eric Yang, Soliton fractional charges in graphene nanoribbon and polyacetylene: Similarities and differences, *Nanomaterials* **9**, 885 (2019). •

[5] S. Ryu and Y. Hatsugai, Topological origin of zero-energy edge states in particle–hole symmetric systems, *Phys. Rev. Lett.* **89**, 077002 (2002).

[6] C. L. Kane, "Topological band theory and the Z2 invariant," in *Contemporary Concepts of Condensed Matter Science Series Vol. VI: Topological Insulators*, edited by M. Franz, L. Molenkamp (Elsevier, Oxford, 2013).•

[7] P. Delplace, D. Ullmo, and G. Montambaux, Zak phase and the existence of edge states in graphene, *Phys. Rev. B* **84**, 195452 (2011).

[8] S. M. Girvin, "The Quantum Hall Effect: Novel excitations and broken symmetries," in *Les Houches Lecture Notes: Topological Aspects of Low Dimensional Systems*, edited by A. Comtet, T. Joliceur, S. Ouvry, and F. David (Springer-Verlag, Berlin and Les Editions de Physique, Les Ulis, 2000). •

[9] S.-R. Eric Yang. M. C. Cha, H. J. Lee, and Y. H. Kim, Topologically ordered zigzag nanoribbon: $e/2$ fractional edge charge, spin–charge separation, and ground–state degeneracy, *Phys. Rev. Res.* **2**, 033109 (2020).

[10] N. F. Colaneri, R. H. Friend, H. E. Schaffer, and A. J. Heeger, Mechanism for photogeneration of metastable charges solitons in polyacetylene, *Phys. Rev. B* **38**, 3960 (1988).

[11] R. Jackiw and C. Rebbi, Solitons with fermion number 1/2, *Phys. Rev. D* **13**, 3398 (1976).

[12] W. P. Su, J. R. Schrieffer, and A. J. Heeger, Solitons in polyacetylene, *Phys. Rev. Lett.* **42**, 1698 (1979).

[13] T.–C. Chung, F. Moraes, J. D. Flood, and A. J. Heeger, Topologically ordered zigzag nanoribbon: Solitons at high density in trans–$(CH)_x$: Collective transport by mobile, spinless charged solitons, *Phys. Rev. B* **29**, 2341 (1984). •

[14] F. Wilczek, *Fantastic Realities* (World Scientific, Singapore, 2006). •

[15] W. P. Su and J. R. Schrieffer, Fractionally charged excitations in charge-density-wave systems with commensurability 3, *Phys. Rev. Lett.* **46**, 738 (1981).

Chapter 5

Anomalous Velocity, Polarization, Zak Phase, and Chern Number

"I remember once going to see him when he was ill at Putney. I had ridden in taxi cab number 1729 and remarked that the number seemed to me rather a dull one, and that I hoped it was not an unfavourable omen. "No," he replied, "it is a very interesting number; it is the smallest number expressible as the sum of two cubes in two different ways."

Hardy-Ramanujan number[a]

"I can't tell you where Ramanujan got his mathematical desires – I should call them desires, because that's what you really require first. Not so much the ability, but desire to play around, and to notice."

Richard P. Feynman

In this chapter, we will study topological properties of electronic band structures. The first Brillouin zone of a band structure is a manifold in k-space and a non-zero Chern number may be associated with it. It is based on the concept of the anomalous velocity, which may give rise to quantized topological numbers, such as the Zak phase and the Chern number. The Zak phase can be related to electric polarization of a periodic crystalline system. In this chapter, we give a short introduction of the modern theory of polarization which explains how the polarization is related to the Zak

---

[a]The two different ways are: $1729 = 1^3 + 12^3 = 9^3 + 10^3$.

phase of the band structure of a crystalline structure.[b] We will also learn that often the symmetry of a physical system leads to quantization of the Zak phase. (For a comprehensive review of this subject, see Refs. [1–4].) A band structure may also possess an anomalous velocity that leads to a Chern number, such as a quantized off-diagonal conductivity. We will learn in a later chapter that the integer quantum Hall effect is related to the Chern number of the Brillouin torus.

5.1. Some Basics Results of Band Structure Theory

This chapter requires a solid understanding of basic band structure theory. So we first refresh some basic results of band structure theory [5]. Students should also reread how the band structure of polyacetylene was derived, see Sec. 4.1.

The Bloch wave function is

$$\psi_{j\vec{k}}(\vec{r}) = e^{i\vec{k}\vec{r}} u_{j\vec{k}}(\vec{r}), \tag{5.1}$$

where the periodic part $u_{j\vec{k}}(\vec{r})$ is an eigenstate of the Hamiltonian

$$h_{\vec{k}} = \frac{(\vec{p} + \hbar\vec{k})^2}{2m} + U(\vec{r}). \tag{5.2}$$

From this it follows

$$h_{\vec{k}+\vec{q}} = h_{\vec{k}} + \frac{\hbar^2}{m}\vec{q} \cdot \left(\frac{1}{i}\nabla + \vec{k}\right) + \frac{\hbar^2}{2m}q^2,$$

$$\frac{1}{\hbar}\frac{\partial h_{\vec{k}}}{\partial \vec{k}} = \frac{\hbar}{m}\left(\frac{1}{i}\nabla + \vec{k}\right). \tag{5.3}$$

In the first equation we may ignore the last term since $\vec{q}$ is small, and the difference $V = h_{\vec{k}+\vec{q}} - h_{\vec{k}}$ may be treated as a small perturbation. The perturbed band energy is

$$\frac{\partial \epsilon_{j\vec{k}}}{\partial k_i} q_i = \int d\vec{r}\, u_{j\vec{k}}^*(\vec{r}) V u_{j\vec{k}}(\vec{r}), \tag{5.4}$$

---

[b]This chapter requires a solid understanding of basic band structure theory. Some basic band structure calculations are explained in Chapter 4.

where $\epsilon_{j\vec{k}}$ is the **eigenvalue**. From this, we can obtain

$$\frac{1}{\hbar}\frac{\partial \epsilon_{j\vec{k}}}{\partial \vec{k}} = \frac{\hbar}{m}\int d\vec{r}\, u_{j\vec{k}}^* \left(\frac{1}{i}\nabla + \vec{k}\right) u_{j\vec{k}} = \frac{\hbar}{m}\int d\vec{r}\, \psi_{j\vec{k}}^* \frac{1}{i}\nabla \psi_{j\vec{k}}. \qquad (5.5)$$

Since $\frac{1}{m}\frac{\hbar}{i}\nabla$ is the velocity operator (see Exercise 5.2) we find the following basic relation between the velocity and the band energy dispersion

$$\vec{v} = \frac{1}{\hbar}\frac{\partial \epsilon_{j\vec{k}}}{\partial \vec{k}}. \qquad (5.6)$$

Exercise 5.1. Show Eq. (5.2).

Exercise 5.2. Show that the velocity operator is

$$\vec{v} = \vec{p}/m, \qquad (5.7)$$

where $\vec{p}$ is the momentum operator. Hint: Use $\frac{d\vec{r}}{dt} = \frac{1}{i\hbar}[\vec{r}, H]$.

An electric field can be incorporated through a time-dependent vector potential, see Exercise 5.3 below. In this approach, crystal periodic symmetries will remain intact. Note that **time dependence** of the wave function $\psi_{j\vec{k}}$ appears **only through wave vector** $\vec{k}$ of $u_{j\vec{k}}(\vec{r})$ of the Bloch wavefunction Eq. (5.1). See the following Exercise 5.4 for an explanation.

Exercise 5.3. Consider a Bloch electron moving in a static and uniform electric field $\vec{E}$. Adding a term $e\vec{E}\cdot\vec{r}$ to the Hamiltonian will break translational invariance. Show that adding a vector potential $\vec{A} = -c\vec{E}t$ will simulate the effect of the electric field without breaking translational invariance. Note that

$$\vec{E} = -\frac{1}{c}\frac{\partial \vec{A}}{\partial t}. \qquad (5.8)$$

Hint: Show that the new Hamiltonian for the Bloch wave functions is

$$H = \frac{(\vec{p} - e\vec{E}t)^2}{2m} + U(\vec{r}), \qquad (5.9)$$

where $U(r)$ is the crystal periodic potential and m is the electron bare mass.

Exercise 5.4. (a) The Bloch wave function is given by $\psi_{j\vec{k}}(\vec{r}) = e^{i\vec{k}\cdot\vec{r}}u_{j\vec{k}}(\vec{r})$. Show that the periodic part $u_{j\vec{k}}(\vec{r})$ is an eigenstate of the following operator (see [5]):

$$h_k = \frac{(\vec{p} + \hbar\vec{k} - e\vec{E}t)^2}{2m} + U(\vec{r}). \tag{5.10}$$

(b) The time-dependent wave function is given by

$$\psi_{j\vec{k}}(\vec{r}, t) = e^{i\vec{k}\cdot\vec{r}}u_{j\vec{k}(t)}(\vec{r}). \tag{5.11}$$

This means that only k in $u_{j\vec{k}}(\vec{r})$ depends on t. Find $\vec{k}(t)$. Answer:

$$\vec{k}(t) = \vec{k} - \frac{e}{\hbar}\vec{E}t = \vec{q}. \tag{5.12}$$

This answer is consistent with the result of the semiclassical model of electron dynamics [5], in which the time variation of wave vector is given by

$$\dot{\vec{q}} = -\frac{e}{\hbar}\vec{E}. \tag{5.13}$$

5.2. Anomalous Velocity

An external electric field modifies electronic states. It gives an extra contribution to the electron velocity due to the corrections to electron wave functions. We will see below that this anomalous contribution to the velocity leads to a finite value of the off-diagonal conductivity related to the Chern number of the Brillouin zone. We first derive the anomalous velocity.

As mentioned in Exercise 5.3, an electric field can be incorporated through a time-dependent vector potential. This approach is convenient because crystal periodic symmetries will remain intact. Time dependence may then be incorporated naturally in the density matrix formalism. (Before reading the following text, the reader should first study the density matrix formalism Appendix C.) Let us follow the derivation in Ref. [2] and compute the current using the density matrix.

Consider an electron subject to a static, weak, and uniform electric field in a periodic crystal potential. Suppose only **one** band is occupied (we will call this band the zeroth band). We want to compute its velocity under the semiclassical model of electron dynamics [5]. The density matrix of a pure state is

$$\rho(t) = |\psi(t)\rangle\langle\psi(t)| = |\psi_0(t)\rangle\langle\psi_0(t)| + \Delta\rho(t). \tag{5.14}$$

Here the quantum state is divided into the initial $|\psi_0(t)\rangle$ and perturbed $\Delta\psi(t)$ parts. The expectation value of the velocity may be computed by taking trace of this density matrix

$$\begin{aligned}
\vec{v}(t) = \mathrm{Tr}\left\{\rho(t)\vec{\mathbf{v}}\right\} &= \langle\psi_0(t)|\vec{\mathbf{v}}|\psi_0(t)\rangle \\
&+ \langle\psi_0(t)|\Delta\rho(t)\vec{\mathbf{v}}|\psi_0(t)\rangle,
\end{aligned} \tag{5.15}$$

where $\vec{\mathbf{v}}$ is the velocity operator.[c]

Using the commutation relation between the Hamiltonian and the density matrix given by Eq. (C.8), we find

$$\begin{aligned}
[H(t), \Delta\rho(t)] = i\hbar\dot{\rho}(t) &\approx i\hbar\frac{d}{dt}|\psi_0(t)\rangle\langle\psi_0(t)| \\
&= i\hbar(|\dot{\psi}_0(t)\rangle\langle\psi_0(t)| + |\psi_0(t)\rangle\langle\dot{\psi}_0(t)|).
\end{aligned} \tag{5.16}$$

Taking the matrix element of this commutator between $\langle\psi_0(t)|$ and $|\psi_n(t)\rangle$ we find[d]

$$(E_0 - E_n)\langle\psi_0(t)|\Delta\rho(t)|\psi_n(t)\rangle = i\hbar\langle\dot{\psi}_0(t)|\psi_n(t)\rangle. \tag{5.19}$$

Using this and the Hermitivity of $\Delta\rho(t)\vec{\mathbf{v}}$, we find the following representation[e]

$$\begin{aligned}
\Delta\rho(t)\vec{\mathbf{v}} &\to i\hbar\sum_{n\neq 0}|\psi_0(t)\rangle\frac{\langle\dot{\psi}_0(t)|\psi_n(t)\rangle}{E_0 - E_n}\langle\psi_n(t)|\vec{\mathbf{v}} \\
&- i\hbar\vec{\mathbf{v}}\sum_{n\neq 0}|\psi_n(t)\rangle\frac{\langle\psi_n(t)|\dot{\psi}_0(t)\rangle}{E_0 - E_n}\langle\psi_0(t)|.
\end{aligned} \tag{5.21}$$

---

[c]The velocity operator is written in bold letters to distinguish it from $\vec{v}$ which is a number. In this book, whenever such a confusion may arise we use bold letters.

[d]Here the index n is a short hand for $(j, \vec{k})$ which labels a Bloch wave function in a periodic crystal potential. The band index is j and wave vector is $\vec{k}$. A Bloch wave function obeys **Bloch's theorem**

$$\psi_{j\vec{k}}(\vec{r} + \vec{R}) = e^{i\vec{k}\cdot\vec{R}}\psi_{j\vec{k}}(\vec{r}), \tag{5.17}$$

where the function $u_{j\vec{k}}(\vec{r})$ is periodic for all $\vec{R}$ in the Bravais lattice

$$u_{j\vec{k}}(\vec{r} + \vec{R}) = u_{j\vec{k}}(\vec{r}). \tag{5.18}$$

[e]Here we have used the following: If $A = \sum_{nm}|n\rangle A_{nm}\langle m|$ is a Hermitian operator, then

$$A^\dagger = \sum_{nm}(|n\rangle A_{nm}\langle m|)^\dagger = \sum_{nm}|m\rangle A_{nm}^*\langle n| = \sum_{nm}|m\rangle A_{mn}\langle n|. \tag{5.20}$$

It can be easily checked that the right-hand side is indeed Hermitian. Inserting this result into the expression for the velocity, Eq. (5.15) and using $\sum_j |\psi_{j\vec{k}}\rangle\langle\psi_{j\vec{k}}| = 1$, we find

$$\vec{v}_{\vec{k}}(t) = \langle\psi_{0\vec{k}}|\vec{\mathbf{v}}|\psi_{0\vec{k}}\rangle + i\hbar\sum_{j\neq 0}\left[\frac{\langle\dot{\psi}_{0\vec{k}}|\psi_{j\vec{k}}\rangle\langle\psi_{j\vec{k}}|\vec{\mathbf{v}}|\psi_{0\vec{k}}\rangle}{\epsilon_{0\vec{k}} - \epsilon_{j\vec{k}}} - \text{c.c.}\right], \quad (5.22)$$

which gives the velocity of an electron in the state $|0\vec{k}\rangle$ of the zeroth band.

Using the result obtained in Exercise 5.4 and the velocity operator $\vec{\mathbf{v}} = \frac{\hbar}{im}\nabla$, we rewrite

$$\hbar\sum_{j\neq 0}\frac{\langle\dot{\psi}_{0\vec{k}}|\psi_{j\vec{k}}\rangle\langle\psi_{j\vec{k}}|\vec{\mathbf{v}}|\psi_{0\vec{k}}\rangle}{\epsilon_{0\vec{k}} - \epsilon_{j\vec{k}}} = \hbar\sum_{j\neq 0}\langle\dot{u}_{0\vec{k}}|u_{j\vec{k}}\rangle\frac{\langle u_{j\vec{k}}|\vec{\mathbf{v}}|u_{0\vec{k}}\rangle}{\epsilon_{0\vec{k}} - \epsilon_{j\vec{k}}}.$$

$$(5.23)$$

It is easy to know that $\langle\psi_{j\vec{k}}|\vec{\mathbf{v}}|\psi_{0\vec{k}}\rangle = \langle u_{j\vec{k}}|\vec{\mathbf{v}}|u_{0\vec{k}}\rangle$: differentiating $|\psi_{0\vec{k}}\rangle$ with respect to $\vec{r}$ gives two terms but one of them is zero due to the orthogonality of $u_{0\vec{k}}(\vec{r})$ and $u_{j\vec{k}}(\vec{r})$ for $j \neq 0$. Then using the result of Exercise 5.5 we find

$$\hbar\sum_{j\neq 0}\langle\dot{u}_{0\vec{k}}|u_{j\vec{k}}\rangle\frac{\langle u_{j\vec{k}}|\vec{\mathbf{v}}|u_{0\vec{k}}\rangle}{\epsilon_{0\vec{k}} - \epsilon_{j\vec{k}}} = \langle\dot{u}_{0\vec{k}}|\nabla_{\vec{k}}u_{0\vec{k}}\rangle \quad (5.24)$$

This leads to the following result for the velocity

$$\vec{v}_{\vec{k}} = \frac{1}{\hbar}\frac{d\epsilon_{0\vec{k}}}{d\vec{k}} + \vec{V}(\vec{k}), \quad (5.25)$$

where the second term is

$$\vec{V}(\vec{k}) = i(\langle\dot{u}_{0\vec{k}}|\nabla_{\vec{k}}u_{0\vec{k}}\rangle - \langle\nabla_{\vec{k}}u_{0\vec{k}}|\dot{u}_{0\vec{k}}\rangle). \quad (5.26)$$

It is called the **anomalous** velocity. This term originates from the changes of the electronic wave functions by an external electric field. When integrated over the first Brillouin zone, the first term of Eq. (5.25) vanishes.[f] For one-dimensional systems, we can express the current in terms of the Berry curvature

$$J(t) = -e\int_{\text{BZ}}\frac{dk}{2\pi}V_k. \quad (5.27)$$

---

[f]In a one-dimensional crystal potential the first Brillouin zone is $-\frac{\pi}{a} < k < \frac{\pi}{a}$, where a is the period of the lattice. Other Bloch wave functions with k values outside the first Brillouin zone are not independent from those in the first Brillouin zone.

Exercise 5.5. The following result will also be used in the derivation of the anomalous velocity. Show

$$|\nabla_{\vec{k}} u_{0\vec{k}}\rangle = \hbar \sum_{j\neq 0} |u_{j\vec{k}}\rangle \frac{\langle u_{j\vec{k}} | \vec{\mathbf{v}} | u_{0\vec{k}}\rangle}{\epsilon_{0\vec{k}} - \epsilon_{j\vec{k}}}. \tag{5.28}$$

Hint: Show the perturbative result

$$u_{0\vec{k}+\Delta\vec{k}} - u_{0\vec{k}} = \sum_{j\neq 0} |u_{j\vec{k}}\rangle \frac{\langle u_{j\vec{k}} | h_{\vec{k}+\Delta\vec{k}} - h_{\vec{k}} | u_{0\vec{k}}\rangle}{\epsilon_{0\vec{k}} - \epsilon_{j\vec{k}}}. \tag{5.29}$$

Use Eq. (5.3) and the first-order perturbative correction to the wave function. Also use how the velocity operator is defined in Exercise 5.2.

5.3. Chern Number of Band Structure

Using the anomalous velocity let us show that the off-diagonal conductivity is related to the Chern number. In a periodic solid, the relevant topological manifold is the surface of the Brillouin zone.

The expectation value of the electron velocity, Eq. (5.25), is

$$\vec{v}_{\vec{k}} = \frac{1}{\hbar} \nabla_{\vec{k}} \epsilon_{\vec{k}} + i \left[\left\langle \frac{\partial u_{0\vec{k}}}{\partial t} \Big| \frac{\partial u_{0\vec{k}}}{\partial \vec{k}} \right\rangle - \left\langle \frac{\partial u_{0\vec{k}}}{\partial \vec{k}} \Big| \frac{\partial u_{0\vec{k}}}{\partial t} \right\rangle \right]. \tag{5.30}$$

Here we have suppressed time dependence of $\vec{k}$. Suppose an electric field is applied along the x-axis: $\vec{E} = \frac{1}{c}\frac{\partial \vec{A}}{\partial t} = (E_x, 0, 0)$. To compute the off-diagonal conductivity we need to replace the time derivatives with

$$\frac{\partial}{\partial t} = \frac{\partial \vec{q}}{\partial t} \cdot \frac{\partial}{\partial \vec{q}} = -\frac{e}{\hbar} \vec{E} \cdot \frac{\partial}{\partial \vec{q}} = -\frac{e}{\hbar} E_x \cdot \frac{\partial}{\partial q_x}. \tag{5.31}$$

Here we have used the result from Exercise 5.4:

$$\hbar\vec{q} = \hbar\vec{k} + \frac{e}{c}\vec{A}(t) = \left(\frac{e}{c}A(t), \hbar k_y, 0\right). \tag{5.32}$$

In this way, the derivative with respect to t appears. There is also the derivative with respect to $\vec{k}$ in the expression for the anomalous velocity. To get a more symmetric expression, it is desirable to replace it with the derivative with respect to k_y

$$\frac{\partial}{\partial \vec{k}} = \frac{\partial \vec{q}}{\partial \vec{k}} \cdot \frac{\partial}{\partial \vec{q}} = \frac{\partial}{\partial \vec{q}} = \left(0, \frac{\partial}{\partial q_y}, 0\right). \tag{5.33}$$

Using Eqs. (5.31) and (5.33), we can write the anomalous velocity (the second term in Eq. (5.30)) as

$$\vec{V}(\vec{k}) = \frac{e}{\hbar}\vec{E} \times \vec{b} = (0, V_y(\vec{k}), 0), \tag{5.34}$$

where

$$V_y(\vec{k}) = -\frac{e}{\hbar}b_z(\vec{k})E_x \tag{5.35}$$

with the z-component of the Berry curvature

$$b_z(\vec{k}) = \Omega_{yx}(\vec{k}) = i\left[\left\langle \frac{\partial u_{0\vec{k}}}{\partial k_x}\Big|\frac{\partial u_{0\vec{k}}}{\partial k_y}\right\rangle - \left\langle \frac{\partial u_{0\vec{k}}}{\partial k_y}\Big|\frac{\partial u_{0\vec{k}}}{\partial k_x}\right\rangle\right]$$

$$= i\langle\nabla_{\vec{k}}u_{0\vec{k}}| \times |\nabla_{\vec{k}}u_{0\vec{k}}\rangle. \tag{5.36}$$

(Here we have made a change of variables $\vec{q} \to \vec{k}$.) The last equality follows from two different expressions for the Berry curvature, see Eqs. (2.34) and (2.35). (It is useful to remember $\Omega_{ij} = \epsilon_{ijk}b_k$, where ϵ_{ijk} is the Levi-Civita antisymmetric tensor.) From the expression for the velocity, Eq. (5.35), we see that an electric field in the x-direction gives rise to a current in the y-direction. By summing over the contribution from all the electrons in the Brillouin zone, we get the following expression for the off-diagonal conductivity

$$\sigma_{yx} = \frac{e}{L^2}\frac{e}{\hbar}\sum_{\mathrm{BZ}}\Omega_{yx} = \frac{e^2}{h}\int_{\mathrm{BZ}}\frac{d\vec{k}}{2\pi}\Omega_{yx}(\vec{k}) = \frac{e^2}{h}\text{integer}. \tag{5.37}$$

Here we have used

$$\frac{i}{2\pi}\oint\langle u_{0\vec{k}}|\nabla_{\vec{k}}|u_{0\vec{k}}\rangle d\vec{k} = \frac{1}{2\pi}\oint\nabla_k\theta_0(\vec{k}) = n, \tag{5.38}$$

where $\theta_0(\vec{k})$ is the Berry phase. The line integral is along the edges of the BZ. (This result agrees with the Kubo result for the off-diagonal conductivity Eq. (10.81).) The off-diagonal conductivity thus represents a Chern number.

Examining the anomalous velocity Eq. (5.34), one can show the following results for the Berry curvature. In the presence of time reversal symmetry

$$\vec{b}(\vec{k}) = -\vec{b}(-\vec{k}). \tag{5.39}$$

(It transforms like a real magnetic field). If inversion symmetry is also present,

$$\vec{b}(\vec{k}) = \vec{b}(-\vec{k}). \tag{5.40}$$

When both symmetries are present then $\vec{b}(\vec{k}) = 0$, i.e., the Chern number is **zero**. When one of them is absent then $\vec{b}(\vec{k}) \neq 0$.

5.4. One-Dimensional Zak Phase and Polarization

The anomalous velocity may be used to define a topological phase called the Zak phase, which is related to the polarization of a periodic system. The concept of polarization of a periodic system is subtle since it is multivalued and unit cell dependent. Only the difference in polarization is physically meaningful. Classical intuition about polarization fails when it is applied to periodic lattice systems. In the modern theory of polarization of periodic systems, spontaneous polarization or strain induced polarization is studied in the **absence** of an applied electric field,[g] see Fig. 5.1. In this theory, polarization is given in terms of **the Zak phase** [6], which is the Berry phase that an electron acquires as it moves across the Brillouin zone. (The wave vector acts as an adiabatic parameter.) The Zak phase exhibits topological properties of a band structure. Also, if quantum fluctuations can be ignored, the Zak phase can be related to the localized boundary charge of a finite system through the **bulk polarization**. This is an example of **bulk edge correspondence**. Moreover, symmetry can lead to integer or fractional quantized values of polarization. Here we will restrict our discussion to the Zak phase of one-dimensional **insulators** with a band gap, see Fig. 5.2.

• Unit Cell Dependent Polarization

In classical physics values of the polarization of a dielectric material is unique and continuous. However, quantum mechanical treatment of polarization of a dielectric material shows that its value is not unique and depends on the choice of unit cell.

---

[g]Before reading this chapter students should study Chapters 8, 9, and 12 of Ref. [5]. These chapters explain Bloch's theorem and semiclassical model of electron dynamics.

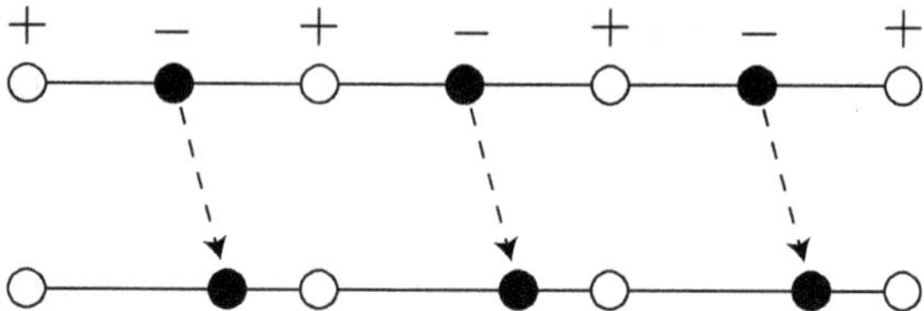

Fig. 5.1. Distortion of a one–dimensional ionic crystal as temperature is lowered. Centrosymmetry (inversion symmetry) is broken.

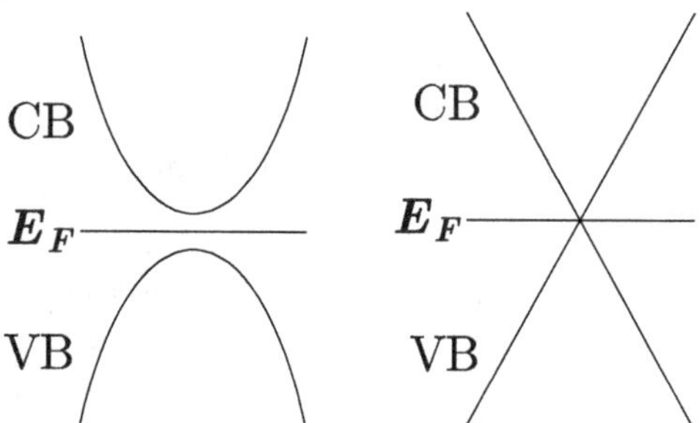

Fig. 5.2. **Left**: we discuss the properties of the Zak phase of a one-dimensional band structure with a gap at the Fermi level. CB and VB stand for conduction and valence bands, respectively. **Right**: the Berry phase is not well defined in a semimetal with the degenerate Fermi energy.

Consider a one-dimensional insulator with a finite length L. The polarization can be defined as

$$P = \frac{1}{L} \int x\rho(x)dx, \qquad (5.41)$$

where $\rho(x)$ is the charge density. Here we are interested in the intrinsic (bulk) polarization whose value is independent of the surface termination. For this purpose, the definition of the classical polarization given above is inapplicable since the polarization value **depends** on the choice of the unit cell of the underlying lattice. This is because different unit cells may have different values of polarization. The polarization of the left cell in Fig. 5.3 is

$$P = \frac{1}{a} \sum_i q_i x_i, \qquad (5.42)$$

where the sum is over all the charges in the cell. For the right unit cell it is $-P$. In fact, the value of P can vary continuously depending on the position of the unit cell. This can be seen by realizing that the charge distribution

Fig. 5.3. Two different unit cells are shown. They have the opposite values of polarization. Unit cell length is a.

is actually continuous and peaked where ions are located. The **absolute value** of polarization of a **periodic** crystal thus does **not** have any physical meaning.

• Multivalued Polarization

The change of polarization can be computed using the basic relation between polarization P and polarization current density J_p

$$\frac{dP}{dt} = J_p(t). \tag{5.43}$$

(The unit of J_p is charge/time in one-dimensional systems.) From this, we see that the polarization change ΔP can be obtained from the integrated current $\int dt J_p(t)$ which flows during an adiabatic transformation. One cannot measure the absolute value of polarization of a periodic system. Only the polarization difference after an adiabatic structural transformation is considered to be meaningful [1, 3]. The following thought experiment is helpful in clarifying these ideas. Consider an adiabatic process in which each anion moves M lattice constants, see Fig. 5.4 (other examples of similar cyclic adiabatic processes can be found in Ref. [1]). The initial and final states cannot be distinguished. Since the total current flowed during this process is Mq, the **difference** ΔP is

$$\Delta P = Mq. \tag{5.44}$$

Consequently, the polarization of the initial state is changed by ΔP. Since any integer value of M is allowed the polarization of the final state is multivalued

$$P_f = P_i + Mq, \tag{5.45}$$

where M is any integer and P_i is the polarization of the initial state. One can only determine **possible values** of how much P is changed, i.e., the absolute values of P_i and P_f are not known.

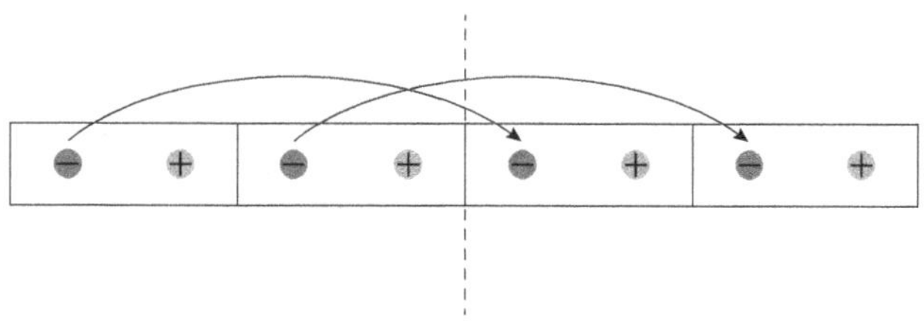

Fig. 5.4. Suppose each anion with charge q moves two lattice constants during an adiabatic process. In this case, two anions move through the cross section of the system (shown as the dashed line in the figure), no matter how the unit cell is chosen. The initial and final states are indistinguishable. Suppose now each anion moves only one lattice constant. In this case one anion moves through the cross section of the system. Again the initial and final states are indistinguishable. In these two processes, the polarization changes are different.

• Modern Theory of Polarization

Here we relate the polarization of a system to the Zak phase using the anomalous velocity. Consider a one-dimensional band structure. Note that, for one-dimensional systems the polarization P is defined as the dipole moment per length. This means P has the same units as e. Let us relate the polarization to a topological quantity called the Zak phase. It is a Berry phase that an electron acquires as it moves through the Brillouin zone. Let us first compute the current from the anomalous velocity and show that it is related to the Berry curvature [2, 7]. After this we will show that the polarization is related to the Berry phase.

If time dependence appears through the parameter $\lambda(t)$ with $\lambda(0) = 0$ and $\lambda(1) = T$, we can write the anomalous current, given by Eq. (5.27), as

$$\int_0^T dt J(t) = -ie \int_0^1 d\lambda \int_{\text{BZ}} \frac{dk}{2\pi} 2i \operatorname{Im} \left\langle \frac{du_{0k}}{d\lambda} \middle| \nabla_k u_{0k} \right\rangle, \qquad (5.46)$$

where the integration domain is shown in Fig. 5.5. In the derivation of Eq. (5.46), we have used

$$\langle \phi | \psi \rangle - \langle \psi | \phi \rangle = 2i \operatorname{Im} \langle \phi \,|\, \psi \rangle \qquad (5.47)$$

and $\frac{du_{0k}}{dt} = \frac{du_{0k}}{d\lambda} \frac{d\lambda}{dt}$. Using the identity

$$2i \operatorname{Im} \left\langle \frac{du_{0k}}{d\lambda} \middle| \nabla_k u_{0k} \right\rangle = \frac{d}{d\lambda} \left\langle u_{0k} \middle| \frac{du_{0k}}{dk} \right\rangle - \frac{d}{dk} \left\langle u_{0k} \middle| \frac{u_{0k}}{d\lambda} \right\rangle, \qquad (5.48)$$

we find

$$\int_0^T dt\, J(t) = \frac{-ie}{2\pi} \int d\lambda \int_{\text{BZ}} dk \left(\frac{d}{d\lambda} \left\langle u_{0k} \middle| \frac{du_{0k}}{dk} \right\rangle - \frac{d}{dk} \left\langle u_{0k} \middle| \frac{du_{0k}}{d\lambda} \right\rangle \right).$$

$$(5.49)$$

We define the Berry connection

$$\vec{A} = (A_\lambda, A_k) = \left[i \left\langle u_{0k} \middle| \frac{du_{0k}}{d\lambda} \right\rangle , i \left\langle u_{0k} \middle| \frac{du_{0k}}{dk} \right\rangle \right]. \qquad (5.50)$$

Note that the Berry connections is defined in terms of the periodic part of the Bloch wave functions, $u_{nk}(x)$, and not in terms of the Bloch wave functions, $\psi(x) = e^{ikx} u_{nk}(x)$. The function $\psi(x) = e^{ikx} u_{nk}(x)$ is **not** periodic in x but $u_{nk}(x)$ is (here n stands for the band index). The curl of $\vec{A}$ gives the **Berry curvature** b_z in the $(\lambda, \vec{k})$ domain, see Eq. (2.33). Using this definition, we find that the total current that flowed through the system can be written as

$$\int_0^T dt\, J(t) = \frac{-e}{2\pi} \int_0^1 d\lambda \int_{\text{BZ}} dk \left(\frac{dA_k}{d\lambda} - \frac{dA_\lambda}{dk} \right)$$

$$= \frac{-e}{2\pi} \int_S b_z dS = \frac{-e}{2\pi} \oint_C \vec{A} \cdot d\vec{l}, \qquad (5.51)$$

where S is the area confined by C. The areal integration over S is converted into a line integral along C, see Fig. 5.5 (the contour is followed **clockwise**). We find that only A_k gives a non-zero contribution since the contributions of A_λ from the opposite vertical paths cancel out. This can be seen by noting that $A_\lambda|_{k=-\frac{\pi}{a}} = A_\lambda|_{k=+\frac{\pi}{a}}$ which follows from $u_{k+K}(x) = e^{-iKx} u_k(x)$, where K is a reciprocal lattice vector. Thus we find that the total current flowing through the system is

$$\int_0^T dt\, J(t) = -\frac{e}{2\pi} \int_{\text{BZ}} dk \left[A_k(\lambda = 1) - A_k(\lambda = 0) \right]. \qquad (5.52)$$

Since this total current is also equal to the polarization change [4]

$$\int_0^T dt\, J(t) = \int_0^T dt \frac{dP}{dt} = \int_0^1 \frac{dP}{d\lambda} d\lambda = P(1) - P(0), \qquad (5.53)$$

we can relate the polarization to the one-dimensional Zak phase

$$P(\lambda) = -\frac{e}{2\pi} \int_{\text{BZ}} dk\, i \langle u_{0k} | \nabla_k u_{0k} \rangle = -\frac{e}{2\pi} Z, \qquad (5.54)$$

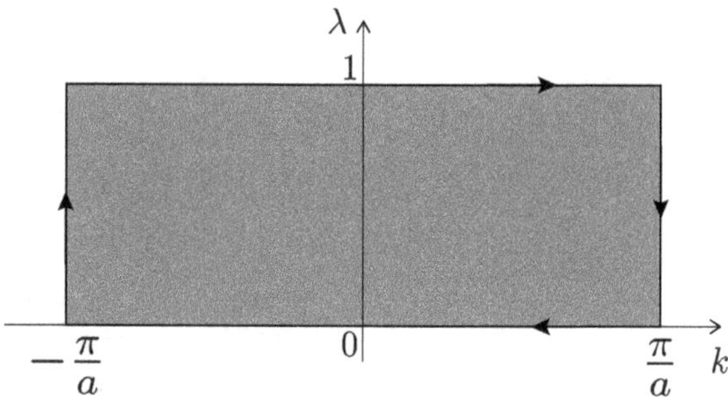

Fig. 5.5. An adiabatic parameter is denoted by λ, which represents the dimensionless lattice distortion shown in Fig. 5.1. Another adiabatic parameter is wave vector k. Adiabatic path goes around the edges of the rectangular in the (λ, k)-plane.

where the Zak phase is defined as

$$Z = i \int_{\mathrm{BZ}} dk \left\langle u_{0k} \left| \frac{du_{0k}}{dk} \right. \right\rangle = \int_{\mathrm{BZ}} dk\, A_k(\lambda). \tag{5.55}$$

(Note that, to compute the Berry connection $A_k(\lambda)$, the band structure must be computed at each λ.) From this result, we see that polarization is given by a Berry phase with the wave vector k serving as an adiabatic parameter. It is the Berry phase that an electron acquires as it moves across the Brillouin zone. (One-dimensional first Brillouin zone is equivalent to a closed circle.) Polarization is **not** related to an integer Chern number. Here the Brillouin zone is only one-dimensional. One needs a surface to define a Chern number. For a three-dimensional system with **several** filled bands without a band crossing the αth component of the polarization is

$$P_\alpha = \frac{-ie}{(2\pi)^3} \sum_{j\ \mathrm{filled}} \int_{\mathrm{BZ}} \left\langle u_{j\vec{k}} \left| \frac{\partial}{\partial k_\alpha} \right| u_{j\vec{k}} \right\rangle d^3k. \tag{5.56}$$

Note that, in three dimensions, the dimension of $[P] = \mathrm{Charge}/\mathrm{Length}^2$, unlike the one-dimensional case.

Here one might ask why the polarization is not given in terms of the Bloch wave functions? Why is it defined in terms of the cell periodic functions $u_{jk}(x)$? This is because the Bloch wave functions lead to vanishing Berry connections in the limit $\Delta k \to 0$ [1]

$$\langle \psi_{jk} | \psi_{jk+\Delta k} \rangle = \int_{-\infty}^{\infty} dx\, e^{ix\Delta k} u_{jk}^*(x) u_{jk+\Delta k}(x) \to 0. \tag{5.57}$$

On the other hand,

$$\langle u_{jk}|u_{jk+\Delta k}\rangle = N_s \int_0^a dx\, u_{jk}^*(x) u_{jk+\Delta k}(x) \tag{5.58}$$

is finite and well-behaved (N_s is the number of unit cells).

The polarization change $\Delta\vec{P}$ is related to the difference between the Zak phases of an insulator and a vacuum. The edge (boundary) charge is thus

$$Q = e\frac{Z}{2\pi}, \tag{5.59}$$

where Z is the Zak phase of the insulator. (The Zak phase of a vacuum is zero.) However, since the value of the Zak phase of the insulator is multi-valued, it gives **only possible** values of the boundary charge and not the actual value of a finite system. To determine which value is actually realized one must perform a calculation incorporating the structure of the unit cell **near** the boundary. Even if the Zak phase is 0 mod 2π the system may actually have non-trivial topological properties, such as an integer value of edge charge. An example of such a system is a zigzag edge of a graphene nanoribbon (see Sec. 17.3). Moreover, in some cases a boundary edge can affect the bulk property. This happens, for example, in a zigzag graphene nanoribbon in the presence of disorder or staggered potential, which induces bulk spin-splitting, see Ref. [8]. This effect is singular since even a small perturbation can significantly induce spin-splitting. If a system has the Zak phase value of π, the system is topologically non-trivial and may lead to a fractional edge charge. Whether a fractional charge actually materializes in a system is related to quantum charge fluctuations, as we already discussed in Sec. 3.8.

5.5. Basic Properties of the One-Dimensional Zak Phase

In this section, some fundamental properties of the one-dimensional Zak phase Z of the filled valence bands are given (the Fermi energy is between the valence and conduction bands). When inversion or chiral symmetry (particle–hole symmetry) is present the Zak phase takes values of 0 or π mod 2π. The Zak phase of graphene zigzag nanoribbons is 0 mod 2π while it is π mod 2π for polyacetylene. When the intra cell hopping is smaller than the inter cell hopping. The values π mod 2π represent the topologically non-trivial case. However, even if the Zak phase takes values of 0 mod 2π, the system may have non-trivial topological properties. An example is zigzag nanoribbons with an integer edge charge.

• Zak Phase of Polyacetylene

In the previous Chapter 4, we studied polyacetylene. Here we will show that the chiral symmetry of polyacetylene yields a non-trivial topological number called the Zak phase. It is instructive to explicitly compute the Zak phase, which is defined as which is defined in Eq. (5.55). In this equation, the BZ may be thought of as a circle (loop) with the length equal to the reciprocal lattice vector $K = 2\pi/a$. However, the periodic part of the Bloch wave function $u_k(x)$ is not equal to $u_{k+K}(x)$ since $u_{k+K}(x) = e^{-iKx}u_k(x)$. The Zak phase is a one-dimensional generalization of the Berry phase. (Note that the Berry phase vanishes in one dimension.)

First it is useful to define a pseudospin. The pseudospin of a band state is defined as the expectation value of the spin operator $\vec{\sigma}$

$$\vec{\Sigma}(k) = \langle u(k)|\vec{\sigma}|u(k)\rangle = \langle \vec{\sigma} \rangle = \begin{pmatrix} \cos\phi(k) \\ \sin\phi(k) \end{pmatrix}. \tag{5.60}$$

The pseudospin of a valence band state is opposite to that of the conduction band state, $-\vec{\Sigma}(k)$. As k varies across the Brillouin zone, the vector $\vec{\Sigma}(k)$ rotates on a circle. Using the eigenvector given in Eq. (4.17), one can show that, for both conduction and valence bands, the Zak phase Z_{1D} is related to the rotation angle $\phi(k)$ of $\vec{\Sigma}(k)$, and that its value is (see Fig. 5.6)

$$Z_{1D} = \frac{1}{2}\oint_{\text{B.Z.}} \frac{d\phi(k)}{dk}dk = \frac{1}{2}[\phi(\pi/a) - \phi(-\pi/a)] = \begin{cases} \pi & \text{mod } 2\pi, \ t'/t < 1, \\ 0 & \text{mod } 2\pi, \ t'/t > 1. \end{cases} \tag{5.61}$$

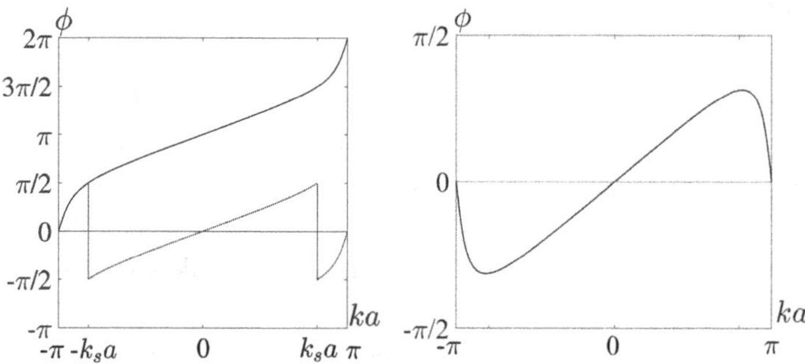

Fig. 5.6. **Left:** Plot of $\phi(k)$ vs. k for $t'/t < 1$. Solid curve is obtained from the dotted curve by adding mod π, the period of the tangent function. The resulting curve is a continuous function of k. (The dotted curve is obtained from the expression for the phase given by Eq. (4.16). It is discontinuous at $\pm k_s$.) **Right:** Plot of $\phi(k)$ vs. k for $t'/t > 1$.

In other words, as k varies across the Brillouin zone, Z_{1D} is given by the difference $\phi(\pi/a) - \phi(-\pi/a)$. When the unit cell of a periodic polyacetylene specimen contains a long bond $(t'/t < 1$), the phase difference is 2π, i.e., Z_{1D} is π mod 2π. In the case of a short-bond unit cell $(t'/t > 1$), Z_{1D} is 0 mod 2π. A **topological phase transition** occurs when the intra- and inter-cell tunneling coefficients are equal, i.e., $t' = t$. This reflects the fact that the bulk polarization can depend on the choice of unit cell [1]. In other words, long-bond polyacetylene is topologically non-trivial whereas short-bond polyacetylene is topologically trivial.

> **Exercise 5.6.** What is the sum of the Zak phases of the conduction and valence bands? Answer: It is $\pi + \pi = 2\pi$, which is equivalent with 0 mod 2π. This result is consistent with the result that the sum of the Berry curvatures should be 0, see Eq. (2.37).

• Absolute Value of Zak Phase is Meaningless

Let us show that a translation of a coordinate system gives a different value of the Zak phase. Suppose there is only one band so that we may ignore the band index j. A Bloch wave function in a coordinate system S is

$$\psi_k(x) = \sum_R e^{ikR}\phi(x - R), \tag{5.62}$$

where R denotes ion positions. The atomic wave function $\phi(x-R)$ is peaked at $x = R$. The periodic part of the Bloch wave function is

$$u_k(x) = e^{-ikx}\sum_R e^{ikR}\phi(x - R). \tag{5.63}$$

Consider a new coordinate system S' that is shifted such that $x' = x - d$. Seen from S' the Bloch wave function takes the form

$$\psi'_k(x') = \sum_R e^{ikR}\phi(x' + d - R), \tag{5.64}$$

and the atoms are located at $x' = R - d$. Note that $\psi'_{k+K}(x') = \psi'_k(x')$, where K is a reciprocal vector. The periodic part of the new Bloch wave function is

$$u'_k(x') = e^{-ikx'} \sum_R e^{ikR} \phi(x' + d - R). \tag{5.65}$$

Note that

$$u'_k(x' + R') = e^{-ik(x'+R')} \sum_R e^{ikR} \phi(x' + R' + d - R)$$

$$= e^{-ikx'} \sum_{R''} e^{ikR''} \phi(x' + d - R'') = u'_k(x'), \tag{5.66}$$

where $R'' = R - R'$.

The periodic part of the new Bloch wave function $u'_k(x')$ is related to the old $u_k(x)$

$$u'_k(x') = e^{ikd} e^{-ikx} \sum_R e^{ikR} \phi(x - R) = e^{ikd} u_k(x). \tag{5.67}$$

Since

$$i\left\langle u'_k(x') \left| \frac{d}{dk} \right| u'_k(x') \right\rangle = -d + i\left\langle u_k(x) \left| \frac{d}{dk} \right| u_k(x) \right\rangle, \tag{5.68}$$

the new Zak phase is

$$Z' = Z - 2\pi \frac{d}{a}, \tag{5.69}$$

where a is the unit cell length. The absolute of the Zak phase thus depends on the choice of the coordinate system.

- **Gauge Transformation**

One can multiply a Bloch wave function with a phase factor $e^{i\chi(k)}$. Since the transformed wave function must satisfy $\psi_{k+K}(x) = \psi_k(x)$, the phase $\chi(k)$ must be a periodic function with the period K: $e^{i\chi(k+K)} = e^{i\chi(k)} \rightarrow \chi(k+K) = \chi(k) + 2\pi n$ (K is a reciprocal lattice vector). The periodic part changes as

$$u_k(x) = e^{i\chi(k)} e^{-ikx} \sum_R e^{ikR} \phi(x - R). \tag{5.70}$$

This factor gives rise to an extra term in the Zak phase

$$Z \rightarrow Z + \int_{BZ} \frac{d\chi(k)}{dk} dk = Z + 2\pi n. \tag{5.71}$$

So in one dimension, the Zak phase is defined mod 2π. This is consistent with the previous result that the polarization is multivalued.

• Inversion Symmetry

Under inversion symmetry $x \to -x$, the polarization of the filled valence bands changes sign $P \to -P$. This implies that the Zak phase also changes sign $Z \to -Z$. By symmetry, the same Z values must also appear in the values of $-Z$: $\gamma + 2\pi n = -\gamma + 2\pi m$, where n and m are integers. This means that $2\gamma = 2\pi(n - m)$, i.e.,

$$\gamma = \pi p, \tag{5.72}$$

where p is any integer. This result combined with the result that the Zak phase is given mod 2π gives

$$Z = \begin{cases} 0 & \mathrm{mod}\ 2\pi, \\ \pi & \mathrm{mod}\ 2\pi. \end{cases} \tag{5.73}$$

So in the presence of inversion symmetry the values of Z/π take even integer or odd integer values, i.e., it is **quantized**.

• Particle–hole Symmetry

When all the valence and conduction band states are filled, the Zak phase is 0 mod 2π (see the gauge invariant expression for the Berry curvature, Eq. (2.37)). Suppose valence bands are fully occupied and their Zak phase is Z, because of particle–hole symmetry, the Zak phase of the conduction bands must be $-Z$. Similar arguments used in the case of inversion symmetry implies that the Zak phase of the valence bands is again **quantized** with the values 0 or π mod 2π.

5.6. Numerical Computation of Gauge Invariant Zak Phase⋆

Numerical evaluations of the eigenstates can give rise to wildly fluctuating phases. Also the band structure can have band crossings, see Fig. 5.7. In such cases, we use a gauge invariant method to compute the Zak phase [9]. (This method was applied to graphene nanoribbons in Ref. [8].) In the following we explain this method.

For simplicity, we will use a one-dimensional model to elucidate the main ideas of this method. Let us first divide the first BZ into small intervals. We

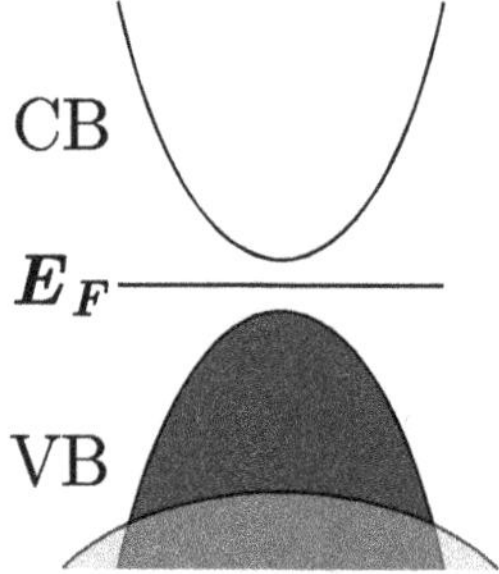

Fig. 5.7. Multi valence bands with degenerate crossings.

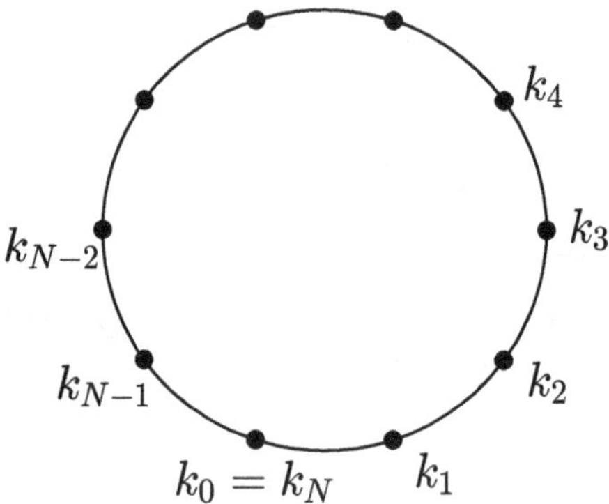

Fig. 5.8. The first Brillouin zone of a one-dimensional system is divided into small intervals.

then find that the total spin-independent **Zak phase** is[h]

$$\mathcal{Z}/2\pi = \frac{1}{2\pi} \sum_{i=0}^{N-1} \mathrm{Arg}\left[\det \mathbf{S}(k_i, k_{i+1})\right], \tag{5.74}$$

where the discrete points in the BZ are k_i with $i = 0, \ldots, N$, see Fig. 5.8. Here the overlap between the **periodic** part of the Bloch wave functions is

$$S_{l,l'}(k_i, k_{i+1}) = \langle u_{l,k_i} | u_{l',k_{i+1}} \rangle = \sum_{p \in \mathrm{Cell}} C^*_{l,k_i,p} C_{l',k_{i+1},p}, \tag{5.75}$$

---

[h] For any complex number z, Arg is defined by $z = |z|\, e^{i\,\mathrm{Arg}\,z}$.

where p denotes lattice sites in the unit cell, $|p\rangle$ is the atomic orbital wave function at site p, and

$$|u_{lk}\rangle = \sum_{p \in \text{Cell}} C_{l,k,p}|p\rangle. \tag{5.76}$$

The expansion coefficients $C_{l,k_i,p}$ are obtained from the **tight-binding** Hamiltonian (an example of such a Hamiltonian is given in Sec. 17.1). Note that the periodic part of the Bloch wave function satisfies

$$u_{l,k+G}(x) = e^{iGx}u_{l,k}(x), \tag{5.77}$$

where $G = \frac{2\pi}{a}$ is a reciprocal vector. This implies that the periodic part of the Bloch wave function at $i = N$ (corresponding to the wave vector $k = \pi/a$) is not identical to the periodic part of the Bloch wave function [5] at $i = 0$ (corresponding to the wave vector $k = -\pi/a$).

• Derivation

The result of Eq. (5.74) can be shown as follows. In the presence of multi-bands without band crossings, the definition of the Zak phase is given by Eq. (5.56), which contains the quantity $-i\sum_n \langle u_{nk}|\nabla_k u_{nk}\rangle$. In the presence of band crossings, we can extend this quantity using the overlap matrix

$$S_{mn}(k, k') = \langle u_{mk}|u_{nk'}\rangle. \tag{5.78}$$

Note that, in this equation, the matrix indices m, n denote **occupied** bands. In the following, we will suppress these matrix indices and write matrices in a bold letter. Then

$$-i\sum_n \langle u_{nk}|\nabla_k u_{nk}\rangle = \text{Im Tr}\left[\nabla_{k'}\mathbf{S}(k, k')\right]\big|_{k'=k}$$

$$= \text{Im Tr}\left[\mathbf{S}^{-1}(k, k')\nabla_{k'}\mathbf{S}(k, k')\right]\big|_{k'=k}$$

$$= \text{Im } \nabla_{k'} \log \det\mathbf{S}(k, k')\big|_{k'=k}$$

$$= \nabla_{k'}\phi(k, k')\big|_{k'=k}. \tag{5.79}$$

Here we have used $\mathbf{S}^{-1}(k, k)$, an identity matrix, and the relation $\text{Tr}(\mathbf{S}^{-1}(k, k')\nabla_k\mathbf{S}(k, k')) = \nabla_k \log \det\mathbf{S}(k, k')$ (see Exercise 5.7). Here

$\phi(k, k')$ denotes the phase $\mathrm{Im}\log\det\mathbf{S}(k, k')$. Thus

$$d\phi = -i \sum_n \langle u_{nk}|\nabla_k u_{nk}\rangle\, dk. \tag{5.80}$$

Then the change of the phase over one discrete interval is

$$\Delta\phi = \mathrm{Im}\log\det\mathbf{S}(k_i, k_{i+1}) = \mathrm{Arg}[\det\mathbf{S}(k_i, k_{i+1}]. \tag{5.81}$$

Since $\mathrm{Im}\log z = \mathrm{Arg}(z)$ this results gives the Zak phase in Eq. (5.74), which is gauge invariant. In Chapter 17, we will compute the Zak phase of a zigzag nanoribbon using Eq. (5.74).

Exercise 5.7. Show $\mathrm{Tr}[\mathbf{S}^{-1}(k, k')\nabla_k\mathbf{S}(k, k')] = \nabla_k\log\det\mathbf{S}(k, k')$. Answer: The matrix $\mathbf{S}(k, k')$ can be diagonalized as $\mathbf{S}(k, k') = \mathbf{U}^{-1}(k, k')\mathbf{D}(k, k')\mathbf{U}(k, k')$. Let the diagonal element be denoted λ_n. Then,

$$\begin{aligned}
\mathrm{Tr}&[\mathbf{S}^{-1}(k, k')\nabla_k\mathbf{S}(k, k')] \\
&= \mathrm{Tr}[\mathbf{U}^{-1}(k, k')\mathbf{D}^{-1}(k, k')\mathbf{U}(k, k')\nabla_k\mathbf{U}^{-1}(k, k')\mathbf{D}(k, k')\mathbf{U}(k, k')] \\
&= \mathrm{Tr}[\mathbf{U}^{-1}(\mathbf{k}, \mathbf{k'})\mathbf{D}^{-1}(\mathbf{k}, \mathbf{k'})\nabla_\mathbf{k}\mathbf{D}(\mathbf{k}, \mathbf{k'})\mathbf{U}(\mathbf{k}, \mathbf{k'})] \\
&= \mathrm{Tr}[\mathbf{D}^{-1}(k, k')\nabla_k\mathbf{D}(k, k')] \\
&= \sum_n \lambda_n^{-1}\nabla\lambda_n = \sum_n \nabla_k\log\lambda_n \\
&= \nabla_k\log\prod_n \lambda_n = \nabla\log\det\mathbf{S}(k, k').
\end{aligned} \tag{5.82}$$

Therefore,

$$\mathrm{Tr}[\mathbf{S}^{-1}(k, k')\nabla_k\mathbf{S}(k, k')] = \nabla_k\log\det\mathbf{S}(k, k'). \tag{5.83}$$

• Single Band Case

When there is **only one occupied band**, the result for the Zak phase reduces to

$$\mathcal{Z} = -\mathrm{Im}\log\left[\langle u_{k_0}|u_{k_1}\rangle\ldots\langle u_{k_{N-1}}|u_{k_N}\rangle\right]. \tag{5.84}$$

In this result, the product $|u_{k_i}\rangle\langle u_{k_i}|$ appears. Any arbitrary phase $e^{i\theta_i}$ of $|u_{k_i}\rangle$ will be canceled by the phase $e^{-i\theta_i}$ of $\langle u_{k_i}|$. So the Zak phase computed in this way is phase independent. Using this result, we will explicitly compute the Zak phase of polyacetylene in Chapter 4.

Bibliography

[1] D. Vanderbilt, *Berry Phases in Electronic Structure Theory* (Cambridge University Press, Cambridge, 2018). •

[2] R. Resta, *Geometry and Topology in Electronic Structure Theory*, Unpublished Notes. •

[3] R. Resta, Theory of the electric polarization in crystals, *Ferroelectrics* **136**, 51 (1992); R. Resta, Polarization as a Berry phase, *Europhys. News* **28**, 18 (1997). •

[4] R. D. King-Smith and D. Vanderbilt, Theory of polarization of crystalline solid, *Phys. Rev. B* **47**, 1651 (1993).

[5] N. W. Ascroft and N. David Mermin, *Solid State Physics* (Thomson Learning, London, 1976). •

[6] J. Zak, Berry's phase for energy bands in solids, *Phys. Rev. Lett.* **62**, 2747 (1989).

[7] D. Xiao, M. C. Chang, and Q. Niu, Berry phase effects on electronic properties, *Rev. Mod. Phys.* **82**, 1959 (2010).

[8] Y. H. Jeong and S.-R. Eric Yang, Topological end states and Zak phase of rectangular armchair ribbon, *Ann. Phys.* **385**, 688 (2017).

[9] R. Resta, Macroscopic polarization in crystalline dielectrics: the geometric phase approach, *Rev. Mod. Phys.* **66**, 899 (1994).

$$\text{Chapter 6}$$

Hartree–Fock Method: Self-consistent Field Method

"A method is more important than a discovery, since the right method will lead to new and even more important discoveries."

Lev Landau

"For the perishable, every additional day in its life translates into a shorter additional life expectancy. For the nonperishable, every additional day imply a longer life expectancy."

Nassim Nicholas Taleb on Lindy effect

In later chapters, we will use many-body techniques to study superconductivity, quantum Hall effect, and disordered interacting zigzag nanoribbons. It is important to understand these techniques before we proceed. In this chapter, we will explain the second quantization, which forms the basis of many-body techniques.

After this, we will explain the self-consistent Hartree–Fock approximation and go through some applications of it. There are two types of the Hartree–Fock methods. In the **unrestricted** Hartree–Fock method, the ground state breaks a symmetry of the Hamiltonian. In this case, the Hartree–Fock approach is said to **include correlation effects** [1]. The unrestricted Hartree–Fock theory uses **different** orbitals for the spin-up and -down electrons. If the ground state does not break any symmetry then the method is called a **restricted** Hartree–Fock method. In the restricted Hartree–Fock theory, spin-up and -down states have the same orbital wave function. Even in the restricted Hartree–Fock approach, there is some correlation between electrons with the same spin as they cannot occupy the same state (this is the Pauli principle and is also a type correlation [1]). This type of correlation becomes more important with localized electrons in disordered systems.

6.1. Indistinguishable Particles

Quantum particles are assumed to be indistinguishable, i.e., they are all identical! (Nature is a perfect factory — its products have no defects.) But what does it mean mathematically? Suppose there are N particles in single-particle states $\alpha, \beta, \ldots$ and suppose you arbitrarily label particles $1, 2, \ldots$. That they are identical means that the probability density is unchanged upon exchange of any two particles (see Fig. 6.1)

$$|\psi(\vec{r}_1, \ldots, \vec{r}_i, \ldots, \vec{r}_j, \ldots, \vec{r}_N)|^2 = |\psi(\vec{r}_1, \ldots, \vec{r}_j, \ldots, \vec{r}_i, \ldots, \vec{r}_N)|^2, \quad (6.1)$$

which implies

$$\psi(\vec{r}_1, \ldots, \vec{r}_i, \ldots, \vec{r}_j, \ldots, \vec{r}_N) = \pm\psi(\vec{r}_1, \ldots, \vec{r}_j, \ldots, \vec{r}_i, \ldots, \vec{r}_N). \quad (6.2)$$

(We do not consider anyons.) Plus (minus) sign corresponds to bosons (fermions).

6.2. Boson Second Quantization

It is inconvenient to work directly with many-particle wave functions. Instead we introduce creation and destruction operators. To explain them, it is instructive to consider quantum theory of one-dimensional crystals. We can draw an analogy between one-dimensional crystals and quantum harmonic oscillators with many different frequencies. The Hamiltonian of

Fig. 6.1. Consider particles occupying single-particle states α, β, and γ. Particles labeled 2 and 3 are exchanged but the probability density remains unchanged, see Eq. (6.1).

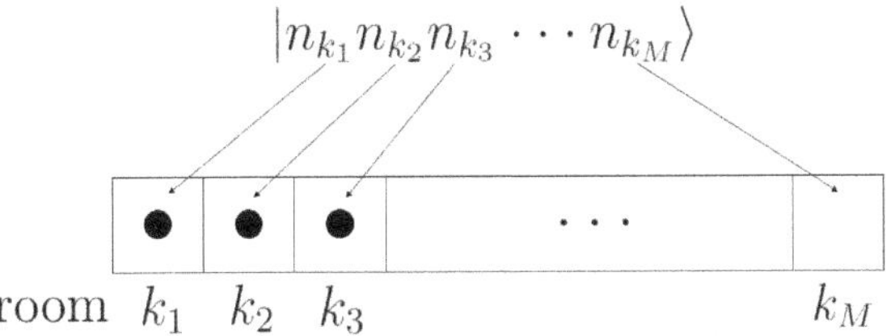

Fig. 6.2. A Fock state is analogous to hotel rooms occupied by guests. Each room is occupied by a number of guests n_k. For bosons n_k are integers with $n_k \geq 0$ and for fermions $n_k = 0$ or 1. The total number of particles (rooms) is $N(M)$.

vibrational modes of a crystal is

$$H = \sum_k \hbar\omega(k) \left(b_k^\dagger b_k + \frac{1}{2} \right), \tag{6.3}$$

where b_k is a boson destruction operator and $\hbar\omega(k)$ is the phonon energy with wave vector k (b_k corresponds to the destruction operator of a harmonic oscillator). Note that each boson k has a different energy $\hbar\omega(k)$. The boson Fock space[a] consists of states

$$|n_{k_1}, n_{k_2}, \ldots\rangle, \tag{6.4}$$

where the occupation numbers n_k are integers with $n_k \geq 0$ (see Fig. 6.2). A level can be occupied by numerous bosons.

The boson destruction and creation operators are defined as

$$b_k |\ldots, n_k, \ldots\rangle = \sqrt{n_k} |\ldots, n_k - 1, \ldots\rangle \tag{6.5}$$

and

$$b_k^\dagger |\ldots, n_k, \ldots\rangle = \sqrt{n_k + 1} |\ldots, n_k + 1, \ldots\rangle. \tag{6.6}$$

It is easy to show that the operators $b_k^\dagger$ and b_k satisfy the commutation relation

$$[b_k, b_{k'}^\dagger] = \delta_{k,k'}. \tag{6.7}$$

This commutation relation is consistent with the symmetric property of the total wave function when two bosons are exchanged. Moreover, it is

---

[a]One should distinguish the Hilbert space (in which all states have the same number of particles) from the Fock space (which contains states with **all possible** particle numbers).

compatible with multiple occupation of a level by many bosons. A many boson state can be written as

$$|\{n_k\}\rangle = \frac{1}{\sqrt{\prod_k n_k!}} \prod_k (b_k^\dagger)^{n_k} |0\rangle. \tag{6.8}$$

Note the normalization factor.

6.3. Fermion Second Quantization

The Hartree–Fock method can be presented using Slater determinant wave functions (see Ref. [2])

$$\Psi = \begin{vmatrix} \psi_\alpha(\vec{r_1}) & \psi_\alpha(\vec{r_2}) & \psi_\alpha(\vec{r_3}) & \cdots \\ \psi_\beta(\vec{r_1}) & \psi_\beta(\vec{r_2}) & \psi_\beta(\vec{r_3}) & \cdots \\ \vdots & \vdots & \vdots & \vdots \end{vmatrix}, \tag{6.9}$$

where $\psi_\alpha(\vec{r}), \psi_\beta(\vec{r}), \ldots$ are single electron wave functions. The Pauli exclusion principle (antisymmetry under particle exchange) is obeyed by the Slater determinant wave functions. The many-particle wave function approach is intuitive and insightful. However, it is not convenient to manipulate with Slater determinant wave functions. It is more convenient to use fermion second quantization, and here we give a quick review of it.

A many-body state in the Fock space can be defined as $|\psi\rangle = |\{n_k\}\rangle$, where the occupation number of each single electron state is $n_k = 0$ or 1 (see Fig. 6.2). It may be helpful to imagine a series of hotel rooms that is either singly occupied or empty. (The Pauli exclusion principle forbids double occupancy.) Let us define the destruction and creation operators of a single particle state p as

$$c_p|1\rangle = |0\rangle, \quad c_p^\dagger|0\rangle = |1\rangle, \tag{6.10}$$

where $|0\rangle(|1\rangle)$ is the empty occupied state of p. The Pauli exclusion principle is implemented by postulating fermion anticommutation relations

$$\{c_j, c_l^\dagger\} = \delta_{jl}, \quad \{c_j^\dagger, c_l^\dagger\} = \{c_j, c_l\} = 0. \tag{6.11}$$

From this, it follows that $c_p^\dagger c_p^\dagger|0\rangle = 0$, which is just the Pauli exclusion principle. Moreover, these anticommutation relations take care of the sign change of many electron wave functions when two electrons are exchanged.

It is useful to know the commutation relations between the occupation number operator $\hat{n}_i = c_i^\dagger c_i$ of site i and site creation and destruction operators

$$[\hat{n}_i, c_i] = -c_i, \quad [\hat{n}_i, c_i^\dagger] = c_i^\dagger. \tag{6.12}$$

A Slater determinant state for N fermions, Eq. (6.9), can be compactly written in terms of creation operators

$$|\psi\rangle = |\{n_p\}\rangle = \prod_p c_p^\dagger|0\rangle = c_{p_1}^\dagger \cdots c_{p_N}^\dagger|0\rangle. \tag{6.13}$$

For example, for $N = 2$

$$|\Psi\rangle = c_\alpha^\dagger c_\beta^\dagger|0\rangle = \begin{vmatrix} \psi_\alpha(\vec{r_1}) & \psi_\alpha(\vec{r_2}) \\ \psi_\beta(\vec{r_1}) & \psi_\beta(\vec{r_2}) \end{vmatrix}. \tag{6.14}$$

(The normalization factor is omitted.) These Slater determinant wave functions form the basis states for constructing many-body states. Any fermion many-body state can be written as a linear combination of Slater determinant wave functions. It can be written compactly using creation operators

$$|\psi\rangle = \sum_{p_1,\ldots,p_N} C_{p_1 \ldots p_N} c_{p_1}^\dagger \cdots c_{p_N}^\dagger|0\rangle. \tag{6.15}$$

Physical quantities of a many-body system can be computed using these states.

We have to determine the action of $c_j^\dagger$ and c_j on a Slater determinant state $|\psi\rangle$. The fermion anticommutation relations dictate the following rules. When the jth occupation number is $n_j = 0$, the action of creating a fermion in the jth state must give

$$c_j^\dagger|\psi\rangle = (-1)^{S_j}|\ldots, n_j = 1, \ldots\rangle, \tag{6.16}$$

where the sum S_j is defined as

$$S_j = \sum_{i<j} n_i. \tag{6.17}$$

(n_i are the occupation numbers of $|\psi\rangle$.) When $n_j = 1$ in the initial state $|\psi\rangle$ the resulting state is the vacuum state since two electrons cannot occupy a single level. When the destruction operator is acted on the same state we get

$$c_j|\psi\rangle = (-1)^{S_j}|\ldots, n_j = 0, \ldots\rangle. \tag{6.18}$$

Let us demonstrate that these results are consistent with the fermion anti-commutation relations. Suppose that $n_j = 1$ and $n_l = 0$ in $|\psi\rangle$. If $j > l$

then

$$c_l^\dagger c_j |\psi\rangle = (-1)^{S_j + S_l} | \ldots, n_l = 1, \ldots, n_j = 0, \ldots \rangle,$$
$$c_j c_l^\dagger |\psi\rangle = (-1)^{S_l + (S_j + 1)} | \ldots, n_l = 1, \ldots, n_j = 0, \ldots \rangle, \qquad (6.19)$$

where S_j and S_l are computed from $|\psi\rangle$. From this follows the anticommutation relation

$$(c_l^\dagger c_j + c_j^\dagger c_l)|\psi\rangle = (-1)^{S_j + S_l}(1 - 1)| \ldots, n_l = 1, \ldots, n_j = 0, \ldots \rangle = 0.$$
$$(6.20)$$

Other anticommutation relations can be shown similarly. The sign factors $(-1)^{S_j}$ are important in implementing **exact diagonalization codes** of many-electron systems. These are clever booking devices to track of antisymmetry under exchange of two fermions.

6.4. Field Operators

We introduce boson and fermion field operators in terms of fermion and boson destruction operators c_k and b_k, respectively,

$$\Psi(\vec{r}) = \sum_k c_k \psi_k(\vec{r}) \qquad (6.21)$$

and

$$\Psi(\vec{r}) = \sum_k b_k \psi_k(\vec{r}), \qquad (6.22)$$

where the single particle wave functions $\{\psi_k(\vec{r})\}$ form a complete set (one often choose them as plane waves). These can be thought of as unitary transformations of c_k and b_k at each $\vec{r}$. The density matrix $|\Psi\rangle\langle\Psi|$ (see Eq. (C.4)) suggests the following definition of the density operator:

$$\rho(\vec{r}) = \Psi^\dagger(\vec{r})\Psi(\vec{r}). \qquad (6.23)$$

In classical physics, physical quantities may be written in terms of the density function. This suggests how the corresponding quantum operators are to be defined [3], (see Exercise 6.1).

Exercise 6.1. The following classical results are useful in deriving the field operator versions of various quantities. (a) Using the density function $\bar{\rho}(\vec{r}) = \sum_i \delta(\vec{r} - \vec{r}_i)$ show that one-particle quantities can be written as

$$\sum_i \bar{A}(\vec{r}_i) = \int d\vec{r}\,\bar{\rho}(\vec{r})\bar{A}(\vec{r}). \tag{6.24}$$

(b) Show that two-particle quantities can be written as

$$\sum_{ij} \bar{B}(\vec{r}_i, \vec{r}_j) = \int d\vec{r}\,d\vec{r}\,'\,\bar{\rho}(\vec{r})\bar{\rho}(\vec{r}\,')\bar{B}(\vec{r}, \vec{r}\,'), \tag{6.25}$$

and that two-particle interaction without self-interaction can be written as

$$\begin{aligned}
\bar{V} &= \frac{1}{2}\sum_{i \neq j} V(\vec{r}_i - \vec{r}_j) \\
&= \frac{1}{2}\int d\vec{r}_1\,d\vec{r}_2\,\bar{V}(\vec{r}_1 - \vec{r}_2)\bar{\rho}(\vec{r}_1)[\bar{\rho}(\vec{r}_2) - \delta(\vec{r}_1 - \vec{r}_2)].
\end{aligned}$$

$$\tag{6.26}$$

Note that these quantities are functions and **not** quantum operators. To distinguish them from quantum operators they are denoted by an overbar.

The expression for the density operator, Eq. (6.23), and the classical result, Eq. (6.24), suggests the following form for the one-particle operator of the Schrödinger equation

$$H_1 = \int d\vec{r}\,\Psi^\dagger(\vec{r})\left(-\frac{\hbar^2\nabla^2}{2m} + v(\vec{r})\right)\Psi(\vec{r}), \tag{6.27}$$

where $v(\vec{r})$ is an external potential. Similarly, the classical expression for two-particle interaction, Eq. (6.26), suggests the following operator representation of two-particle interaction

$$V = \frac{1}{2}\int d\vec{r}_1\,d\vec{r}_2\,V(\vec{r}_1 - \vec{r}_2)\Psi^\dagger(\vec{r}_1)\Psi^\dagger(\vec{r}_2)\Psi(\vec{r}_2)\Psi(\vec{r}_1), \tag{6.28}$$

where $V(\vec{r}_1 - \vec{r}_2)$ is the two-particle interaction. Note that this operator is normal ordered, i.e., the creation operators all appear to the left; this will eliminate infinities in the expectation values of V, see Exercise 6.2.

> **Exercise 6.2.** Show that, as in Eq. (6.26), the self-interaction term is subtracted from the two-particle interaction given in Eq. (6.28). Hint: Change the order of the field operators so that $\rho(\vec{r}_1)\rho(\vec{r}_2)$ appears in the interaction term.

Plugging the field operator (6.21) into the Hamiltonians equations (6.27) and (6.28) we find the many-electron Hamiltonian

$$
H = \sum_{\nu\lambda} \langle \nu | \left(-\frac{\hbar^2 \nabla^2}{2m} + v(\vec{r}) \right) | \lambda \rangle c_\nu^\dagger c_\lambda
$$

$$
+ \frac{1}{2} \sum_{\nu\lambda\alpha\beta} \langle \nu\lambda \, | V(\vec{r}_1 - \vec{r}_2) | \alpha\beta \rangle c_\nu^\dagger c_\lambda^\dagger c_\beta c_\alpha, \tag{6.29}
$$

where the two-particle interaction is

$$
\langle \nu\lambda \, | V(\vec{r}_1 - \vec{r}_2) | \alpha\beta \rangle = \int d\vec{r}_1 d\vec{r}_2 \phi_\nu^*(\vec{r}_1) \phi_\lambda^*(\vec{r}_2) V(\vec{r}_1 - \vec{r}_2) \phi_\alpha(\vec{r}_1) \phi_\beta(\vec{r}_2).
$$

$$
\tag{6.30}
$$

Indices $\nu, \lambda, \alpha, \beta$ include spin quantum numbers $\sigma = \uparrow, \downarrow$. The creation operator $c_\nu^\dagger$ creates an electron with the wave function $\phi_\nu(\vec{r})$. Note that the order of indices in $c_\nu^\dagger c_\lambda^\dagger c_\beta c_\alpha$ is different from that appearing in $\langle \nu\lambda \, | V(\vec{r}_1 - \vec{r}_2) | \alpha\beta \rangle$. This Hamiltonian is generally difficult to solve exactly. In this book, we will study three well-known approaches one can use to find solutions. First, one can try to guess a good Ansatz for the ground state that is essentially exact; some examples are the Laughlin states (Chapter 11) and the Ansatz state for the toric code (Chapter 13). Second, one can use bosonization method (Chapter 7). These approaches are essentially exact but they cannot be universally applied and one often has to apply an approximate method, such as the Hartree–Fock self-consistent approximation.

6.5. Basic Idea of Hartree–Fock Approximation

The art of the self-consistent field method is to find a **good ground state Ansatz**[b] (not necessarily exact) with an appropriate **self-consistent mean field** [4] that replaces electron–electron interactions. This means

---

that an effective single particle picture is approximately correct. Suppose that the interaction term in the Hamiltonian can be written as a product of two bilinear operators AB. One guesses the ground state Ansatz $|\psi_0\rangle$ made of some as yet undetermined variational single-particle states. One can replace AB in the Hamiltonian with

$$AB \to A\langle B\rangle + B\langle A\rangle - \langle A\rangle\langle B\rangle, \tag{6.31}$$

where the expectation value $\langle \ldots \rangle$ is with respect to $|\psi_0\rangle$. This decoupling scheme constitutes the Hartree–Fock approximation (there may be more than one way to perform the decoupling). Since the interaction term can be exactly rearranged as

$$\begin{aligned} AB &= (A - \langle A\rangle + \langle A\rangle)(B - \langle B\rangle + \langle B\rangle) \\ &= A\langle B\rangle + B\langle A\rangle + (A - \langle A\rangle)(B - \langle B\rangle) - \langle A\rangle\langle B\rangle, \end{aligned} \tag{6.32}$$

the Hartree–Fock approximation amounts to ignoring quantum fluctuations, represented by the third term. In the Hartree–Fock approximation one tries to approximately replace the quartic term AB in the Hamiltonian with **bilinear** terms[c] because a bilinear Hamiltonian can be exactly **diagonalized**, which is well-known from linear algebra. The energy variational principle is used to derive equations that determine the self-consistent single-particle states. One then computes these Hartree–Fock single-particle eigenstates and occupies the N lowest-energy states. The resulting many-electron state yields the ground state Ansatz $|\psi_0\rangle$.

An example of this method from classical physics is the Weiss mean field theory of the nearest-neighbor Ising model, see Exercise 6.3. One of the best known examples from quantum physics is the BCS trial wave function (Chapter 8). It should be noted that there is no unique way of dividing the interaction term into bilinear terms, in other words, different physical systems, for example, antiferromagnetism and superconductivity, have different schemes. A deep and intuitive understanding of the essential physics is needed to find the correct scheme.

Some examples of self-consistent fields are given in Table 6.1. In the case of superconductivity, the self-consistent field is a pairing field $\Delta(\vec{r})$: it scatters a spin-up electron with wave vector $\vec{k}$ to another spin-down electron

---

method of obtaining a good ground state Ansatz. It is more of an art than a science. Good examples are the BCS and Laughlin wave functions.

[c] Examples of a bilinear operator are $A = \sum_{ij} A_{ij} a_i^\dagger a_j$ and $A = \sum_{ij} A_{ij} a_i^\dagger a_j^\dagger$, where $a_i^\dagger$ is a creation operator.

Table 6.1.   Examples of self-consistent fields.

Condensed systems	Self-consistent fields
Ferromagnet	Magnetic field
Crystal	Periodic field
Superconductor	Paring field
Antiferromagnet	Alternating magnetic field
Disorder	Random field

with wave vector $-\vec{k}$. The non-local exchange mean field of an electron gas is another example. It couples an electron at position $\vec{r}$ to an electron at position $\vec{r}\,'$ with the same spin. Third example is the effective magnetic field of a ferromagnet that polarizes electron spins. In disordered systems, many self-consistent **random fields** are introduced. In this book, we will apply this method to topologically ordered systems with disorder, such as integer quantum Hall bars and interacting disordered graphene zigzag nanoribbons, see Chapters 10 and 18.

6.6.   Gaussian Fluctuations

Let us solve a concrete problem to understand better what a mean field approach is and how to include fluctuations neglected in the approach. It is instructive to study a classical model because it is simpler to treat than quantum problems. We find the mean-field solution of the one-dimensional Ising model and incorporate thermal fluctuations.

Consider the (classical) nearest-neighbor Ising Hamiltonian in one dimension

$$H = -\sum_{\langle ij \rangle} J_{ij} s_i s_j - h \sum_i s_i, \tag{6.33}$$

where h is an external magnetic field, J_{ij} is the ferromagnetic spin coupling constant, $\langle ij \rangle$ is the sum over nearest neighbors, and s_i is spin at site i. The Hamiltonian can be written as

$$H = -\sum_i s_i h_i, \tag{6.34}$$

where the self-consistent field at site i is

$$h_i = h + \sum_j J_{ij} \langle s_j \rangle + \sum_j J_{ij} (s_j - \langle s_j \rangle), \tag{6.35}$$

and $\langle\ldots\rangle$ stands for the statistical average. If the fluctuations are small[d] the last term can be ignored and the effective mean field at site i is

$$h_i \approx h + 2Jm, \tag{6.36}$$

where the nearest neighbor coupling is $J_{ij} = J$ and the average site spin is $m = \langle s_i \rangle$. This means that the system is equivalent to non-interacting spins in an effective magnetic field h_i. (The ground state Ansatz is thus a uniformly spin-polarized state.) The partition function is given by

$$Z = \mathrm{Tr}(e^{-\frac{H_{\mathrm{MF}}}{k_B T}}) = \prod_{i=1}^{N_s} \sum_{s_i = \pm 1} e^{\frac{h_i s_i}{k_B T}}, \tag{6.37}$$

where the mean field Hamiltonian is $H_{\mathrm{MF}} = -\sum_i s_i h_i$ (N_s is the total number of sites). We can find the site magnetization using $m = \langle s_i \rangle = -\frac{1}{N_s}\frac{\partial F}{\partial h_i}$ where $F = -k_B T \log Z$ is the free energy of the chain. We find that m satisfies a self-consistent equation

$$m = \tanh\left(\frac{h + 2Jm}{k_B T}\right), \tag{6.38}$$

where k_B is the Boltzmann factor and T is temperature. Solutions show that even in the absence of an external magnetic field h, the system is spin polarized below a critical temperature.

How can one include fluctuations neglected in the mean field theory? This can be done using a path integral approach. The exact partition function is

$$Z = \mathrm{Tr}\{e^{\frac{\beta}{2}\sum_{ij} J_{ij} s_i s_j + \beta \sum_i h_i s_i}\}, \tag{6.39}$$

where the inverse temperature is $\beta = 1/k_B T$. Using an auxiliary field functional integral approach, namely the Hubbard–Stratonovich transformation (see Eq. (G.11)), we can rewrite

$$e^{\frac{1}{2}\sum_{ij} J_{ij} s_i s_j} = \sqrt{\frac{\det J}{2\pi}} \int \prod_{k=1}^{N_s} d\phi_k e^{-\frac{1}{2}\sum_{ij} \phi_i (J^{-1})_{ij} \phi_j + \sum_i \phi_i s_i}. \tag{6.40}$$

This result gives the following expression for the partition function

$$Z = \int \prod_{k=1}^{N_s} d\psi_k e^{-\beta S}, \tag{6.41}$$

---

[d]This assumption is valid when T is not near the critical temperature T_c.

where S is the action given by

$$S = \frac{1}{2}\sum_{ij}(\psi_i - h_i)J_{ij}^{-1}(\psi_j - h_j) - \frac{1}{\beta}\sum_{i}\log(2\cosh\beta\psi_i). \tag{6.42}$$

(Here we have set $\psi_i = \phi_i + h_i$. We have ignored the overall multiplicative constant $\sqrt{\frac{\det J}{2\pi}}$.) The action in the saddle point (Gaussian) approximation[e] is

$$S \approx S[\overline{\psi}_i] + \frac{1}{2}\sum_{pk}(\psi_p - \overline{\psi}_p)\frac{\delta^2 S}{\delta\psi_p\delta\psi_k}(\psi_k - \overline{\psi}_k), \tag{6.43}$$

where $\{\overline{\psi}_i\}$ is the mean field solution, satisfying $\frac{\delta S}{\delta\psi_k} = 0$, and

$$\frac{\delta^2 S}{\delta\psi_p\delta\psi_k} = J_{pk}^{-1} - \beta[1 - \tanh^2(\beta\psi_p)]\delta_{pk} \tag{6.44}$$

represents a functional differentiation (see Sec. G.1 in Appendix). Then the free energy is determined[f] from $Z = De^{\beta F}$ with $D \propto [\det J^{-1}]^{-1/2}$

$$F \approx S[\overline{\psi}_i] + \frac{1}{2\beta}\log\det[\delta_{ij} - \beta(1 - \tanh^2(\beta\overline{\psi}_i)J_{ij}]. \tag{6.45}$$

(See Ref. [5].) Here, the determinant of a matrix $C = AB$ is

$$\det C = \det A \, \det B. \tag{6.46}$$

Exercise 6.3. Derive the action Eq. (6.42). Hint: Use

$$\mathrm{Tr}e^{\beta\psi_i s_i} = e^{\sum_i \log[2\cosh(\beta\psi_i)]}. \tag{6.47}$$

---

[e] In the Gaussian approximation fluctuations do not interact with each other. We will encounter the quantum version of this approach, namely the random phase approximation, in Sec. 6.14, where we explain the time-dependent Hartree–Fock approximation.

[f] It can be shown that, when the spatial dimension is less than or equal to 4, the correction terms in the expansion of S become important near the critical temperature T_c. The physics near T_c is thus dominated by fluctuations. Onsager found the exact solution of two-dimensional systems.

Further reading. The derivation of the Hartree–Fock approximation in path integral approach is subtle. One cannot just apply the Hubbard–Stratonovich transformation. However, one can use the freedom in formulating the auxiliary field functional integral, see Refs. [5,6].

6.7. Restricted Hartree–Fock Approximation of Atoms

Let us try to describe electrons in an atom using the restricted Hartree–Fock approximation. Let us apply the variational principle to the many-body Hamiltonian equation (6.29) using the ground state Ansatz

$$|\psi_0\rangle = c_{\nu_1}^\dagger c_{\nu_2}^\dagger \dots c_{\nu_N}^\dagger |0\rangle. \tag{6.48}$$

Spin quantum numbers $\sigma = \pm 1$ may be included through indices $\nu = (j,\sigma)$ (j labels atomic orbitals). An operator $c_\nu^\dagger$ creates an electron in an atomic orbital with a wave function $\phi_\nu(\vec{r})$. How is this wave function to be determined in the presence of electron–electron interactions? The wave function $\phi_\nu(\vec{r})$ must be determined **variationally** by minimizing the total energy. This leads to a set of equations that are to be solved **self-consistently**.

We have to minimize the total Hartree–Fock energy $\langle\psi_0|H|\psi_0\rangle$ under the constraint $\int d\vec{r}\,|\phi_\nu(\vec{r})|^2 = 1$. (The many-body Hamiltonian is given in Eq. (6.29).) According to the Lagrange multiplier method, we have to minimize the following functional:

$$\langle\psi_0|H|\psi_0\rangle - \sum_\nu \epsilon_\nu \left[\int d\vec{r}\,|\phi_\nu(\vec{r})|^2 - 1 \right]. \tag{6.49}$$

The first term $\langle\psi_0|H|\psi_0\rangle$ contains the interaction energy $\langle\psi_0|V|\psi_0\rangle$. It can be evaluated using the result for the expectation value $\langle\psi_0|c_\nu^\dagger c_\lambda^\dagger c_\beta c_\alpha|\psi_0\rangle$, which can be easily computed because the Ansatz allows decoupling of the quartic interaction term into bilinear terms, see Exercise 6.4. It has two contributions. The first contribution is the Hartree energy term

$$\frac{1}{2}\sum_{\nu\lambda} n_\nu n_\lambda \phi_\nu^*(\vec{r}_1)\phi_\lambda^*(\vec{r}_2)\frac{e^2}{|\vec{r}_1 - \vec{r}_2|}\phi_\nu(\vec{r}_1)\phi_\lambda(\vec{r}_2) \tag{6.50}$$

and the second is the exchange energy term

$$-\frac{1}{2}\sum_{\nu\lambda} n_\nu n_\lambda \phi_\nu^*(\vec{r}_1)\phi_\lambda^*(\vec{r}_2)\frac{e^2}{|\vec{r}_1 - \vec{r}_2|}\phi_\lambda(\vec{r}_1)\phi_\nu(\vec{r}_2). \tag{6.51}$$

Here the **occupation function** is $n_\gamma = 1(0)$ for occupied (unoccupied) states. The sum is thus over only occupied states ν and λ. Other terms in the functional can be easily evaluated.

Exercise 6.4. Evaluate $\langle \psi_0 | c_\nu^\dagger c_\lambda^\dagger c_\beta c_\alpha | \psi_0 \rangle = \langle c_\nu^\dagger c_\lambda^\dagger c_\beta c_\alpha \rangle$. Answer: Use

$$\langle c_\nu^\dagger c_\lambda^\dagger c_\beta c_\alpha \rangle = \langle \psi_0 | c_\nu^\dagger c_\lambda^\dagger c_\beta c_\alpha | \psi_0 \rangle = \langle \{n_k\} | c_\nu^\dagger c_\lambda^\dagger c_\beta c_\alpha | \{n_k\} \rangle. \tag{6.52}$$

A non-zero term exists when $\nu = \beta$, $\lambda = \alpha$, and $\nu \neq \lambda$. In this case we find

$$\langle c_\nu^\dagger c_\lambda^\dagger c_\beta c_\alpha \rangle = -\langle \{n_k\} | c_\nu^\dagger c_\nu c_\lambda^\dagger c_\lambda | \{n_k\} \rangle$$
$$= -\langle \{n_k\} | c_\nu^\dagger c_\nu | \{n_k\} \rangle n_\lambda = -n_\nu n_\lambda. \tag{6.53}$$

This term gives the exchange energy in Eq. (6.51). Another non-zero term exists when $\nu = \alpha$, $\lambda = \beta$, and $\nu \neq \lambda$, which gives

$$\langle c_\nu^\dagger c_\lambda^\dagger c_\beta c_\alpha \rangle = -\langle \{n_k\} | c_\nu^\dagger c_\lambda^\dagger c_\nu c_\lambda | \{n_k\} \rangle$$
$$= \langle \{n_k\} | c_\nu^\dagger c_\nu c_\lambda^\dagger c_\lambda | \{n_k\} \rangle = n_\nu n_\lambda. \tag{6.54}$$

This term gives the third term in Eq. (6.50). These results are equivalent to the following **decoupling scheme** used in the Hartree–Fock method

$$\langle c_\nu^\dagger c_\lambda^\dagger c_\beta c_\alpha \rangle = \langle c_\nu^\dagger c_\alpha \rangle \langle c_\lambda^\dagger c_\beta \rangle - \langle c_\nu^\dagger c_\beta \rangle \langle c_\lambda^\dagger c_\alpha \rangle. \tag{6.55}$$

Functional differentiating (see Appendix G.2) the functional with respect to $\phi_{j\sigma}^*(\vec{r})$, we find the Hartree–Fock equation

$$\left[-\frac{\hbar^2}{2m} \nabla^2 + V_{\text{ext}}(\vec{r}) + V_{\text{H}}(\vec{r}) \right] \phi_{i\sigma}(\vec{r}) + \int d\vec{r}\,' V_{\text{X}\sigma}(\vec{r}, \vec{r}\,') \phi_{i\sigma}(\vec{r}\,')$$

$$= \epsilon_{i\sigma} \phi_{i\sigma}(\vec{r}). \tag{6.56}$$

We have included an external potential $V_{\text{ext}}(\vec{r})$, which includes the ionic positive background potential. Note that we have displayed the explicit spin dependence ($\sigma = \uparrow$ or $\downarrow$). The local Hartree potential is

$$V_{\text{H}}(\vec{r}) = -e \int d\vec{r}\,' \rho(\vec{r}\,') \frac{1}{|\vec{r} - \vec{r}\,'|}, \tag{6.57}$$

where the total electron charge density is the sum of spin-up and -down densities

$$\rho(\vec{r}) = -e \sum_{j,\sigma} n_{j\sigma} |\phi_{j\sigma}(\vec{r})|^2. \tag{6.58}$$

An electron with spin σ experiences a non-local exchange potential

$$V_{X\sigma}(\vec{r}, \vec{r}\,') = -e^2 \sum_j \frac{n_{j\sigma}}{|\vec{r} - \vec{r}\,'|} \phi_{j\sigma}^*(\vec{r}\,')\phi_{j\sigma}(\vec{r}) = -e\frac{\rho(\vec{r}, \sigma; \vec{r}\,', \sigma)}{|\vec{r} - \vec{r}\,'|}, \tag{6.59}$$

where the density matrix elements are

$$\rho(\vec{r}, \sigma; \vec{r}\,', \sigma) = e \sum_j n_{j\sigma} \phi_{j\sigma}^*(\vec{r}\,')\phi_{j\sigma}(\vec{r}). \tag{6.60}$$

(Here we have used that the density matrix is $\rho = e \sum_{j,\sigma} n_{j\sigma} |\phi_{j\sigma}\rangle\langle\phi_{j\sigma}|$, see Eqs. (C.4) and (C.3) in Sec. 5.2.) The exchange potential originates from the antisymmetry of the Slater determinant wave function and its sign is negative (in contrast the Hartree potential is positive). The Hartree–Fock single particle energy is $\epsilon_{i\sigma}$. The Hartree–Fock equations can also be written as follows:

$$\left[-\frac{\hbar^2}{2m}\nabla^2 + V_{\text{ext}}(\vec{r}) \right] \phi_{i\sigma}(\vec{r}) + \int d\vec{r}\,' \Sigma_\sigma(\vec{r}, \vec{r}\,')\phi_{i\sigma}(\vec{r}\,') = \epsilon_{i\sigma}\phi_{i\sigma}(\vec{r}), \tag{6.61}$$

where the **self-energy** is

$$\Sigma_\sigma(\vec{r}, \vec{r}\,') = V_{\text{H}}(\vec{r})\delta(\vec{r} - \vec{r}\,') + V_{X\sigma}(\vec{r}, \vec{r}\,'). \tag{6.62}$$

> **Exercise 6.5.** Derive the Hartree–Fock equation (6.56). Hint: Either $(\phi_{j\sigma}(\vec{r}), \phi_{j\sigma}^*(\vec{r}))$ or $(\text{Re}\phi_{j\sigma}(\vec{r}), \text{Im}\phi_{j\sigma}(\vec{r}))$ can be chosen as variational parameters. We functional differentiate the energy functional with respect to $\phi_{j\sigma}^*(\vec{r})$ (see Appendix G.2). $\phi_{j\sigma}(\vec{r})$ is independent of $\phi_{j\sigma}^*(\vec{r})$.

It is convenient to rewrite the Hartree–Fock equations in a more compact form using the second quantization method. From the Hartree–Fock equation, we find that the mean field version of the Hamiltonian equation (6.29) is bilinear

$$H_{\mathrm{MF}} = \sum_{\alpha\beta} \left\{ \langle\alpha| \left(-\frac{\hbar^2\nabla^2}{2m} \right) |\beta\rangle \right.$$

$$\left. + \langle\alpha|V_{\mathrm{ext}}|\beta\rangle + \langle\alpha|V_{\mathrm{H}}|\beta\rangle + \langle\alpha|V_{\mathrm{X}}|\beta\rangle \right\} c_\alpha^\dagger c_\beta + \mathrm{constant}.$$

$$(6.63)$$

Here α and β include spin quantum numbers σ.

The Hartree–Fock equation must be solved self-consistently: (a) One guesses good initial states $\{\phi_{j\sigma}\}$. (b) From them the Hartree and exchange potentials are computed. (c) Solve the Hartree–Fock equation. (d) The Hartree–Fock eigenstates with N **lowest** energies $\epsilon_{i\sigma}$ are chosen as the new states $\{\phi_{j\sigma}\}$. (e) Go back to step (b) and repeat until solutions do not change.

Exercise 6.6. Show that the total Hartree–Fock energy is

$$E_{\mathrm{tot}} = \sum_\alpha n_\alpha \epsilon_\alpha - \frac{1}{2} \sum_{\alpha,\beta} n_\alpha n_\beta [\langle\alpha\beta|V|\alpha\beta\rangle - \langle\alpha\beta|V|\beta\alpha\rangle]. \quad (6.64)$$

The Hartree–Fock single-particle state and energy are $|\alpha\rangle$ and ϵ_α, respectively. The Hartree–Fock occupation number is n_α. Hint: Use

$$\langle\alpha|T|\beta\rangle + \sum_\delta n_\delta [\langle\alpha\delta|V|\beta\delta\rangle - \langle\alpha\delta|V|\delta\beta\rangle] = \epsilon_\alpha \delta_{\alpha\beta} \quad (6.65)$$

and

$$E_{\mathrm{tot}} = \sum_\alpha \langle\alpha|T|\alpha\rangle + \frac{1}{2} \sum_{\alpha,\beta} n_\alpha n_\beta [\langle\alpha\beta|V|\alpha\beta\rangle - \langle\alpha\beta|V|\beta\alpha\rangle].$$

$$(6.66)$$

6.8. Hartree–Fock Tunneling Density of States

The tunneling DOS is an important quantity that can be readily measured experimentally. We will compute it several times for different physical systems in this book.

At zero temperature, the tunneling DOS of interacting electrons is computed using

$$D(\epsilon) = \sum_k D_k(\epsilon),$$

where

$$D_k(\epsilon) = \sum_k ([|\langle \psi_j^{N+1}|a_k^\dagger|\psi_0^N\rangle|^2 \delta[\epsilon - (E_j^{N+1} - E_0^N)]$$

$$+ [|\langle \psi_j^{N-1}|a_k|\psi_0^N\rangle|^2 \delta[\epsilon + (E_j^{N-1} - E_0^N)]]). \tag{6.67}$$

Here j labels many-body states with $N+1$ or $N-1$ particles and k labels any complete set of single-particle states. The interacting ground state of N electrons is $|\psi_0^N\rangle$ and its energy is E_0^N. We describe the Hartree–Fock single particles by the creation and destruction operators $c_k^\dagger$ and c_k. Eq. (6.67) simplifies in the Hartree–Fock approximation. The Koopmans' theorem [7] (see Exercise 6.7) gives

$$\begin{aligned}
E_k^{N+1} &= E_0^N + \epsilon_k, \\
E_k^{N-1} &= E_0^N - \epsilon_k,
\end{aligned} \tag{6.68}$$

where ϵ_k is the Hartree–Fock single-particle energy. The excited many-body states are given by

$$\begin{aligned}
|\psi_k^{N+1}\rangle &= c_k^\dagger|\psi_0^N\rangle, \\
|\psi_k^{N-1}\rangle &= c_k|\psi_0^N\rangle.
\end{aligned} \tag{6.69}$$

If we choose

$$a_k^\dagger = c_k^\dagger, \tag{6.70}$$

the result simplifies

$$D(\epsilon) = \sum_k (|\langle \psi_0|c_k c_k^\dagger|\psi_0\rangle|^2 \delta[\epsilon - \epsilon_k] + |\langle \psi_0|c_k^\dagger c_k|\psi_0\rangle|^2 \delta[\epsilon + \epsilon_k]). \tag{6.71}$$

From this result, we find that the Hartree–Fock tunneling **DOS consists of delta functions at the Hartree–Fock single-particle energies.** The first (second) term gives the relevant DOS for tunneling events from the environment (system) into the system (environment). This result will be used several times in later chapters.

Exercise 6.7. Show Koopmans' theorem

$$E_k^{N+1} - E_0^N = \epsilon_k, \tag{6.72}$$

where E_0^N is the Hartree–Fock ground state energy, ϵ_k is the Hartree–Fock single-particle energy, and E_k^{N+1} is the Hartree–Fock energy of $c_k^\dagger |\psi_0^N\rangle$. Hint: Use the result of Exercise 6.6.

6.9. Restricted Hartree–Fock Approximation of Electron Gas

We apply the second quantization method to a homogeneous electron gas and derive the self-consistent Hartree–Fock equations, which determine electron wave functions. In this approximation, electrons move in self-consistent Hartree and Fock (exchange) potentials. Before reading this section, the reader should study Chapter 17 of Ref. [2], which gives a nice derivation of the Hartree–Fock approximation using Slater determinant wave functions. In an electron gas, translational invariance is present. So we can use plane wave functions, i.e., wave vector $\vec{k}$ is a good quantum number.

• Ground State Ansatz

The ground state is assumed to exhibit translational invariance, which suggests the following Hartree–Fock ground state Ansatz consisting of plane wave states:

$$|\psi_0\rangle = c_{\vec{k}_1\uparrow}^\dagger c_{\vec{k}_1\downarrow}^\dagger \cdots c_{\vec{k}_N\uparrow}^\dagger c_{\vec{k}_N\downarrow}^\dagger |0\rangle = |\vec{k}_{1\uparrow}, \vec{k}_{1\downarrow}, \ldots\rangle. \tag{6.73}$$

Here states with $|\vec{k}| \le k_F$ are all occupied, where k_F is the Fermi wave vector. Note that the ordering of indices in the creation operators and the ordering of states in the ket are the same. In this state, all the available states below the Fermi energy E_F are filled.

The mean field Hamiltonian is (see Exercise 6.8)

$$H_{\mathrm{MF}} = \sum_{\vec{k}\vec{k}'}\{\epsilon_{\vec{k}}\delta_{\vec{k},\vec{k}'} + \langle\vec{k}|V_{\mathrm{ext}}|\vec{k}'\rangle + \langle\vec{k}|V_H|\vec{k}'\rangle + \langle\vec{k}|V_X|\vec{k}'\rangle\}c_{\vec{k}}^{\dagger}c_{\vec{k}'} + \text{constant},$$

$$(6.74)$$

where $\epsilon_{\vec{k}}$ is the free electron dispersion (here, for convenience, spin index σ is included in wave vector quantum number $\vec{k}$ so that $\vec{k}$ really **stands** for $(\vec{k},\sigma)$). The self-energy matrix elements $\langle\vec{k}|V_H|\vec{k}'\rangle$ and $\langle\vec{k}|V_X|\vec{k}'\rangle$ are

$$\langle\vec{k}|V_H|\vec{k}'\rangle = \sum_{\vec{p}} n_{\vec{p}}\langle\vec{k},\vec{p}|V|\vec{k}',\vec{p}\rangle \qquad (6.75)$$

and

$$\langle\vec{k}|V_X|\vec{k}'\rangle = -\sum_{\vec{p}} n_{\vec{p}}\langle\vec{k},\vec{p}|V|\vec{p},\vec{k}'\rangle. \qquad (6.76)$$

Here states $\{|\vec{p}\rangle\}$ denote the eigenstates of the Hartree–Fock Hamiltonian. Note that all the states in the Fock matrix element $\langle\vec{k},\vec{p}|V|\vec{p},\vec{k}'\rangle$ must have the **same** spin values. The effect of an external potential may be included in the Hartree term. By taking the matrix elements of the Hartree–Fock Hamiltonian $\langle\vec{k}|H_{\mathrm{MF}}|\vec{k}'\rangle = H_{\vec{k}\vec{k}'}$, we find the Hamiltonian matrix equation

$$\sum_{\vec{k}'} H_{\vec{k}\vec{k}'}a_{\vec{k}'} = \epsilon(\vec{k})a_{\vec{k}}, \qquad (6.77)$$

where the matrix elements are

$$H_{\vec{k},\vec{k}'} = \epsilon_{\vec{k}}\delta_{\vec{k},\vec{k}'} + \langle\vec{k}|V_H|\vec{k}'\rangle + \langle\vec{k}|V_X|\vec{k}'\rangle. \qquad (6.78)$$

The eigenvalues $\epsilon(\vec{k})$ represent the renormalized single-particle energy. In the absence of an external potential, the Hamiltonian matrix (6.78) is **diagonal** in the plane wave basis $\{|\vec{k}\rangle\}$, i.e., only terms with $\vec{k} = \vec{k}'$ survive (this is a consequence of translational invariance).

Exercise 6.8. Show Eq. (6.74) from Eq. (6.63). Hint: Use the following unitary transformations (see Appendix B) on Hamiltonian equation (6.63)

$$|\alpha\rangle = \sum_{\vec{k}} U_{\alpha,\vec{k}}|\vec{k}\rangle, \quad \langle\alpha| = \sum_{\vec{k}}\langle\vec{k}|U_{\vec{k},\alpha}^{\dagger},$$
$$c_{\alpha}^{\dagger} = \sum_{\vec{k}} U_{\alpha,\vec{k}}c_{\vec{k}}^{+}, \quad c_{\alpha} = \sum_{\vec{k}} c_{\vec{k}}U_{\vec{k},\alpha}^{\dagger}.$$

$$(6.79)$$

The old and new basis vector sets are, respectively, $\{|\vec{k}\rangle\}$ and $\{|\alpha\rangle\}$. Note that

$$\langle\alpha| = (|\alpha\rangle)^\dagger = \left(\sum_{\vec{k}} U_{\alpha,\vec{k}}|\vec{k}\rangle\right)^\dagger = \sum_{\vec{k}}\langle\vec{k}|U^*_{\alpha,\vec{k}} = \sum_{\vec{k}}\langle\vec{k}|U^\dagger_{\vec{k}\alpha}$$

$$(6.80)$$

and

$$\sum_\beta U^*_{\vec{k}'\beta}U_{\vec{q}\beta} = \sum_\beta U^*_{\vec{k}'\beta}U^T_{\beta\vec{q}} = \sum_\beta (U_{\vec{k}'\beta}U^+_{\beta\vec{q}})^* = \delta_{\vec{k}'\vec{q}}. \quad (6.81)$$

The unitary matrix satisfies

$$UU^\dagger = U^\dagger U = I. \qquad (6.82)$$

Note that the transformed Hamiltonian is given by $U^\dagger HU$.

In a translationally invariant electron gas, the Hartree self-energy cancels with the contribution from the uniform positive background charge. However, the negative exchange self-energy matrix elements are diagonal and are given by

$$\langle\vec{k}|V_X|\vec{k}\rangle = -\sum_{\vec{k}'} n_{\vec{k}'}\langle\vec{k},\vec{k}'|V|\vec{k}'\vec{k}\rangle. \qquad (6.83)$$

It renormalizes the bare electron energy $\epsilon(\vec{k})$, see Fig. 6.3. Note that the Fermi velocity is divergent, which is unphysical. It may be removed by screening the Coulomb interaction $V(\vec{r}_1 - \vec{r}_2)$. However, one must not screen the Coulomb interaction appearing in the Hartree self-energy, which will lead to double counting since the Hartree effect already includes screening effect. A dynamically screened version of the Hartree–Fock approximation is called the GW approximation [1], which is widely used in band structure calculations.

Exercise 6.9. Show that, in a homogeneous electron gas, plane waves are solutions of the self-consistent Hartree–Fock equations. Hint: Use $\frac{e^2}{|\vec{r}-\vec{r}'|} = 4\pi e^2 \int \frac{d\vec{q}}{(2\pi)^3} \frac{1}{q^2} e^{i\vec{q}\cdot(\vec{r}-\vec{r}')}$, see Ref. [2]. Note that the Hartree term cancels out with the contribution from the positively charged ion density.

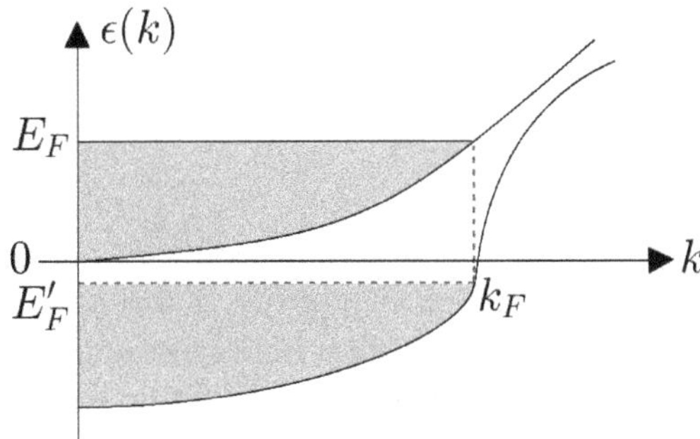

Fig. 6.3. In the absence of electron interactions, the band dispersion $\epsilon(\vec{k})$ is parabolic. Exchange self-energy renormalizes the electron band dispersion. The Fermi energy is shifted to a new value E'_F. The new slope at k_F is singular, which implies that the Fermi velocity is infinite. This result is unphysical and is artifact of the Hartree–Fock approximation.

Exercise 6.10. Compute the exchange self-energy equation (6.83). Hint: See Ref. [2].

The self-consistent eigenstates of the Hartree–Fock Hamiltonian can be used to compute the Hartree–Fock total energy, see Exercise 6.6.

6.10. Hubbard Model

Consider an electron moving in a two-dimensional square lattice. Electron interactions are included in the Hubbard model as follows [8]:

$$H = -t \sum_{\langle ij \rangle, \sigma} c^{\dagger}_{i,\sigma} c_{j,\sigma} + U \sum_{i} n_{i\uparrow} n_{i\downarrow}. \tag{6.84}$$

The second term originates from the two-body interactions. The general expression for the two-body interaction term is

$$\frac{1}{2} \sum_{i\sigma_i, j\sigma_j, k\sigma_k, l\sigma_l} \langle i\sigma_i, j\sigma_j | V | k\sigma_k l\sigma_l \rangle c^{\dagger}_{i\sigma_i} c^{\dagger}_{j\sigma_j} c_{l\sigma_l} c_{k\sigma_k}. \tag{6.85}$$

Assume that the two-body interaction V is zero when the electrons are not on the same site. The Pauli exclusion principle dictates that the spins of two electrons on the same site must have opposite values. When they are on the same site, the on-site repulsion is U. There are thus two non-zero contributions

$$\tfrac{1}{2}U c_{i\uparrow}^{\dagger} c_{i\downarrow}^{\dagger} c_{i\downarrow} c_{i\uparrow} = \tfrac{1}{2}U n_{i\uparrow} n_{i\downarrow},$$

$$\tfrac{1}{2}U c_{i\downarrow}^{\dagger} c_{i\uparrow}^{\dagger} c_{i\uparrow} c_{i\downarrow} = \tfrac{1}{2}U n_{i\downarrow} n_{i\uparrow}. \tag{6.86}$$

The electron energy dispersion in the absence of on-site repulsion U is computed in Exercise 6.11 and plotted in Fig. 6.4. The Hubbard Hamiltonian has SU(2) symmetry, i.e., spin rotational invariance, see Exercise 6.12. However, a magnetic ground state may spontaneously break this symmetry.

Exercise 6.11. Derive the energy dispersion $\epsilon_{\vec{k}} = -2t(\cos k_x a + \cos k_y a)$. Hint: Use

$$c_{i,\sigma}^{\dagger} = \sqrt{\frac{1}{N_s}} \sum_{\vec{k}} e^{i\vec{k}\cdot\vec{R}_i} c_{\vec{k},\sigma}^{\dagger}, \tag{6.87}$$

to transform the non-interacting Hamiltonian $H = -t\sum_{\langle ij\rangle,\sigma} c_{i,\sigma}^{\dagger} c_{j,\sigma}$ into k-space ($\vec{R}_i$ denotes the position of a lattice site, and only the nearest-neighbor sites matter, see Fig. 6.7.) Then construct the matrix Hamiltonian $\langle \vec{k}|H|\vec{k}'\rangle = H_{\vec{k}\vec{k}'}$, where $|\vec{k}\rangle = c_{\vec{k}}^{\dagger}|0\rangle$.

Exercise 6.12. Show that one can rewrite the on-site repulsion term as follows:

$$U\sum_{i} n_{i\uparrow} n_{i\downarrow} = -\frac{2U}{3}\sum_{i} \vec{s}_i^{\,2} + \frac{U}{2}\sum_{i}(n_{i\uparrow} + n_{i\downarrow}). \tag{6.88}$$

The spin term $\vec{s}_i^{\,2}$ is SU(2) invariant (i.e., spin rotation invariant). Hint: Use

$$s_{i,\alpha} = \frac{1}{2}\sum_{\sigma\sigma'} c_{i\sigma}^{\dagger} (\sigma_\alpha)_{\sigma\sigma'} c_{i\sigma'}, \tag{6.89}$$

where $\alpha = x, y, z$ and $\hbar = 1$ is dimensionless. Use also the following result for the Pauli spin matrices

$$\sum_{\alpha=x,y,z} \sigma_{ij}^{\alpha}\sigma_{kl}^{\alpha} = 2\delta_{il}\delta_{jk} - \delta_{ij}\delta_{kl}. \tag{6.90}$$

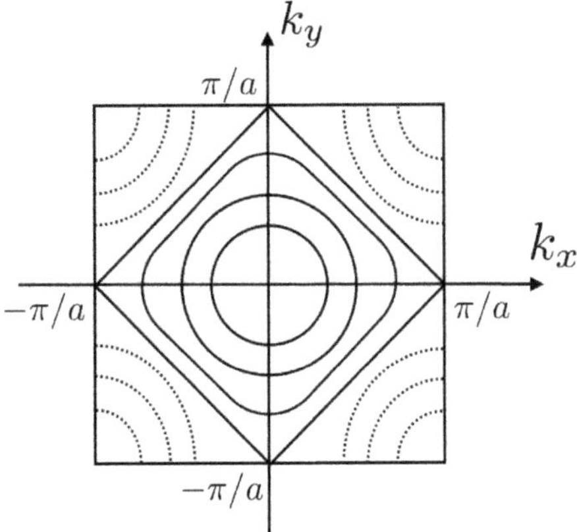

Fig. 6.4. The big square is the Brillouin zone. Contour plot of the energy dispersion $\epsilon_{\vec{k}} = -2t(\cos k_x a + \cos k_y a)$ is displayed. Energy dispersion is $\epsilon_{\vec{k}} < 0$ for the straight line and $\epsilon_{\vec{k}} > 0$ for the dotted line. The Brillouin zone is half-filled. The lattice constant is a. The small square is the **magnetic Brillouin zone** (MBZ), whose meaning will be explained in Sec. 6.12.

6.11. Unrestricted Hartree–Fock Approximation

Let $|\psi_0\rangle$ be the ground state Ansatz of the Hubbard model. The decoupling scheme for the on-site interaction term in Eq. (6.84) is assumed to be (see Fig. 6.5)

$$\langle\psi_0|n_{i,\uparrow}n_{i,\downarrow}|\psi_0\rangle = \langle\psi_0|c^\dagger_{i,\uparrow}c_{i,\uparrow}c^\dagger_{i,\downarrow}c_{i,\downarrow}|\psi_0\rangle = \langle c^\dagger_{i,\uparrow}c_{i,\uparrow}c^\dagger_{i,\downarrow}c_{i,\downarrow}\rangle$$

$$= \langle c^\dagger_{i,\uparrow}c^\dagger_{i,\downarrow}c_{i,\downarrow}c_{i,\uparrow}\rangle \rightarrow \langle n_{i,\uparrow}\rangle\langle n_{i,\downarrow}\rangle - \langle s^+_i\rangle\langle s^-_i\rangle, \tag{6.91}$$

where the spin raising and lowering operators are

$$s^+_i = s_{x,i} + is_{y,i}, \quad s^-_i = s_{x,i} - is_{y,i},$$
$$s^+_i = c^\dagger_{i,\uparrow}c_{i,\downarrow}, \quad s^-_i = c^\dagger_{i,\downarrow}c_{i,\uparrow} \tag{6.92}$$

with site spin operators

$$s_{x,i} = \frac{1}{2}[c^\dagger_{i,\uparrow}c_{i,\downarrow} + c^\dagger_{i,\downarrow}c_{i,\uparrow}],$$

$$s_{y,i} = -\frac{i}{2}[c^\dagger_{i,\uparrow}c_{i,\downarrow} - c^\dagger_{i,\downarrow}c_{i,\uparrow}], \tag{6.93}$$

$$s_{z,i} = \frac{1}{2}[c^\dagger_{i,\uparrow}c_{i,\uparrow} - c^\dagger_{i,\downarrow}c_{i,\downarrow}].$$

(Here $\hbar = 1$.)

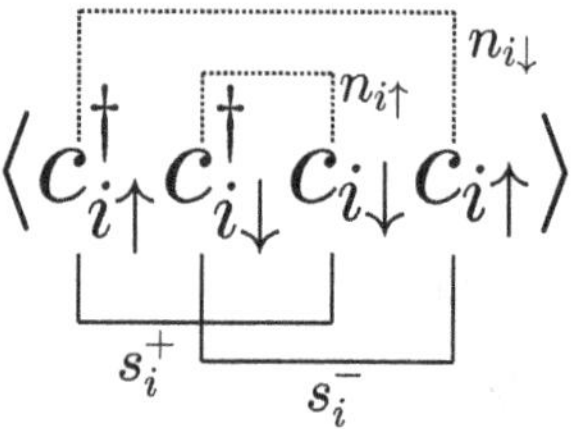

Fig. 6.5. Decoupling scheme leading to the mean field Hamiltonian equation (6.91). It gives rise to two terms. Note that only those ground state Ansatzs satisfying this decoupling scheme is used.

Exercise 6.13. Show that $s_{x,i}$, $s_{y,i}$, and $s_{z,i}$ satisfy the usual spin commutation relations.

The decoupling scheme of Eq. (6.91) gives the following mean-field version of the Hubbard Hamiltonian

$$
\begin{aligned}
H_{\mathrm{MF}} &= T + V_1 + V_2, \\
T &= -t \sum_{\langle ij\rangle,\sigma} c_{i,\sigma}^\dagger c_{j,\sigma}, \\
V_1 &= U \sum_i [n_{i,\uparrow}\langle n_{i,\downarrow}\rangle + n_{i,\downarrow}\langle n_{i,\uparrow}\rangle - \langle n_{i,\downarrow}\rangle\langle n_{i,\uparrow}\rangle], \quad (6.94) \\
V_2 &= -U \sum_i [s_i^+\langle s_i^-\rangle + s_i^-\langle s_i^+\rangle - \langle s_i^-\rangle\langle s_i^+\rangle].
\end{aligned}
$$

The symbol $\langle ij\rangle$ means that the sum is over pairs of nearest-neighbor sites.

The mean-field Hamiltonian equation (6.94) can be written in a more intuitive way (see Exercise 6.14)

$$
\begin{aligned}
H_{\mathrm{MF}} = -t \sum_{\langle ij\rangle,\sigma} c_{i,\sigma}^\dagger c_{j,\sigma} + U \sum_i [n_{i,\uparrow}\langle n_{i,\downarrow}\rangle + n_{i,\downarrow}\langle n_{i,\uparrow}\rangle - \langle n_{i,\downarrow}\rangle\langle n_{i,\uparrow}\rangle] \\
+ \sum_i [s_{ix}\langle h_{ix}\rangle + s_{iy}\langle h_{iy}\rangle], \quad (6.95)
\end{aligned}
$$

where the self-consistent "magnetic fields" are $\langle h_{ix}\rangle = -2U\langle s_{ix}\rangle$ and $\langle h_{iy}\rangle = -2U\langle s_{iy}\rangle$. (Imagine that there is an external magnetic field $\langle \vec{h}\rangle$ whose components at site i are $\langle h_{ix}\rangle$ and $\langle h_{iy}\rangle$.)

> **Exercise 6.14.** Derive Eq. (6.95) from Eq. (6.94). Hint: Write s_i^+ and s_i^- in terms of $s_{x,i}$ and $s_{y,i}$.

The mean-field Hamiltonian equation (6.95) is a one-particle Hamiltonian. Its eigenstates are given by

$$\psi_\alpha = \left(A_{1\uparrow}^\alpha \ A_{2\uparrow}^\alpha \ \cdots \ A_{N_s\uparrow}^\alpha \ A_{1\downarrow}^\alpha \ \cdots \ A_{N_s\downarrow}^\alpha \right)^T, \qquad (6.96)$$

where N_s denotes the number of sites. The self-consistent fields appearing in the mean-field Hamiltonian can be found using these eigenstates, see Exercise 6.15.

> **Exercise 6.15.** Using the eigenstates (6.96), show that the spin expectation values for site i are
>
> $$\langle \sigma_i^x \rangle = \sum_\alpha n_\alpha (A_{i\uparrow}^{\alpha*} A_{i\downarrow}^\alpha + A_{i\downarrow}^{\alpha*} A_{i\uparrow}^\alpha),$$
>
> $$\langle \sigma_i^y \rangle = i \sum_\alpha n_\alpha (-A_{i\uparrow}^{\alpha*} A_{i\downarrow}^\alpha + A_{i\downarrow}^{\alpha*} A_{i\uparrow}^\alpha), \qquad (6.97)$$
>
> $$\langle \sigma_i^z \rangle = \langle n_{i\uparrow} \rangle - \langle n_{i\downarrow} \rangle,$$
>
> where
>
> $$\langle n_{i\uparrow} \rangle = \sum_\alpha n_\alpha (A_{i\uparrow}^{\alpha*} A_{i\uparrow}^\alpha),$$
>
> $$\langle n_{i\downarrow} \rangle = \sum_\alpha n_\alpha (A_{i\downarrow}^{\alpha*} A_{i\downarrow}^\alpha), \qquad (6.98)$$
>
> and $n_\alpha = 1(0)$ for occupied (unoccupied) Hartree–Fock states. Hint: For each site i use the Pauli spin matrices for σ_i^x, σ_i^y, and σ_i^z.

The particle–hole symmetric version of the Hubbard Hamiltonian can be given in the grand canonical ensemble formalism

$$H = -t \sum_{\langle ij \rangle, \sigma} c_{i,\sigma}^\dagger c_{j,\sigma} + U \sum_i (n_{i\uparrow} - 1/2)(n_{i\downarrow} - 1/2) - \mu \sum_i (n_{i\uparrow} + n_{i\downarrow}),$$

$$(6.99)$$

where μ is the chemical potential. Under particle–hole transformation, the filling factor and chemical potential transform as follows:

$$n_{\mathrm{f}} \to 1 - n_{\mathrm{f}}, \quad \text{and} \quad \mu \to -\mu, \qquad (6.100)$$

where the filling factor (the average occupation number per site) is defined as $n_{\mathrm{f}} = N/N_s$ (N is the number of electrons; $n_{\mathrm{f}} = 1$ at half-filling). This implies $n_{\mathrm{f}}(\mu) = 2 - n_{\mathrm{f}}(-\mu)$.

Exercise 6.16. Find the total mean-field energy in the presence of on-site disorder potential $v_{\mathrm{ext}} = \sum_{j\sigma} v_i c_{j,\sigma}^{\dagger} c_{j,\sigma}$, which breaks translational invariance. Answer: The contribution from the on-site interaction is given by (see Eq. (6.91))

$$V = U \sum_i [\langle n_{i,\uparrow}\rangle\langle n_{i,\downarrow}\rangle - \langle s_i^-\rangle\langle s_i^+\rangle]. \tag{6.101}$$

Suppose that the operator a_α^+ creates a Hartree–Fock eigenstate

$$a_\alpha = \sum_j A_{\alpha,j}^\sigma c_{j,\sigma}. \tag{6.102}$$

(α includes spin index.) The expectation values of $\langle n_{i,\sigma}\rangle$ and $\langle s_i^\pm\rangle$ can be expressed in terms of $A_{\alpha j}^\sigma$, see Exercise 6.15. The contribution from the hopping term is given by the inverse matrix $B = A^{-1}$

$$T = -t \sum_{\langle ij\rangle} \sum_\alpha n_\alpha B_{\alpha,i}^{\sigma*} B_{j,\alpha}^\sigma, \tag{6.103}$$

where $n_\alpha = 1(0)$ for occupied (unoccupied) Hartree–Fock states. The contribution from disorder is

$$D = \sum_{j,\sigma} v_j \langle n_{j,\sigma}\rangle. \tag{6.104}$$

The total energy is

$$E_{\mathrm{tot}} = T + D + V. \tag{6.105}$$

(Note that in this result the chemical potential is not zero while in Eq. (6.94) it is set to zero.) At half-filling, the spin-flip terms are absent and the total energy can be also given in a different but equivalent form

$$E_{\mathrm{tot}} = \sum n_\alpha \epsilon_\alpha - U \sum_i \langle n_{i,\uparrow}\rangle\langle n_{i,\downarrow}\rangle, \tag{6.106}$$

where ϵ_α is the Hartree–Fock single-particle energy.

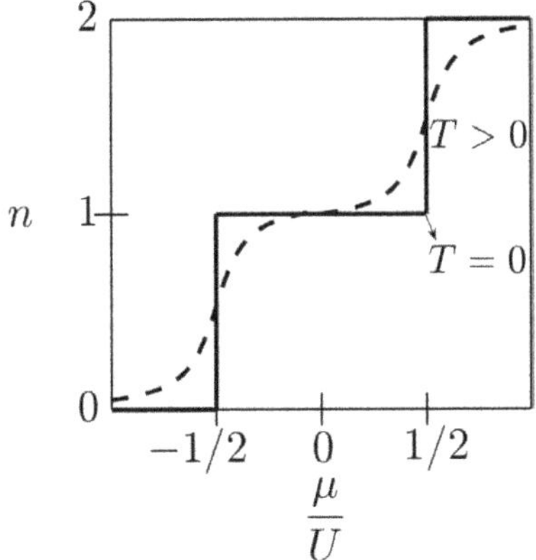

Fig. 6.6.   Using the single-site Hubbard model, the dependence of site occupation number on the chemical potential is computed. The single-site Hubbard model is an excellent toy model for understanding the Coulomb blockade where an electron may tunnel onto an artificial atom [9] by overcoming the electrostatic charging energy (the Coulomb repulsion energy $\sim U$). Notice that at the first step in the figure an electron enters the artificial atom. As μ increases, the charging energy is overcome and another electron may enter the artificial atom and the second step develops.

Exercise 6.17. Consider a single-site Hubbard Hamiltonian

$$H = U(n_{i\uparrow} - 1/2)(n_{i\downarrow} - 1/2) - \mu(n_{i\uparrow} + n_{i\downarrow}). \qquad (6.107)$$

It has four possible states $|0\rangle$, $|\uparrow\rangle$, $|\downarrow\rangle$, and $|\uparrow\downarrow\rangle$. Find the partition function $Z = \text{Tr}[e^{-\beta H}]$ and the site occupation number $n = \langle n_\uparrow + n_\downarrow \rangle = Z^{-1}\text{Tr}[(n_\uparrow + n_\downarrow)e^{-\beta H}]$. For several values of β plot n as a function of μ (see Fig. 6.6). Hint: Consult Ref. [8].

6.12.   Unrestricted Hartree–Fock Approximation and Antiferromagnetic Order in Hubbard Model

• Antiferromagnetic Order at Half-Filling

The Hubbard model may look deceptively simple but no exact solution is known in two dimensions. Some insight about the ground state may be gained from the following consideration. In the limit of very strong on-site repulsion, $U \to \infty$, one can show that the Hubbard model at half-filling can be written as the nearest-neighbor antiferromagnetic spin Hamiltonian [10]

$$H = \frac{1}{2} \sum_{\langle i,j \rangle} J \vec{s}_i \cdot \vec{s}_j, \tag{6.108}$$

where $\vec{s}_i$ are site spin operators and the antiferromagnetic coupling constant is

$$J = \frac{4t^2}{U} > 0. \tag{6.109}$$

(Note that in this case $|\langle s_i^z \rangle| = 1/2$.) The main physics behind antiferromagnetic coupling between the nearest-neighbor sites is the virtual hopping of, for example, a spin-up electron to a neighboring site occupied by a spin-down electron and then hopping back to the original site (the energy conservation law may be violated by amount ΔE during the hopping time Δt such that $\Delta E \Delta t > \hbar$). Hopping of a spin-down electron to a neighboring site occupied by a spin-down electron is forbidden by the Pauli exclusion principle. The virtual process will lower the total energy, which may be shown using the second-order perturbation theory.

Even if U is not strong, the ground state of the two-dimensional Hubbard at half-filling has an antiferromagnetic order: if the average spin of a site of a square lattice is up then the average spin values of the neighboring sites are down, see Fig. 6.7. (Note that $|\langle s_i^z \rangle| \leq 1/2$, where $\vec{s}_i$ is the ith site spin-$\frac{1}{2}$ operator.) The Hamiltonian has SU(2) symmetry but the antiferromagnetic ground state does not. This is an example of a **spontaneously broken symmetry** with an accompanying **local-order parameter** (see below). In addition, time-reversal symmetry is also broken due to the presence of magnetism.

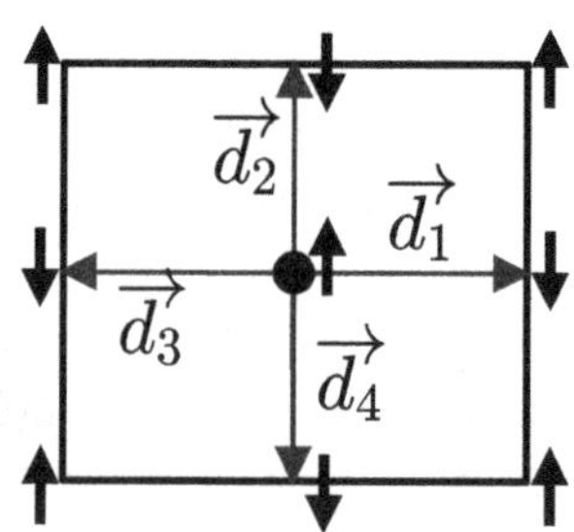

Fig. 6.7. Antiferromagnetic order in a periodic square lattice is depicted. Site spin expectation values are shown. A site $\vec{R}_i$ has four nearest-neighbor sites $\vec{R}_j = \vec{R}_i + \vec{d}_p$ with $p = 1, 2, 3, 4$. The lattice constant is a.

Below we will examine, using the unrestricted Hartree–Fock approximation, antiferromagnetism of square lattice away from the limit $U \to \infty$. The Ansatz for the **periodic** and **self-consistent** site occupation numbers [11] is

$$\langle n_{i,\sigma} \rangle = \frac{1}{2}(1 + m e^{-i\vec{Q}\cdot\vec{R}_i} \lambda_{\sigma,\uparrow}), \tag{6.110}$$

where $\vec{R}_i$ is the position vector of site i, $\vec{Q} = (\frac{\pi}{a}, \frac{\pi}{a})$ is the spin modulation vector, and

$$\lambda_{\sigma,\sigma} = 1, \quad \lambda_{\sigma,-\sigma} = -1. \tag{6.111}$$

The second term of Eq. (6.110) represents the spin modulation and is present both in the x and y directions. Its magnitude is

$$m = |\langle n_{i,\uparrow} \rangle - \langle n_{i,\downarrow} \rangle| \tag{6.112}$$

with the condition $\langle n_{i,\uparrow} \rangle + \langle n_{i,\downarrow} \rangle = 1$ maintained. (Here m is not the electron mass.) The ground state expectation value of $\vec{s}_i$ is

$$\langle \vec{s}_i \rangle = \langle \Psi_0 | \vec{s}_i | \Psi_0 \rangle. \tag{6.113}$$

Note that $\langle \vec{s}_i \rangle$ is spatially modulated, described by the vector $\vec{Q}$. The quantity m serves as the **local order parameter** of a spontaneously broken symmetry state. $e^{i\vec{Q}\cdot\vec{R}_i}$ changes sign when $\vec{R}_i$ changes to a nearest-neighbor site. The ground state has a non-zero **staggered** magnetization, meaning that the direction of $\vec{s}_i$ alternates from site to site. Sites with dominantly spin-up (-down) form a lattice and we will call it an $A(B)$-lattice. We thus have a bipartite square lattice. The origin of this antiferromagnetic coupling is due to a nesting effect in two-dimensional Hubbard model, i.e., numerous points on the magnetic zone boundary can be connected by the modulation wave vector $\vec{Q}$ (see Fig. 6.8). We will see below that this effect induces, at half-filling, a gap at the Fermi energy.

- **Antiferromagnetic Ground State Ansatz**

The variational ground state Ansatz is given by

$$|\Psi\rangle = \left(\prod_{\vec{k} \in F_\uparrow} a^\dagger_{\vec{k},\uparrow} |0\rangle \right) \left(\prod_{\vec{k} \in F_\downarrow} b^\dagger_{\vec{k},\downarrow} |0\rangle \right), \tag{6.114}$$

where $F_{\uparrow,\downarrow}$ stands for two Fermi seas of two magnetic bands. The new fermion operators $a^\dagger_{\vec{k},\sigma}$ and $b^\dagger_{\vec{k},\sigma}$ are related to the old ones via a canonical

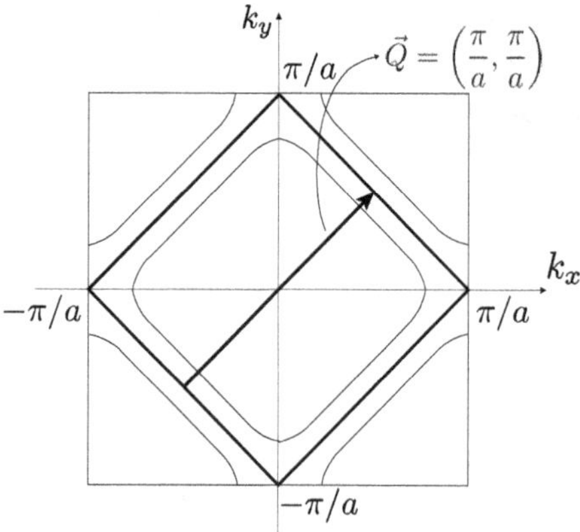

Fig. 6.8.　First Brillouin zone of two-dimensional Hubbard model of the half-filled square lattice. Nesting effect is present: numerous points on the MBZ boundary can be connected by the same wave vector $\vec{Q} = (\frac{\pi}{a}, \frac{\pi}{a})$. In the presence of antiferromagnetism a gap opens up at the Fermi energy (the half-filled square lattice). Gap edges are drawn as thin lines. Note that the electron energy dispersion in the absence of U is $\epsilon(\vec{k}) < 0$ in the MBZ.

transformation:

$$
\begin{bmatrix} c_{\vec{k},\sigma} \\ c_{\vec{k}+\vec{Q},\sigma} \end{bmatrix} = \begin{pmatrix} \cos\theta_{\vec{k},\sigma} & -\sin\theta_{\vec{k},\sigma} \\ \sin\theta_{\vec{k},\sigma} & \cos\theta_{\vec{k},\sigma} \end{pmatrix} \begin{bmatrix} a_{\vec{k},\sigma} \\ b_{\vec{k},\sigma} \end{bmatrix},
\tag{6.115}
$$

where

$$
\tan 2\theta_{\vec{k},\sigma} = -\frac{Um\lambda_{\sigma,\uparrow}}{\epsilon_{\vec{k}} - \epsilon_{\vec{k}+\vec{Q}}},
$$

$$
\sin^2\theta_{\vec{k},\sigma} = \frac{1}{2}\left(1 - \lambda_{\sigma,\uparrow}\frac{\epsilon_{\vec{k}} - \epsilon_{\vec{k}+\vec{Q}}}{\sqrt{[\epsilon_{\vec{k}} - \epsilon_{\vec{k}+\vec{Q}}]^2 + (Um)^2}}\right).
\tag{6.116}
$$

Its inverse transformation is

$$
\begin{bmatrix} a_{\vec{k},\sigma} \\ b_{\vec{k},\sigma} \end{bmatrix} = \begin{pmatrix} \cos\theta_{\vec{k},\sigma} & \sin\theta_{\vec{k},\sigma} \\ -\sin\theta_{\vec{k},\sigma} & \cos\theta_{\vec{k},\sigma} \end{pmatrix} \begin{bmatrix} c_{\vec{k},\sigma} \\ c_{\vec{k}+\vec{Q}\sigma} \end{bmatrix}.
\tag{6.117}
$$

The justification of this Ansatz is given below (we follow the derivation given in Ref. [10]).

• Mean Field Hamiltonian

Let us explain how this ground state Ansatz rises. At half-filling, the spin-flip terms in the mean Hubbard Hamiltonian Eq. (6.94) may be neglected. We get the following Hamiltonian at half-filling

$$H_{\mathrm{MF}} = -t \sum_{<ij>,\sigma} c^\dagger_{i,\sigma} c_{j,\sigma},$$

$$+ U \sum_i [n_{i,\uparrow}\langle n_{i,\downarrow}\rangle + n_{i,\downarrow}\langle n_{i,\uparrow}\rangle - \langle n_{i,\downarrow}\rangle\langle n_{i,\uparrow}\rangle]. \tag{6.118}$$

Imagine that there is a self-consistent magnetic field with alternating sign from site to site. Such a field will produce antiferromagnetic site spins, i.e., the net site spin $\langle n_{i,\uparrow}\rangle - \langle n_{i,\downarrow}\rangle$ will change sign from site to site.

We can rewrite the mean-field Hamiltonian in k-space as follows (see Exercise 6.18):

$$H_{\mathrm{MF}} = \sum_{\vec{k}\in\mathrm{MBZ},\sigma} \left[\epsilon_{\vec{k}} n_{\vec{k},\sigma} + \epsilon_{\vec{k}+\vec{Q}} n_{\vec{k}+\vec{Q},\sigma} - \frac{Um\lambda_{\sigma,\uparrow}}{2}(c^\dagger_{\vec{k}+\vec{Q},\sigma} c_{\vec{k},\sigma} + \mathrm{h.c.}) \right]$$

$$+ \frac{UN_s(1+m^2)}{4}, \tag{6.119}$$

where the sum over the Brillouin zone is split into two sums over the magnetic Brillouin zone. As yet we do not know the value of the order parameter m. Note that when $\vec{k} \in \mathrm{MBZ}$ then

$$\epsilon_{\vec{k}+\vec{Q}} = -\epsilon_{\vec{k}}. \tag{6.120}$$

Exercise 6.18. Using the mean Hamiltonian Eq. (6.118) derive Eq. (6.119). Hint: Use Eq. (6.110) and the positions of the nearest neighbor sites (see Fig. 6.7). Wave vectors $\vec{k} \in \mathrm{BZ}$ are divided into those $\vec{k}$ belonging to the magnetic Brillouin zone and those belonging to outside of the magnetic Brillouin zone (equivalent points $\vec{k}+\vec{Q}$ outside the Brillouin zone are actually included), see Fig. 6.9.

In terms of the new transformed particle operators $a^\dagger_{\vec{k},\sigma}$ and $b^\dagger_{\vec{k},\sigma}$ (see Eq. (6.117)), the **bilinear** mean field Hamiltonian is **diagonal**

$$H_{\mathrm{MF}} = \sum_{\vec{k}\in\mathrm{MBZ}} \left[E^-_{\vec{k}}(a^\dagger_{\vec{k},\uparrow} a_{\vec{k},\uparrow} + b^\dagger_{\vec{k},\downarrow} b_{\vec{k},\downarrow}) + E^+_{\vec{k}}(a^\dagger_{\vec{k},\downarrow} a_{\vec{k},\downarrow} + b^\dagger_{\vec{k},\uparrow} b_{\vec{k},\uparrow}) \right]$$

$$+ \frac{UN_s(1+m^2)}{4}, \tag{6.121}$$

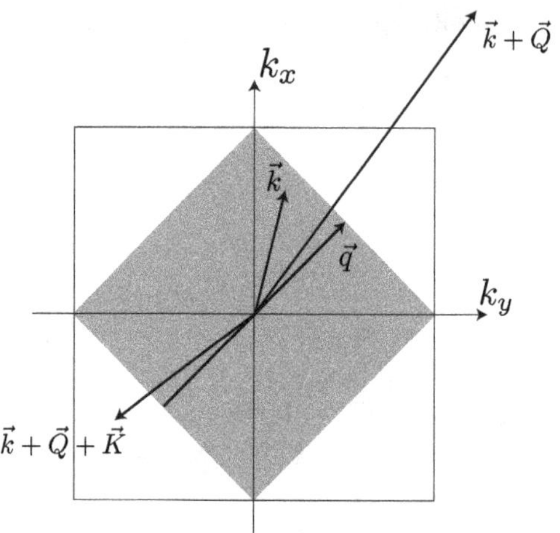

Fig. 6.9. Wave vectors $\vec{k}$ in the Brillouin zone can be divided into two sums, one in the MBZ and the other outside it: $\sum_{\vec{k}\in\mathrm{BZ}} = \sum_{\vec{k}\in\mathrm{MBZ}} + \sum_{\vec{k}\in\overline{\mathrm{MBZ}}}$, where $\overline{\mathrm{MBZ}} = \mathrm{BZ} - \mathrm{MBZ}$. In the figure, a reciprocal vector $\vec{K} = (-\frac{2\pi}{a}, -\frac{2\pi}{a})$ brings $\vec{k} + \vec{Q}$ into $\overline{\mathrm{MBZ}}$.

where the new renormalized electron energy is

$$E_{\vec{k}}^{\pm} = \frac{1}{2}\left[\epsilon_{\vec{k}} + \epsilon_{\vec{k}+\vec{Q}} \pm \sqrt{[\epsilon_{\vec{k}} - \epsilon_{\vec{k}+\vec{Q}}]^2 + (Um)^2}\, \right]$$
$$= \pm\sqrt{\epsilon_{\vec{k}}^2 + (Um/2)^2}. \tag{6.122}$$

Here we have used $\epsilon_{\vec{k}+\vec{Q}} = -\epsilon_{\vec{k}}$. Each eigenenergy of $E_{\vec{k}}^{\pm}$ is spin degenerate. Note that $E_{\vec{k}}^{-} < 0$ and $E_{\vec{k}}^{+} > 0$ and there is a gap between them.

Exercise 6.19. Derive the diagonalization condition (6.116). Hint: Use Eqs. (6.115) and (6.119), $\epsilon_{\vec{k}} = -\epsilon_{\vec{k}+\vec{Q}}$, $\cos^2 x = 1/(1 + \tan^2 x)$, and $\sin^2 x = (1 - \cos 2x)/2$.

- **Minimization of Total Energy and Gap Equation**

To find the magnetization m, Eq. (6.112), we must minimize the total energy at zero temperature. We find

$$m = \frac{2}{N_s} \sum_{\vec{k}\in\mathrm{MBZ}} \frac{Um}{\sqrt{4\epsilon_{\vec{k}}^2 + (Um)^2}}. \tag{6.123}$$

Factor 2 is due to spin degeneracy.

Exercise 6.20. Derive the gap equation (6.123). Hint: Use that the total energy is

$$E_{\text{tot}} = \frac{U N_s (1 + m^2)}{4} + 2 \sum_{\vec{k} \in \text{MBZ}} E_{\vec{k}}^- . \qquad (6.124)$$

The first originates from the constant term in the mean-field Hamiltonian equation (6.119). The second term originates from the occupied band. Minimize E_{tot} with respect to the order parameter m.

The band structure is displayed in Fig. 6.10. We have plotted the weighting factors $\sin^2 \theta_{\vec{k},\uparrow}$ and $\sin^2 \theta_{\vec{k},\downarrow}$, see Fig. 6.11. From these results, we infer that the creation operator for **the new particle with spin-up is** $a_{\vec{k},\uparrow}^\dagger \sim c_{\vec{k}+\vec{Q},\uparrow}^\dagger \sim c_{\vec{k},\uparrow}^\dagger$, **which creates a spin $\uparrow$ particle predominantly on the A-lattice sites**, see Table 6.2. Note that $b_{\vec{k},\downarrow}^\dagger \sim c_{\vec{k}+\vec{Q},\downarrow}^\dagger$ creates a new particle with spin-down predominantly on the B-lattice sites, see Table 6.2. The wave functions of these particles are thus **different**, consistent with the fact that the ground state is antiferromagnetically ordered. Note that we have used the **unrestricted** Hartree–Fock method. States with $E_{\vec{k}}^- < 0$ are occupied in the new ground state.

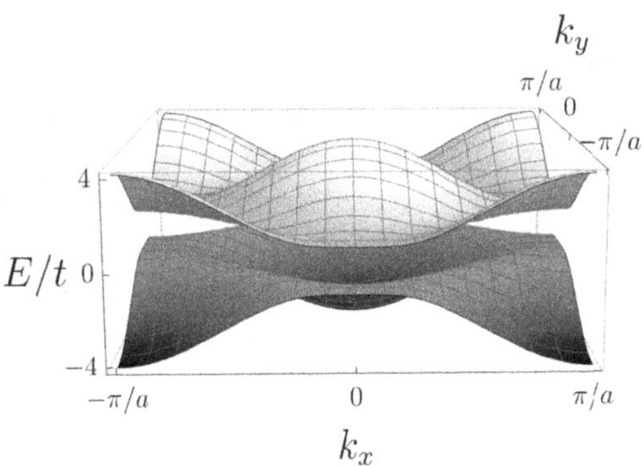

Fig. 6.10. Schematic display of the band structure with a gap. The Fermi level is in the gap. The trough of the conduction and the crest of the valence band represent the edges of the MBZ.

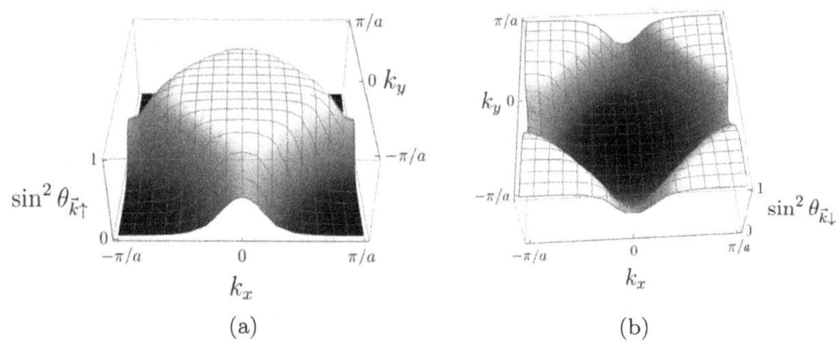

Fig. 6.11. (a) Weighting factor $\sin^2\theta_{\vec{k},\uparrow} \approx 1$ in the MBZ. (b) However, $\sin^2\theta_{\vec{k},\downarrow} \approx 0$.

Table 6.2. Character of new particles.

Occupied	Unoccupied
$a^\dagger_{\vec{k},\uparrow} \sim c^\dagger_{\vec{k}+\vec{Q},\uparrow}$	$a^\dagger_{\vec{k},\downarrow} \sim c^\dagger_{\vec{k},\downarrow}$
$b^\dagger_{\vec{k},\downarrow} \sim c^\dagger_{\vec{k}+\vec{Q},\downarrow}$	$b^\dagger_{\vec{k},\uparrow} \sim c^\dagger_{\vec{k},\uparrow}$

6.13. Unrestricted Hartree–Fock Approximation and Spin Density Wave in Hubbard Model⋆

• Ground State Ansatz

Away from half-filling, the Hartree–Fock approximation of the Hubbard can give several competing candidates for the ground state (these states have nearly degenerate energies).[g] One such possible state represents a spin density wave [12]. (A ground state spin density wave should not be confused with a spin wave, which is an elementary excitation of the system with energy higher than the ground state energy.) To describe such a state, a term that was neglected at half-filling must be included, see Eq. (6.94). It represents spin flip terms, which will allow spins to flip spiral, see Fig. 6.12. Such a state may be described by introducing a Hartree–Fock self-consistent spiraling magnetic field.

The following variational ground state Ansatz can be used (we follow the derivation given in Ref. [10])

$$|\Psi\rangle = \prod_{\vec{k}\in F^\pm,\lambda=\pm} a^\dagger_{\vec{k},\lambda}|0\rangle, \qquad (6.125)$$

---

[g]Often a better approximation is needed to determine the true ground state.

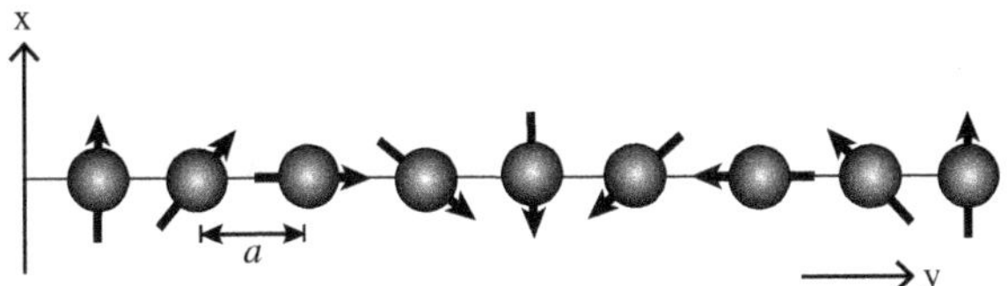

Fig. 6.12. Example of a spiraling spin density wave in the xy-plane with $\vec{q} = (0, 2\pi/8a, 0)$. Note the expectation value $\langle s_i^+ \rangle = \langle s_{x,i} \rangle + i\langle s_{y,i} \rangle$ has both x- and y-components. Another example of a spin density wave is the following spin configuration: the wave vector may point along the z-axis and spins spiral in the xy-plane as z-coordinate varies.

where $F^\pm$ stands for two Fermi seas of two magnetic bands and the new fermion operators $a_{\vec{k},\lambda}^+$ and $a_{\vec{k},\lambda}$ are related to the old ones via a canonical transformation: Here, unlike in the case of antiferromagnetism, a linear combination of **opposite** spins is used

$$\begin{bmatrix} c_{\vec{k},\uparrow} \\ c_{\vec{k}+\vec{q},\downarrow} \end{bmatrix} = \begin{pmatrix} \cos\theta_{\vec{k}} & -\sin\theta_{\vec{k}} \\ \sin\theta_{\vec{k}} & \cos\theta_{\vec{k}} \end{pmatrix} \begin{bmatrix} a_{\vec{k},+} \\ a_{\vec{k},-} \end{bmatrix} \tag{6.126}$$

and

$$\begin{bmatrix} a_{\vec{k},+} \\ a_{\vec{k},-} \end{bmatrix} = \begin{pmatrix} \cos\theta_{\vec{k}} & \sin\theta_{\vec{k}} \\ -\sin\theta_{\vec{k}} & \cos\theta_{\vec{k}} \end{pmatrix} \begin{bmatrix} c_{\vec{k},\uparrow} \\ c_{\vec{k}+\vec{q},\downarrow} \end{bmatrix}, \tag{6.127}$$

where $\cos\theta_{\vec{k}}$ depends on $\vec{q}$. The unknown quantities $\cos\theta_{\vec{k}}$ and $\vec{q}$ can be determined using the variational ground state Ansatz.

• Minimization of Total Energy

Using the Ansatz equation (6.125), we minimize the total energy including the kinetic term. The expectation value of the interaction term with respect to the Ansatz $|\Psi\rangle$ is

$$U\left\langle \sum_i n_{i\uparrow} n_{i\downarrow} \right\rangle = U \sum_i \langle n_{i\uparrow} \rangle \langle n_{i\downarrow} \rangle - U \sum_i \langle s_i^+ \rangle \langle s_i^- \rangle$$

$$= N_s U n_{\mathrm{f}}^2/4 - U m_z^2 - U m_{\vec{q}}^2, \tag{6.128}$$

where m_z and $m_{\vec{q}}$ describe, respectively, spin polarization in the z direction and spiraling spins in the xy-plane. They are, respectively, related to $\langle s_i^+ \rangle$

and $\langle s_i^z \rangle$,

$$\langle s_i^+ \rangle = N^{1/2} m_{\vec{q}}\, e^{-i\vec{q}\cdot\vec{R}_i}, \quad \langle s_i^z \rangle = N^{1/2} m_z. \tag{6.129}$$

Exercises 6.21 and 6.22 show how these may be calculated. Note that the total number of electrons is $n_{\mathrm{f}} N_s$.

Exercise 6.21. Show that

$$m_{\vec{q}} = \frac{1}{2N_s} \sum_{\vec{k}} \sin(2\theta_{\vec{k}})(n_{\vec{k}}^+ - n_{\vec{k}}^-), \tag{6.130}$$

where $n_{\vec{k}}^{\pm}$ are the occupation numbers: $n_{\vec{k}}^{\pm} = 1$ when $\vec{k} \in F^{\pm}$. Hint: We can write the expectation value of s_i^+ as

$$\langle s_i^+ \rangle = \langle c_{i,\uparrow}^\dagger c_{i,\downarrow} \rangle = \frac{1}{N_s} \sum_{\vec{p}} e^{-i\vec{p}\cdot\vec{R}_i} \sum_{\vec{k}} \langle c_{\vec{k},\uparrow}^\dagger c_{\vec{k}+\vec{p},\downarrow} \rangle. \tag{6.131}$$

Note $c_{i,\sigma}^\dagger = \sqrt{\frac{1}{N_s}} \sum_{\vec{k}} e^{i\vec{k}\cdot\vec{R}_i} c_{\vec{k},\sigma}^\dagger$. Use the transformation equation (6.126) and $\langle a_{\vec{k},\pm}^\dagger a_{\vec{k}',\pm} \rangle = n_{\vec{k}}^{\pm} \delta_{\vec{k},\vec{k}'}$.

Exercise 6.22. Show

$$m_z = \frac{1}{2N_s} \sum_{\vec{k}} \cos(2\theta_{\vec{k}})(n_{\vec{k}}^+ - n_{\vec{k}}^-). \tag{6.132}$$

Hint: Use $c_{i,\sigma}^\dagger = \sqrt{\frac{1}{N_s}} \sum_{\vec{k}} e^{i\vec{k}\cdot\vec{R}_i} c_{\vec{k},\sigma}^\dagger$. Show that the expectation value of the z-component of spin is

$$\langle s_i^z \rangle = \frac{1}{2} \langle [c_{i,\uparrow}^\dagger c_{i,\uparrow} - c_{i,\downarrow}^\dagger c_{i,\downarrow}] \rangle$$

$$= \frac{1}{2N_s} \sum_{\vec{p}} e^{-i\vec{p}\cdot\vec{R}_i} \sum_{\vec{k}} \langle c_{\vec{k},\uparrow}^\dagger c_{\vec{k}+\vec{p},\uparrow} - c_{\vec{k},\downarrow}^\dagger c_{\vec{k}+\vec{p},\downarrow} \rangle. \tag{6.133}$$

Apply $\langle a_{\vec{k},\pm}^\dagger a_{\vec{k}',\pm} \rangle = n_{\vec{k}}^{\pm} \delta_{\vec{k},\vec{k}'}$.

The expectation value of the kinetic term is

$$T = \left\langle \sum_{\vec{k}\sigma} \epsilon_{\vec{k}} c_{\vec{k}\sigma}^{\dagger} c_{\vec{k}\sigma} \right\rangle$$

$$= \sum_{\vec{k}\in F^{+}} \left[\frac{\epsilon_{\vec{k}} + \epsilon_{\vec{k}+\vec{q}}}{2} + \left(\frac{\epsilon_{\vec{k}} - \epsilon_{\vec{k}+\vec{q}}}{2} \right) \cos(2\theta_{\vec{k}}) \right] n_{\vec{k}}^{+}$$

$$+ \sum_{\vec{k}\in F^{-}} \left[\frac{\epsilon_{\vec{k}} + \epsilon_{\vec{k}+\vec{q}}}{2} - \left(\frac{\epsilon_{\vec{k}} - \epsilon_{\vec{k}+\vec{q}}}{2} \right) \cos(2\theta_{\vec{k}}) \right] n_{\vec{k}}^{-}.$$

$$(6.134)$$

Here $n_{\vec{k}}^{\pm}$ are the occupation numbers of the new transformed particles created by $a_{\vec{k},\pm}$. By subtracting $-2(N_s U m_z^2 - N_s U m_q^2)$ from T and adding it to the total energy, we find that the total energy can be written as

$$E_{\text{tot}}(m_z, m_{\vec{q}}, \theta_{\vec{k}}, \vec{q}, F^{\pm}) = \sum_{\vec{k}\in F^{+}} E_{\vec{k}}^{+} n_{\vec{k}}^{+} + \sum_{\vec{k}\in F^{-}} E_{\vec{k}}^{-} n_{\vec{k}}^{-}$$

$$+ U n_{\text{f}}^2/4 + U m_z^2 + U m_{\vec{q}}^2, \qquad (6.135)$$

where

$$E_{\vec{k}}^{\pm} = \frac{\epsilon_{\vec{k}} + \epsilon_{\vec{k}+\vec{q}}}{2} \pm \left[\left(\frac{\epsilon_{\vec{k}} - \epsilon_{\vec{k}+\vec{q}}}{2} - U m_z \right) \cos(2\theta_{\vec{k}}) - U m_{\vec{q}} \sin(2\theta_{\vec{k}}) \right].$$

$$(6.136)$$

(Note that $E_{\vec{k}}^{\pm}$ are not true quasiparticle energies. See Eq. (6.121) and Sec. 8.4 for how the quasiparticle energy may be determined.)

Note that there are several variational parameters: the modulation vector $\vec{q}$, the angle $\theta_{\vec{k}}$,[h] the magnetizations m_z and $m_{\vec{q}}$, and the Fermi seas $F^{\pm}$. They can be determined from the minimization conditions. For example, minimizing E_{tot} with respect to $\cos(2\theta_{\vec{k}})$ gives

$$\cos(2\theta_{\vec{k}}) = \frac{(\epsilon_{\vec{k}} - \epsilon_{\vec{k}+\vec{q}})/2 - U m_z}{\sqrt{((\epsilon_{\vec{k}} - \epsilon_{\vec{k}+\vec{q}})/2 - U m_z)^2 + (U m_q)^2}}. \qquad (6.137)$$

---

[h]Note that $\theta_{\vec{k}}$ is not a geometric angle of $\vec{k}$. It is defined by the canonical transformation defined in Eq. (6.126).

> **Exercise 6.23.** Show Eq. (6.137).

Inserting this result into Eq. (6.136) we find

$$E_{\vec{k}}^{\pm}(m_z, m_{\vec{q}}) = \frac{\epsilon_{\vec{k}} + \epsilon_{\vec{k}+\vec{q}}}{2} \pm \sqrt{\left(\frac{\epsilon_{\vec{k}} - \epsilon_{\vec{k}+\vec{q}}}{2} - U m_z\right)^2 + U^2 m_{\vec{q}}^2}.$$

$$(6.138)$$

Note that when $m_z \neq 0$ and $m_{\vec{q}} = 0$, the gap vanishes and itinerant ferromagnetism appears. Minimizing E_{tot} with respect to m_z and $m_{\vec{q}}$, respectively, reproduces Eqs. (6.130) and (6.132) (m_z, $m_{\vec{q}}$, and $\cos(2\theta_{\vec{k}})$ are independent variational parameters). In addition, the Fermi surfaces may be determined by minimizing E_{tot}. Finally, minimizing E_{tot} with respect to m_z and $m_{\vec{q}}$ and expanding $E_{\vec{k}}^{\pm}$ to linear order in m_z and $m_{\vec{q}}$ (see Exercise 6.24), one can determine the Stoner's criteria for a ferromagnetic instability

$$U\chi(0) = 1 \tag{6.139}$$

and spin density wave instability

$$U\chi(\vec{q}) = 1. \tag{6.140}$$

The function $\tilde{\chi}(\vec{q})$ is defined in Exercise 6.24. From the spin density wave instability, Eq. (6.140), one can **determine the value of the wave vector $\vec{q}$**.

> **Exercise 6.24.** Show Eqs. (6.139) and (6.140). Hint: Using the total energy equation (6.135) show
>
> $$\frac{dE}{dm_{\vec{q}}} = \frac{dE^0}{dm_{\vec{q}}} - 2U m_{\vec{q}} = 0,$$
>
> $$\frac{dE}{dm_z} = \frac{dE^0}{dm_z} - 2U m_z = 0, \tag{6.141}$$
>
> where $E^0 = \sum_{\vec{k}\in F+} E_{\vec{k}}^+ n_{\vec{k}}^+ + \sum_{\vec{k}\in F-} E_{\vec{k}}^- n_{\vec{k}}^-$. Use
>
> $$\frac{dE^0}{dm_{\vec{q}}}\Big|_{m_z=0} = 2U^2 m_{\vec{q}} \chi(\vec{q}),$$
>
> $$\tag{6.142}$$
>
> $$\frac{dE^0}{dm_z}\Big|_{m_{\vec{q}}=0} = 2U^2 m_z \chi(0),$$

and

$$\chi(\vec{q}) = \sum_{\vec{k}} \frac{n_{\vec{k}+\vec{q}} - n_{\vec{k}}}{\epsilon_{\vec{k}} - \epsilon_{\vec{k}+\vec{q}}},$$

$$\chi(0) = \sum_{\vec{k}} \frac{dn_{\vec{k}}}{d\epsilon_{\vec{k}}} = g(E_F).$$

(6.143)

Here $g(E_F)$ is the DOS per spin multiplied by the volume of the system (the DOS, $D(E)$, is defined as number of states per energy per volume). Hint: Consult Ref. [10]. Use that when $m_{\vec{q}} = 0$ we have $E_{\vec{k}}^{+} = \epsilon_{\vec{k}} + Um_z$ and $E_{\vec{k}}^{-} = \epsilon_{\vec{k}+\vec{q}} - Um_z$. In this case the volume of $+$ and $-$ Fermi seas are equal. Note that the number of electrons in the energy interval ΔE is $\Delta N = g(E_F)\Delta E$.

6.14. Time-Dependent Hartree–Fock Method⋆

How does an interacting electron system respond to a time-dependent perturbation? Hartree–Fock approximation provides a conceptually simple scheme to describe such a response. One introduces **time-dependent** self-consistent Hartree and Fock fields. Using these fields one can compute response functions, for example the susceptibility, from which the collective modes (e.g., plasmons) can be found. Note that the time-dependent Hartree–Fock method describes collective modes consisting of two-particle excitations, i.e., electron–hole pair excitations across the Fermi surface.

• Time-Dependent Density Matrix

Consider a non-interacting system with a set of eigenstates $\{|i\rangle\}$. Suppose one applies a time-dependent perturbation $V(t) \sim e^{i\omega t}$. The following result for the change of the density matrix (see Exercise 6.25) is very useful

$$\delta\rho_{ij}(\omega) = \left[\frac{n(\epsilon_i) - n(\epsilon_j)}{\epsilon_i - \epsilon_j - \hbar(\omega + i\eta)} \right] V_{ij}(\omega),$$

(6.144)

where $\langle i|V(t)|j\rangle = V_{ij}e^{i\omega t}$ and $n(\epsilon_i)$ is the Fermi distribution function.

Exercise 6.25. Derive the Fourier transform of the time-dependent density matrix given by Eq. (6.144). Answer: We start from the von Neumann equation (C.8)

$$i\hbar \frac{\partial \rho}{\partial t} = [H, \rho]. \tag{6.145}$$

The Hamiltonian is $H + V(t)$, where $V(t)$ is the time-dependent perturbation. The density operator can be divided into two parts $\rho(t) = \rho_0 + \delta\rho(t)$. From these we get

$$i\hbar \frac{\partial}{\partial t} \langle i|\delta\rho(t)|j\rangle = \langle i|[H_0, \delta\rho(t)]|j\rangle + \langle i|[V(t), \rho_0]|j\rangle$$

$$= (\epsilon_i - \epsilon_j)\langle i|\delta\rho(t)|j\rangle + [n_j - n_i]\langle i|V(t)|j\rangle. \tag{6.146}$$

Note that $\rho_0|i\rangle = n_i|i\rangle$. Using $\delta\rho(t) = \delta\rho(\omega)e^{-i\omega t}$ and $V_{ij}(t) = V_{ij}(\omega)e^{-i\omega t}$ we find the desired result of Eq. (6.144). Consult Ref. [13].

When electron interactions are present, ϵ_i should be replaced by the **renormalized Hartree–Fock** quasiparticle energy $\tilde{\epsilon}_i$, which includes the self-energy corrections containing Hartree and exchange self-energies. The Hartree self-energy is given by

$$\langle i'|V_{\mathrm{H}}|i\rangle = \sum_\alpha n_\alpha \langle i'\alpha|V|i\alpha\rangle = \sum_{jj'} \langle i'j'|V|ij\rangle \rho_{j'j}, \tag{6.147}$$

where the Hartree–Fock density matrix element is

$$\rho_{j'j} = \sum_\alpha n_\alpha \langle j'|\alpha\rangle\langle\alpha|j\rangle. \tag{6.148}$$

Likewise, the exchange self-energy can be written as

$$\langle i'|V_{\mathrm{X}}|i\rangle = -\sum_{jj'} \langle i'j'|V|ji\rangle \rho_{j'j}. \tag{6.149}$$

• Time-Dependent Hartree–Fock Potentials

In the presence of electron interactions, the effect of a time-dependent potential $V_{\mathrm{ext}}(x, t) = V_{\mathrm{ext}}e^{i(qx-\omega t)}$ also gets renormalized. The resulting self-consistent Hartree–Fock potential is

$$\langle i'|\delta H_{\mathrm{HF}}(t)|i\rangle = \langle i'|\delta V_{\mathrm{H}}(t)|i\rangle + \langle i'|\delta V_{\mathrm{X}}(t)|i\rangle + V_{\mathrm{ext}}(t)\langle i'|e^{iqr}|i\rangle. \tag{6.150}$$

The change in the Hartree potential is

$$\langle i'|\delta V_{\mathrm{H}}(t)|i\rangle = \sum_{jj'}\langle i'j'|V|ij\rangle \delta\rho_{j'j}(t). \tag{6.151}$$

The change in the exchange potential is

$$\langle i'|\delta V_{\mathrm{X}}(t)|i\rangle = -\sum_{jj'}\langle i'j'|V|ji\rangle \delta\rho_{j'j}(t). \tag{6.152}$$

Note that $\delta V_{\mathrm{H}}(t) \sim \delta V_{\mathrm{X}}(t) \sim e^{-i\omega t}$ because $\delta\rho_{j'j}(t) \sim e^{-i\omega t}$. Plugging these results into the change of the density matrix (6.144), we find

$$\delta\rho_{ii'} = \frac{n_i - n_{i'}}{\tilde\epsilon_i - \tilde\epsilon_{i'} - (\omega + i\eta)}\langle i|\delta H_{\mathrm{HF}}(\omega)|i'\rangle$$

$$= \frac{n_i - n_{i'}}{\tilde\epsilon_i - \tilde\epsilon_{i'} - (\omega + i\eta)}(\langle i|\delta V_{\mathrm{H}}|i'\rangle + \langle i|\delta V_{\mathrm{X}}|i'\rangle + V_{\mathrm{ext}}\langle i|e^{iqr}|i'\rangle).$$

$$\tag{6.153}$$

This equation gives the following inhomogeneous equation:

$$\sum_{jj'}((\tilde\epsilon_i - \tilde\epsilon_{i'} - \omega)\delta_{ij}\delta_{i'j'} - (n_i - n_{i'})[\langle i'j'|V|ij\rangle - \langle i'j'|V|ji\rangle])\delta\rho_{j'j}$$

$$= (n_i - n_{i'})V_{\mathrm{ext}}\langle i|e^{iqr}|i'\rangle. \tag{6.154}$$

For translationally invariant systems, i and i' stand for wave vectors $\vec{k}$ and $\vec{k}'$. These states are connected through e^{iqx} (this implies that j and j' are also related). The energy of the collective modes are the eigenvalues of the homogeneous part of this equation. For the application of this time-dependent Hartree–Fock method to integer quantum Hall systems, see Refs. [14, 15].

An alternative method to compute the dispersion of a collective mode is as follows. Consider a translationally invariant electron gas. A collective mode consists of a linear combination of particle-hole excitations

$$|\Psi_{\vec{q}}\rangle = \sum_{\vec{k}'} A_{\vec{k}'} c^{\dagger}_{\vec{k}'+\vec{q}} c_{\vec{k}'}|\Psi_0\rangle = \sum_{\vec{k}'} A_{\vec{k}'}|\psi_{\vec{k}'}\rangle, \tag{6.155}$$

where $|\Psi_0\rangle$ is the **Hartree–Fock** ground state and $|\psi_{\vec{k}'}\rangle$ represents an excited state consisting of a hole with wave vector $\vec{k}'$ and an electron with wave vector $\vec{k}' + \vec{q}$. The eigenvalues of the Hamiltonian matrix

$$H_{\vec{k}\vec{k}'} = \langle \psi_{\vec{k}}|H|\psi_{\vec{k}'}\rangle \tag{6.156}$$

give the energy dispersion $E(\vec{q})$ of the collective mode. Note that H is the full Hamiltonian and not the Hartree–Fock Hamiltonian. The matrix elements are [16]

$$\langle \psi_{\vec{k}} | H | \psi_{\vec{k}'} \rangle = (\tilde{\epsilon}_{\vec{k}+\vec{q}} - \tilde{\epsilon}_{\vec{k}}) \delta_{\vec{k}\vec{k}'} - [\langle \vec{k}+\vec{q}, \vec{k}' | V | \vec{k}' + \vec{q}, \vec{k} \rangle$$
$$- \langle \vec{k}+\vec{q}, \vec{k}' | V | \vec{k}, \vec{k}' + \vec{q} \rangle]. \tag{6.157}$$

Note that for off-diagonal elements, $\vec{k} \neq \vec{k}'$, the first term is absent.

Exercise 6.26. Evaluate the matrix elements, Eq. (6.156), for a spinless one-dimensional electron gas.

Further reading. For a derivation of the time-dependent Hartree–Fock method using Feynman diagrams, see Ref. [15]. It is instructive to compare the Feynman diagram method to that of density matrix. Let us first analyze the Feynman diagrams. The response function is depicted in Fig. 6.13(a). The self-consistent Green's function of Fig. 6.13(b) is computed in the Hartree–Fock approximation. The vertex corrections are depicted in Fig. 6.13(c). It has two contributions from electron interactions. The third diagram represents excitonic effects (exchange local-field corrections). The fourth diagram describes depolarization effects (Coulomb local-field corrections). The excitonic effect corresponds to the term $-\langle i'j' | V | ji \rangle$ in the time-dependent Hartree–Fock equation (6.154). This term represents the attractive interaction between an electron and a hole. The depolarization effect corresponds to the term $\langle i'j' | V | ij \rangle$ in Eq. (6.154), and represents the well-known random phase approximation. The two-dimensional magnetoplasmon excitation energy $\hbar\omega(q)$ may be calculated using the time-dependent Hartree–Fock method. In the presence of translational symmetry or a parabolic potential, the many-body corrections cancel each other in the limit $q \to 0$ (this is called Kohn's theorem). This is because, in this limit, the depolarization effect vanishes and the excitonic and self-energy corrections cancel exactly. As a result, $\hbar\omega(q)$ approaches the cyclotron energy $\hbar\omega_c$, consistent with Kohn's theorem.

It should be emphasized that if the Hartree and exchange corrections are included in the calculation of the self-energy then excitonic and depolarization must be also included in the calculation of the vertex function.[i]

---

[i]In a homogenous system the Hartree self-energy and the contribution from the uniform positive background charge cancel each other, see Exercise 6.10.

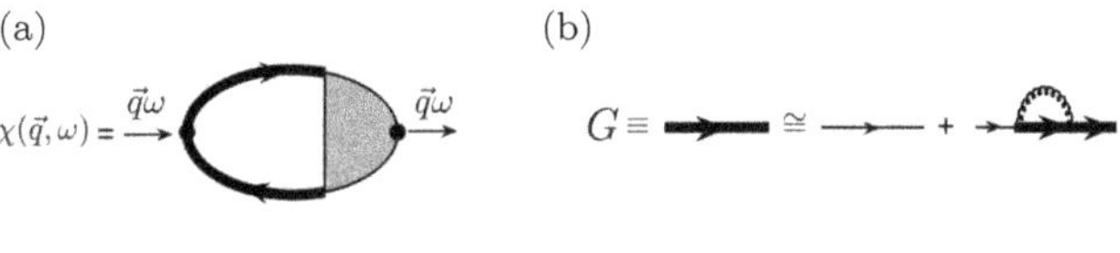

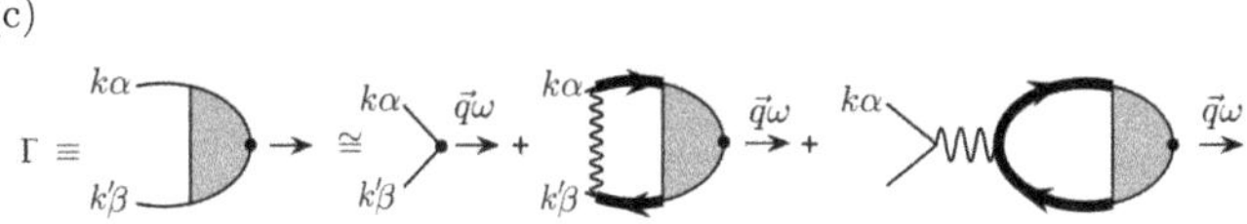

Fig. 6.13. Diagrammatic representation of the time-dependent Hartree–Fock method. (a) Response function χ consists of the self-consistent Greens functions (thick lines) and the vertex corrections. (b) The self-consistent Green's functions is computed using unscreened Coulomb interaction (wavy line). The thin lines represent the bare Green's function. (c) The vertex corrections consist of excitonic (the second diagram) and depolarization (the third diagram) effects. Note these corrections must be computed using the self-consistent Hartree–Fock Green's functions.

If only the Hartree self-energy correction is included then only the depolarization vertex correction should be included. In path integral approach of many-body physics, the random phase approximation corresponds to an approximate scheme which includes Gaussian fluctuations near the saddle point (an example is given in connection with the Anderson–Higgs mechanism, see Sec. 8.8).

Bibliography

[1] R. M. Martin, L. Reining, and D. M. Ceperley, *Interacting Electrons: Theory and Computational Approaches* (Cambridge University Press, Cambridge, 2016).

[2] N. W. Ascroft and N. David Mermin, *Solid State Physics* (Thomson Learning, London, 1976). •

[3] Y. V. Nazarov and J. Danon, *Advanced Quantum Mechanics* (Cambridge University Press, Cambridge, 2013). •

[4] R. D. Mattuck, *A Guide to Feynman Diagrams in the Many-Body Problem* (Dover Publications, Mineola, 1992). •

[5] J. W. Negele and H. Orland, *Quantum Many-Particle Systems* (CRC Press, Boca Raton, 1998).

[6] A. K. Kerman, S. Levit, and T. Troudet, Functional integral mean field expansions for nuclear many fermion systems, *Ann. Phys.* **148**, 436 (1983).

[7] E. Merzbacher, *Quantum Mechanics* (Wiley, Hoboken, 1997). •

[8] R. T. Scalettar, "An introduction to the Hubbard Hamiltonian," *Quantum Materials: Experiments and Theory Modeling and Simulation Vol. VI*, edited by E. Pavarini, E. Koch, J. van den Brink, and G. Sawatzky, (Forschungszentrum Julich, Julich, 2016). •

[9] M. Kastner, Artificial atoms, *Phys. Today* **46**, 1, 24 (1993). •

[10] A. Auerbach, *Interacting Electrons and Quantum Magnetism* (Springer, New York, 1994).

[11] F. Gebhard, *The Mott Metal-Insulator Transition* (Springer, New York, 2013).

[12] S. Brown and G. Grüner, Charge and spin density waves, *Sci. American* **270**, 4, 50 (1994). •

[13] H. Ehrenreich and M. H. Cohen, Self-consistent field approach to the many-electron problem, *Phys. Rev.* **115**, 786 (1959).

[14] A. H. MacDonald, Hartree–Fock approximation for response functions and collective excitations in a two-dimensional electron gas with filled Landau levels, *J. Phys. C: Solid State Phys.* **18**, 1003 (1985).

[15] C. Kallin and B. I. Halperin, Excitations from a filled Landau level in the two-dimensional electron gas, *Phys. Rev. B* **30**, 5655 (1984).

[16] D. S. Koltun and J. M. Eisenberg, *Quantum Mechanics of Many Degrees of Freedom* (Wiley, Hoboken, 1988).

Chapter 7

Bosonization*

Weinberg's Laws of Progress in Theoretical Physics:

- First Law: "The conservation of Information" (You will get nowhere by churning equations)
- Second Law: "Do not trust arguments based on the lowest order of perturbation theory"
- Third Law: "You may use any degrees of freedom you like to describe a physical system, but if you use the wrong ones, you'll be sorry!"

S. Weinberg [1]

Consider the electron susceptibility[a] at frequency ω and wave vector $\vec{q}$

$$\chi(\vec{q}, \omega) = \sum_{\vec{k}} \frac{n_{\vec{k}} - n_{\vec{k}+\vec{q}}}{\omega + \epsilon_{\vec{k}} - \epsilon_{\vec{k}+\vec{q}} + i\eta}. \tag{7.1}$$

(Here $n_{\vec{k}}$ is the occupation number, $\epsilon_{\vec{k}}$ is the dispersion energy, $i\eta$ is a convergence factor, and $\hbar = 1$.) In a one-dimensional free electron gas, the Fermi surface consists of only two points. This leads to a logarithmic singularity of the static susceptibility at $q = 2k_F$, see Fig. 7.1 (k_F is the Fermi wave vector). This suggests that our naive zeroth-order picture of independent electrons is incorrect in one dimension.

The main physics behind this breakdown is that the creation of an electron amounts to creating numerous bosonic density excitations (plasmons) in interacting one-dimensional systems. In this chapter, we briefly introduce bosonization method in which fermion degrees of freedom are

---

[a]The susceptibility relates the induced charge to the full physical potential, see Ref. [2].

"

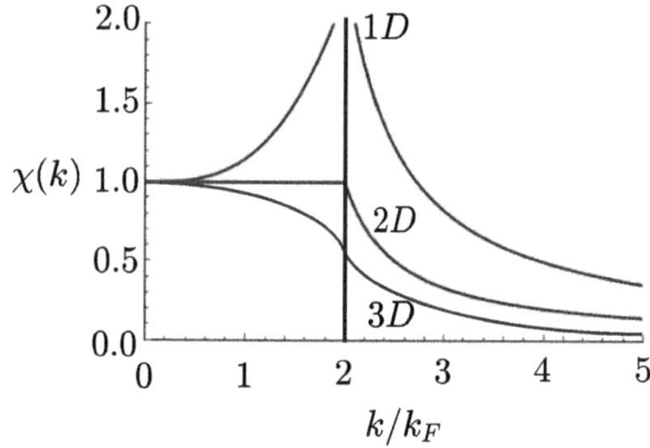

Fig. 7.1. Static susceptibilities in one-, two- and three-dimensional systems. Note that susceptibility measures the response of charge density in the electron gas to the total electric potential, see Ref. [2].

described by boson degrees of freedom. It is valid for linearized spectrum near the Fermi wave vector.[b] Unlike the Hartree–Fock method, bosonization approach is essentially exact because quantum fluctuations are properly included. Bosonization theory is based on the density operators that obey the Kac–Moody commutation relations.

Later in Chapter 12, we will see that this approach is vital in describing one-dimensional edge excitations of quantum Hall systems. In this chapter, we give only a short introduction to the very basic ideas of bosonization and study **spinless** fermions. (We will consider in Sec. 12.5 how spinful electrons may display spin–charge separation.) A good introduction of bosonization is given in Refs. [4, 5] and a short historical account of the development of bosonization can be found in Ref. [6]. For in-depth treatment of this important method, see Refs. [6–8].

7.1. Tomonaga and Luttinger Models

In most condensed matter systems at low temperatures, only small wave vector and energy excitations near the Fermi energy are important, and

---

[b]The results presented in this chapter are based on a linearization of the energy spectrum. However, Haldane showed that bosonization theory is generally applicable to low energy physics of interacting one-dimensional systems and coined the term Luttinger liquid for these systems [3]. In his generalization of the bosonization formula the density has oscillating terms containing all the higher harmonics (see also Sec. 12.4). Moreover, a spinless fermion model under periodic boundary conditions contains two types of leftand right-moving topological excitations in addition to bosonic modes.

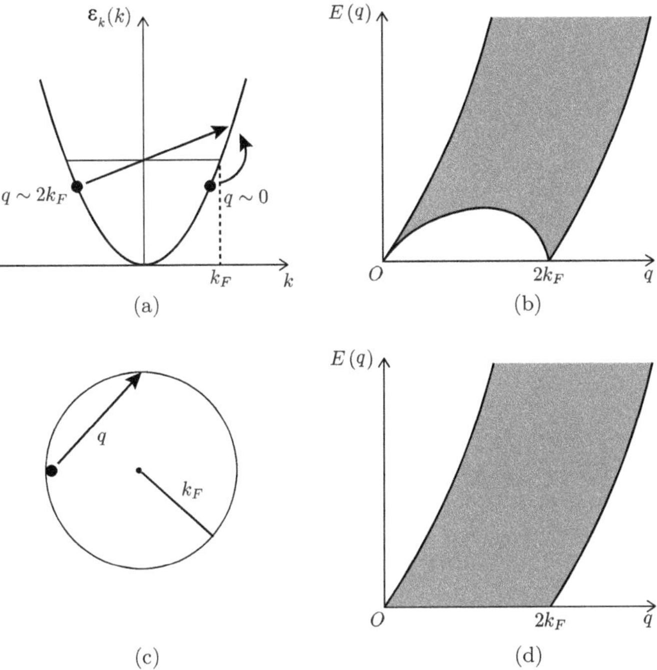

Fig. 7.2. (a) Wave vector transfer $q \approx 0$ or $2k_F$ in electron–hole pair excitations in one-dimensional systems. (b) In the low energy and low wave vector limits, these excitations, depicted in Figs. 7.3, form well-defined intra-band density modes. (c) The wave vector transfer in electron–hole pair excitations can take continuous values in two-dimensional systems. (d) These pair excitations do not lead to well-defined low-energy density excitations. (There can be plasmon modes but they do not have a linear energy dispersion.)

they form the basis for bosonization. The full electron–hole pair excitation spectra in one- and two dimensional systems are shown in Fig. 7.2. We see that, in one-dimensional systems, a well-defined collective mode can be formed when the wave vector transfer of electron–hole pair excitations is $q \approx 0$. We will try to rewrite the Hamiltonian in terms of these boson modes. In contrast, in two-dimensional systems, the wave vector transfer of electron–hole pair excitations takes continuous values and low-energy collective modes with a linear dispersion are ill-defined.[c]

In the Tomonaga model of one-dimensional systems, one linearizes the energy spectrum: $\epsilon_k = v_F k$ for $k > 0$ and $\epsilon_k = v_F k$ for $k < 0$, see Figs. 7.3

---

[c] However, in the presence of electron interactions, a plasmon mode with energy $\propto \sqrt{q}$ can be formed.

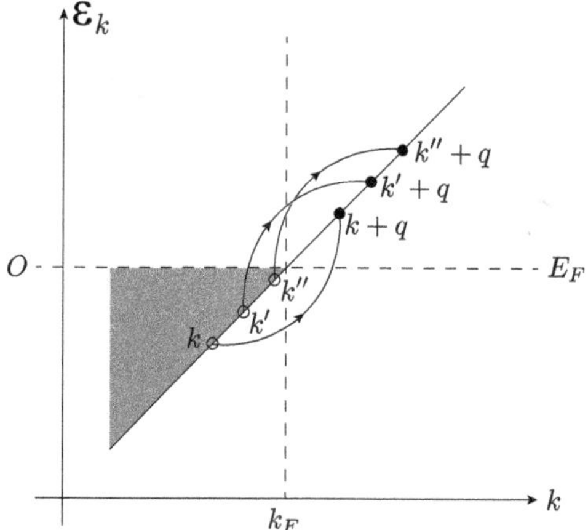

Fig. 7.3.   In the one-dimensional Tomonaga model, all the electron–hole pair excitations near the Fermi energy with a given wave vector q have **same** energy, provided that a linear energy dispersion $\epsilon_k \approx \hbar v_F (k - k_F)$ is used. (v_F and k_F are the Fermi velocity and Fermi wave vector.) In this model the electron energy is $\epsilon_k > 0$.

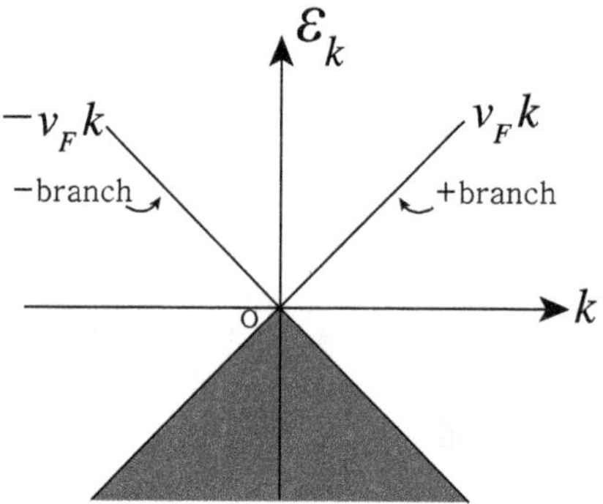

Fig. 7.4.   In the Luttinger model two branches of fermions originate from the linearized spectrum (the Fermi wave vector is $k_F = 0$). Only states below the Fermi energy ($E_F = 0$) are occupied.

(v_F is the Fermi velocity). In the Luttinger model (see Fig. 7.4), the Fermi energy is set to zero $\epsilon_k = 0$ and the fermions are separated into two linear energy branches $\alpha = r, l$ (or equivalently $\alpha = +1, -1$).

It is rather difficult to solve the Tomonaga–Luttinger model using the fermion picture. One has to resum the various divergences in the perturbation theory or resort to the renormalization group method [7]. Mattis and Lieb showed that electrons of the Tomonaga–Luttinger liquid model could be treated by the bosonization method. We will see in the next section that bosonization is a powerful technique that can solve the problem exactly.

7.2. Bosonic Excitations of One-Dimensional Free Fermi Gas

Let us first show that electron–hole excitations of a free Fermi gas are bosons. The Hamiltonian is

$$H_0 = \sum_k \epsilon_k c_k^\dagger c_k, \tag{7.2}$$

where k is wave vector and $c_k^\dagger$ is the electron creation operator. One can write the fermion field operator as

$$\psi(x) = \frac{1}{\sqrt{L}} \sum_k e^{ikx} c_k \approx \frac{1}{\sqrt{L}} \sum_{k \approx -k_F} e^{ikx} c_k + \frac{1}{\sqrt{L}} \sum_{k \approx k_F} e^{ikx} c_k, \tag{7.3}$$

where L is the system length.

In the Luttinger model, the total density operator is[d]

$$\rho_q \approx \rho_{1,q} + \rho_{-1,q}, \tag{7.4}$$

where the density operator of a branch α is defined as

$$: \rho_{\alpha,q} := : \sum_k c_{\alpha,k}^+ c_{\alpha,k+q} :, \tag{7.5}$$

with normal ordering

$$: AB := AB - \langle 0|AB|0 \rangle. \tag{7.6}$$

This is equivalent to placing destruction operators to the right of creation operators: in the case of fermions we have $: c_\alpha^\dagger c_{\alpha'} := c_\alpha^\dagger c_{\alpha'}$ and $: c_{\alpha'} c_\alpha^\dagger :$ $= -c_\alpha^\dagger c_{\alpha'}$.[e] The density operators satisfy the Kac–Moody commutation

---

[d]The density components with $2k_F$ and higher harmonics are ignored.
[e]Normal ordering is sometimes needed to avoid meaningless infinite terms. Consider the boson Hamiltonian

$$H = \frac{1}{2} \sum_k E_k (b_k b_k^\dagger + b_k^\dagger b_k). \tag{7.7}$$

Normal ordering of this Hamiltonian gives the sensible result

$$H = \frac{1}{2} \sum_k E_k b_k^\dagger b_k = \frac{1}{2} \sum_k E_k n_k. \tag{7.8}$$

Note that for bosons $: b_{k'} b_k^\dagger := b_k^\dagger b_{k'}$, in contrast to the case of fermions.

relations

$$[\rho_{\alpha,q}, \rho_{\alpha',-q'}] = \alpha \delta_{\alpha,\alpha'} \delta_{q,q'} \frac{qL}{2\pi}. \tag{7.9}$$

For same branches, $\alpha = \alpha'$ it is trivial to show this result. However, $\alpha \neq \alpha'$ is non-trivial and one must use normal ordering (see Exercise 7.1).

Exercise 7.1. Show the commutation relation (7.9) for the same branches $\alpha = \alpha'$. Answer:

$$: [\rho_{\alpha,q}, \rho_{\alpha,-q'}] : = : \sum_p (c^\dagger_{\alpha,p+q-q'} c_{\alpha,p} - c^\dagger_{\alpha,p-q'} c_{\alpha,p-q}) :$$

$$= \sum_p (\langle 0|c^\dagger_{\alpha,p+q-q'} c_{\alpha,p}|0\rangle - (\langle 0|c^\dagger_{\alpha,p-q'} c_{\alpha,p-q}|0\rangle))$$

$$= \delta_{q,q'} \sum_p (\langle 0|c^\dagger_{\alpha,p} c_{\alpha,p}|0\rangle - \langle 0|c^\dagger_{\alpha,p-q} c_{\alpha,p-q}|0\rangle).$$

$$\tag{7.10}$$

This result is ambiguous. This can be avoided by adopting the Luttinger model where the linear spectrum is assumed for all values of k, see Fig. 7.4. We shift the Fermi energy E_F to zero and set $k_F = 0$. Note that right moving electrons (with velocity $v > 0$) in the Fermi sea have negative wave vectors $k < 0$ and left moving electrons ($v < 0$) have positive wave vectors $k > 0$. In the Luttinger model, the summation interval in the first term is $[-\infty, k_F]$ while the second term is $[-\infty, k_F + q]$. So the final result is given by the difference $q \times \frac{L}{2\pi}$.

The free-boson Hamiltonian is

$$H_B = \frac{2\pi v_F}{L} \sum_{q>0,\alpha} : \rho_{\alpha,q} \rho_{\alpha,-q} :, \tag{7.11}$$

where

$$: \rho_{\alpha,q} \rho_{\alpha,-q} : \equiv \begin{cases} \rho_{R,-q} \rho_{R,q} & \text{if } \alpha = R, \\ \rho_{L,q} \rho_{L,q} & \text{if } \alpha = L. \end{cases} \tag{7.12}$$

From the commutation relations of density operators, we find the following commutation relation

$$[H_B, \rho_{\alpha,q}] = v_F \alpha q \rho_{\alpha,q}. \tag{7.13}$$

Here the density operators are normal ordered, i.e., $\rho_{R,q}$ and $\rho_{L,q}$ are placed to the right because they are destruction operators

$$\rho_{L,q<0}|0\rangle = 0, \ \rho_{R,q>0}|0\rangle = 0, \tag{7.14}$$

where $|0\rangle$ is the non-interacting ground state. We will use H_B in place of H_0. This procedure for forming H_B is called the Sugawara construction. Only deviations from the density of ground state are meaningful in this construction. As far as the **density fluctuations** are concerned, these two Hamiltonians are equivalent. Note that the velocity of the boson mode of a free Fermi gas is just the Fermi velocity v_F. This type of Hamiltonian is equally applicable to the left and right edges of the integer quantum Hall state, see Fig. 12.7.

7.3. Collective Modes of Interacting Electrons

Let us show that excitations of an interacting spinless electron gas may also be described as bosons. The interaction term is

$$V = \frac{1}{2L} \sum_{kk'q} V(q) c^\dagger_{k-q} c^\dagger_{k'+q} c_{k'} c_k. \tag{7.15}$$

In terms of density operators, one may rewrite this interaction operator as follows

$$V =\ :\frac{1}{2L} \sum_{\alpha,q} [g_2 \rho_{\alpha,q}\rho_{-\alpha,-q} + g_4 \rho_{\alpha,q}\rho_{\alpha,-q}]\ :, \tag{7.16}$$

where the g_2 term describes the inter branch electron–hole excitations and the g_4 term describes the intra branch electron–hole excitations, see Fig. 7.5.

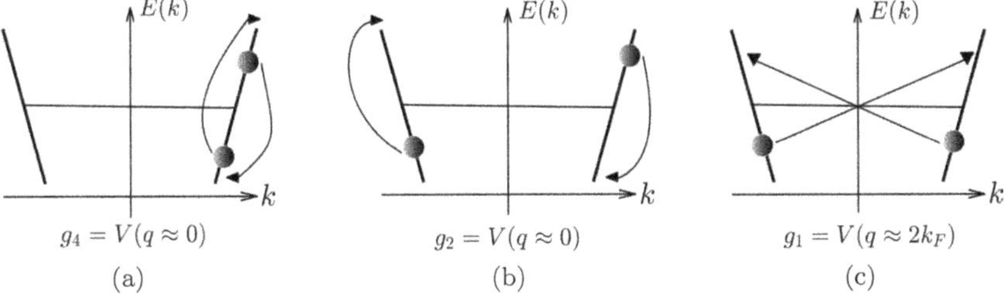

Fig. 7.5. Two particle scattering processes and excitations across the Fermi level are depicted. Process (c) represents a backscattering. For a spinless electrons processes labeled g_1 and g_2 are identical. However, for spinful electrons they can be different.

Let us define new operators for $q > 0$. For the left branch, we define the following operators:

$$b_{-q} = \frac{1}{\sqrt{L_q}}\rho_{l,-q}, \quad b^\dagger_{-q} = \frac{1}{\sqrt{L_q}}\rho_{l,q}, \tag{7.17}$$

where $L_q = qL/2\pi$. For the right branch,

$$b_q = \frac{1}{\sqrt{L_q}}\rho_{r,-q}, \quad b^\dagger_q = \frac{1}{\sqrt{L_q}}\rho_{r,q}. \tag{7.18}$$

> **Exercise 7.2.** Show that these operators b_q and $b^\dagger_q$ satisfy the Bose commutation relations. Use the commutation relation given in Eq. (7.9).

Inverting these operators, for example, $\rho_{rq} = (\frac{qL}{2\pi})^{1/2}b^\dagger_q$, we can write the interaction equation (7.16) as follows:

$$V =\, : \frac{1}{2\pi}\sum_{q>0}(b_q, b_{-q})v(q)\begin{bmatrix}b^\dagger_q \\ b^\dagger_{-q}\end{bmatrix} :, \tag{7.19}$$

where the 2×2 matrix is

$$v(q) = q\begin{pmatrix} g_4 & g_2 \\ g_2 & g_4 \end{pmatrix}. \tag{7.20}$$

Diagonalization of the total Hamiltonian matrix of $H_B + V$ gives two new renormalized boson modes. The renormalized velocity is higher for repulsive interactions and lower for attractive interactions. We will apply this method to find edge modes of quantum Hall liquids in Chapter 12.

Further reading. What about g_1 processes (see Fig. 7.5)? They are more complicated to analyze. In the presence of spin rotation invariance, there are two cases, $g_1 > 0$ and $g_1 < 0$. For $g_1 > 0$, various massless phases are possible. However, for attractive interactions with $g_1 < 0$, the spin modes become massive. When spin rotation invariance is broken, there are two processes depending on whether the spins σ and σ' of the interacting fermions are equal ($g_{1\parallel}$) or opposite ($g_{1\perp}$). Even in this case, there can be massive phases for $g_{1\parallel} < 0$ and $g_{1\perp} > 0$ or $g_{1\parallel} < 0$ and $g_{1\perp} < 0$. See Ref. [7] for details (in this book we do not need to study these phases). There are also g_3 processes that we have not mentioned. They represent Umpklapp scattering processes in which the total momentum is only conserved up to the addition (or subtraction) of a reciprocal lattice vector (remember that only discrete translational invariance is present in a crystal). These processes are not important in the problems treated in this book and will be ignored.

7.4. Bosonization of a Fermion Field

Let us show that the electron field operator may be written in terms of boson operators. A fermion field can be written as

$$\psi_r(x) = \frac{1}{\sqrt{L}} \sum_k c_{r,k} e^{i(k+k_F)x}, \quad \psi_l(x) = \frac{1}{\sqrt{L}} \sum_k c_{l,k} e^{i(k-k_F)x}. \quad (7.21)$$

The dominant contributions stem from $k \sim 0$ (corresponding to wave vectors near k_F). We can write the commutation relation as

$$[b_q, \psi_r(x)] = -\frac{1}{\sqrt{L_q}} e^{-iqx} \psi_r(x), \quad [b_q^\dagger, \psi_r(x)] = -\frac{1}{\sqrt{L_q}} e^{iqx} \psi_r(x). \tag{7.22}$$

One can show that ψ_r can be written in terms of b_q and $b_q^\dagger$ (see Exercise 7.3)

$$\psi_r(x) e^{-ik_F x} = F_r e^{-\sum_{q>0} \frac{1}{\sqrt{L_q}} b_q^\dagger e^{-iqx}} e^{-\sum_{q>0} \frac{1}{\sqrt{L_q}} b_q e^{iqx}}. \quad (7.23)$$

Here one must attach the so-called Klein operators F_R and F_L, whose function is to destroy an electron. They obey the following commutation relations:

$$[F_\alpha, N_\alpha] = F_\alpha, \quad [F_\alpha^\dagger, N_\alpha] = -F_\alpha^\dagger, \tag{7.24}$$

where N_α is the total number of fermions on branch α. They are present because the exponential operators do not change fermion number. For processes conserving particle number, they can be omitted.

> **Exercise 7.3.** Show that the result of the fermion field (7.23) satisfies the commutation relation (7.22). Hint: Use the identity
>
> $$[b_{q'}, e^{\alpha b_q^\dagger}] = \delta_{q,q'} \alpha e^{\alpha b_q^\dagger}. \tag{7.25}$$
>
> See also Appendix E.

We will express fermion field operators in terms of **chiral phase operators**

$$\phi_r(x) = \frac{1}{L} \sum_{q \neq 0} \rho_r(q) \frac{e^{iqx}}{iq}, \quad \phi_l(x) = \frac{1}{L} \sum_{q \neq 0} \rho_l(q) \frac{e^{iqx}}{iq}. \tag{7.26}$$

It is easy to show that these operators satisfy

$$\nabla \phi_\alpha = 2\pi \rho_\alpha(x). \tag{7.27}$$

The Kac–Moody commutation relations (7.9) may be expressed in terms of the phase operators

$$\left[\frac{1}{2\pi} \frac{\partial \phi_\alpha(x)}{\partial x}, \phi_\beta(x') \right] = i\alpha \delta_{\alpha,\beta} \delta(x - x'). \tag{7.28}$$

Note that $\alpha = l$ and r have different signs -1 and 1, respectively. Then we use the identity

$$e^{\alpha b_q^\dagger + \gamma b_q} = e^{\alpha b_q^\dagger} e^{\gamma b_q + \alpha\gamma/2} = e^{\gamma b_q} e^{\alpha b_q^\dagger - \alpha\gamma/2} \tag{7.29}$$

and the relation between b_q and b_{-q} and ρ_α (given in Eqs. (7.17) and (7.18)) to show the desired representation of the fermion destruction operators. They can be written in terms of the chiral phase operators

$$\psi_r(x) = \frac{1}{2\pi\delta} F_r e^{i\phi_r(x) + ik_F x}, \quad \psi_l(x) = \frac{1}{2\pi\delta} F_l e^{i\phi_l(x) - ik_F x}. \tag{7.30}$$

In the expression $\sum_{q>0}[b_q, b_q^\dagger]/L_q$, the summation is replaced by the integral over $[\frac{2\pi}{L}, \infty]$. To suppress the divergence, a convergence factor $e^{-q\delta}$ is introduced, which gives δ in the denominator of Eq. (7.30). Time dependence may be included by replacing $ix \to i(x - vt)$ for the right moving electrons and by $ix \to i(x + vt)$ for the left moving electrons.[f] We will use these results in Secs. 12.3 and 12.4 to compute the electron Green's function.

Further reading. Let us give motivation behind Eq. (7.27) to gain physical understanding of the phase field $\phi_\alpha(x) = \phi_r(x)$ or $\phi_l(x)$ [7]. Below we will consider them as functions and not operators. Let us define a function $\tilde{\phi}_l(x)$ such that $\tilde{\phi}_l(x) = n$ (n are integers) when x is equal to position of the nth electron $x = x_n$ (see Fig. 7.6). It is easy to show that

$$\tilde{\rho}_l(x) = \sum_n \delta(x - x_n) = \sum_n |\nabla \tilde{\phi}_l(x)| \delta(\tilde{\phi}_l(x) - n). \tag{7.31}$$

---

[f]This is a consequence of the Lorentz invariance of the linear dispersion. Electron interactions may break the Lorentz invariance.

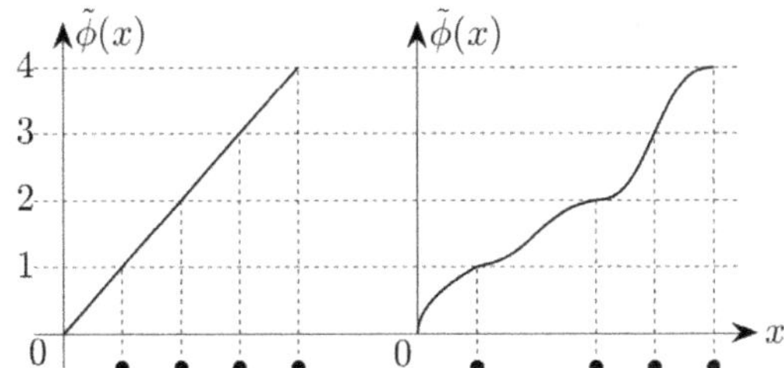

Fig. 7.6. A new quantity $\tilde{\phi}_l(x_n) = n$ can be used to determine the position of the nth Electron x_i. Left: Imagine electrons forming a periodic chain. Right: Electron positions deviate from the equilibrium positions.

Using

$$\delta(y - y') = \frac{1}{L}\sum_p e^{\frac{2\pi}{L}p(y-y')} \tag{7.32}$$

we find

$$\tilde{\rho}_l(x) = \frac{\nabla\tilde{\phi}_l(x)}{2\pi}\sum_p e^{ip\tilde{\phi}_l(x)}, \tag{7.33}$$

where p are integers.

Introducing a new phase field $\phi_l(x)$

$$\tilde{\phi}_l(x) = 2\pi\rho_0 x - \phi_l(x), \tag{7.34}$$

the density can be written as

$$\tilde{\rho}_l(x) = \left[\rho_0 - \frac{1}{2\pi}\nabla\phi_l(x)\right]\sum_p e^{ip(2\pi\rho_0 x - \phi_l(x))}. \tag{7.35}$$

(See Exercise 7.4.) The uniform density is ρ_0 while the average density over long distances is

$$\tilde{\rho}_l(x) \approx \rho_0 - \frac{1}{2\pi}\nabla\phi_l(x). \tag{7.36}$$

We thus find that the density modulation is given by

$$\rho_l(x) = \tilde{\rho}_l(x) - \rho_0 = -\frac{1}{2\pi}\nabla\phi_l(x). \tag{7.37}$$

Similarly we obtain

$$\rho_r(x) = \tilde{\rho}_r(x) - \rho_0 = +\frac{1}{2\pi}\nabla\phi_l(x). \tag{7.38}$$

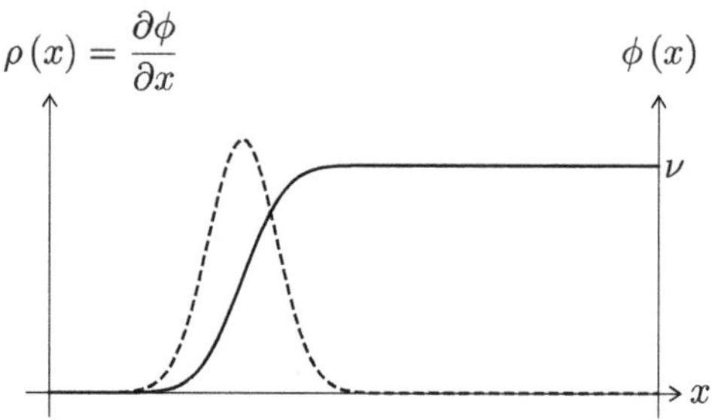

Fig. 7.7.　The phase field (solid line) and a kink excitation (dashed line).

The signs of $\rho_{l,r}(x)$ are chosen so that the commutation relation (7.28) is satisfied. In the bosonization approach, only density deviation from the uniform density of the ground state matters, as commented below Eq. (7.14). This means that whenever $\phi_\alpha(x)$ changes rapidly, a density modulation is present, see Fig. 7.7. This result is analogous to a poly-acetylene domain-wall soliton (kink) that is located at the position where dimerization abruptly changes, see Figs. 4.1 and 4.7.

Exercise 7.4.　Show Eq. (7.33) using

$$\delta(f(x)) = \sum_i \frac{1}{|f'(x_i)|} \delta(x - x_i), \tag{7.39}$$

where $f(x_i) = 0$. Use also

$$\delta(y - y') = \frac{1}{L} \sum_n e^{i \frac{2n\pi}{L}(y - y')} \tag{7.40}$$

and Poisson summation formula

$$\sum_{n=-\infty}^{\infty} s(n) = \sum_{k=-\infty}^{\infty} S(k), \tag{7.41}$$

where the Fourier transform is

$$S(k) = \int_{-\infty}^{\infty} dx\, s(x) e^{-i2\pi k x}. \tag{7.42}$$

7.5. Lagrangians and Conjugate Variables

The Kac–Moody commutation relation (7.28) suggests that $\frac{\partial \phi_r(x)}{\partial x}$ and $\phi_r(x)$ are **canonically conjugate** variables. From the Lagrangian $L = p\dot{q} - H(p,q)$ we find the following Lagrangian densities in the absence of interactions effects:

$$\mathcal{L}_r = -\frac{1}{4\pi}\frac{\partial \phi_r(x)}{\partial x}\left[\frac{\partial \phi_r(x)}{\partial t} + v_F \frac{\partial \phi_r(x)}{\partial x}\right],$$

$$\mathcal{L}_l = -\frac{1}{4\pi}\frac{\partial \phi_l(x)}{\partial x}\left[\frac{\partial \phi_l(x)}{\partial t} + v_F \frac{\partial \phi_l(x)}{\partial x}\right]. \tag{7.43}$$

The Hamiltonians are

$$H_r = \frac{v_F}{4\pi}\int dx \left(\frac{\partial \phi_r(x)}{\partial x}\right)^2, \tag{7.44}$$

$$H_l = \frac{v_F}{4\pi}\int dx \left(\frac{\partial \phi_l(x)}{\partial x}\right)^2, \tag{7.45}$$

which are of the form given in the Hamiltonian equation (7.11).

The total Lagrangian [4] is

$$\mathcal{L} = \mathcal{L}_r + \mathcal{L}_l. \tag{7.46}$$

We introduce new fields $\phi(x)$ and $\theta(x)$

$$\phi_l = \phi - \theta, \tag{7.47}$$

$$\phi_r = \phi + \theta. \tag{7.48}$$

With these new fields, we can write the total Lagrangian as

$$\mathcal{L} = \frac{1}{\pi}\frac{\partial \theta(x)}{\partial x}\frac{\partial \phi(x)}{\partial t} - \frac{v_F}{2\pi}\left[\left(\frac{\partial \theta(x)}{\partial x}\right)^2 + \left(\frac{\partial \phi(x)}{\partial x}\right)^2\right]. \tag{7.49}$$

Note that $\frac{1}{\pi}\frac{\partial \theta(x)}{\partial x}$ and ϕ are canonically conjugate variables. The fermion fields can be written as

$$\psi_r^\dagger(x) \sim e^{i(\phi(x)+\theta(x))}, \tag{7.50}$$

$$\psi_l^\dagger(x) \sim e^{i(\phi(x)-\theta(x))}. \tag{7.51}$$

Here we have suppressed the Klein factors. In the presence of interaction effects, the value of v_F in Eq. (7.49) renormalizes differently for the two terms in the bracket: $v_F K$ and v_F/K, respectively.

Further reading. For in-depth treatment of bosonization in interacting one-dimensional systems, see Refs. [6, 7].

Bibliography

[1] S. Weinberg, "Why the renormalization group is a good thing," in *Asymptotic Realms of Physics*, edited by. A. Guth, K. Huang, R.L. Jaffe (MIT Press, Cambridge, 1983).

[2] N. W. Ascroft and N. David Mermin, *Solid State Physics* (Thomson Learning, London, 1976). •

[3] F. D. M. Haldane, Luttinger liquid theory of one-dimensional quantum fluids. I. Properties of the Luttinger model and their extension to the general 1D interacting spinless Fermi gas, *J. Phys. C: Solid State Phys.* **14**, 2585 (1981); F. D. M. Haldane, Effective harmonic-fluid approach to low-energy properties of one-dimensional quantum fluids, *Phys. Rev. Lett.* **47**, 1840 (1981).

[4] C. L. Kane, "Lectures on bosonization," in *Boulder School 2005* (Boulder School for Condensed Matter and Materials Physics, 2005).

[5] M. P. A. Fisher and L. I. Glazman, "Transport in a one-dimensional Luttinger liquid," in *Mesoscopic Electron Transport: NATO ASI Series E*, edited by L. Kowenhoven, G. Schoen, and L. Sohn (Springer, New York, 1997). •

[6] A. O. Gogolin, A. A. Nersesyan, and A. M. Tsvelik, *Bosonization and Strongly Correlated Systems* (Cambridge University Press, Cambridge, 1998).

[7] T. Giamarchi, *Quantum Physics in One Dimension* (Oxford University Press, Oxford, 2003).

[8] Y. Kuramoto, *Quantum Many-Body Physics: A Perspective on Strong Correlations* (Springer, Tokyo, 2020).

PART 2

Topologically Ordered Phases

Chapter 8

Superconductivity

"A complex trait like intelligence is not only partly determined by many interacting genes, it changes across the lifespan as some genes are automatically turned on and some turned off. The most appreciated abilities in society, such as creativity and leadership, rarely fully present themselves early on."

Scott B. Kaufman

Superconductivity is a singular phenomenon arising from a macroscopic number of particles condensing into a single quantum state. A three-dimensional superconductor has several characteristic features of a topologically ordered phase. It displays vortices, spin–charge fractionalization [1], and ground state degeneracy in torus geometry. Although not generally appreciated, it is topologically ordered and cannot be described by a local order parameter [2]. In fact, the correct order parameter of a superconductor is **non-local** and there is **no** spontaneously broken symmetry [3,4], consistent with the general result that it is not possible to spontaneously break a local symmetry [5].

Before we study p-type superconductors with non-Abelian braiding statistics (which will be discussed in the next chapter) it is important to understand first the basic topological properties of s-wave superconductivity. In this chapter, we briefly mention some basic results and concepts of s-wave superconductivity and lay the groundwork for the study of p-type superconductors. There is a vast literature on s-wave superconductivity,[a] and here we will focus on subjects that are relevant to topological insulators.

---

[a]For an accessible introduction, see Refs. [6,7]

193

8.1. Cooper Pair

The electron–phonon interaction profoundly affects electrons in the presence of a sharp Fermi surface. It is a singular perturbation and leads to formation of Cooper pairs. There is a special correlation between electrons with quantum numbers $k \uparrow$ and $-k \downarrow$ near the Fermi surface. They form a weak bound state by exchanging phonons, see Fig. 8.1. A net interaction between two electrons is of the form [8]

$$V(\vec{q}, \omega) = \frac{4\pi e^2}{q^2 + k_0^2} \frac{\omega^2}{\omega^2 - \omega_{q^2}},\tag{8.1}$$

where k_0 is the Thomas–Fermi wave vector and ω_q is the frequency of a phonon of wave vector $\vec{q}$, see Fig. 8.2(a). It consists of a product between a repulsive term and frequency-dependent term. We see when $\omega < \omega_q$, the net interaction can be negative. Cooper made a toy model of two interacting electrons near the Fermi energy. The change in electron energy due to phonons is $\delta E \sim \hbar \omega_D$, where ω_D is the Debye cutoff frequency of phonons. This implies that the paring interaction takes place in a spherical shell with radius k_F and thickness $\delta k = m\omega_D / \hbar k_F$, where k_F is the Fermi wave vector, see Fig. 8.2. This means that the Fermi surface is not sharp but blurred near the Fermi wave vector k_F, i.e., the occupation number is not a sharp step function.

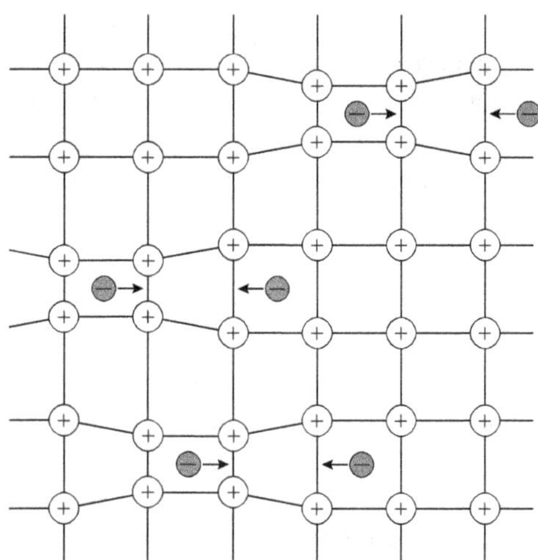

Fig. 8.1.   A passing electron attracts ions nearby which in turn attract other electrons.

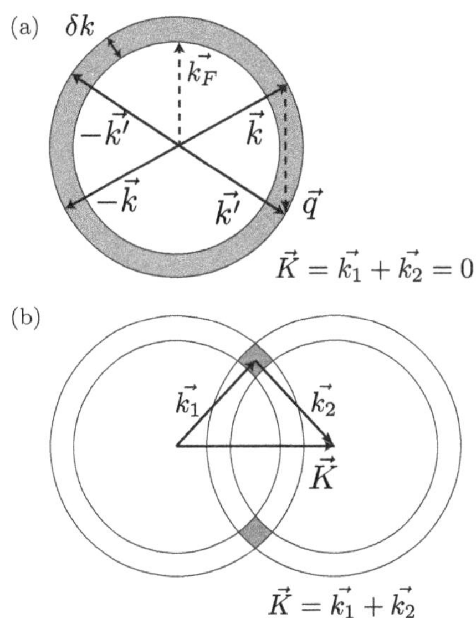

Fig. 8.2. The paring interaction between two electrons interacting outside the Fermi sea. Their wave vectors $\vec{k}_1 = k$ and $\vec{k}_2 = -\vec{k}$ belong to a spherical shell with radius k_F and thickness δk, shown as the shaded area in (a). Both wave vectors are outside the Fermi sea. Available phase space of two-particle scattering (shaded region) is large when the total momentum is $\vec{K} = 0$ (a) while it is small when $\vec{K} \neq 0$ (b).

Consider two interacting electrons with wave vectors $|\vec{k}| > k_F$. Multiple scattering processes between two electrons with zero total momentum $\vec{K} = 0$ and reversed spins are $(\vec{k} \uparrow, -\vec{k} \downarrow) \rightarrow (\vec{k}' \uparrow, -\vec{k}' \downarrow) \rightarrow (\vec{k}'' \uparrow, -\vec{k}'' \downarrow)$, etc., see Fig. 8.2. (Here the magnitude of the wave vectors are all larger than that of the Fermi wave vector.) These scattering events give rise to a bound state. The interaction is attractive $V_{\vec{k},\vec{k}} = -V$ only when $\vec{k}$ and $\vec{k}'$ both belong to the shell as shown in Fig. 8.2(a). (Note that these processes are not possible inside the Fermi sea due to Pauli exclusion principle. So both $|\vec{k}|$ and $|\vec{k}'|$ are larger than k_F.) Despite the presence of a Fermi sea, a stable bound state can be formed, see Exercise 8.1:

$$E_b = -\hbar \omega_D e^{-\frac{1}{VD(E_F)}}, \tag{8.2}$$

where $D(E_F)$ is the density of states at the Fermi energy. This is an example of the Fermi surface instability in the presence of an attractive interaction between electrons. Note that the binding energy non-analytic in V, i.e., even an infinitely small attractive interaction can generate a bound state. If there is no Fermi sea then $D(E_F) = 0$ and a bound state is not guaranteed to exist. In one and two dimensions, $D(E)$ is finite at zero energy and gives rise to bound states even in the absence of a Fermi sea (do Exercise 8.2).

The resulting bound state can be described as

$$|\psi(\vec{r}_1 - \vec{r}_2)\rangle = \sum_{\vec{k}} f_{\vec{k}} e^{i\vec{k}(\vec{r}_1 - \vec{r}_2)} (|\uparrow,\downarrow\rangle - |\downarrow,\uparrow\rangle), \tag{8.3}$$

where the spin wave function forms a spin singlet and the spatial wave function is in the s-orbital. In momentum space, the creation operator of a bound Cooper pair can be described as

$$b^{\dagger}_{\text{pair}} = \sum_{\vec{k}} f(\vec{k}) c^{\dagger}_{\vec{k}\uparrow} c^{\dagger}_{-\vec{k}\downarrow} = \sum_{\vec{k}} f(\vec{k}) b^{\dagger}_{\vec{k}},$$
$$b^{\dagger}_{\vec{k}} = c^{\dagger}_{\vec{k}\uparrow} c^{\dagger}_{-\vec{k}\downarrow}. \tag{8.4}$$

The orbital part of the Fourier transform of this will give the wave function of the relative motion. The size of a Cooper pair is much larger than atomic distances.

Exercise 8.1. Show that the binding energy of a single Cooper pair is given by Eq. (8.2). Hint: In the relative coordinate system, the orbital part of the pair wave function is

$$\psi(\vec{r}) = \sum_{\vec{k}} f_{\vec{k}} e^{i\vec{k}\vec{r}}. \tag{8.5}$$

The expansion coefficients satisfy

$$\sum_{\vec{k}} V_{\vec{k},\vec{k}'} f_{\vec{k}} = (E - 2\epsilon_k) f_{\vec{k}},$$
$$-V \sum_{\vec{k}' \in \delta k} f_{\vec{k}'} = (E - 2\epsilon_k) f_{\vec{k}}. \tag{8.6}$$

From this, we find

$$-\sum_{\vec{k} \in \delta k} \frac{V}{E - 2\epsilon_k} = 1. \tag{8.7}$$

Introducing the density of states $D(\epsilon)$, we find

$$-V \int d\epsilon \frac{D(\epsilon)}{E - 2\epsilon_{\vec{k}}} = -D(E_F) V \int_{E_F}^{E_F + \hbar\omega_D} \frac{d\epsilon}{E - 2\epsilon} = 1. \tag{8.8}$$

From this, we can find the binding energy.

Exercise 8.2. Compute the density of states of one- and two-dimensional systems. Hint: In one-dimension, $D(E = 0)$ diverges and in two dimensions, $D(E)$ is discontinuous at $E = 0$. The density of states in 1, 2, and 3 dimensions are, respectively, $D_{1d}(E) = \frac{1}{\pi\hbar}(2m/E)^{1/2}$, $D_{2d}(E) = \frac{m}{\pi\hbar^2}$, and $D_{3d}(E) = \frac{m}{\pi^2\hbar^3}(2mE)^{1/2}$. Spin degeneracy 2 is included. Do also Exercise 8.3.

Exercise 8.3. Consider a particle with the energy dispersion $\epsilon_k = \epsilon_0 + d_p k^p$. Show that the DOS per spin is

$$D_n(\epsilon) = \frac{1}{(2\pi)^d}\frac{d\Omega_n(k)}{d\epsilon} = \frac{1}{(2\pi)^d}\frac{nV_n}{pd_p^{n/p}}(\epsilon - \epsilon_0)^{\frac{n}{p}-1}. \qquad (8.9)$$

Here the volume of an n-dimensional k-space sphere with radius k is $\Omega_n(k) = V_n k^n$, where $V_1 = 2$, $V_2 = \pi$, and $V_3 = 4\pi/3$. Hint: Use

$$\Omega_n(\epsilon) = \frac{V_n}{d_p^{n/p}}(\epsilon - \epsilon_0)^{n/p}. \qquad (8.10)$$

Further reading. Kohn and Luttinger demonstrated superfluidity and superconductivity in three-dimensional Fermi systems with purely repulsive interactions. This mechanism has nothing to do with the conventional electron–phonon attractive interaction in metals. It is due to the sharpness of the Fermi surface, which may produce a long-range oscillatory effective interaction between electrons. The ground state of such a system has an instability that destroys the sharp Fermi surface [9].

8.2. BCS Trial Wave Function

Although a Cooper pair consists of two electrons with opposite momentum and spin, it is only approximately a boson. Despite this, one can boldly imagine that Cooper pairs Bose condensate and form a macroscopically coherent wave function: The many-body ground state consists of numerous overlapping Cooper pairs. (before reading further students are encouraged to study a chapter by Feynman on superconductivity in Ref. [10].) Bound Cooper pairs are a part of the Fermi sea. Bardeen, Cooper, and Schrieffer

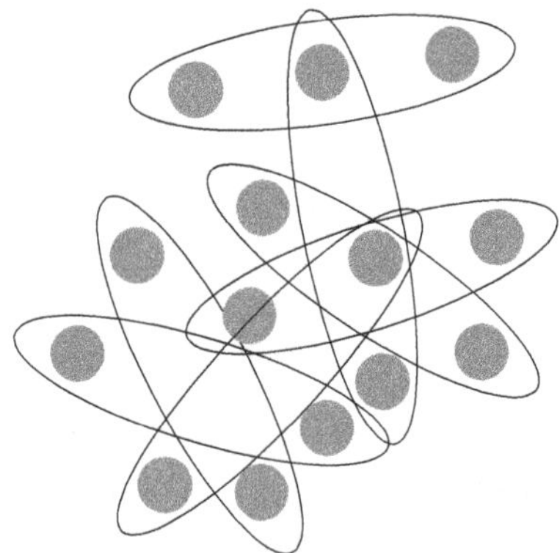

Fig. 8.3. Overlapping Cooper pairs. Electrons near the Fermi energy form numerous Cooper pairs.

(BCS) constructed a trial wave function[b] with many **overlapping** bound Cooper pairs in the presence of a Fermi sea,[c] see Fig. 8.3.

Let us explain the BCS trial wave function. The ground state consists of a product of the following pair states. Consider a superposition of two states, one without a pair and one with a pair with wave vector k

$$|g_k\rangle = (u_k + v_k b_k^\dagger)|0\rangle. \tag{8.11}$$

(Here we consider, for simplicity, a one-dimensional superconductor.) The probability amplitude that a pair is in the momentum state k is v_k and the probability amplitude that it is not is u_k. Now we make a product state consisting of all possible values of k

$$|\Psi_{\text{BCS}}\rangle = \prod_k |g_k\rangle = \prod_k (u_k + v_k b_k^\dagger)|0\rangle \propto \prod_k (1 + d_k b_k^\dagger)|0\rangle = \prod_k e^{d_k b_k^\dagger}|0\rangle. \tag{8.12}$$

(Note that $(b_k^\dagger)^2 = (c_{\vec{k}\uparrow}^\dagger c_{-\vec{k}\downarrow}^\dagger)^2 = 0$ since $(c_\alpha^\dagger)^2 = 0$ for any fermion operator.[d]) The probability amplitudes u_k and v_k are determined by minimizing the total energy. This ground state is called the BCS state. Note that this

---

[b]The trial wave function approach is important not only in studying superconductivity but also in describing fractional quantum Hall states and toric codes. In this approach, one guesses a good ground state trial wave function and use it to build quasiparticles and collective modes.

[c]John Bardeen, Leon N. Cooper, and John Robert Schrieffer won a Nobel Prize for developing the BCS theory of superconductivity.

[d]Note the following commutation relations

$$[b_k, b_k^\dagger] \neq 1 \tag{8.13}$$

Fock state consists of superpositions of $0, 1, 2, 3, \ldots$ number of Cooper pairs, i.e., the number of Cooper pairs is not well-defined.[e]

The BCS wave function can be rewritten as [11]

$$|\Psi_{\text{BCS}}\rangle = \prod_k u_k \left(1 + \frac{v_k}{u_k} b_k^\dagger\right) |0\rangle \propto e^{\gamma b_{\text{pair}}^\dagger} |0\rangle, \tag{8.15}$$

where the pair operator is

$$b_{\text{pair}}^\dagger = \frac{1}{\gamma} \sum_k \frac{v_k}{u_k} b_k^\dagger, \tag{8.16}$$

with

$$\gamma = \sqrt{\sum_k \left(\frac{v_k}{u_k}\right)^2}. \tag{8.17}$$

This result for $b_{\text{pair}}^\dagger$ is identical to the result of a bound Cooper pair, see Eq. (8.4). We interpret the operator $b_{\text{pair}}^\dagger$ as the creation operator of a bound Cooper pair.

> **Exercise 8.4.** Show, from the BCS wave function of Eq. (8.12), the exponential form of the BCS wave function given in Eq. (8.15). Hint: Use $(b_k^\dagger)^2 = 0$.

The BCS trial wave function for the ground state of a superconductor consists of a linear combination of states with different number of Cooper pairs

$$|\Psi_{\text{BCS}}\rangle \propto e^{\gamma b_{\text{pair}}^\dagger} |0\rangle \sim [1 + \gamma b_{\text{pair}}^\dagger + \gamma^2 (b_{\text{pair}}^\dagger)^2 + \cdots] |0\rangle. \tag{8.18}$$

and

$$[b_k, b_{k'}^\dagger] = 0 \tag{8.14}$$

for $k \neq k'$.

[e]Note that in the Hilbert space, all states have the same number of particles while in the Fock space states can have all possible particle numbers.

It is a linear combination of $0, 1, 2, \ldots, N$ bound pairs.[f] An important point is that all the bound states are identical. Note that the Pauli exclusion principle is automatically satisfied. Let us explain this by writing

$$(b^\dagger_{\text{pair}})^2 |0\rangle = \sum_{k \neq p} f(k) f(p) c^\dagger_{k\uparrow} c^\dagger_{-k\downarrow} c^\dagger_{p\uparrow} c^\dagger_{-p\downarrow} |0\rangle. \tag{8.20}$$

The product $b^\dagger_k b^\dagger_p$ in this expression does not vanish when $k \neq p$, but when $k = p$ they vanish due to the Pauli exclusion principle. What is thus crucial about the BCS Ansatz is that it **precludes two Cooper pairs from occupying the same momentum state**. This is the key correlation.

Further reading. There are several spectacularly successful trial wave functions in condensed matter physics.[g] The first one is of course the BCS wave function. The other one is the Laughlin wave function. The third one is the valence bond states that are variational wave functions for antiferromagnetic Heisenberg models. Then there is the Sutherland wave function that describes fractional charges in an artificial one-dimensional model. Gutzwiller variational Ansatz, incorporating local correlation effects, correctly predicts Mott transitions. It is worthwhile to study these trail wave functions.

8.3. Gap Equation

From the variation principle, one can determine the parameters $\{u_k\}$ and $\{v_k\}$.

- Model Hamiltonian

The formation of bound Cooper pairs suggests that the most important two-particle scattering processes[h] are $(k + q, \uparrow, -k - q, \downarrow) \rightarrow (k, \uparrow, -k, \downarrow)$

---

[f]This is in contrast to the wave function of Bose–Einstein condensed particles

$$\Psi_G(r_1, \ldots, r_N) = \phi(r_1), \ldots, \phi(r_N), \tag{8.19}$$

which has exactly N particles.

[g]It is unclear whether an advanced AI of distant future could match the human ability to guess useful trial wave functions. However, there are some recent developments in this direction, see Ref. [12]. But it seems that what is required is artificial general intelligence (AGI). What AGI means for the future of creativity and why humans have miserably failed to develop it so far, see, for example, Ref. [13].

[h]This process is called a relevant perturbation in renormalization group theory.

(see Fig. 8.2). The simplest possible momentum space Hamiltonian with such an interaction term is[i]

$$H = \sum_{k,\sigma} \epsilon_k n_{k\sigma} - \frac{U}{2} \sum_{k,q,\sigma} c^\dagger_{k+q,\sigma} c^\dagger_{-k-q,-\sigma} c_{-k,-\sigma} c_{k,\sigma}$$

$$= \sum_{k\sigma} \epsilon_k n_{k\sigma} - U \left(\sum_k b^\dagger_k \right) \left(\sum_{k'} b_{k'} \right), \tag{8.21}$$

where

$$b^\dagger_k = c^\dagger_{k,\uparrow} c^\dagger_{-k,\downarrow} \quad \text{and} \quad b_{k'} = c_{-k',\downarrow} c_{k',\uparrow}. \tag{8.22}$$

There are other two-particle scattering processes besides the one included in the Hamiltonian equation (8.21) but they are not responsible for superconductivity. The operator $b^\dagger_k$ creates an unbound electron pair in momentum space, i.e., a pair state consisting of one electron with quantum numbers $k \uparrow$ and the other with $-k \downarrow$ (note that the Fermi wave vector is set to $k_F = 0$ for convenience.)

- Total Energy

We compute the total energy of the model Hamiltonian using the BCS wave function. There is a nice way to compute with the BCS wave function. It is convenient to introduce the Nambu box representation [14]. The BCS trail wave function consists of states $|g_k\rangle$ given in Eq. (8.11). We represent it as

$$|g_k\rangle = v_k |2\rangle_k + u_k |0\rangle_k, \tag{8.23}$$

where states $|2\rangle_k$ and $|0\rangle_k$ are defined through the relations

$$b^\dagger_k |0\rangle_k = |2\rangle_k, \; b_k |2\rangle_k = |0\rangle_k. \tag{8.24}$$

In the basis $\{|2\rangle_k, |0\rangle_k\}$ these operators can be represented as

$$b^\dagger_k = \begin{pmatrix} 0 & 1 \\ 0 & 0 \end{pmatrix} \; b_k = \begin{pmatrix} 0 & 0 \\ 1 & 0 \end{pmatrix}, \tag{8.25}$$

and the occupation number in the state k is

$$n_{k\uparrow} + n_{-k\downarrow} = \begin{pmatrix} 2 & 0 \\ 0 & 0 \end{pmatrix}. \tag{8.26}$$

---

[i]The electron interaction is given by a δ-function in real space.

Here

$$|2\rangle_k = (1,0)_k^T \text{ and } |0\rangle_k = (0,1)_k^T \tag{8.27}$$

with two electrons and no electron in the state $|k\rangle$, respectively. It is straightforward to show $(b_k^\dagger)^2 = 0$.

Using the Nambu boxes, we can easily compute various quantities. We find

$$\langle b_k^\dagger \rangle = \langle g_k | b_k^\dagger | g_k \rangle = u_k v_k^*. \tag{8.28}$$

Here the expectation value is $\langle \ldots \rangle = \langle \Psi_{\mathrm{BCS}} | \ldots | \Psi_{\mathrm{BCS}} \rangle$. Using the Hamiltonian equation (8.21) we find the mean-field ground state energy

$$\langle H \rangle = E_{\mathrm{tot}} = E_{\mathrm{single}} + E_{\mathrm{two}} = 2 \sum_k \epsilon_k |v_k|^2 - U | \sum_k u_k v_k^* |^2. \tag{8.29}$$

The first term is the contribution from the single particle operators of H and the second term represents the contribution from the two-particle operators. We have used

$$\left\langle \left(\sum_k b_k^\dagger \right) \left(\sum_{k'} b_{k'} \right) \right\rangle = \sum_k \langle b_k^\dagger \rangle \sum_{k'} \langle b_{k'} \rangle. \tag{8.30}$$

(see Exercise 8.5).

Exercise 8.5. Show Eq. (8.30). Answer: Using Eqs.(8.25) and (8.27) we find for $k \neq k'$

$$\left\langle \sum_{k \neq k'} b_k^\dagger b_{k'} \right\rangle = \sum_{k \neq k'} \prod_p (u_p^* \, _p\langle 0| + v_p^* \, _p\langle 2|) b_k^\dagger b_{k'}$$

$$\times \prod_{p'} (u_{p'} |0\rangle_{p'} + v_{p'} |2\rangle_{p'}) = \sum_{k \neq k'} u_k v_k^* u_{k'}^* v_{k'}. \tag{8.31}$$

When $k = k'$

$$\left\langle \sum_k b_k^\dagger b_k \right\rangle = \sum_k \prod_p (u_p^* \, _p\langle 0| + v_p^* \, _p\langle 2|) b_k^\dagger b_k \prod_{p'} (u_{p'} |0\rangle_{p'} + v_{p'} |2\rangle_{p'})$$

$$= \sum_k v_k^* v_k. \tag{8.32}$$

The final result is

$$\left(\sum_{k,k'} u_k v_k^* u_{k'}^* v_{k'} - \sum_k |v_k|^2 |u_k|^2 \right) + \sum_k |v_k|^2$$

$$= \sum_{k,k'} u_k v_k^* u_{k'}^* v_{k'} + \sum_k |v_k|^2 \left(1 - |u_k|^2\right)$$

$$\approx \sum_{k,k'} u_k v_k^* u_{k'}^* v_{k'} = \sum_k u_k v_k^* \sum_{k'} u_{k'}^* v_{k'} = \sum_k \langle b_k^\dagger \rangle \sum_{k'} \langle b_{k'} \rangle. \tag{8.33}$$

In the second equation, the second term, in comparison to the first term, vanishes in the thermodynamic limit (note that $V k_F^3 \sim N \gg 1$, where N is the number of electrons and V is the volume of the superconductor).

- Self-Consistent Equations

We minimize E_{tot} by varying u_k^* and v_k^* independently, and we find that

$$\delta E_{\text{tot}} = 2\epsilon_k v_k \delta v_k^* - \Delta u_k \delta v_k^* - \Delta^* v_k \delta u_k^*, \tag{8.34}$$

where the gap is

$$\Delta^* = U \sum_k \langle b_k^\dagger \rangle = U \sum_k u_k^* v_k. \tag{8.35}$$

The normalization constraint $u_k u_k^* + v_k v_k^* = 1$ is used to find a relation between δu_k^* and δv_k^*. This leads to

$$\frac{\delta E}{\delta u_k^*} = 2\epsilon_k v_k - \Delta u_k + \Delta^* v_k^2 / u_k = 0. \tag{8.36}$$

From the condition of the minimum total energy, Eq. (8.36), we find

$$\frac{v_k}{u_k} = \frac{-\epsilon_k \pm E_k}{\Delta} = \frac{\epsilon_k^2 - E_k^2}{(-\epsilon_k \mp E_k)\Delta^*} = \frac{\Delta}{\epsilon_k \pm E_k}, \tag{8.37}$$

where the quasiparticle energy is

$$E_k = \sqrt{\epsilon_k^2 + |\Delta|^2}. \tag{8.38}$$

We must choose plus sign in $\frac{v_k}{u_k}$ since it must go to zero for large k. From the normalization condition $1 = |u_k|^2 + |v_k|^2 = |u_k|^2 + |u_k|^2 \frac{|v_k|^2}{|u_k|^2}$ it can be shown that

$$|v_k|^2 = \frac{1}{2}\left(1 - \frac{\epsilon_k}{\sqrt{\epsilon_k^2 + |\Delta|^2}}\right),$$

$$|u_k|^2 = \frac{1}{2}\left(1 + \frac{\epsilon_k}{\sqrt{\epsilon_k^2 + |\Delta|^2}}\right). \tag{8.39}$$

The coefficients $u_k \to 1$ and $v_k \to 0$ as k increases away from the Fermi wave vector, see Fig. 8.4. The opposite is true when k decreases away from the Fermi wave vector. By assuming that both u_k and v_k are real, we find from these results that

$$u_k^* v_k = \frac{1}{2}\frac{\Delta}{\sqrt{\epsilon_k^2 + |\Delta|^2}}. \tag{8.40}$$

This function is called the **coherence factor** and is approximately a delta function near $\epsilon_k = 0$, see Fig. 8.5. In a normal metal it is zero.

The gap equation

$$\Delta = \frac{U}{2}\sum_k \frac{\Delta}{\sqrt{\epsilon_k^2 + |\Delta|^2}} \tag{8.41}$$

is derived using its definition given in Eq. (8.35). The gap Δ is determined self-consistently from this equation. One must know the gap to compute other quantities of physical relevance.

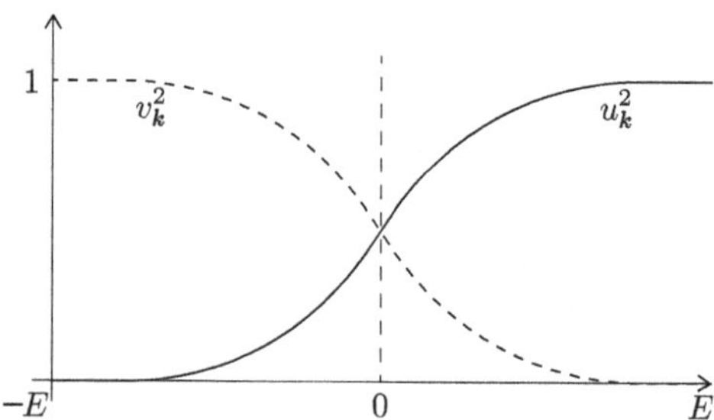

Fig. 8.4.　Coherence factors u_k and v_k.

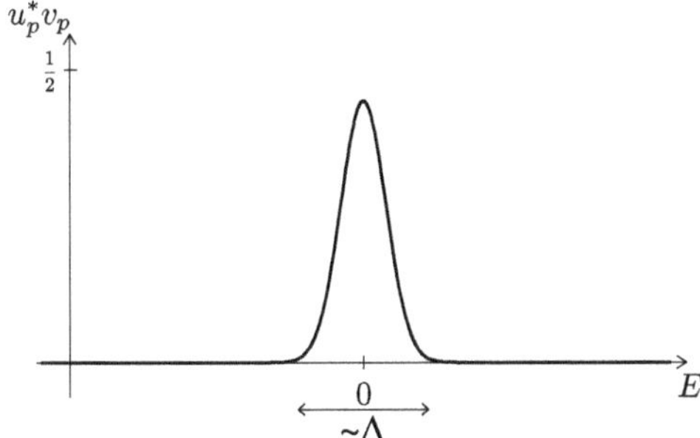

Fig. 8.5. Product of the coherence factors $u_p^* v_p$ is a sharply peaked at the Fermi energy.

8.4. Quasiparticle and Spin–Charge Separation

Let us adopt a different but equivalent microscopic approach. This approach will allow us to study the excitation spectrum consisting of quasi-particles, in addition to ground state properties. We consider the toy model of **one-dimensional** s-wave superconductor, consisting of the following Hubbard model with on-site **attraction**

$$H = \sum_{i\sigma}[-t(c_{i,\sigma}^\dagger c_{i+1,\sigma} + c_{i+1,\sigma}^\dagger c_{i,\sigma}) - \mu c_{i\sigma}^\dagger c_{i\sigma}] - U\sum_{j} n_{j\uparrow}n_{j\downarrow}. \quad (8.42)$$

Here periodic boundary conditions are assumed. Note that the on-site interaction $-U < 0$ represents an attractive interaction mediated by phonons. It is also instructive to rederive the gap equation using this Hamiltonian.

- Mean Field Hamiltonian

  The BCS Ansatz suggests the following decoupling scheme:

$$n_{i\uparrow}n_{i\downarrow} = c_{i,\uparrow}^\dagger c_{i,\uparrow} c_{i,\downarrow}^\dagger c_{i,\downarrow} = c_{i,\uparrow}^\dagger c_{i,\downarrow}^\dagger c_{i,\downarrow} c_{i,\uparrow} \to c_{i,\uparrow}^\dagger c_{i,\downarrow}^\dagger \overline{c_{i,\downarrow} c_{i,\uparrow}} + \overline{c_{i,\uparrow}^\dagger c_{i,\downarrow}^\dagger} c_{i,\downarrow} c_{i,\uparrow}.$$

$$(8.43)$$

The resulting mean field Hamiltonian is

$$H_{\mathrm{MF}} = \sum_{i\sigma}[-t(c_{i,\sigma}^\dagger c_{i+1,\sigma} + c_{i+1,\sigma}^\dagger c_{i,\sigma}) + \mu c_{i\sigma}^\dagger c_{i\sigma}] - \sum_{i}(\Delta c_{i\uparrow}^\dagger c_{i\downarrow}^\dagger + \Delta^* c_{i\downarrow} c_{i\uparrow})$$
$$+ \Delta^2/U, \quad (8.44)$$

where the gap is

$$\Delta = NU\langle c_{i,\downarrow} c_{i,\uparrow}\rangle, \quad \Delta^* = NU\langle c_{i,\uparrow}^\dagger c_{i,\downarrow}^\dagger\rangle. \quad (8.45)$$

(We will assume that Δ is uniform and real). Pairs are only allowed to be created or destroyed at the same site. The pair wave function is symmetric and spin wave function is antisymmetric (in accordance with Pauli principle).

The k-space mean Hamiltonian be can written as follows:

$$H_{\mathrm{MF}} = \sum_k [\epsilon_k (c^\dagger_{k,\uparrow} c_{k,\uparrow} + c^\dagger_{-k,\downarrow} c_{-k\downarrow}) - (\Delta c^\dagger_{k,\uparrow} c^\dagger_{-k,\downarrow} + \Delta^* c_{-k,\downarrow} c_{k,\uparrow})] + \frac{\Delta^2}{U}.$$

(8.46)

The gap is now given by

$$\Delta^* = U \sum_k \langle c^\dagger_{k\uparrow} c^\dagger_{-k\downarrow} \rangle = U \sum_k \langle b^\dagger_k \rangle.$$

(8.47)

Note that $\epsilon_k = \epsilon_{-k}$ due to inversion symmetry. Under the global gauge transformation $c_{k\sigma} \to e^{i\phi} c_{k\sigma}$, the Hamiltonian is not invariant due to the pairing terms. According to Noether's theorem, this Hamiltonian thus does not conserve particle number. The mean-field Hamiltonian can also be written as

$$H_{\mathrm{MF}} = \sum_{k,\sigma} \epsilon_k n_{k,\sigma} - \sum_k (\Delta b^\dagger_k + \Delta^* b_k) + \frac{\Delta^2}{U},$$

(8.48)

which may be written in the canonical form

$$H_{\mathrm{MF}} = \sum_k (c^\dagger_{k\uparrow}, c_{-k\downarrow}) \begin{pmatrix} \epsilon_k - \mu & -\Delta \\ -\Delta^* & -\epsilon_{-k} + \mu \end{pmatrix} \begin{bmatrix} c_{k\uparrow} \\ c^\dagger_{-k,\downarrow} \end{bmatrix} + \frac{\Delta^2}{U}.$$

(8.49)

Here we have restored the constant term in the mean-field Hamiltonian. The value of the gap must be determined by minimizing the ground state energy.

Exercise 8.6. Derive the momentum space Hamiltonian (Eq. (8.46)) from the real space Hamiltonian. Hint: Use the following relations between momentum space and the real space quantities:

$$\sum_j c^\dagger_j c_j = \sum_k c^\dagger_k c_k, \quad \sum_j c^\dagger_j c_{j+1} = \sum_k e^{ika} c^\dagger_k c_k,$$
$$\sum_j c^\dagger_j c_{j+1} = \sum_k c^\dagger_k c_k e^{ika}.$$

(8.50)

Here a is the lattice constant.

- Diagonalization and Quasiparticle

We diagonalize the 2×2 Hamiltonian matrix h of Eq. (8.49) using a unitary matrix[j] U (Appendix H)

$$\tilde{h} = U^{\dagger} h U, \tag{8.53}$$

where

$$U = \begin{bmatrix} u_k & -v_k \\ v_k & u_k \end{bmatrix}. \tag{8.54}$$

(u_k and v_k can be assumed to be real.) Here matrix $U^{\dagger}$ is the conjugate transpose of U. The appropriate unitary matrix U with $U^{-1} = U^{\dagger}$ has the following matrix elements:

$$u_k = \sqrt{\frac{\epsilon_k + E_k}{2E_k}} = \frac{1}{\sqrt{2}}\left(1 + \frac{\epsilon_k}{E_k}\right)^{1/2} = \frac{1}{\sqrt{2}}(1 + \cos\theta_k)^{1/2} = \cos\theta_k/2,$$
$$\tag{8.55}$$
$$v_k = \sqrt{\frac{E_k - \epsilon_k}{2E_k}} = \frac{1}{\sqrt{2}}\left(1 - \frac{\epsilon_k}{E_k}\right)^{1/2} = \frac{1}{\sqrt{2}}(1 - \cos\theta_k)^{1/2} = \sin\theta_k/2.$$

Note that these quantities are identical to those probability amplitudes introduced in Eq. (8.39).

> **Exercise 8.7.** Show that the unitary transformation U of Eq. (8.56) diagonalizes the Hamiltonian h given in Eq. (8.49), i.e., the diagonalized matrix $\tilde{h}$ is given by the unitary transformation (8.53).

Introducing the operators $\tilde{\psi}_k = (\alpha_{k\sigma}, \beta^{\dagger}_{-k})$ where

$$\begin{bmatrix} \alpha_k \\ \beta^{\dagger}_{-k} \end{bmatrix} = U^{\dagger}\begin{bmatrix} c_{k\uparrow} \\ c^{\dagger}_{-k,\downarrow} \end{bmatrix} = \begin{bmatrix} u_k & v_k \\ -v_k & u_k \end{bmatrix}\begin{bmatrix} c_{k\uparrow} \\ c^{\dagger}_{-k,\downarrow} \end{bmatrix}, \tag{8.56}$$

---

[j]It is helpful to remind ourselves that the unitary matrix U has the following structure:

$$U = e^{i\varphi/2}\begin{bmatrix} e^{i\varphi_1}\cos\eta & e^{i\varphi_2}\sin\eta \\ -e^{-i\varphi_2}\sin\eta & e^{-i\varphi_1}\cos\eta \end{bmatrix}, \tag{8.51}$$

with

$$\det(U) = e^{i\varphi}. \tag{8.52}$$

The absolute value of the determinant of a unitary matrix is one.

we can write each term in the mean-field Hamiltonian H_{MF} as

$$\psi_k^\dagger \begin{pmatrix} \epsilon_k & -\Delta \\ -\Delta^* & -\epsilon_{-k} \end{pmatrix} \psi_k = \tilde{\psi}_k^\dagger \begin{pmatrix} E_k & 0 \\ 0 & -E_k \end{pmatrix} \tilde{\psi}_k, \tag{8.57}$$

where $\psi_k = (c_{k\uparrow}, c_{-k,\downarrow}^\dagger)^T$. The new transformed operators can be written in terms of the electron operators and the unitary matrix U. The Bogoliubov operators $\alpha_k^\dagger$ and $\beta_k^\dagger$ diagonalize the mean-field Hamiltonian:

$$H_{\mathrm{MF}} = \sum_k E_k [\alpha_k^\dagger \alpha_k + \beta_k^\dagger \beta_k] + \frac{\Delta^2}{U}. \tag{8.58}$$

These operators satisfy the usual anticommutation relations of fermions. When these operators act on the BCS ground state,

$$\alpha_k |\Psi\rangle = 0. \tag{8.59}$$

The eigenenergy of this Hamiltonian is (see Fig. 8.6)

$$E_k = \sqrt{\epsilon_k + |\Delta|^2}. \tag{8.60}$$

We measure energies from the chemical potential $\mu = 0$. One can show that

$$H_{\mathrm{MF}} \alpha_k^\dagger |\Psi\rangle = E_k \alpha_k^\dagger |\Psi\rangle, \tag{8.61}$$

which implies that $\alpha_k^\dagger |\Psi\rangle$ is an eigenstate of H_{MF}.

The new operators thus describe the Bogoliubov quasiparticle excitations. For $|k| \sim 0$, a quasiparticle state represents a **hybridized state of a negatively charged electron and a positively charged hole,** as can be seen from the energy dispersion displayed in Fig. 8.6(b). Note that $|u_k|^2 \to 1$ as $k \to \infty$ and $|v_k|^2 \to 1$ as $k \to -\infty$. When $k \to \infty$ the operator $\alpha_k^\dagger = u_k^* c_{k\uparrow}^\dagger + v_k^* c_{-k\downarrow}^\dagger \to c_{k\uparrow}^\dagger$ and creates a spin-up electron. Similarly, as $k \to -\infty$ the operator $\beta_{-k}^\dagger = -v_k c_{k\uparrow} + u_k c_{-k\downarrow}^\dagger \to c_{k\uparrow}$ and creates a spin-down hole with $k > 0$ (a hole is an antiparticle of an electron with opposite spin and momentum).

One can find the total ground state energy from the quasiparticle energy dispersion

$$E_{\mathrm{tot}} = -\sum_{k \in \mathrm{BZ}} E_k + \frac{\Delta^2}{U}. \tag{8.62}$$

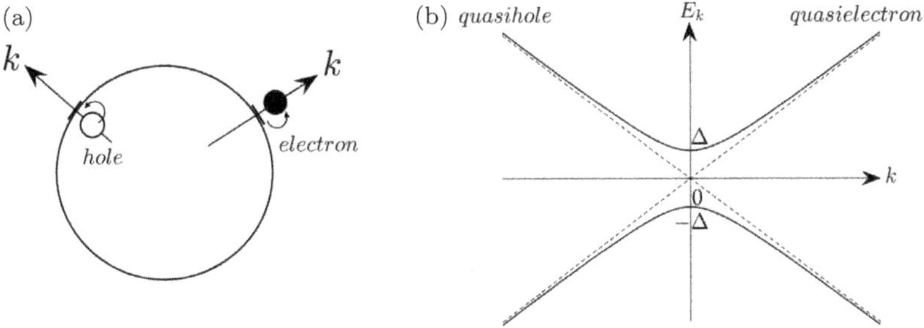

Fig. 8.6. (a) Consider a superconductor connected to an external reservoir whose chemical potential is set to zero. The figure depicts two processes. In the first process, an electron from the Fermi surface is transferred to a state with $k > 0$ (note that we have set $k_F = 0$ and the Fermi energy $E_F = 0$). In the second process, an electron in a state with $k < 0$ and negative energy is transferred to the Fermi surface (thereby a hole with $k > 0$ and positive energy is created in the Fermi sea). (b) Quasiparticle energy dispersion (solid lines) E_k is shown near the Fermi energy. For $k > 0$, the energy $E_k > 0$ is the cost of creating an electron. For $k < 0$, the energy $E_k > 0$ is the cost of creating a hole. (The energy dispersion with $E_k < 0$ corresponds to the reverse processes.) The dashed lines show the electron and hole energy levels in the normal metal. In the superconducting state, these states become **hybridized**. The tunneling DOS is sharply peaked at $E = \pm\Delta$ (the van Hove singularity) while it is zero in the gap $-\Delta < E < \Delta$.

(The energy of occupied quasiparticles is $-E_k$.) Minimizing with respect to Δ, we find the gap equation

$$\Delta = \frac{U}{2} \sum_{k \in \mathrm{BZ}} \frac{\Delta}{\sqrt{\epsilon_k^2 + \Delta^2}}. \tag{8.63}$$

Exercise 8.8. Show that the mean-field Hamiltonian of Eq. (8.46) and the Bogoliubov–de Gennes Hamiltonian of Eq. (8.58) are identical.

- Spin–Charge Separation

Unlike an electron, a quasiparticle is **charge neutral** (this effect is an example of spin–charge separation [1]; we will see that the edges of the integer quantum Hall state also exhibit spin–charge separation, see 12.5.). In superconductors, the absolute distinction between electrons and holes is blurred. Consider the following hand waving argument. An electron can

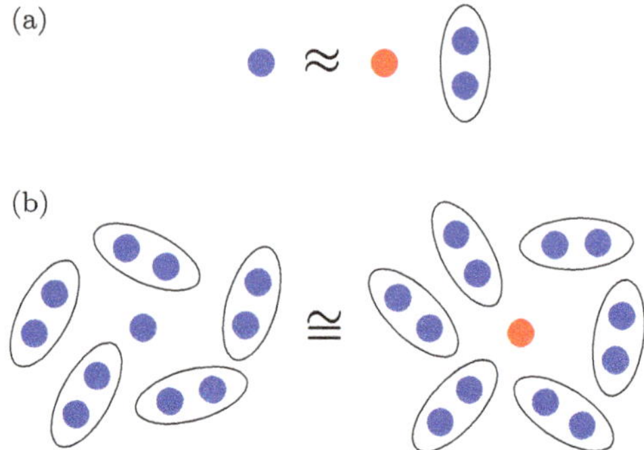

Fig. 8.7. An electron (blue circle) is approximately a hole (red circle) plus a Cooper pair (a) Because of screening by Cooper pairs, this means that an electron and a hole cannot be distinguished in a superconductor, i.e., they are effectively charge neutral (b) Reprinted with permission from Ref. [15].

be thought of a hole and a Cooper pair

$$e^- = e^+ + (e^- e^-). \tag{8.64}$$

Since there are many other Cooper pairs in the ground state, the hole e^+ will be screened [15], see Fig. 8.7.

The same argument can be applied to a positively charged hole. So this suggests that a neutral particle may be created in a superconductor. But is it its own antiparticle? This does not actually work for s-wave superconductors: the quasiparticle operators of Eq. (8.56) do not obey $\alpha_{k\sigma}^\dagger = \alpha_{k\sigma}$: because of spin degree of freedom is the problem.

Since a quasiparticle is charge neutral made up of an electron **and** hole, one might think that no Aharonov–Bohm phase can exist. This is actually incorrect. A quasiparticle is **not** an electron plus a hole. It is **either** an electron **or** hole as it is in a quantum superposition of them. (A hole and an electron live in different worlds according to the many worlds interpretation of quantum mechanics. For an easy introduction to the Everettian quantum mechanics, see Refs. [16, 17].) Thus an electron or hole that goes round a vortex picks up a phase

$$e^{\pm i2\pi \frac{\Phi_S}{\Phi_0}} = -1. \tag{8.65}$$

(Any flux in a superconductor must appear as an integer multiple of the elementary flux Φ_S of a charge $2e$.)

8.5. Coherent State, Phase, and Number Uncertainty

Here we discuss the BCS state from the perspective of coherent states. A best example of a coherent state is laser light. The BCS state is **approximately** a coherent state.

A coherent state [14] is an eigenstate of a boson **destruction operator** b with a **complex** eigenvalue λ

$$b|\psi\rangle = \lambda|\psi\rangle. \tag{8.66}$$

A coherent state can be written in terms of a boson creation operator b^+ with the commutation relation $[b, b^\dagger] = 1$

$$|\psi\rangle = Ce^{\lambda b^\dagger}|0\rangle. \tag{8.67}$$

Properties of a coherent state can be elucidated by considering the one-dimensional quantum harmonic oscillator [7]. A coherent state can be written as

$$|\psi\rangle = C\left[\psi_0(x) + \frac{\lambda}{(1!)^{1/2}}\psi_1(x) + \frac{\lambda^2}{(2!)^{1/2}}\psi_2(x) + \cdots\right], \tag{8.68}$$

where $\psi_n(x)$ is an eigenstate of the oscillator. In the number representation, the coherent state is

$$|\psi\rangle = e^{-\frac{1}{2}|\lambda|^2}\sum_n \frac{\lambda^n}{\sqrt{n!}}|n\rangle, \tag{8.69}$$

where $(b^\dagger)^n \propto |n\rangle$. A coherent state does not have a definite value of quantum number n. The probability of observing n is

$$P_n = \frac{|\lambda|^{2n}}{n!}e^{-|\lambda|^2}. \tag{8.70}$$

It is easy to show that the average number of bosons in a coherent state is

$$\bar{n} = |\lambda|^2. \tag{8.71}$$

With $\lambda = |\lambda|e^{i\phi}$, a coherent state has a definite phase

$$|\psi\rangle = C\left[\psi_0(x) + e^{i\phi}\frac{|\lambda|}{(1!)^{1/2}}\psi_1(x) + e^{i2\phi}\frac{|\lambda|^2}{(2!)^{1/2}}\psi_2(x) + \cdots\right], \tag{8.72}$$

Differentiating this state, we find

$$\frac{1}{i}\frac{\partial}{\partial\phi}|\psi\rangle = \hat{n}|\psi\rangle, \qquad (8.73)$$

where $\hat{n}$ is the number operator. This result suggests

$$\frac{1}{i}\frac{\partial}{\partial\phi} = \hat{n}. \qquad (8.74)$$

It implies that phase and number are conjugate variables. Since any linear combination of $\psi_n(x)$ can be written as a linear combination of coherent states ($\{|\psi\rangle\}$ forms an overcomplete set), one can show the following uncertainty relation between the phase and number

$$\delta n\delta\phi \geq 1/2. \qquad (8.75)$$

Let us give a heuristic argument for this result. The overlap between two coherent states is

$$|\langle\psi|\psi'\rangle|^2 = e^{-|\lambda-\lambda'|^2}. \qquad (8.76)$$

Thus two coherent states cannot be distinguished if the difference between their parameters λ is sufficiently small. This suggests that the uncertainty in λ is

$$\delta|\lambda| = 1/2. \qquad (8.77)$$

Thus the uncertainty in the boson number is

$$\delta n = 2|\lambda||\delta\lambda| = |\lambda|. \qquad (8.78)$$

and the uncertainty in the phase is

$$\delta\phi = \frac{\delta|\lambda|}{|\lambda|} = \frac{1}{2\sqrt{\bar{n}}}. \qquad (8.79)$$

These results indicate that the uncertainty relation between particle number and phase is

$$\delta n\delta\phi \geq 1/2. \qquad (8.80)$$

A squeezed state with minimum uncertainty relation $\delta n\delta\phi = 1/2$ is shown in Fig. 8.8.

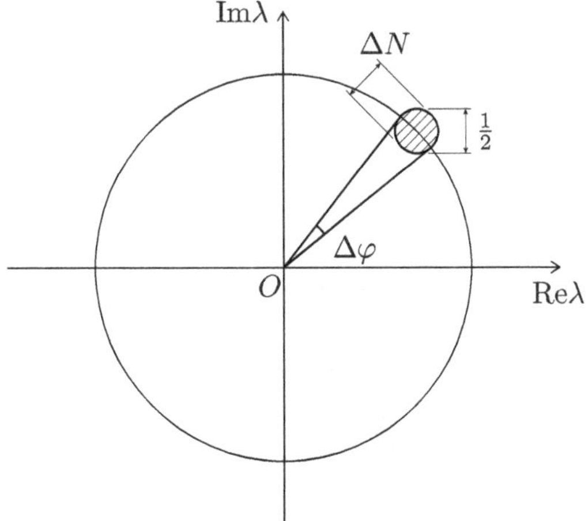

Fig. 8.8. The large circle shows a state with a fixed number of particles but indefinite phase. A coherent state which minimizes the uncertainty relation between number is shown as a small circle. Here it is represented by a dot of size $1/2$.

Let us put $\gamma = |\gamma|e^{i\phi}$ in the BCS state (8.15). The resulting BCS state has a **definite** phase ϕ

$$\prod_k (u_k + v_k e^{i\phi} c^\dagger_{k\uparrow} c^\dagger_{-k\downarrow} |0\rangle. \tag{8.81}$$

The quantity ϕ is the phase of the bound Cooper pair wave function. By identifying γ with λ, we see that the BCS state given in Eq. (8.15) is identical to the coherent state of Eq. (8.67). However, this identification is not exact. This is because the following commutation relation is not satisfied

$$[b_{\text{pair}}, b^\dagger_{\text{pair}}] \neq 1. \tag{8.82}$$

Nonetheless, it is a good approximation to treat the superconducting state as a coherent state. What really matters is off-diagonal long-range order, which is one of the essential features of a coherent state. Superfluid has off-diagonal long-range order in the one-particle density matrix while a superconductor has off-diagonal long-range order in the pair density matrix

$$\langle \Psi^\dagger(\vec{r})\Psi(\vec{r}')\rangle \to \text{constant} \quad \text{as} \quad |\vec{r} - \vec{r}'| \to \infty, \tag{8.83}$$

where the pairing field operator is (see Eq. (8.45))

$$\Psi^\dagger(\vec{r}) = \psi^\dagger_\uparrow(\vec{r})\psi^\dagger_\downarrow(\vec{r}). \tag{8.84}$$

($\psi_\sigma^\dagger(\vec{r})$ is the electron field operator.) The concept of off-diagonal long-range order is well explained in Ref. [7].

Exercise 8.9. Show that the BCS wave function has the same structure as the coherent state, Eq. (8.67).

Exercise 8.10. Show Eq. (8.82). Hint: Use Eqs. (8.16) and (8.25).

A standard deviation in the number of Cooper pairs of the BCS state is $\Delta N \propto \sqrt{N}$. One can also construct a BCS wave function with definite number of Cooper pairs by a Fourier transform

$$|\Psi_{2m}\rangle \propto \int_0^{2\pi} d\phi\, e^{-i2m\phi} \prod_k (u_k + v_k e^{2i\phi} b_k^\dagger)|0\rangle. \tag{8.85}$$

The phase of this state is completely uncertain. However, the number of particle is certain. The Fourier transform implies that there is an uncertainty relation between N and ϕ:

$$\Delta N \Delta \phi > 1. \tag{8.86}$$

An important physical implication of this uncertainty relation is the following. In a Josephson junction [10], the phase difference between two superconductors leads to current. If you can count the number of electrons precisely in a superconductor, ϕ is not a good quantum number, and there will be no current. If you can define a phase difference between two superconductors, a current will flow but you cannot count the electrons in superconductors.

One can also formally define phase and number operators as follows:

$$b = e^{-i\hat{\phi}}\sqrt{\hat{N}}, \quad b^\dagger = \sqrt{\hat{N}}e^{i\hat{\phi}}. \tag{8.87}$$

where $\hat{\phi}$ is defined as the phase operator and $\hat{N} = b^\dagger b$ as the number operator. The following commutation relation holds (see Exercise 8.12):

$$[\hat{N}, \hat{\phi}] = -i, \tag{8.88}$$

Exercise 8.11.   (a) Show the commutator

$$[\hat{N}, e^{i\hat{\phi}}] = e^{i\hat{\phi}}. \tag{8.89}$$

Hint: Use

$$\Psi_0^\dagger = \hat{N}^{1/2} e^{i\hat{\phi}}, \quad \Psi_0 = e^{-i\hat{\phi}} \hat{N}^{1/2} \tag{8.90}$$

and the commutation relation of the boson field operator

$$\Psi_0 \Psi_0^\dagger - \Psi_0^\dagger \Psi_0 = 1. \tag{8.91}$$

(b) Show also that the number operator $\hat{N} = -i\frac{\partial}{\partial\phi}$ satisfies this commutation relation.

Exercise 8.12. (a) Show the commutation relation

$$[\hat{N}, \hat{\phi}] = -i. \tag{8.92}$$

Hint: Use $e^{i\hat{\phi}} \approx 1 + i\hat{\phi}$ in the commutation relation (8.89). (b) Show that the number operator $\hat{N}$ is $-i\frac{\partial}{\partial\phi}$ satisfies the same commutation relation. Hint: Let the commutator act on a wave function $\Psi(\phi)$.

Exercise 8.13.   Using the commutator $[\hat{N}, e^{i\hat{\phi}}] = e^{i\hat{\phi}}$ show that

$$e^{i\hat{\phi}}|n\rangle = |n+1\rangle. \tag{8.93}$$

Further reading. Read about the Josephson effect (B. Josephson suggested this effect when he was only a 22-year old graduate student. He won the Nobel Prize in 1973). He predicted that, even in the absence of a voltage difference, a current can flow between two weakly coupled superconductors. Any two phase coherent weakly coupled systems should exhibit this effect. Feynman's book [10] has one of the best presentations on it. The equation that describes the nonlinear dynamics of Josephson junctions is the same as that for a classical pendulum [18].

8.6. Charge Neutral Superfluid

The Anderson–Higgs mechanism is one of the most interesting properties of a superconductor. Before we discuss the Anderson–Higgs mechanism for a charged superconductor, we will first show how a Goldstone mode arises in a neutral superfluid, i.e., in a Bose–Einstein condensate.[k] The Lagrangian of the system has U(1) global symmetry[l] and particle number is conserved (see Noether's theorem in Appendix H). However, the ground state breaks the symmetry — it is a spontaneously broken symmetry state.

- Goldstone Mode in Charge Neutral Superfluid

Since neutral particles do not couple to electromagnetic fields, the Lagrangian density is ($\hbar = 1$ and $c = 1$)

$$\mathcal{L} = i\Psi^* \frac{\partial \Psi}{\partial t} - \frac{1}{2m^*} \nabla\Psi^* \cdot \nabla\Psi - \frac{g}{2}(\rho_0 - \Psi^*\Psi)^2, \tag{8.94}$$

where Ψ is the order parameter. (It is a Ginzberg–Landau free-energy density, see Sec. 8.8.) Here the two-particle interaction term is represented by the contact interaction g and the particle mass is m^*.

We rewrite the order parameter as $\Psi = \rho e^{i\phi}$ with broken U(1) symmetry. Writing the Lagrangian density in terms of amplitude ρ and phase θ of the order parameter, we find

$$\mathcal{L} = -\rho\frac{\partial\phi}{\partial t} - \frac{1}{2m^*}\left[\frac{1}{4\rho}(\nabla\rho)^2 + \rho(\nabla\phi)^2\right] - \frac{g}{2}(\rho_0 - \rho)^2. \tag{8.95}$$

Let us denote fluctuations around the average amplitude $\rho = \rho_0$ by the amplitude mode a (see Fig. 8.9)

$$\sqrt{\rho} = \sqrt{\rho_0} + a. \tag{8.96}$$

In terms of the amplitude mode a and phase mode ϕ, the Lagrangian density has the following form:

$$\mathcal{L} = -\frac{1}{2m^*}(\nabla a)^2 - 2g\rho_0 a^2 - (2\sqrt{\rho_0}\partial_0\phi)a - \frac{\rho_0}{2m^*}(\nabla\phi)^2. \tag{8.97}$$

---

[k]A good explanation Bose–Einstein condensation for non-interacting particles is given in Refs. [7, 19]. It should be noted that an ideal Bose–Einstein condensate is **not superfluid**. The critical velocity for superfluid flow is non-zero only when particle interactions are included.

[l]U(1) global symmetry is not the local gauge symmetry of electromagnetism because the vector potential is not involved in it.

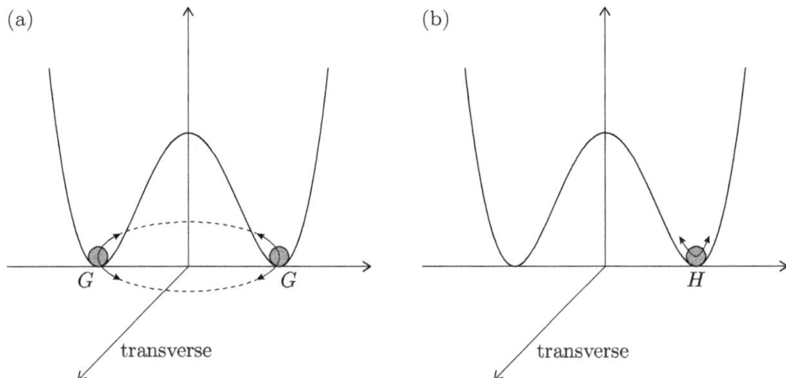

Fig. 8.9. (a) Goldstone mode (G) represents the phase mode moving in the valley of a Mexican hat potential. (b) Massive Higgs mode (H) represents the massive amplitude mode (longitudinal).

Using integration by parts, we can make the following replacement in $\int d\vec{r}\mathcal{L}$

$$-\frac{1}{2m^*}(\nabla a)^2 - 2g\rho_0 a^2 \rightarrow -a\left(\frac{-1}{2m^*}\nabla^2 + 2g\rho_0\right)a. \tag{8.98}$$

We then integrate out the amplitude mode a using functional integration $\int d[a]e^{\int d\vec{r}\mathcal{L}}$ (see Appendix G.2). We find [20]

$$-a\left(-\frac{1}{2m^*}\nabla^2 + 2g\rho_0\right)a - (2\sqrt{\rho_0}\partial_0\phi)a$$
$$\rightarrow \sqrt{\rho_0}\partial_0\phi\frac{1}{(-1/2m^*)\nabla^2 + 2g\rho_0}\sqrt{\rho_0}\partial_0\phi. \tag{8.99}$$

This procedure modifies the Lagrangian density for the phase modes. The resulting effective Lagrangian describing the phase degrees of freedom is

$$\mathcal{L}_{\text{eff}} = \rho_0\frac{\partial\phi}{\partial t}\frac{1}{2g\rho_0 - 1/2m^*\nabla^2}\frac{\partial\phi}{\partial t} - \frac{\rho_0}{2m^*}(\nabla\phi)^2. \tag{8.100}$$

Ignoring the gradient term in the denominator, we obtain the following low-energy effective Lagrangian:

$$\mathcal{L}_{\text{eff}} = \frac{1}{2g}\left(\frac{\partial\phi}{\partial t}\right)^2 - \frac{\hbar^2\rho_0}{2m^*}(\nabla\phi)^2. \tag{8.101}$$

This describes a phase mode with the linear energy dispersion

$$E_k = \left(\frac{\rho_0 g}{m^*}\right)^{1/2}k, \tag{8.102}$$

which is the Goldstone mode (see Fig. 8.10).

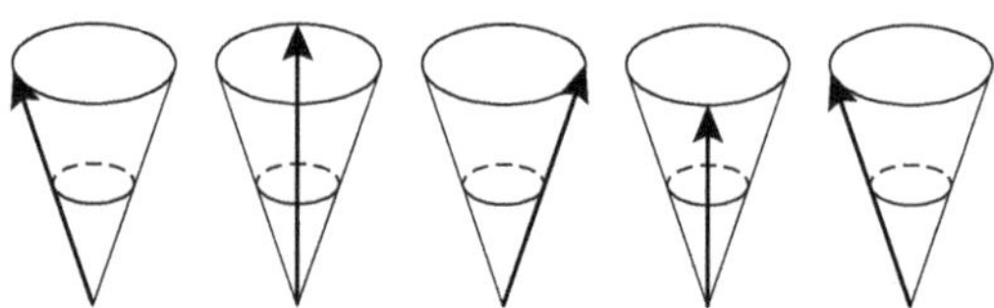

Fig. 8.10. Goldstone mode in a superfluid is analogous to a spin wave of a ferromagnet. Spin rotational symmetry is spontaneously broken and a low-energy Goldstone mode appears as a spin wave. As one moves from site to site, spins rotate slightly on a cone. Energy vanishes linearly as wave vector approaches 0, which is a hallmark of the Goldstone mode.

- Topological Excitations in Ring of Charge Neutral Superfluid

The superfluid number density ρ satisfies the continuity equation (as we mentioned before this is a consequence of Noether's theorem)

$$\frac{d\rho}{dt} + \nabla \vec{j} = 0, \tag{8.103}$$

where the current density is defined as

$$\vec{j}(\vec{r}, t) = -\frac{i\hbar}{2m^*}(\Psi^* \nabla \Psi - \Psi \nabla \Psi^*). \tag{8.104}$$

Plugging the condensate wave function $\Psi = \sqrt{\rho}e^{i\varphi}$ into the expression for the current density and using that the current density is $j = \rho v$, we find the velocity of the superfluid in terms of the phase gradient

$$v = \frac{\hbar}{m^*}\nabla\phi. \tag{8.105}$$

Here we have assumed that only the phase of the wave function varies with position.

Now consider a ring of a superfluid. The accumulated phase as one moves around the ring must be an integer [14] multiple of 2π

$$\oint dl \frac{d\phi}{dl} = 2\pi M. \tag{8.106}$$

The winding number M plays an important role in a neutral superfluid ring. When the integer $M = 0$, the superfluid phase is independent of position. When $M = 1$, the phase changes by 2π as one goes around the ring. A state with a certain value of topological charge M cannot change into another state with a different value of M. Such a state is topologically stable. A supercurrent in a narrow ring does not decay because it is a topological effect of a multiply connected domain, i.e., of a ring.

- Gross–Pitaevskii Equation

An atomic Bose–Einstein condensate is a weakly interacting Bose gas and is a superfluid, unlike the ideal Bose–Einstein condensate. We consider an atomic Bose–Einstein condensate in a finite confinement potential.[m] From the Ginzberg–Landau free-energy density (8.94) one can find the Hamiltonian of a weakly interacting Bose condensate[n]

$$H = \sum_i \frac{\vec{p}_i^2}{2m^*} + \sum_{i<j} U(\vec{r}_i - \vec{r}_j) + \sum_i V(\vec{r}_i), \tag{8.107}$$

where $V(r)$ is the external potential that confines bosons and $U(\vec{r}_i - \vec{r}_j) = \lambda\delta(\vec{r}_i - \vec{r}_j)$ is the particle interaction. The particle mass is m^*. The ground state Ansatz is

$$\psi(\vec{r}_1, \ldots, \vec{r}_N) = \prod_{i=1}^{N} \phi(\vec{r}_i), \tag{8.108}$$

where $\phi(\vec{r})$ is a single particle wave function to be determined. This state has a definite number of bosons. We define the condensate wave function as

$$\Psi(r) = \sqrt{N}\phi(r) \tag{8.109}$$

so that it is normalized to give the total number of particles N. Using a Lagrange multiplier ϵ to minimize the ground state energy expectation value $E[\Psi]$, we find

$$\frac{\delta}{\delta\Psi^*(r)} \left[E[\Psi] - \epsilon \left(\int dr |\Psi|^2 - N \right) \right] = 0, \tag{8.110}$$

which gives the time-independent Gross–Pitaevskii equation [21]

$$-\frac{\hbar^2 \nabla^2}{2m^*} \Psi(\vec{r}) + V(\vec{r})\Phi + \lambda\Psi(\vec{r})|\Psi(\vec{r})|^2 = \epsilon\Psi. \tag{8.111}$$

---

[m] Strictly speaking there is no phase transition in a finite system. However, the number of trapped atoms is large enough and the transition temperature is sharply defined.
[n] A weakly interacting Bose gas does not give rise to the roton minimum (see Sec. 11.8) in the excitation spectrum, in contrast strongly interacting ^{4}He.

An interacting charge neutral Bose–Einstein condensate can be described by the time-dependent Gross–Pitaevskii equation (see Exercise 8.15)

$$i\hbar\frac{\partial\Psi}{\partial t} = -\frac{\hbar^2\nabla^2}{2m^*}\Psi(\vec{r}) + V(\vec{r})\Phi + \lambda\Psi(\vec{r})|\Psi(\vec{r})|^2. \tag{8.112}$$

Exercise 8.14. Derive the time-independent Gross–Pitaevskii equation from Eq. (8.110). Hint: First compute the ground state energy $E[\Psi] = \langle\Psi|H|\Psi\rangle$ in terms of $\phi(r)$ and $\phi^*(r)$. Then functional differentiate with respect to $\phi^*(r)$.

Exercise 8.15. Derive the time-dependent Gross–Pitaevskii equation. Hint: In the Heisenberg picture the field operator $\hat{\Psi}$ satisfies

$$i\hbar\frac{\partial\hat{\Psi}}{\partial t} = [\hat{\Psi}(\vec{r}), \hat{H}]. \tag{8.113}$$

The Hamiltonian $\hat{H}$ contains the following contact interaction term:

$$\hat{V} = \frac{\lambda}{2}\int \hat{\Psi}(\vec{r})^{\dagger 2}\hat{\Psi}(\vec{r})^2. \tag{8.114}$$

First evaluate the commutator. Then replace the expectation value of the field operator with the condensate wave function, $\langle\hat{\Psi}(\vec{r})\rangle = \Psi(\vec{r})$ and $\langle\hat{\Psi}^\dagger(\vec{r})\rangle = \Psi^*(\vec{r})$.

8.7. Breaking Local Gauge Transformation?⋆

The BCS wave function has a definite phase, which means that local gauge transformation is absent, implying that local charge conservation is broken (see also Exercise 8.16). But the true order parameter of a charged system must be local gauge invariant. We discuss this issue here. (There is a difference between local and global charge conservation laws, see Exercise 8.17.)

What is a local gauge transformation? (We prefer to use this term over "local gauge symmetry".) It implies that change of the phase of an electron wave function by a different amount at each point in spacetime will not change physics. Consider a charged particle in electromagnetic potentials $(\phi, \vec{A})$. Suppose we make a local phase transformation to the wave function of the particle

$$\psi(\vec{r}, t) \rightarrow e^{i \frac{q}{\hbar c} \lambda(\vec{r}, t)} \psi(\vec{r}, t), \tag{8.115}$$

where q is the charge of the particle. The transformed wave function does not satisfy the same Schrödinger equation. It satisfies a transformed Schrödinger equation

$$H = \frac{1}{2m} \left(\frac{\hbar}{i} \nabla - \frac{q}{c} \vec{A}' \right)^2 + q\phi' \tag{8.116}$$

with the following changes

$$\begin{aligned} \vec{A}'(\vec{r}, t) &= \vec{A}(\vec{r}, t) + \nabla \lambda(\vec{r}, t), \\ \phi'(\vec{r}, t) &= \phi(\vec{r}, t) - \frac{1}{c} \frac{\partial \lambda(\vec{r}, t)}{\partial t}. \end{aligned} \tag{8.117}$$

This is the well-known gauge transformation of electrodynamics. Note that the transformation has no effect on $\vec{B}$ because $\nabla \times \nabla \lambda = 0$. When $\lambda(\vec{r}, t)$ is a constant, the Schrödinger equation does not change at all (it is a global gauge transformation). Invariance under local phase transformation of the wave function requires the existence of the electromagnetic fields. Electromagnetism is called a gauge theory because the gauge transformation requires the existence of electromagnetic forces.

A local gauge symmetry acts trivially on all observables and is merely a redundancy in our description. Let us quote what Wen [22] said about gauge symmetry breaking: "Usually when the same thing has the same properties, we do not say that there is a symmetry. Thus the terms 'gauge symmetry' and 'gauge symmetry breaking' are two of the most misleading terms in theoretical physics ... We will say there is a gauge structure (instead of a gauge 'symmetry') when we use many labels to label the same state. When we change our labeling scheme, we will say that there is a change of gauge structure (instead of gauge 'symmetry breaking')".

Exercise 8.16. Using locally gauge invariant Maxwell's equations derive the continuity equation (local charge conservation)

$$\nabla \vec{j} + \frac{\partial \rho}{\partial t} = 0, \tag{8.118}$$

where $\vec{j}$ and ρ are current and charge densities.

> **Exercise 8.17.** What is the difference between local and global charge conservation laws? Answer: global charge conservation law allows some charge to instantaneously move from place to place. Charge is conserved locally if it leaves a region by moving with some velocity. Consult Ref. [23].

- Global Phase Transformation

One should distinguish between global phase transformation and local gauge transformation. The global phase invariance is a subgroup of local gauge transformations. A global gauge group is responsible for a conservation law and is a real symmetry (see Noether's theorem in Appendix H). Global U(1) symmetry may be spontaneously broken in a condensed matter system in the **absence** of an electromagnetic field. However, as mentioned above, local gauge symmetry cannot spontaneously break down as a matter of principle, since it is not a physical symmetry of the system to begin with, but merely an invariance of description. Local gauge transformation is analogous to a rotation of coordinate axis while global phase transformation is analogous to a rotation of the physical system.° Physics cannot change because one rotates the coordinate axis. On the other hand, if for example an antiferromagnetic state is rotated, another degenerate state is obtained.

A good example of a global phase invariance is superfluidity. Under global U(1) transformation, a creation operator transforms as

$$a_{\vec{k}}^{\dagger} \to e^{i\lambda} a_{\vec{k}}^{\dagger}, \tag{8.119}$$

which leaves the Lagrangian of the superfluid unchanged (the invariance of the Lagrangian leads to conservation of charge, see Noether's theorem, Appendix H). However, the local order parameter of a superfluid transforms as

$$\Psi \to e^{-i\lambda} \Psi, \tag{8.120}$$

and does not leave it invariant. Thus the Lagrangian/Hamiltonian equation (8.94) is invariant but the ground state order parameter is not — this is an example of a spontaneously broken symmetry. A superconductor also breaks global gauge symmetry.

°On this point, students are encouraged to read Ref. [24]. It is a good pedagogical paper.

- ● Local Gauge Transformation and Non-local Order Parameter

Local gauge transformation may be applied to a charged superconductor. The BCS wave function may look different in a different gauges, but the state itself will remain the same. This can be seen as follows. A general N-electron state can be written as

$$|\chi\rangle = \int d\vec{r}_1 \dots d\vec{r}_N \chi(\vec{r}_1 \dots \vec{r}_N)\psi^\dagger(\vec{r}_1)\dots\psi^\dagger(\vec{r}_N)|0\rangle, \qquad (8.121)$$

where $\psi^\dagger(\vec{r})$ is the electron creation field operator. Here spin variables are suppressed. Under a local gauge transformation, the **wave function** transforms as

$$\chi(\vec{r}_1 \dots \vec{r}_N) \to \prod_i^N e^{-i\frac{e}{\hbar c}\lambda(\vec{r}_i)}\chi(\vec{r}_1 \dots \vec{r}_N) \qquad (8.122)$$

and the field operators transform as

$$\psi^\dagger(\vec{r}_i) \to e^{i\frac{e}{\hbar c}\lambda(\vec{r}_i)}\psi^\dagger(\vec{r}_i). \qquad (8.123)$$

Thus a local gauge transformation leaves the state, including the **BCS state** $|\Psi\rangle$ of Eq. (8.121), invariant. One can also show directly from the BCS state

$$|\chi\rangle = \prod_{\vec{k}}(u_{\vec{k}} + v_{\vec{k}}e^{i\phi}c^\dagger_{\vec{k}\uparrow}c^\dagger_{-\vec{k}\downarrow})|0\rangle \qquad (8.124)$$

that it is invariant under a local gauge transformation. This is because under a local gauge transformation we have

$$c^\dagger_{\vec{k}\sigma} \to e^{i\frac{e}{\hbar c}\lambda}c^\dagger_{\vec{k}\sigma}, \quad \phi \to \phi - \frac{2e}{\hbar c}\lambda. \qquad (8.125)$$

Here ϕ is the phase of the order parameter, i.e., the phase of the wave function of the bound Cooper pair. The order parameter

$$\Psi(\vec{r}) = |\Psi(\vec{r})|e^{i\phi} = \langle\chi|\psi_\uparrow(\vec{r})\psi_\downarrow(\vec{r})|\chi\rangle = \frac{1}{V}\sum_{\vec{k}}v_{\vec{k}}u^*_{\vec{k}}e^{i\phi} \qquad (8.126)$$

transforms as

$$\phi \to \phi - \frac{2e}{\hbar c}\lambda(\vec{r}). \qquad (8.127)$$

It appears that the "order parameter" $\Psi(\vec{r})$ is **not** invariant under local gauge transformation. But the real order parameter should be, as we discussed before. Moreover, Elitzur's theorem [5] dictates that a quantity that transforms as $f(\vec{r}) \to e^{-i\frac{2e}{\hbar c}\lambda(\vec{r})} f(\vec{r})$ under local gauge transformation is zero when averaged over all possible gauge fields. It turns out that, in reality, the true order parameter is a **non-local** quantity [2] which remains invariant under local gauge transformation (a non-local order parameter is consistent with the expectation that a superconductor is topologically ordered). Strictly speaking, the traditional definition of the order parameter of superconductivity is thus not the correct one. However, the results obtained from the Ginzburg–Landau theory are all correct, including the Higgs mechanism and the Meissner effect (in a certain gauge, namely the Coulomb gauge $\nabla \vec{A} = 0$, the non-local order parameter reduces to the usual local one [3–5]). In the following, we will assume that the order parameter transforms according to local gauge transformation in the usual way.

8.8. Anderson–Higgs Mechanism⋆

In superconductors, global U(1) symmetry is broken but local gauge symmetry should be present. One may think that a Goldstone mode exists but does it really exist in the presence of local gauge invariance? Anderson solved this problem by maintaining gauge invariance at every stage of his theory [25]. He included quantum fluctuations using the random phase approximation, which obeys local charge conservation. (We briefly encountered the random phase approximation at the end of Sec. 6.14. See Ref. [6] for the role of conserving approximations [26] in the gauge invariant formulation of superconductors.) Anderson's approach showed that the Goldstone mode is replaced by another mode: the gauge photon absorbs the Goldstone boson originating from the ground state with a broken symmetry. As a result, the photons become massive. (This mode is not the massive Higgs boson, i.e., it is not the amplitude mode that absorbs the Goldstone boson.) The inverse mass of the new mode is proportional to the range λ of the modified electromagnetic force, which is short-ranged. In 3D, the Goldstone mode acquires a gapped dispersion due to the long-range character of Coulomb interactions. The Goldstone mode is lifted to the plasma frequency. This is the famous Anderson-Higgs mechanism.[P]

[P]The Nobel Prize in Physics 2013 was awarded jointly to François Englert and Peter W. Higgs for this theoretical discovery. Higgs himself referred it as the "ABEGHHKtH" mechanism, the full acronym referring to Anderson, Brout, Englert, Guralink, Hagen, Higgs, Kibble, and t'Hooft. Read the fascinating story about how these physicists were all involved in the discovery of the mechanism, see "The infinity Puzzle" by F. Close [27].

• Massive Photon Mode in Charged Superconductor

Let us derive the Anderson–Higgs mechanism. Instead of using the Green's function approach, we follow a different but equivalent derivation using the path integral method (see Ref. [20]). The classical action of a superconductor in the presence of an electromagnetic field is assumed to be given by the Ginzburg–Landau free-energy density.[q] In this approach, one pretends that Ψ is a local order parameter with the appropriate transformation property. The Lagrangian is

$$\mathcal{L} = i\hbar\Psi^*\frac{\partial\Psi}{\partial t} + \frac{1}{2m^*}|(\vec{p} - \frac{q}{c}\vec{A}(\vec{r}))\Psi|^2 + \alpha|\Psi|^2 + \frac{\beta}{2}|\Psi|^4 + \frac{|\vec{B}(\vec{r})|^2}{8\pi},$$

$$(8.128)$$

where $\vec{p}$ is the momentum operator, $\vec{A}(\vec{r})$ is a vector potential, and $\vec{B}(\vec{r})$ is a magnetic field. The mass of a Cooper pair m^* is $2m$ and its charge $q = -2e$ (m is the electron mass and $e > 0$). Because of the covariant derivative, $\mathcal{L}$ is invariant under local gauge transformation. The order parameter

$$\Psi(\vec{r}) = \sqrt{\rho}e^{i\theta} \tag{8.129}$$

could be interpreted as the center of mass wave function of Cooper pairs of electrons. The current density is

$$\vec{j}(\vec{r}) = -c\frac{\partial\mathcal{L}}{\partial\vec{A}(\vec{r})} = -\frac{|\Psi|^2 q^2}{m^* c}(\vec{A}(\vec{r}) - \frac{\hbar c}{q}\nabla\theta(\vec{r})). \tag{8.130}$$

In the absence of electromagnetic fields, the Ginzburg–Landau free-energy density has the same form as the Lagrangian density of a neutral superfluid given in Eq. (8.94), except for an unimportant constant (this can be verified by doing Exercise 8.18).

Exercise 8.18. Show that the superfluid density is $\rho_0 = \frac{-\alpha}{\beta}$. Hint: Minimizing the Ginzburg–Landau free energy when $\vec{A} = 0$. Consult Ref. [28].

---

[q]A good introduction to the Ginzburg–Landau theory can be found in Ref. [28].

The total Lagrangian density is ($\hbar = 1$ and $c = 1$)

$$\mathcal{L} = -\rho(\partial_0\theta + qA_0) - \frac{1}{2m^*}\left[\frac{1}{4\rho}(\nabla\rho)^2 + \rho(\nabla\theta - q\vec{A})^2\right]$$

$$-\frac{g}{2}(n - \rho)^2 - \frac{1}{4}F_{\mu\nu}F^{\mu\nu}, \tag{8.131}$$

where $F_{\mu\nu}$ is the electromagnetic field tensor. It is convenient to form a gauge invariant quantity

$$A'_\mu = \frac{1}{q}\partial_\mu\theta(x) + A_\mu(\vec{r}). \tag{8.132}$$

(Here $\partial_\mu = (\frac{\partial}{\partial t}, \frac{\partial}{\partial x}, \frac{\partial}{\partial y}, \frac{\partial}{\partial z})$ and $\partial^\mu = (\frac{\partial}{\partial t}, -\frac{\partial}{\partial x}, -\frac{\partial}{\partial y}, -\frac{\partial}{\partial z})$.) Note that this is a **gauge transformation of $A_\mu(\vec{r})$, and is facilitated by the phase $\theta(\vec{r})$ of the order parameter**. The electromagnetic field tensor is not affected by this transformation

$$F_{\mu\nu} = \partial_\mu A_\nu - \partial_\nu A_\mu = \partial_\mu A'_\nu - \partial_\mu A'_\nu. \tag{8.133}$$

The Lagrangian density can be written as

$$\mathcal{L} = -\rho qA'_0 - \frac{1}{2m^*}\left[\frac{1}{4\rho}(\nabla\rho)^2 + \rho q^2 \vec{A}'^2\right] - \frac{g}{2}(n - \rho)^2 - \frac{1}{4}F_{\mu\nu}F^{\mu\nu}. \tag{8.134}$$

The fluctuations a of the amplitude of the order are separated as follows:

$$\sqrt{\rho} = \sqrt{\rho_0} + a. \tag{8.135}$$

Plugging this into the Lagrangian density, we find

$$\mathcal{L} = -\frac{\rho_0 q^2}{2m^*}\vec{A}'^2 - \frac{1}{2m^*}(\nabla a)^2 - 2g\rho_0 a^2 - 2q\sqrt{\rho_0}A'_0 a - \frac{1}{4}F_{\mu\nu}F^{\mu\nu}. \tag{8.136}$$

Using functional integration, one integrates out the fluctuating amplitude mode a and finds

$$-a\left(-\frac{1}{2m^*}\nabla^2 + 2g\rho_0\right)a - (2q\sqrt{\rho_0}A'_0)a$$

$$\rightarrow q\sqrt{\rho_0}A'_0\frac{1}{(-1/2m^*)\nabla^2 + 2g\rho_0}q\sqrt{\rho_0}A'_0. \tag{8.137}$$

The effective Lagrangian density can now be written as

$$
\begin{aligned}
\mathcal{L}_{\text{eff}} &= -\frac{\rho_0 q^2}{2m^*}\vec{A}'^2 + q\sqrt{\rho_0}A'_0 \frac{1}{(-1/2m)\nabla^2 + 2g\rho_0} q\sqrt{\rho_0}A'_0 - \frac{1}{4}F_{\mu\nu}F^{\mu\nu} \\
&= \frac{q^2}{2g}A'^2_0 - \frac{\rho_0 q^2}{2m^*}\vec{A}'^2 - \frac{1}{4}F_{\mu\nu}F^{\mu\nu} \\
&= \frac{q^2}{2g}(A'^2_0 - v^2\vec{A}'^2) - \frac{1}{4}F_{\mu\nu}F^{\mu\nu}.
\end{aligned}
\tag{8.138}
$$

where $v^2 = \rho_0 g/m^*$. (Here we have ignored again the gradient term in the denominator.) This is precisely the Lagrangian for the **massive** electromagnetism. It represents the new relevant field of a longitudinal plasmon-like mode[r] (the photon mode is not present anymore). Using this Lagrangian density, one can derive the Euler–Lagrange equation for A'_μ,

$$
(\partial^2 + M^2)A'_\mu = 0,
\tag{8.139}
$$

This is the Klein–Gordon equation of a massive field. It shows that a longitudinal plasmon-like mode exists whose energy goes to a finite value as wave vector approaches zero (the photon mode is not present anymore). Here M is the new mass. Metaphorically speaking, the gauge field A_μ has eaten the Nambu–Goldstone boson and has gained weight [19].

Exercise 8.19. (a) Derive the Euler–Lagrange equation

$$
\partial_\mu F^{\mu\nu} + M^2 A'_\mu = 0,
\tag{8.140}
$$

Hint: Read Section 13.2 in Ref. [20]. (b) From this, derive the Klein–Gordon equation (8.139). Hint: Use $\partial_\mu\partial_\nu F_{\mu\nu} = 0$ and $\partial_\mu A'_\mu = 0$.

---

[r]As noted by Anderson, this phenomenon is somewhat analogous to what is happening within plasma, where a longitudinal electromagnetic wave exists besides two transverse ones, in contrast to vacuum with only two transverse electromagnetic modes. Gauge invariance and a massive carrier of a force can coexist in a plasma.

8.9. Meissner Effect and Flux Quantization

There are two hallmarks of superconductivity, namely the Meissner effect and flux quantization. According to the Meissner effect, a superconductor expels magnetic fields. Flux quantization means that a superconductor ring can only trap a quantized value of magnetic flux.

Inside a superconductor, one may assume $\rho \approx$ constant. With the assumption $\theta =$ constant, we have

$$\vec{j} = -\frac{q^2 \rho}{2m^* c}\vec{A}, \tag{8.141}$$

where the charge is $q = -2e$. From the Maxwell's equations

$$\nabla \times \vec{B} = \frac{4\pi}{c}\vec{j}, \quad \vec{B} = \nabla \times \vec{A}, \tag{8.142}$$

one finds in the Landau gauge

$$\nabla \cdot \vec{A} = 0, \quad \nabla \times \nabla\vec{A} = \nabla(\nabla \cdot \vec{A}) - \nabla^2\vec{A}. \tag{8.143}$$

We find that the vector potential of a superconductor satisfies

$$\nabla^2\vec{A} = \frac{\vec{A}}{\lambda^2}, \tag{8.144}$$

where the penetration length λ is given by

$$\frac{1}{\lambda^2} = \frac{8\pi\rho e^2}{m^* c^2}. \tag{8.145}$$

The vector potential decays over the penetration depth near the surface of a superconductor. Because of this vector potential, a surface current flows on the surface (see Eq. (8.141)) that counteracts the external magnetic field. For the Meissner effect to exist, it is crucial that a superconductor is charged. This Meissner effect is depicted in Fig. 8.11.

Now let us show that magnetic flux threading a superconducting ring is quantized. Thread the hole of the ring with a magnetic flux. Then consider a loop lying well within the interior of the ring. The circulation of the current along this loop must be zero. The expression for the current,

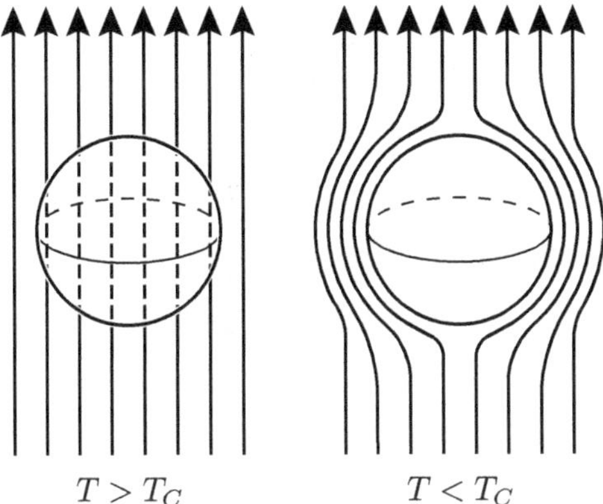

Fig. 8.11. Below the critical temperature T_c magnetic fields are expelled from a super-conductor. A supercurrent flows on the surface of the superconductor that counteracts the external magnetic field.

Eq. (8.130), has two contributions. One from the vector potential is

$$\int \vec{A} \cdot d\vec{l} = \int \vec{B} d\vec{S} = \Phi.$$

(8.146)

The other contribution from the phase gradient term is an integer multiple of 2π

$$\oint \nabla \theta \cdot d\vec{l} = 2\pi N$$

(8.147)

since the order parameter is single-valued, i.e., the wave function is periodic in the angle ϕ. (Here it must be noted that a twist boundary condition (see, Sec. 3.2) cannot be imposed on the wave function of a superconductor because the phase of the order parameter exhibits a phase stiffness.) Combining these results it follows from the expression for the current equation (8.130) that the trapped flux is quantized in units $\Phi_0/2$

$$\Phi_S = \Phi_0 \frac{N}{2}.$$

(8.148)

8.10. Ground State Degeneracy

Unlike a neutral superfluid, a charged superconductor does not have a long-range order. Instead it may have topological order. In this section, we will explain ground state degeneracy of a superconductor [11], which is indicative of the presence of topological order. Consider a torus with solenoids

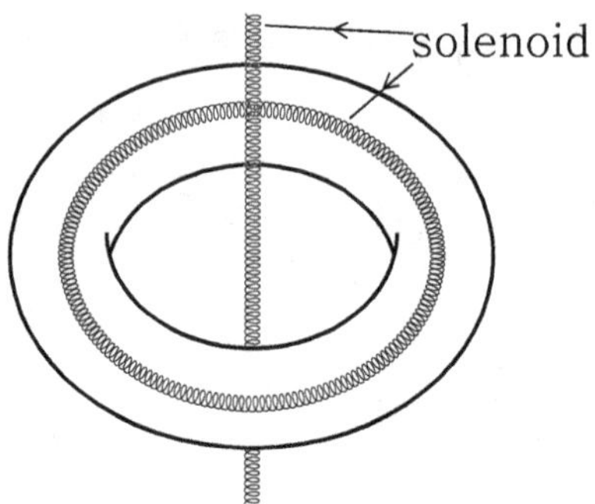

Fig. 8.12.   Imagine Cooper pairs living on the surface of a torus threaded by two magnetic fluxes.

that wind through the two holes of it. Imagine electrons living on the surface of the torus, see Fig. 8.12. Note the following important point: electrons on the surface do **not "feel"** the magnetic field of the solenoids. This means that the electron ground state of such a system should be four-fold degenerate. This can be explained as follows. Let the flux through ith hole be $\Phi_i = n_i \Phi_S$, where $\Phi_S = \frac{hc}{2e}$ and n_i is an integer and $i = 1, 2$ label the solenoids. Then states $(n_1, n_2) = (0, 1), (1, 0), (0, 0), (1, 1)$ represent different degenerate states, implying a degeneracy of four. Note that since Φ_S is half of Φ_0 it cannot be gauged way (see Sec. 3.2). We will see later that the topologically ordered state of the toric code has also a fourfold degeneracy on the torus (see Chapter 13).

Bibliography

[1] S. A. Kivelson and D. S. Rokhsar, Bogoliubov quasiparticles, spinons, and spin-charge decoupling in superconductors, *Phys. Rev. B* **41**, 11693 (1990).

[2] T. K. Kvorning, C. Spanslatt, A. P. O. Chan, and S. Ryu, Nonlocal order parameters for the states with topological electromagnetic response, *Phys. Rev. B* **101**, 205101 (2020).

[3] T. H. Hansson, V. Organesyan, and S. L. Sondhi, Superconductors are topologically ordered, *Ann. Phys.* **313**, 497 (2004).

[4] E. Fradkin, *Field Theories of Condensed Matter Physics* (Cambridge University Press, Cambridge, 2013).

[5] S. Elitzur, Impossibility of spontaneously breaking local symmetries, *Phys. Rev. D* **12**, 3978 (1975).

[6] J. R. Schrieffer, *Theory of Superconductivity* (CRC Press, Taylor & Francis Group, New York, 2018). •

[7] J. F. Annett, *Superconductivity, Superfluids, and Condensates* (Oxford University Press, Oxford, 2004). •

[8] N. W. Ascroft and N. David Mermin, *Solid State Physics* (Thomson Learning, London, 1976). •

[9] W. Kohn and J. M. Luttinger, New mechanism for superconductivity, *Phys. Rev. Lett.* **15**, 524 (1965).

[10] R. P. Feynman, R. B. Leighton, and M. Sands, *The Feynman Lectures on Physics Vol. III: Quantum Mechanics* (Addison-Wesley, Boston, 1977). •

[11] S. M. Girvin and K. Yang, *Modern Condensed Matter Physics* (Cambridge University Press, Cambridge, 2019).

[12] G. Carleo and M. Troyer, Solving the quantum many-body problem with artificial neural networks, *Science* **355**, 602 (2017).

[13] David Deutsch, *Creative Blocks*, Aeon, 3 Oct. (2012). •

[14] Y. V. Nazarov and J. Danon, *Advanced Quantum Mechanics* (Cambridge University Press, Cambridge, 2013).

[15] F. Wilczek, Majorana returns, *Nature Phys.* **5**, 614 (2009). •

[16] B. Greene, *The Hidden Reality: Parallel Universes and the Deep Laws of the Cosmos* (Vintage, New York, 2011). •

[17] M. Tegmark, *Our Mathematical Universe: My Quest for the Ultimate Nature of Reality* (A. A. Knopf, New York, 2014). •

[18] S. H. Strogatz, *Nonlinear Dynamics and Chaos* (CRC Press, Boca Raton, 2015). •

[19] A. Zee, *Fly by Night Physics: How Physicists Use the Backs of Envelopes* (Princeton University Press, Princeton, 2020). •

[20] T. Lancaster and S. J. Blundell, *Quantum Field Theory for the Gifted Amateur* (Oxford University Press, Oxford, 2014). •

[21] K. Konishi and G. Paffuti, *Quantum Mechanics* (Oxford University Press, Oxford, 2009).•

[22] X.-G. Wen, *Quantum Field Theory of Many-Body Systems: From the Origin of Sound to an Origin of Light and Electrons* (Oxford University Press, Oxford, 2004).

[23] R. P. Feynman, *The Character of Physical Law* (The MIT Press, 2017). •

[24] M. Greiter, Is electromagnetic gauge invariance spontaneously violated in superconductors? *Ann. Phys.* **319**, 217 (2005). •

[25] P. W. Anderson, Random-phase approximation in the theory of superconductivity, *Phys. Rev.* **112**, 1900 (1958).

[26] L. Kadanoff and G. Baym, *Quantum Statistical Methods* (CRC Press, Boca Raton, 1962).

[27] F. Close, *The Infinity Puzzle: Quantum Field Theory and the Hunt for an Orderly Universe* (Basic Books, New York, 2011). •

[28] N. Goldenfeld, *Lectures on Phase Transitions and the Renormalization Group* (CRC Press, Boca Raton, 1992). •

Chapter 9

p-Wave Superconductors and Majorana Zero Modes

"There are known knowns; there are things we know we know. We also know there are known unknowns; that is to say we know there are some things we do not know. But there are also unknown unknowns–the ones we don't know we don't know."

Donald Rumsfeld

We saw in the previous chapter that the ground state of a s-wave superconductor is degenerate in torus geometry and that it displays spin–charge separation. These are some characteristic features of a topologically ordered phase. In a p-type superconductor, a neutral quasiparticle can split into two Majorana[a] zero modes and exchanging these objects gives rise to non-Abelian statistics, which is a property unique to a topological ordered phase [2].[b] Wilczek has succinctly explained what a Majorana zero mode is [5]: "Majorana modes have a statistic that is different and more complex than conventional anyons. Specifically, the statistic is inherently non-Abelian–exchanges of particles associated with Majorana modes result not only in a change of the phase of the quantum mechanical wavefunction, but also in the change of the internal states of the modes". It should be stressed that a zero mode is not really a particle but a part of a quasiparticle.

---

[a]Majorana disappeared mysteriously from a passenger ship and was never to be seen again. Read about his life in "A Brilliant Darkness" [1].

[b]The 5/2 quantum Hall state also exhibits non-Abelian anyons [3,4], unlike the Laughlin states that display Abelian anyons.

233

In this chapter, our goal is not ambitious. We will try to understand the basic concepts and ideas of Majorana half modes using a one-dimensional Kitaev toy model for spinless fermions exhibiting a topologically ordered odd pairing superconductivity. One-dimensional Majorana chain is called an invertible topologically ordered system because it has no fractionalized excitations. Then we will explain the rudiments of a more realistic model of two-dimensional p-wave superconductivity. For a more in depth and lengthy treatment of this exciting field, the reader should read other articles.

9.1. One-Dimensional Periodic p-Wave Superconductor: Kitaev Model

Before we study a two-dimensional p-wave superconductor, let us consider a simpler one-dimensional p-wave superconductor.

- Hamiltonian and Topological Phase

The Kitaev model exhibits a finite topological winding number. Let us show this. Kitaev's Hamiltonian [6] of a lattice with N sites is

$$H = -\sum_{j=1}^{N}[t(c_j^\dagger c_{j+1} + c_{j+1}^\dagger c_j) + \mu c_j^\dagger c_j + \Delta(c_j^\dagger c_{j+1}^\dagger + c_{j+1}c_j)], \quad (9.1)$$

where $c_j^\dagger$ creates a fermion at site i and the third term is the chemical potential term. This mean-field Hamiltonian is consistent with a p-wave pairing (compare this with the Hamiltonian of a s-wave pairing, Eq. (8.44)).[c] Under periodic boundary condition, one can transform the Hamiltonian into k-space. The mean-field Hamiltonian for p-wave pairing is

$$H = \sum_{k\in\mathrm{BZ}}[\epsilon_k c_k^\dagger c_k - \Delta(k)c_k^\dagger c_{-k}^\dagger - \Delta^*(k)c_{-k}c_k], \quad (9.3)$$

where

$$\epsilon_k = -2t\cos ka - \mu \quad (9.4)$$

---

[c]In the continuum limit the mean-field Hamiltonian of a one-dimensional p-wave super-conductor can be written as

$$H_{1\mathrm{D}} = \int dx\left\{\psi^+(x)\left(-\frac{\nabla_x^2}{2m} - \mu\right)\psi(x) + \psi(x)\Delta' e^{i\phi}ip_x\psi(x) + \mathrm{h.c.}\right\}. \quad (9.2)$$

Here $\hbar = 1$.

and

$$\Delta(k) = i\Delta \sin ka, \tag{9.5}$$

which is an odd function in k (a is the lattice constant). Note that the gap is zero at $k = 0$ and $k = \pm\pi/a$.

Exercise 9.1. We need the following identity for p-wave superconductors

$$\sum_j c_j c_{j+1} = \frac{1}{2}\sum_k (c_k c_{-k} e^{-ika} + c_{-k} c_k e^{ika}) = -\sum_{k \in \mathrm{BZ}} 2i c_k c_{-k} \sin ka. \tag{9.6}$$

Show this. Note that c_j and c_{j+1} anticommute. Hint: Use $c_j = \frac{1}{\sqrt{N}}\sum_k e^{ijak} c_k$, where a is the lattice constant and sites $j = N+1$ and $j = 1$ are equivalent due to periodic boundary conditions.

The mean-field Hamiltonian can be written as

$$H_{\mathrm{MF}} = \sum_k (c_k^\dagger, c_{-k}) \frac{1}{2}\begin{pmatrix} \epsilon_k & -2\Delta(k) \\ -2\Delta^*(k) & -\epsilon_{-k} \end{pmatrix}\begin{bmatrix} c_k \\ c_{-k}^\dagger \end{bmatrix}$$

$$= \sum_k (c_k^\dagger, c_{-k}) H(k) \begin{bmatrix} c_k \\ c_{-k}^\dagger \end{bmatrix}. \tag{9.7}$$

The 2×2 matrix has the following form:

$$H(k) = \vec{h} \cdot \vec{\sigma} = \begin{pmatrix} h_z & h_x - ih_y \\ h_x + ih_y & -h_z \end{pmatrix}, \tag{9.8}$$

where

$$h_x = 0, \; h_y = \Delta \sin ka, \; h_z = \epsilon_k/2. \tag{9.9}$$

Since the Hamiltonian Eq. (9.8) has diagonal terms proportional to h_z, it is not in the desirable form. One can rotate the coordinate system counterclockwise by $\pi/2$ about the y-axis, which gives the new field $\vec{R} = (0, h_y, h_z) \to (h_z, h_y, 0)$. This will bring h_z from diagonal elements to off-diagonal elements

$$\tilde{H}(k) = \vec{R} \cdot \vec{\sigma} = \begin{pmatrix} 0 & h_z - ih_y \\ h_z + ih_y & 0 \end{pmatrix} = h_z \sigma_x + h_y \sigma_y. \tag{9.10}$$

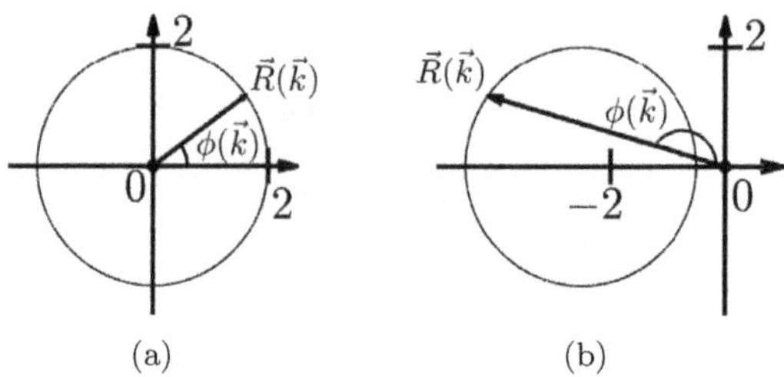

Fig. 9.1. Schematic display of how the vector $\vec{R}(\vec{k})$ rotates as $\vec{k}$ moves across the Brillouin zone. (a) $\mu = 0$, $\Delta/t = 1$ and (b) $\mu = 2.5$, $\Delta/t = 1$.

Note that this Hamiltonian has the same structure as that of topological polyacetylene, see Eq. (4.22). Kitaev model has particle–hole symmetry.

> **Exercise 9.2.** Show that the transformed Hamiltonian $\tilde{H}(k) = UH(k)U^{\dagger}$ is given in Eq. (9.10) (see Appendix H for unitary transformations). Hint: Look up the spin rotation operator $U = e^{-i\frac{\pi}{4}\sigma_y}$.

We have investigated how $\vec{R}$ rotates as k varies across the first Brillouin zone. A finite Berry phase will be present if $\vec{R}$ encircles the origin, which will depend on the values of Δ and μ, see Fig. 9.1. The system is topologically **non-trivial** for $|\mu| < 2t$.

9.2. Majorana Half Modes in Finite-Length One-Dimensional p-Wave Superconductor

Consider Kitaev's toy model of one-dimensional p-wave superconductivity. It exhibits a charge neutral particle that is divided between the two ends of the chain. Particle–hole symmetry plus bulk edge correspondence is consistent with the existence of a Majorana edge state with zero energy.

- Numerical Study of Majorana Edge Modes of Kitaev Model

The existence of such zero modes can be investigated by finding excitations of a **finite length** chain. Its mean-field Hamiltonian can be written as

$$H = [c_1^\dagger, \ldots, c_N^\dagger, c_1, \ldots, c_N] h \begin{bmatrix} c_1 \\ \cdots \\ c_N \\ c_1^\dagger \\ \vdots \\ c_N^\dagger \end{bmatrix} + \text{constant.} \tag{9.11}$$

Consider for a moment a finite length chain consisting of four sites. The Hamiltonian matrix h is

$$h = \begin{pmatrix} -\mu/2 & -t/2 & 0 & 0 & 0 & -\Delta/2 & 0 & 0 \\ -t/2 & -\mu/2 & -t/2 & 0 & \Delta/2 & 0 & -\Delta/2 & 0 \\ 0 & -t/2 & -\mu/2 & -t/2 & 0 & \Delta/2 & 0 & -\Delta/2 \\ 0 & 0 & -t/2 & -\mu/2 & 0 & 0 & \Delta/2 & 0 \\ 0 & \Delta/2 & 0 & 0 & \mu/2 & t/2 & 0 & 0 \\ -\Delta/2 & 0 & \Delta/2 & 0 & t/2 & \mu/2 & t/2 & 0 \\ 0 & -\Delta/2 & 0 & \Delta/2 & 0 & t/2 & \mu/2 & t/2 \\ 0 & 0 & -\Delta/2 & 0 & 0 & 0 & t/2 & \mu/2 \end{pmatrix}.$$
$$\tag{9.12}$$

It is easy to extend this matrix when there are more sites. The matrix is Hermitian and the eigenvalues are all real. When $\mu = 0$ the DOS must be symmetric $D(E) = D(-E)$ (particle–hole symmetry) since the diagonal elements are all zero.

Exercise 9.3. Extend the matrix given in Eq. (9.12) for more sites. Find the eigenvalues and eigenvectors for $t = \Delta$ using Mathematica.

It is recommended that the reader try to find all the eigenvectors and eigenvalues of the Hamiltonian matrix and plot them. Useful insights can be gained. Numerically, we find that there are **two** zero energy states that are well separated from other eigenenergies, see Fig. 9.2. One of them, ϕ_l, is localized at one end of the chain and the other degenerate state ϕ_r at the opposite end. Their eigenvectors, denoted as $\{a_i\}$, should be interpreted as

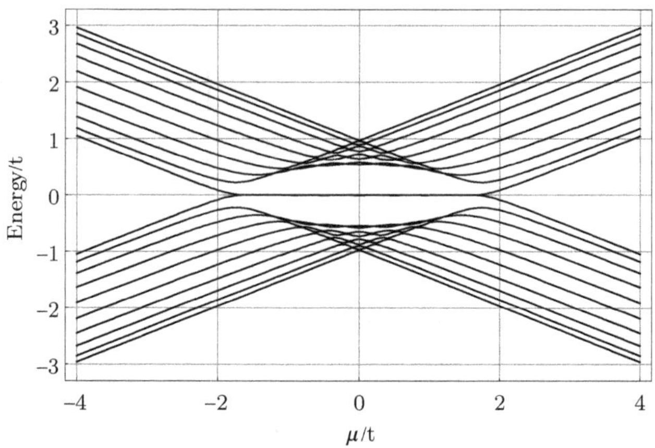

Fig. 9.2. Energy levels vs. μ for 10 sites with $\Delta = t/2$. The model has a degenerate ground state, corresponding to the presence or absence of a zero-energy electron.

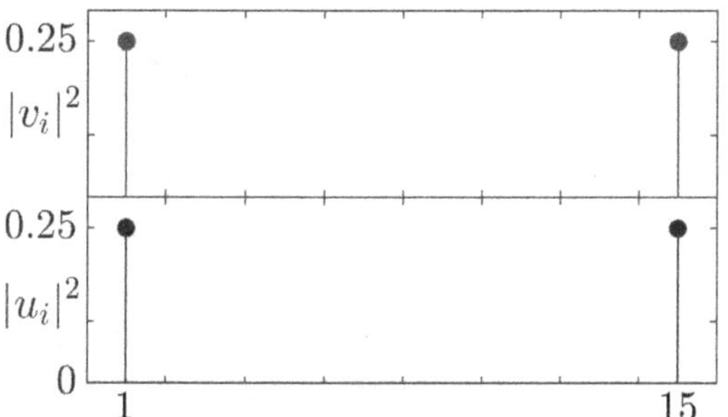

Fig. 9.3. Kitaev's chain has two degenerate zero-energy states. They can be combined to give one bonding and one antibonding states. For each of these states, the probability to find an electron (hole) at site $i = 1, \ldots 15$ is $|u_i|^2$ ($|v_i|^2$). We have set the hopping parameter equal to the value of the gap $t = \Delta$ and the chemical potential $\mu = 0$. A Majorana zero mode (half a real fermion mode) at each end of the chain is displayed. Reprinted with permission from Ref. [7].

follows: if a site is probed, the probability to find an electron is $|a_i|^2$ and the probability to find a hole is $|a_{i+N}|^2$. If there are 15 sites, we find 14 states with energy t, 14 states with energy $-t$, and 2 states with energy zero.[d] The probability density of a degenerate zero mode is computed numerically and is plotted in Fig. 9.3

---

[d]To get mutually orthogonal degenerate states one may insert a very small number in the diagonal elements of the Hamiltonian matrix.

One can form nearly degenerate bonding and antibonding states $\frac{1}{\sqrt{2}}(\phi_l \pm \phi_r)$ (the coupling between the left and right sites is singular perturbation, see Sec. 3.5). If the chain is very long, their energies must be close to zero. Each of these states describes two fractional particles at the left and right ends. For $|\mu| > 2t$, no zero-energy states exist and the system is topologically trivial. Existence of zero-energy end states is very analogous to that of finite length polyacetylene (see Fig. 4.10), except for the charge of the particle. Why is the zero mode charge **neutral**? The zero mode is in an exactly equal superposition of an electron and a hole.

- Majorana Operators

Let us study analytically the Kitaev model. This will provide further insights into Majorana modes. We can define Majorana operators

$$\gamma_A = \frac{1}{\sqrt{2}}(e^{i\phi/2}c + e^{-i\phi/2}c^\dagger), \quad \gamma_B = \frac{1}{i\sqrt{2}}(e^{i\phi/2}c - e^{-i\phi/2}c^\dagger), \quad (9.13)$$

where the phase of the superconductor is ϕ. Then

$$\{\gamma_k, \gamma_p\} = \delta_{kp}. \quad (9.14)$$

When $k = p$ we find

$$\gamma_k^2 = 1. \quad (9.15)$$

Note that

$$\gamma_p = \gamma_p^\dagger. \quad (9.16)$$

One can try to write the "number operator" in terms of Majorana operators, but it is ill-defined since it is always $\gamma_p^\dagger \gamma_p = \gamma_p^2 = 1$. Thus operators γ_p and $\gamma_p^\dagger$ are **not really particle operators**.[e] Any fermion operator can be

---

[e] What is a Majorana fermion? It is a charge neutral fermion and it is its own antiparticle. It is described by the Dirac equation, but, in contrast to a Dirac electron, its field is **real** and not complex (on the other hand, their Dirac γ matrices are all imaginary; this is called the Majorana condition).

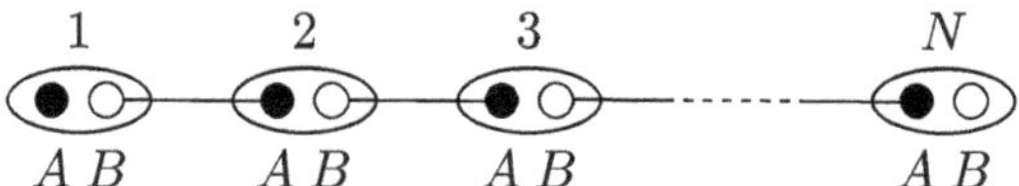

Fig. 9.4. Kitaev chain. There is no intra cell coupling. The end sites $1A$ and NB are uncoupled from the chain. The model displays a topological phase for $-1 < \mu/t < 2$.

written as

$$c = \frac{1}{\sqrt{2}}(e^{-i\phi/2}\gamma_A + ie^{-i\phi/2}\gamma_B), \quad c^\dagger = \frac{1}{\sqrt{2}}(e^{i\phi/2}\gamma_A - ie^{i\phi/2}\gamma_B).$$

(9.17)

We apply these representations of the fermion operators to the Kitaev chain with $j = 1, \ldots, N$ sites for $\phi = 0$, $\mu = 0$ and $\Delta = t$, see Fig. 9.4. At each site, we introduce Majorana operators and write the fermion operators as

$$c_j = \frac{1}{\sqrt{2}}(\gamma_{jA} + i\gamma_{jB}), \quad c_j^\dagger = \frac{1}{\sqrt{2}}(\gamma_{jA} - i\gamma_{jB}).$$

(9.18)

Then the Hamiltonian can be written as

$$H = -2it \sum_{j=1}^{N-1} \gamma_{jB}\gamma_{j+1A}.$$

(9.19)

There are coupling between sites of inter pairs $(jB, j+1A)$ but there is no coupling between sites of intra pairs (jA, jB). The end site operators γ_{1A} and γ_{NB} are not present in the Hamiltonian. This means there is one zero mode. Half of its weight is at the left end site $1A$ and the other half on the right end site NB. (We will see later that zigzag graphene nanoribbons have also a similar zero energy end state with two split $e^-/2$ charges, see Chapter 18.) Such a fermion can be described in terms of Majorana operators at site $i = 1$ and N

$$c = \frac{1}{\sqrt{2}}(\gamma_{1A} + i\gamma_{NB}).$$

(9.20)

9.3. Majorana Edge Modes of Quasi-one-Dimensional Strip

Before we investigate two-dimensional p-wave superconductors, it is instructive to consider an infinite strip of a quasi-one-dimensional p-wave superconductor [8]. Let us start from the mean-field Hamiltonian of two-dimensional *p*-wave superconductors

$$H = \int d^2\vec{r}\left[\psi^+(\vec{r})\left(-\frac{\nabla^2}{2m} - \mu\right)\psi(\vec{r})\right.$$

$$\left. + \psi(\vec{r})\frac{\Delta'}{2}e^{i\phi}i(p_x + ip_y)\psi(\vec{r}) + \text{H.c.}\right], \tag{9.21}$$

where Δ' is the order parameter, ϕ is the phase of the order parameter, and μ is the chemical potential (the absolute magnitude of the order parameter is $\Delta' \sim \text{Energy} \times \text{Length}$ and $\hbar = 1$). We rewrite this Hamiltonian as

$$H = \frac{1}{2}\sum_{\vec{p}}\Psi_{\vec{p}}^\dagger\begin{pmatrix}\frac{p^2}{2m} - \mu(\vec{r}) & -\Delta'e^{-i\phi}i(p_x - ip_y) \\ \Delta'e^{i\phi}i(p_x + ip_y) & -\frac{p^2}{2m} + \mu(\vec{r})\end{pmatrix}\Psi_{\vec{p}}, \tag{9.22}$$

where $\Psi_{\vec{p}} = (\psi^\dagger(\vec{r}), \psi(\vec{r}))^T$ a two-component field operator. The *p*-wave order parameter in the Fourier space is $\Delta'(\vec{p}) = \Delta'e^{i\phi}(p_x+ip_y)$ with $\Delta'(\vec{p}) = -\Delta'(-\vec{p})$.

Focusing on low-energy excitation modes and assuming that the chemical potential $|\mu| \gg p^2/2m$, we can ignore the kinetic term in H. The Bogoliubov–de Gennes equation[f] is

$$\begin{pmatrix}-\mu(\vec{r}) & -i\Delta'(k_x - \frac{\partial}{\partial y}) \\ i\Delta'(k_x + \frac{\partial}{\partial y}) & \mu(\vec{r})\end{pmatrix}\begin{bmatrix}\psi_e(y) \\ \psi_h(y)\end{bmatrix} = E\begin{bmatrix}\psi_e(y) \\ \psi_h(y)\end{bmatrix}. \tag{9.23}$$

(It follows from the Hamiltonian by choosing the phase of the superconductor such that $e^{i\phi} = 1$.) The eigenstates are

$$\Psi = \begin{bmatrix}\psi_e(y) \\ \psi_h(y)\end{bmatrix}e^{ik_x x}. \tag{9.24}$$

Let us find solutions that represent edge modes that are confined to the boundaries. Using the Ansatz $\psi_h = -i\psi_e$, we find right-moving modes

$$\psi_e \sim e^{-\frac{\mu}{\Delta'}(d/2-y)} \quad \text{for} \quad y < d/2 \text{ with } E = \Delta'k_x, \tag{9.25}$$

where d is the width of the strip. For the left-moving modes, we use the Ansatz $\psi_h = i\psi_e$

$$\psi_e \sim e^{\frac{\mu}{\Delta'}(-d/2+y)} \quad \text{for} \quad y > -d/2 \text{ with } E = -\Delta'k_x. \tag{9.26}$$

---

[f]This equation is analogous to the Hartree–Fock equation that determines the eigenstates and eigenenergies of single-particle excitations, see Chapter 6.

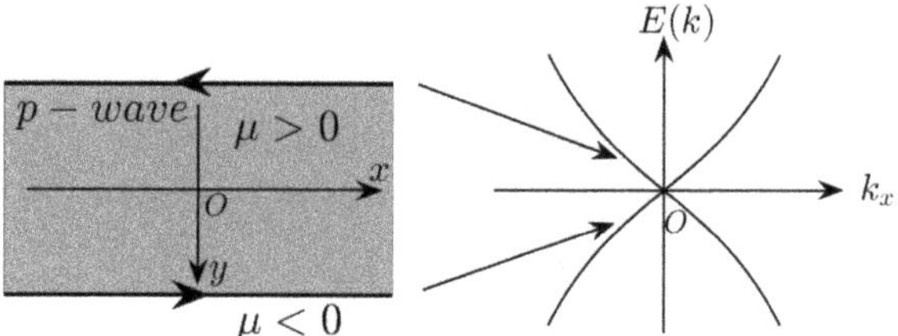

Fig. 9.5. An infinite strip of a p-wave superconductor has two chiral edge modes propagating in opposite directions. The excitation spectrum of a strip of a p-wave superconductor is shown. Gap states representing chiral edge modes are indicated.

When $\mu > 0$, these chiral energy modes are localized on the boundaries, as shown in Fig. 9.5. (Note the similarity of these modes to the edge modes of integer quantum Hall states, see Fig. 10.12.)

Exercise 9.4. Derive the Bogoliubov–de Gennes equation given in Eq. (9.22). Show that it can be written as

$$H = \left(\frac{\vec{p}^2}{2m} - \mu \right) \sigma_z + \Delta'(p_x \sigma_x + p_y \sigma_y). \tag{9.27}$$

Here spin-up and -down states stand for electron and hole states and the phase angle of the gap is set to $e^{-i\phi} = i$. Show that the Hamiltonian anticommutes with the **chiral** operator $P = \sigma_x K$, where K is complex conjugation. Hint: Note that $P \begin{bmatrix} \psi_e \\ \psi_h \end{bmatrix} = \begin{bmatrix} \psi_h^* \\ \psi_e^* \end{bmatrix}$. The transformed state has energy $-E$ if the energy of the original state is E; this is particle–hole symmetry. It is easy to show $\{H, P\} = 0$. A different way to prove it is to show

$$PH\psi = \sigma_x(H\psi)^* = -HP\psi = -H\sigma_x\psi^*. \tag{9.28}$$

Note that $(H\psi)^* = H^*\psi^*$ and $KH = H^*$ (note that H is a matrix).

9.4. Two-Dimensional Chiral p-Wave Superconductor

As we mentioned in the beginning of this chapter, a two-dimensional p-wave superconductor has an invertible topological order. Let us explain this. Its vortex can inhabit a zero mode [9], which can be spatially split into two entities. However, these split objects are not a fractionalized **particle**. Let us quote Wen on this subtle point [2]: "A Majorana zero mode is not a Majorana fermion. In fact, it is not even a particle. It is a property of a particle, just like the mass is a property of a particle. If a mobile particle carriers a Majorana zero mode, then the particle will have a non-Abelian statistics." Nonetheless they seem to represent the right degrees of freedom of the system because counting them gives the correct number of ground state degeneracy. In any case, we explain below how these Majorana zero modes are formed and elucidate their counter-intuitive non-Abelian nature.

- Non-trivial Winding Number

Consider a **spinless**[g] **two-dimensional** p-wave superconductor. The Fourier transform of the two-dimensional mean-field Hamiltonian Eq. (9.21) is

$$H = \frac{1}{2} \int d^2\vec{k}\,\Psi^\dagger(\vec{k}) H(\vec{k}) \Psi(\vec{k}), \tag{9.29}$$

where

$$H(\vec{k}) = \begin{pmatrix} \epsilon_k & \tilde{\Delta}(\vec{k})^* \\ \tilde{\Delta}(\vec{k}) & -\epsilon_k \end{pmatrix} \tag{9.30}$$

with the gap function

$$\tilde{\Delta}(\vec{k}) = i\Delta' e^{i\phi}(k_x + ik_y) \tag{9.31}$$

and the energy dispersion

$$\epsilon(k) = \frac{k^2}{2m} - \mu. \tag{9.32}$$

Compare this gap function with that of the one-dimensional system, see Eq. (9.5).

The Hamiltonian can be written in the standard form (see Sec. 2.4 about magnetic monopoles)

$$H(\vec{k}) = \mathbf{h}(\vec{k}) \cdot \vec{\sigma} \tag{9.33}$$

---

[g]It is sufficient to consider the spinless case. In spinful 2D spin-triplet $p + ip$ superconductivity a half quantum vortex binds a single zero-energy Majorana mode in one spin channel [9].

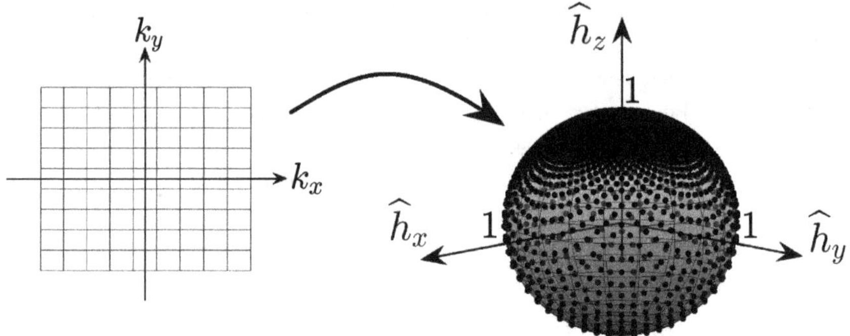

Fig. 9.6. Mapping $\vec{k}$ to $\hat{\mathbf{h}}(\vec{k})$ on the unit sphere. When the infinitesimal grid points cover the entire two-dimensional plane the points on the sphere represented by $\hat{\mathbf{h}}(\vec{k})$ cover the entire sphere.

with the components

$$h_x(\vec{k}) = \Re[\tilde{\Delta}(\vec{k})],$$
$$h_y(\vec{k}) = \Im[\tilde{\Delta}(\vec{k})], \qquad (9.34)$$
$$h_z(\vec{k}) = \epsilon(k).$$

The Chern number can be defined in terms of the unit vector $\hat{\mathbf{h}}(\vec{k}) = \frac{\mathbf{h}(\vec{k})}{|\mathbf{h}(\vec{k})|}$ (see Fig. 9.6)

$$C = \int \frac{d^2\vec{k}}{4\pi} \hat{\mathbf{h}} \cdot [\partial_{k_x}\hat{\mathbf{h}} \times \partial_{k_y}\hat{\mathbf{h}}]. \qquad (9.35)$$

(This is the winding number given in Eq. (2.66). It is identical to the Chern number (Do Exercise 9.5).) It is -1 for $\mu > 0$, i.e., the system is topologically non-trivial.

> **Exercise 9.5.** Using Mathematica, show that the Chern number, Eq. (9.35), is -1 (the accuracy depends on the number of grid points, see Fig. 9.6). Show that for two-component wavefunctions the winding number is identical to the Chern number, defined in Eq. (5.37). Hint: The projection operator is $P = |\psi\rangle\langle\psi| = \frac{1}{2}(1 + \vec{h} \cdot \vec{\sigma})$. This is not an easy problem. Consult Ref. [10].

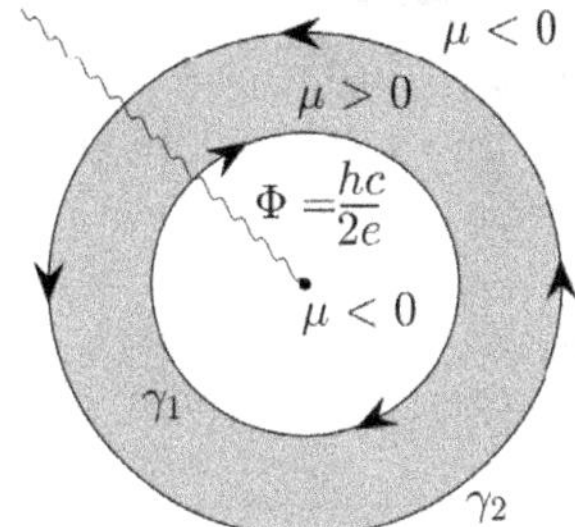

Fig. 9.7. A topological phase occupies the annulus region, consistent with the result of the analysis of the Chern number. We will consider two cases with and without a half flux $\Phi_0/2$ threading the ring. Operators γ_1 and γ_2 describe Majorana half modes at the inner and outer edges, respectively. The wavy line will be explained in Sec. 9.5

- ● Chiral Edge States of an Annulus

Consider a topological superconductor occupying an annulus region, as in Fig. 9.7. First consider the case **without a flux threading** the ring. Focusing on low-energy excitation modes we can discard the kinetic term in H. In addition, we assume that the chemical potential $\mu(r)$ is slowly varying. In polar coordinates (r, θ), the Hamiltonian for the annulus can be written as

$$H_{\text{edge}} = \int d^2\vec{r} \left[-\mu(r)\psi^\dagger\psi + \left[\frac{\Delta'}{2} e^{i\phi} e^{i\theta} \psi \left(\partial_r + \frac{i\partial_\theta}{r} \right) \psi + \text{H.c.} \right] \right].$$

$$(9.36)$$

One can remove the factor $e^{i\theta}$ via the following transformation:

$$\psi \to e^{-i\theta/2}\psi.$$

$$(9.37)$$

Note that the new ψ satisfies antiperiodic boundary conditions. With $\Psi^\dagger(\vec{r}) = [\psi^\dagger(\vec{r}), \psi^\dagger(\vec{r})]$ the Hamiltonian can be rewritten as

$$H_{\text{edge}} = \frac{1}{2} \int d^2\Psi^\dagger(\vec{r}) H(\vec{r}) \Psi(\vec{r}),$$

$$(9.38)$$

where

$$H(\vec{r}) = \begin{pmatrix} -\mu(r) & \Delta' e^{-i\phi}(-\partial_r + \frac{i\partial_\theta}{r}) \\ \Delta' e^{i\phi}(\partial_r + \frac{i\partial_\theta}{r}) & \mu(r) \end{pmatrix}.$$

$$(9.39)$$

> **Exercise 9.6.** Show Eqs. (9.36) and (9.39). Hint: Use the Hamiltonian equation (9.21) and express $\frac{\partial}{\partial x}$ and $\frac{\partial}{\partial y}$ in polar coordinates. Note that in polar coordinate we can write $\partial_x + i\partial_y = i(p_x + ip_y) = e^{i\theta}(\partial_r + i\partial_\theta/r)$ and $-(\partial_x - i\partial_y) = -i(p_x - ip_y) = -e^{-i\theta}(\partial_r - i\partial_\theta/r)$. This can be shown using relations like $x = r\cos\theta$, $\frac{\partial}{\partial x} = \frac{\partial}{\partial r}\frac{\partial r}{\partial x} + \frac{\partial}{\partial \theta}\frac{\partial \theta}{\partial x}$, etc.

The solutions have the form

$$\xi_n(\vec{r}) = e^{in\theta}\begin{bmatrix} e^{-i\phi/2}[f(r) + ig(r)] \\ e^{i\phi/2}[f(r) - ig(r)] \end{bmatrix}, \tag{9.40}$$

where the functions f and g satisfy

$$\begin{aligned} (E + n\Delta'/r)f &= -i[\mu(r) - \Delta'\partial_r]g, \\ (E - n\Delta'/r)g &= i[\mu(r) - +\Delta'\partial_r]f. \end{aligned} \tag{9.41}$$

Note that n is a **half-integer** because of the antiperiodic boundary condition. For modes that are well-localized at $r = R_i$ ($i = $ in or out) it is a good approximation to set $1/r \to 1/R_i$ in Eq. (9.41). It is straightforward to find the eigenstates (see Exercise 9.7)

$$\xi_n^{\mathrm{out}}(\vec{r}) = e^{in\theta}e^{\frac{1}{\Delta'}\int_{R_{\mathrm{out}}}^r dr'\mu(r')}\begin{bmatrix} ie^{-i\phi/2} \\ -ie^{i\phi/2} \end{bmatrix}, \tag{9.42}$$

$$\xi_n^{\mathrm{in}}(\vec{r}) = e^{in\theta}e^{-\frac{1}{\Delta'}\int_{R_{\mathrm{in}}}^r dr'\mu(r')}\begin{bmatrix} e^{-i\phi/2} \\ e^{i\phi/2} \end{bmatrix}. \tag{9.43}$$

These modes are localized at the inner and outer boundaries when $\mu(r) > 0$. In an annulus with a finite width, the inner and outer modes are coupled. One half of the probability is found near the inner boundary and the other half at the outer boundary of the annulus. This is analogous to a bonding or antibonding state of the double-well problem. Their energies are computed in Exercise 9.7.

> **Exercise 9.7.** Find the eigenfunctions of Eq. (9.41) with the following eigenenergies:
>
> $$E_{\mathrm{out}} = \frac{n\Delta'}{R_{\mathrm{out}}}, \quad E_{\mathrm{in}} = -\frac{n\Delta'}{R_{\mathrm{out}}}, \tag{9.44}$$
>
> where n is a half integer. Hint: Consult Ref. [9].

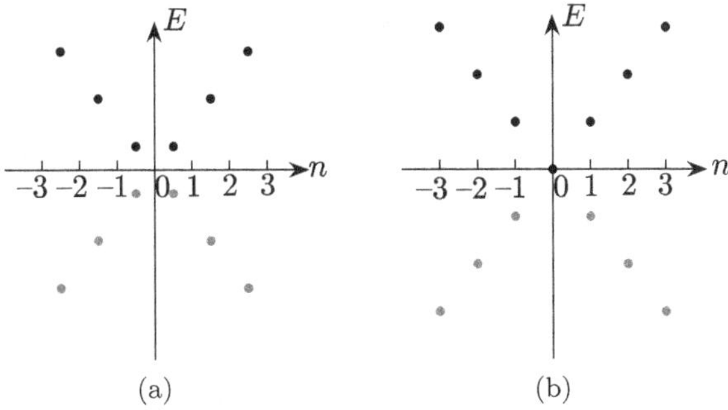

Fig. 9.8. (a) The excitation spectrum of an annulus in the absence a flux: n takes half-integer values and no zero-energy mode exists. (b) Excitation spectrum in the presence of a flux $\Phi_0/2$: n takes integer values and a zero-energy mode exists. Energy E_{out} of chiral edge modes is increasing with increasing n while E_{in} is decreasing, i.e., their velocities are opposite as shown in Fig. 9.7.

- Chiral Edge States of a Vortex

Now suppose a **flux** $\Phi = \frac{hc}{2e}$ threads the center region of the annulus, see Fig. 9.7. This changes the gap to $\Delta(\vec{r}) \to \Delta(\vec{r})e^{-i\theta}$. It is a consequence of the Aharonov–Bohm effect of a Cooper pair with the resulting phase factor $e^{i2\pi\Phi/\Phi_{2e}} = 1$ ($\Phi_{2e} = hc/2e$). Here we choose not to include the corresponding vector potential in the Hamiltonian. Instead, we incorporate its effect in the order parameter as a phase factor, see Sec. 3.2. This new phase factor $e^{-i\theta}$ cancels the already existing phase factor $e^{i\theta}$ in the Hamiltonian equation (9.36). As a result of this, half-integer n values change to **integer** values. The quasiparticle energies before and after the introduction of a vortex are plotted in Fig. 9.8. Note that they represent chiral modes with opposite velocities. In the presence of the flux, two zero modes exist. One half of the quasiparticle is located at the inner boundary while the other half at the outer boundary of the annulus.

- Two Vortices

Consider two nearby vortices, as shown in Fig. 9.9. The ground state of such a system is 2-fold **degenerate**. The degenerate states are

$$|0\rangle \quad \text{and} \quad c^\dagger|0\rangle. \tag{9.45}$$

Let the operators γ_1 and γ_2 represent two Majorana zero modes sitting at vortex 1 and 2, respectively. The fermion operator $c^\dagger$ may be written

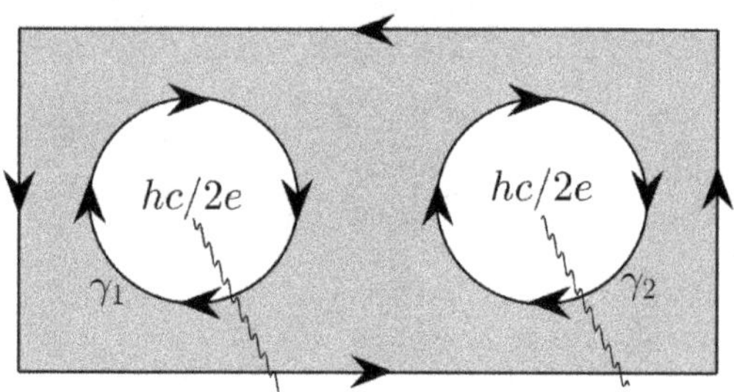

Fig. 9.9. Two nearby vortices are shown. Note that there is now only one outer mode which obeys antiperiodic boundary conditions because it contains **two** half fluxes $\frac{hc}{2e}$. Its excitation mode has a **gap**, see Fig. 9.8(a). Operators γ_1 and γ_2 describe Majorana half modes at the left and right edges, respectively. So the situation is rather different from the case of an annulus in the presence of a half vortex.

as

$$c^\dagger = \frac{1}{2}(\gamma_1 + i\gamma_2). \tag{9.46}$$

$|0\rangle$ and $c^\dagger|0\rangle$ have, respectively, even and odd number of fermions.

9.5. Braiding Through Vortex Encircling and Non-Abelian Statistics

The only way to induce transition among degenerate ground states is by braiding Majorana modes. This may be achieved by swapping vortices [11, 12].

- Non-Abelian Braiding

When a vortex encircles another vortex the phase factor of the superconductor changes by $e^{i2\pi}$**, i.e., the change in the phase factor of a Cooper pair is** $e^{i2\pi}$ [13]**, see Fig. 9.10. Then for a single fermion the phase factor change is** $e^{i\pi}$**. As a result of the change** π **in the phase** $\phi/2$**, we should have**

$$\gamma_1 \to -\gamma_2, \quad \gamma_2 \to +\gamma_1. \tag{9.47}$$

(This follows from Eq. (9.13) with replacement of $1 \to A$ and $2 \to B$. Remember that a fermion is "split" into two Majorana modes.)

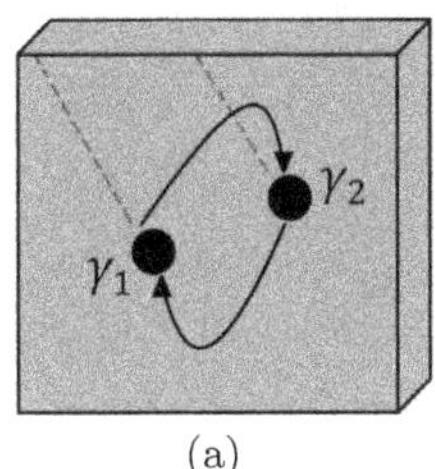
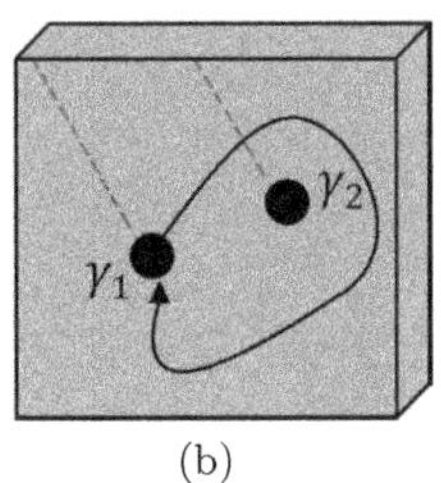

Fig. 9.10. When a branch cut crossed, a **sign change** for the Majorana edge modes occurs. (a) Two Majorana zero modes are exchanged. One Majorana zero mode must cross the branch cut emanating from the other vortex. (b) One Majorana zero mode encircles another Majorana zero mode. Each Majorana zero mode must cross the branch cut of the other Majorana fermion.

Let us show this result differently using the following physical argument. First let us review how the order parameter near vortex i behaves:

$$
\begin{aligned}
\Delta(\vec{r}) &= |\Delta(\vec{r})| e^{i\theta(\vec{r}-\vec{R}_i)+i\Omega_i} \\
\Omega_i &= \sum_{i\neq j} \arg(\vec{R}_i - \vec{R}_j),
\end{aligned}
\tag{9.48}
$$

$\theta(\vec{r} - \vec{R}_i)$ is the polar angle of vector $\vec{r} - \vec{R}_i$. If one vortex moves around another, the separation $\vec{R}_i - \vec{R}_j$ passes through 2π. A zero-energy Bogoliubov quasiparticle of a vortex splits into two Majorana zero modes. The order parameter (9.48) and Majorana operators given in Eq. (9.13) suggest the following wave functions [14] of the half modes located at $\vec{R}_1$ and $\vec{R}_2$

$$
\begin{aligned}
\psi_1(r,\phi) &= e^{i\frac{\pi+\phi}{2}} f(|\vec{r} - \vec{R}_1|) e^{i\theta(\vec{r}-\vec{R}_1)}, \\
\psi_2(r,\phi) &= e^{i\frac{\phi}{2}} f(|\vec{r} - \vec{R}_2|) e^{i\theta(\vec{r}-\vec{R}_2)}.
\end{aligned}
\tag{9.49}
$$

The overall superfluid phase at vortex 1 is $\pi + \phi$ and ϕ at vortex 2, see Fig. 9.11. Note that, in contrast to the case of vortices, the superconducting phase in the Majorana wave functions appears as $\phi/2$, consistent with Majorana operators given in Eq. (9.13). Let us consider moving a Majorana zero mode around another that is placed in a different vortex far away, see Fig. 9.12. The phases may be represented by braiding of Majorana zero modes, see Fig. 9.12. (An accessible and detailed discussion of braiding of Majorana fermions in p-wave superconductors can be found in Refs. [9] and [13]. A simple and beautiful explanation of what the braid groups are, is given for laymen in Chap. 5 of Ref. [15].) After an adiabatic exchange of

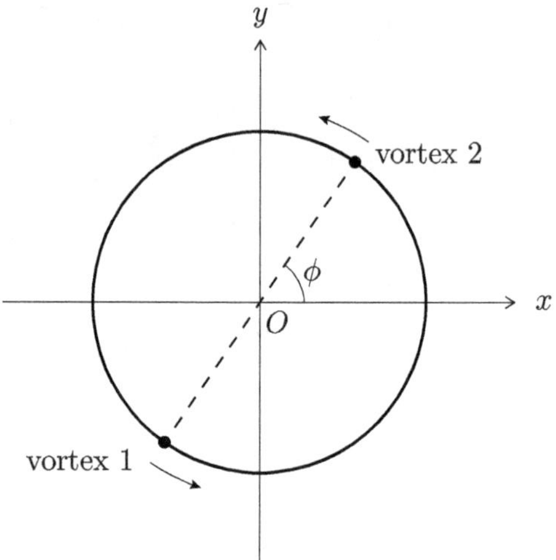

Fig. 9.11.　Exchange of two Majorana half modes: $\phi \to \phi + \pi$.

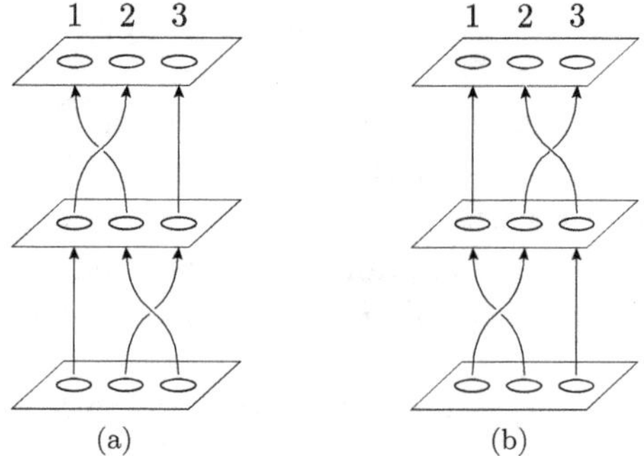

Fig. 9.12.　Three particles are located at sites 1, 2, and 3. In (a), initially, particles at 2 and 3 are braided (U_{23}) and then those at 1 and 2 (U_{12}) (braiding two worldlines corresponds to a particle exchange, see Fig. 3.11). In (b), initially, those at 1 and 2 are braided (U_{12}), and then those at 2 and 3 (U_{23}). Braiding the particles amounts to a rotation among the degenerate states, see Sec. 2.6 about non-Abelian matrix Berry phases. Since these rotations are non-commuting, the first combined operation $U_{12}U_{23}$ is not equivalent to the second combined operation $U_{23}U_{12}$.

vortices 1 and 2

$$\psi_1(r, \pi) \to -\psi_2(r, 0),$$
$$\psi_2(r, \pi) \to \psi_1(r, 0).$$

(9.50)

These results agree with Eq. (9.47).

- Many Body Matrix Berry Phase

Formally, these phase changes may be thought of as the non-Abelian matrix phases, see Sec. 2.6. The only difference is that now a many-body degenerate manifold is present instead of single-particle degeneracy.

Let us explain this. Suppose adiabatic parameters $\vec{R}$ follow a closed loop in the parameter space (vortex positions may be used as the adiabatic parameters). Since a time-dependent many-body wave function can be written as a linear combination of basis states the results obtained thus far also applies to many-body wave functions

$$\Psi(t) = \sum_i c_i(t)\Psi_i(t). \tag{9.51}$$

The αth component of the vector matrix Berry connection between M degenerate ground states $|\Psi_i\rangle$ is

$$A_{ij}^\alpha = i\langle\Psi_i|\frac{\partial}{\partial R_\alpha}|\Psi_j\rangle, \tag{9.52}$$

where R_α is an adiabatic parameter. Then the initial and final states are connected by the matrix Berry phase

$$|\Psi_i\rangle = \sum_{j=1}^{M}\Gamma_{ij}|\Psi_j\rangle, \tag{9.53}$$

where the matrix Berry phase is

$$\mathbf{\Gamma} = Pe^{\oint_C \vec{A}\cdot d\vec{R}}. \tag{9.54}$$

($\vec{\mathbf{A}}$ is defined in terms of A_{ij}^α, see Eq. (2.81).) $\mathbf{\Gamma}$ is given by a $M \times M$ matrix, see Eq. (2.83). So the **internal state of the system changes** during the adiabatic process and a non-Abelian (non-commuting) matrix phase arises. The matrix Berry phase also gives rise to the statistical phase factors upon exchange of two particles

$$\psi_i(\vec{r}_\alpha, \vec{r}_\beta) \to e^{i\sum_a \theta_a T_{ij}^a}\psi_j(\vec{r}_\alpha, \vec{r}_\beta), \tag{9.55}$$

where the parameters θ_a and matrices $\mathbf{T^a}$ may be obtained from $\mathbf{\Gamma}$. This phase can be thought of as the Aharonov–Bohm phase due to a statistical gauge field $\vec{\mathbf{A}}$.

9.6. Number Parity Conservation

Even and odd parity states of a two-dimensional chiral p-wave superconductor may be relevant in braiding Majorana modes. Let us explain these even and odd parity states. In a superconducting state, there are two **degenerate** many-body states with different number of fermions. The first one is the ground state with an even number of particles, i.e., even parity state (it is the BCS state)

$$|\Psi_0\rangle = \sum_{N \text{ even}} C_N |\Phi_N\rangle. \tag{9.56}$$

Although the mean value of N is well-defined, it has fluctuations. The other state $c^\dagger |\Psi_0\rangle$ has two half-fermion modes and the number of electrons is odd

$$|\Psi_1\rangle = c^\dagger |\Psi_0\rangle. \tag{9.57}$$

A finite system usually has a well-defined particle number. In a finite system the energy difference between N or $N \pm 1$ or $N \pm 2$, etc. pairs is non-zero. But in the thermodynamic limit, the difference vanishes.

Let us study a simple model to understand better the degeneracy between the even and odd parity states. Consider a toy model consisting of a Kitaev chain with only two sites (see Ref. [16]). Its Hamiltonian can be solved exactly. Its energy spectrum is shown in Fig. 9.13. Let us write down the eigenstates. Occupy the **two lowest** single-particle energy levels and construct a two-body state $\Psi_e = \alpha_2^\dagger \alpha_1^\dagger |0\rangle$. The single-particle operators are of the form $\alpha_i^\dagger = (u_1 c_1 + u_2 c_2 + v_1 c_1^\dagger + v_2 c_2^\dagger)|0\rangle$ and the coefficients (u_1, u_2, v_1, v_1) are given in Fig. 9.13. We find

$$\Psi_e = \frac{1}{\sqrt{2}}(1 + c_2^\dagger c_1^\dagger)|0\rangle \tag{9.58}$$

with energy $E_e = -t$ (this state has **even** parity). One-electron ground state is

$$\Psi_o = \frac{1}{\sqrt{2}}(c_1^\dagger - c_2^\dagger)|0\rangle \tag{9.59}$$

and it has also energy $E_o = -t$. This state is an **odd** parity state. These even and odd parity states are thus degenerate. It is interesting to note that the Majorana operators γ_{1A} and γ_{2B} each **connect** even and odd parity

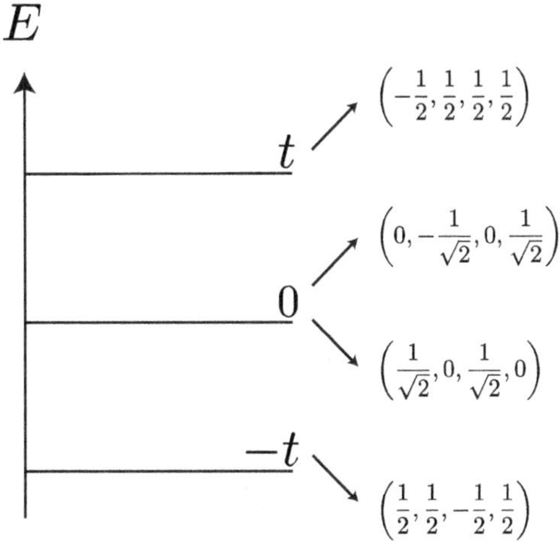

Fig. 9.13. The lowest four mean-field single-particle energy levels of a two-site Kitaev model described by the matrix Hamiltonian given in Eq. (9.12) for $\mu = 0$ and $\Delta = t$. The middle energy state is doubly degenerate. Each state is given by a linear combination $(u_1 c_1 + u_2 c_2 + v_1 c_1^\dagger + v_2 c_2^\dagger)|0\rangle$. The expansion coefficients (u_1, u_2, v_1, v_2) are given in the figure.

states

$$\gamma_{1A}\Psi_e = \frac{i}{\sqrt{2}}(c_1^\dagger - c_2^\dagger)|0\rangle = i\Psi_o \tag{9.60}$$

and

$$\gamma_{2B}\Psi_e = \frac{1}{\sqrt{2}}(c_1^\dagger - c_2^\dagger)|0\rangle = \Psi_o. \tag{9.61}$$

Additionally note that

$$\gamma_{2B}\Psi_e = -i\gamma_{1A}\Psi_e. \tag{9.62}$$

Moreover, we have

$$c\Psi_e = i\Psi_o, \tag{9.63}$$

where the electron destruction operator is

$$c = \frac{1}{\sqrt{2}}(\gamma_{1A} + i\gamma_{2B}). \tag{9.64}$$

Particle number conservation should be obeyed universally in condensed matter physics. Fermion parity operator is defined as

$$P = e^{i\pi \sum_p n_p} = (-1)^{\sum_p n_p}, \qquad (9.65)$$

where the total particle number operator is $\sum_p n_p = \sum_p c_p^\dagger c_p = N$. The conservation of fermion number is

$$[H, P] = 0. \qquad (9.66)$$

Fermionic parity is always conserved as long as fermionic operators come up in pairs in the Hamiltonian, for example $c_p^\dagger c_q$ or $c_p^\dagger c_p^\dagger$, etc. This means, if you start with a system of even (odd) number of fermions, you will always end up with even (odd) number of fermions, i.e., an adiabatic operation does not change the parity number of fermion.

However, the braiding process described above seems to break particle number conservation. Lin and Leggett [14] advocate the need for particle-number conserving theory of Majorana zero modes. This issue is subtle in the case of superconductivity since the BCS wave function does not have a definite particle number. See Ref. [12] for the recent efforts in the experimental investigation of Majorana fermions. There is as yet no generally accepted experimental evidence for non-abelian statistics.

Bibliography

[1] J. Magueijo, *A Brilliant Darkness: The Extraordinary Life and Mysterious Disappearance of Ettore Majorana, the Troubled Genius of the Nuclear Age* (Basic Books, New York, 2009).•

[2] X.-G. Wen, Colloquium: Zoo of quantum-topological phases of matter, *Rev. Mod. Phys.* **89**, 041004 (2017).

[3] G. Moore and N. Read, Nonabelions in the fractional quantum Hall effect, *Nucl. Phys. B* **360**, 362 (1991).

[4] N. Read, Topological phases and quasiparticle braiding, *Phys. Today* **65**, 7, 38 (2012). •

[5] F. Wilczek, Majorana returns, *Nature Phys.* **5**, 614 (2009). •

[6] A. Y. Kitaev, Unpaired Majorana fermions in quantum wires, *Phys.-Usp* **44**, 131 (2001).

[7] S.-R. Eric Yang, M. C. Cha, H. J. Lee, and Y. H. Kim, Topologically ordered zigzag nanoribbon: $e/2$ fractional edge charge, spin–charge separation, and ground-state degeneracy, *Phys. Rev. Res.* **2**, 033109 (2020).

[8] B. A. Bernevig and T. L. Hughes, *Topological Insulators and Topological Superconductors* (Princeton University Press, Princeton, 2013)

[9] J. Alicea, New directions in the pursuit of Majorana fermions in solid state systems, *Rep. Prog. Phys.* **75**, 076501 (2012). •

[10] J. de Lisle, S. De, E. Alba, A. Bullivant, J. J. Garcia-Ripoll, V. Lahtinen, and J. K. Pachos, Detection of Chern numbers and entanglement in topological two-species systems through subsystem winding numbers, *New J. Phys.* **16**, 083022 (2014).

[11] D. A. Ivanov, Non-Abelian statistics of half-quantum vortices in p-wave superconductors, *Phys. Rev. Lett.* **86**, 268 (2001).

[12] C. W. J. Beenakker, Search for Majorana fermions in superconductors, *Annu. Rev. Condens. Matter Phys.* **4**, 113 (2013). •

[13] M. Leijnse and K. Flensberg, Introduction to topological superconductivity and Majorana fermions, *Semicond. Sci. Technol.* **27**, 124003 (2012). •

[14] Y. Lin and A. J. Leggett, Towards a particle-number conserving theory of Majorana zero modes in p+ip superfluids, arXiv:1803.08003v1.

[15] E. Frenkel, *Love and Math* (Basic Books, New York, 2013). •

[16] S.-Q. Shen, *Topological Insulators* (Springer, New York, 2012).

Chapter 10

Integer Quantum Hall Effect

"One cannot understand Nature by pure thought, without
hints from experiment."

Max Born

One may expect that the presence of disorder in two-dimensional systems gives rise to random values of the Hall coefficient, dependent on types of disorder. But these systems display precise values of the quantized Hall conductivity given by the fundamental constants $\hbar$ and e, independent of types of disorder and electron interactions. The quantized Hall effect is thus a truly remarkable phenomenon. We will see that gauge invariance under insertion of an elementary quantum flux through the holes of a multiply connected domain, such as cylinder or torus, plays important role in these systems [1]. Charge transfer during this process is quantized, i.e., it is an integer or fractional charge. Moreover, transitions between quantum Hall plateaus should be viewed as a metal-insulator transition with universal critical exponents. We will learn here that the quantized value of the Hall coefficient of both integer and fractional quantum Hall states is related to the presence of topological order.

In this chapter, we focus on the integer quantum Hall effect. The presence of a magnetic field breaks time-reversal symmetry and the properties of the integer quantum Hall state are not determined by symmetry. The integer quantum Hall state [2] displays edge states [3], which exist with topological reasons, see Fig. 10.1. It can be characterized by the Chern number of the filled energy band.[a] Also spin–charge separation is present [4]. The

---

[a] See also the quantum anomalous Hall effect in Sec. 15.

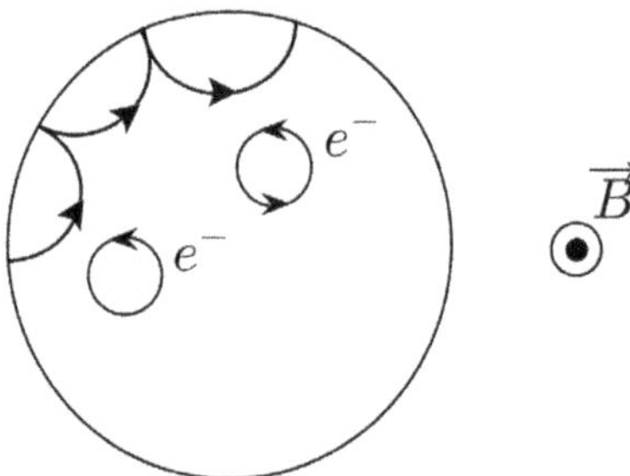

Fig. 10.1. Confined two-dimensional electrons in a magnetic field $\vec{B}$ must have edge states by topological reasons. An edge electron performs a skipping motion and moves clockwise. Electrons inside the system move counterclockwise in a cyclotron motion.

integer quantum Hall system is said to have an invertible topological order because it has no fractionalized excitations.

Although the integer quantum Hall effect is a mature subject, our current understanding of the integer quantum Hall transition is incomplete. We have to better understand the role of the Coulomb interaction between electrons since it is expected to provide a relevant perturbation at the non-interacting fixed point. For a consistent description of the dynamic critical exponent of the metal–insulator transition,[b] the long-range Coulomb interaction must be properly included, which is a difficult problem.

The field of the integer quantum Hall effect is vast, and there are several excellent books [6,8,9]. Here we will only discuss subjects that are relevant to topological insulators, such as edge states and the Chern number.

---

[b]Electron localization leads to the strange effect of shutoff of quantum tunneling through a potential barrier. The conventional one-parameter scaling theory of localization [5] predicts the following (here the dimensionless scale parameter is a measure of the conductance). In three-dimensional systems some states are delocalized and other states are localized, and there is a metal–insulator transition at the boundary between these two types of states (mobility edge). A lower-dimensional system does not undergo a genuine phase transition because the conductance always decreases with system size. All states are eventually localized if the system is large enough. In a disordered two-dimensional system with broken time reversal symmetry the one-parameter scaling theory of localization does not apply as localized and delocalized states may coexist (integer quantum Hall systems belongs to this class). Before studying this chapter students are encouraged to read about localization-delocalization transition; see, for example, Chap. 11 in Ref. [6]. For a short introduction to Anderson localization with a good historical account, see Ref. [7].

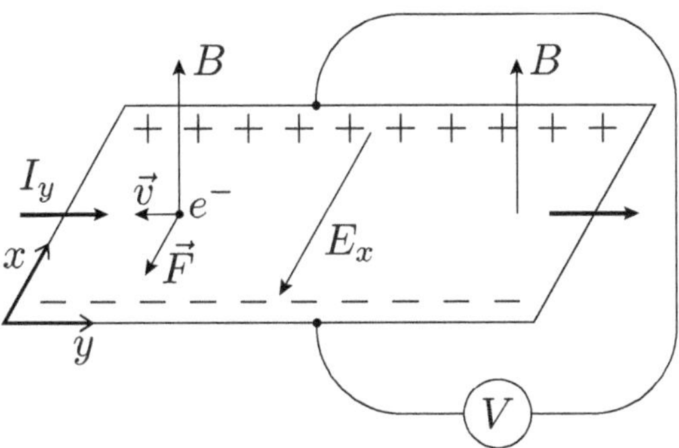

Fig. 10.2. In the quantum Hall experiment, one measures the voltage across the two-dimensional Hall bar system when a longitudinal current flows along the system. Since the bulk states are localized only edge states carry the current. The Lorentz force $\vec{F} = q\frac{\vec{v}}{c} \times \vec{B}$ on electrons make them to pile up at the lower edge. Under a steady state the Lorentz force is compensated by the electric field E_x between the upper and lower edges. The transverse voltage difference is measured with a voltmeter. The Hall bar is a square with length L.

10.1. Phenomenology of Integer Quantum Hall Effect

Let us first explain the two-dimensional Hall effect. Its experimental setup is illustrated in Fig. 10.2. A magnetic field is applied perpendicular to the xy-plane and a longitudinal current I_y flows along the y-axis. There is also a transverse current because, in the presence of both an electric and magnetic field, an electron will move perpendicular to the electric and magnetic fields (its velocity is called the drift velocity)

$$\vec{v}_D = c\frac{\vec{E} \times \vec{B}}{B^2}. \tag{10.1}$$

Electrons will pile on one side of the sample until a steady state is achieved. This charge unbalance between the opposite edges of the sample will create a transverse electric field E_x along the x-direction, i.e., a transverse voltage difference between the opposite edges will appear. One finds the Hall resistance[c] by measuring the transverse voltage and the longitudinal

---

[c]In Hall materials, the conductivity and resistivity are related as follows:

$$\sigma = \begin{pmatrix} \sigma_{xx} & \sigma_{xy} \\ \sigma_{yx} & \sigma_{yy} \end{pmatrix} = \begin{pmatrix} \rho_{xx} & \rho_{xy} \\ \rho_{yx} & \rho_{yy} \end{pmatrix}^{-1}. \tag{10.2}$$

Resistivity is independent of the shape of material. Note that resistance R and resistivity ρ are identical in two dimensions since $R = \rho L^{2-d}$, where d is the dimension of space.

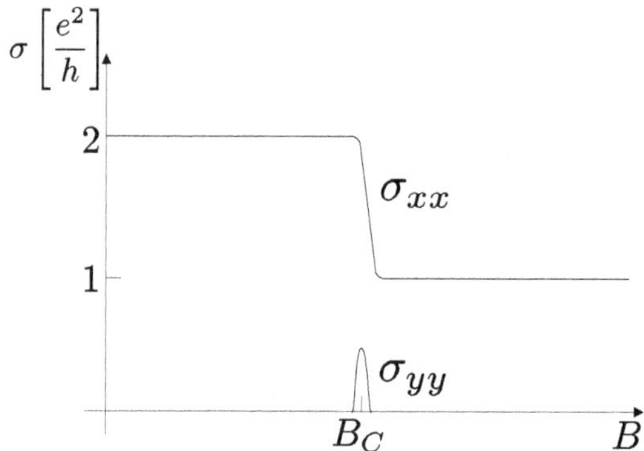

Fig. 10.3. Plateau transition as a function of B is displayed schematically near zero temperature. σ_{xy} becomes a step function and σ_{yy} a delta function in the limit of zero temperature. The critical value of σ_{yy} is expected to be universal and $\sim \frac{1}{2}\frac{e^2}{h}$. Spin degree of freedom is ignored here.

current

$$R_{xy} = \frac{V}{I_y} = \rho_{xy} = \frac{1}{n_t}\frac{h}{e^2}, \tag{10.3}$$

where n_t is the number of filled Landau levels (to be defined below). As long as the Fermi energy is between two Landau level energies σ_{xy}, or ρ_{xy}, remains unchanged. In this regime, σ_{xy} displays a "plateau" as a function of B but the diagonal conductivity is $\sigma_{xx} = 0$, see Fig. 10.3. Moreover, note that no longitudinal bulk current flows and only edge current flows. This effect is in sharp contrast to the classical Hall effect in three dimensions which does not display quantized "plateau" values. (At this point if a student does not know what the classical Hall effect is, he should do Exercise 10.1.) The two-dimensional Hall resistance is measured to ten significant digits $\frac{h}{e^2} = 25812.80745\ldots.$[d] The precise and universal value of the Hall conductivity was also unexpected, considering that there are some 10^{23} electrons that interact with each other and with numerous random impurities in the sample. It was also observed that, as magnetic field varies, the value σ_{xy} suddenly changes from $\frac{e^2}{h}n_t \to \frac{e^2}{h}(n_t \pm 1)$ and at the critical value of B_c, the diagonal conductivity takes a universal value $\sigma_{yy} \sim \frac{1}{2}\frac{e^2}{h}$, see Fig. 10.3.

---

[d]von Klitzing won a Nobel Prize for his discovery of the integer quantum Hall effect

> **Exercise 10.1.** Show that $V/I_y \propto B$ in the classical Hall effect. Show also that the electron density can be measured in the Hall experiment.

10.2. Landau Levels

The integer quantum Hall effect originates from Landau levels in two dimensions. Let us try to understand how the Landau levels are formed.

- Intra and Inter Landau Level Operators

First we will ignore disorder and electron–electron interaction effects and consider the single electron problem in a magnetic field. The two-dimensional Hamiltonian of an electron with the effective mass m^* in a magnetic field is[e]

$$H = \frac{1}{2m^*}\vec{\Pi}^2, \tag{10.4}$$

where $\Pi = -i\hbar\nabla + e\vec{A}/c$ is the mechanical momentum $(e > 0)$ with the commutation relation

$$[\Pi_x, \Pi_y] = -\frac{i\hbar e}{c}B_z = -\frac{i\hbar^2}{\ell^2}, \tag{10.5}$$

where **the magnetic length** is

$$\ell = \sqrt{\frac{\hbar c}{eB}}. \tag{10.6}$$

(The magnetic field points along the z-axis, $\vec{B} = B\hat{z}$.)

The motion of an electron in a magnetic field can be treated as the superposition of a relatively fast circular motion $\vec{\eta}$ around a point called the guiding center $\vec{R} = (X, Y)$ and a relatively slow drift of this point, see Fig. 10.4. The quantum version of the guiding center coordinates are the

---

[e]Since the relevant length scale is much larger than atomic distances one can employ the continuum limit approximation and use the effective mass approximation, see Appendix A.

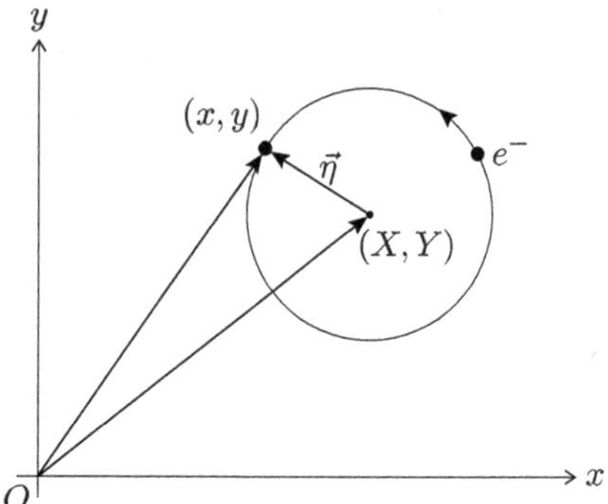

Fig. 10.4. Electron position is given by the sum of relative position $\vec{\eta}$ and guiding center position $\vec{R}$: $\vec{r} = \vec{\eta} + \vec{R}$. The cyclotron motion is circular. Note that the magnetic field points out of the page.

guiding center operators[f] given by

$$(\mathbf{X}, \mathbf{Y}) = \vec{r} - \vec{\eta} = \vec{r} - \frac{\ell^2}{\hbar}(\hat{z} \times \Pi) = \left(x + \frac{\ell^2}{\hbar}\Pi_y, y - \frac{\ell^2}{\hbar}\Pi_x \right). \quad (10.7)$$

Note that the commutation relation between $\mathbf{X}$ and $\mathbf{Y}$ is non-zero

$$[\mathbf{X}, \mathbf{Y}] = -i\ell^2. \quad (10.8)$$

The commutation relation between η_x and η_y is

$$[\eta_x, \eta_y] = i\ell^2. \quad (10.9)$$

The intra Landau level operators can be written in terms of the guiding center coordinate operators

$$b = \frac{1}{\sqrt{2}\ell}(\mathbf{X} - i\mathbf{Y}), \; b^\dagger = \frac{1}{\sqrt{2}\ell}(\mathbf{X} + i\mathbf{Y}),$$
$$[b, b^\dagger] = 1. \quad (10.10)$$

One can define inter Landau level operators in terms of $\vec{\Pi}$

$$a = \frac{\ell}{\sqrt{2}\hbar}(\Pi_x + i\Pi_y), \; a^\dagger = \frac{\ell}{\sqrt{2}\hbar}(\Pi_x - i\Pi_y),$$
$$[a, a^\dagger] = 1. \quad (10.11)$$

The intra and inter Landau level orperators commute with each other.

---

[f]These operators are written in bold letters to distinguish it from coordinates (X, Y) which are numbers. In this book, whenever such a confusion may arise we use bold letters.

These operators $a^\dagger$ and a describe the cyclotron motion of an electron and describe different energy states (this is why they are called the inter Landau level operators). The single electron Hamiltonian can be written in terms of the **inter** Landau level operators $a^\dagger$ and a

$$H = \left(a^\dagger a + \frac{1}{2}\right)\hbar\omega_c, \tag{10.12}$$

where the **cyclotron frequency** is

$$\omega_c = \frac{eB}{m^*c}. \tag{10.13}$$

Note that this Hamiltonian is identical to the Hamiltonian of a harmonic oscillator. Its eigenstates are

$$|nm\rangle = \frac{1}{\sqrt{n!m!}}(a^\dagger)^n(b^\dagger)^m|0\rangle. \tag{10.14}$$

(The operation $(b^\dagger)^m$ does not change the Landau level index. This is why these are called intra Landau level operators.) The eigenenergies are

$$E_n = \hbar\omega_c(n + 1/2), \tag{10.15}$$

where n is the Landau level index. Note that the eigenenergies are independent of the intra Landau level index m. This is the origin of the massive degeneracy of each Landau level, see Fig. 10.5. The operators $b^\dagger$ and b

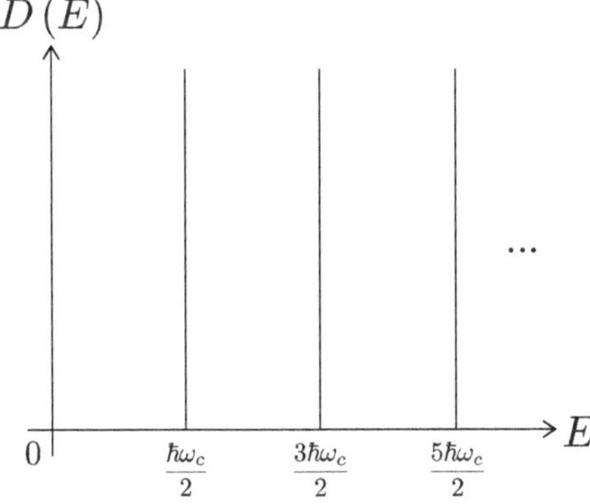

Fig. 10.5. DOS of a quantum Hall system in the absence of disorder and electron interactions. Each Landau level $n = 0, 1, 2$, etc. has a large **degeneracy** $N_\phi = \frac{A}{2\pi\ell^2}$ where A is the two–dimensional area of the system. Here we have ignored the Zeeman spin–splitting.

describe different degenerate states within a Landau level that have different guiding center positions.

The results given above are all **gauge-independent**, i.e., they are valid in any gauge. Now we give gauge-dependent results. For rectangular geometry, it is convenient to use the Landau gauge. In the Landau gauge, $\vec{A} = (0, -Bx, 0,)$ with $\vec{B} = -B\hat{z}$ and the guiding center operators are

$$(\mathbf{X}, \mathbf{Y}) = \left(\ell^2 \frac{1}{i} \frac{\partial}{\partial y}, y - \ell^2 \frac{1}{i} \frac{\partial}{\partial x} \right). \tag{10.16}$$

In this gauge, the eigenstate wave functions of the nth Landau level are

$$\psi_{n,k}(x, y) = \frac{1}{\sqrt{L}} e^{iky} \phi_n(x - k\ell^2), \tag{10.17}$$

where the orbital wave function $\phi_n(x)$ is the nth Hermite polynomial and L is the length of the system consisting of a square. Its probability density along the x-axis is localized at the guiding center position with width ℓ

$$X_k = k\ell^2. \tag{10.18}$$

Along the y-axis it represents a plane wave. For the $n = 0$ Landau level

$$\psi_{0k}(x, y) = \frac{1}{\pi^{1/4} \sqrt{L}} e^{iky} e^{-(x - k\ell^2)^2 / 2\ell^2}. \tag{10.19}$$

Note that it is a Gaussian wave function **localized** at $x = k\ell^2$. The wave vector (i.e., intra Landau level index) is

$$k = \frac{2\pi}{L} j, \quad j = 0, \ldots, N_\phi, \tag{10.20}$$

where the **Landau level degeneracy** is

$$N_\phi = L^2 / 2\pi\ell^2. \tag{10.21}$$

The allowed values of j are restricted by the condition that the guiding center, X, must physically lie within the system, $0 \leq X \leq L$. This gives the following range for j,

$$0 \leq j < \frac{L^2}{2\pi\ell^2} = N_\phi. \tag{10.22}$$

This implies that each Landau level state occupies an area of $2\pi\ell^2$. Note also that

$$N_\phi = \Phi/\Phi_0 = BL^2/\Phi_0. \tag{10.23}$$

From this, it follows that **each Landau level state is pierced by one quantum of magnetic flux**

$$\Phi_0 = B2\pi\ell^2. \tag{10.24}$$

The distance between neighboring centers of the wave function is

$$\Delta X = \frac{L}{N_\phi} = \frac{2\pi\ell^2}{L}. \tag{10.25}$$

The Landau level wave functions in the symmetric gauge will be given later.

10.3. Magnetic Translations

In a magnetic field, the translational operator is not given by the usual translational operator $T = e^{-\frac{i\vec{d}\cdot\vec{p}}{\hbar}}$. (Note $Tf(\vec{r}) = f(\vec{r}+\vec{d})$ and the minus sign in the exponent.) The correct translational operators must commute with the Hamiltonian

$$[T_x, H] = [T_y, H] = 0. \tag{10.26}$$

(T_x and T_y are the translation operators along the x- and y-axes; the amount translations will be denoted by d_x and d_y, respectively.) To define a Brillouin zone, T_x and T_y must commute with each other:

$$[T_x, T_y] = 0. \tag{10.27}$$

Let us examine whether the operators $[10, 11]$

$$\begin{aligned}
T_x &= e^{-\frac{i}{\hbar}d_x(p_x - \frac{e}{c}A_x)}, \\
T_y &= e^{-\frac{i}{\hbar}d_y(p_y - \frac{e}{c}A_y)}
\end{aligned} \tag{10.28}$$

satisfy the conditions for translation operators in the presence of a magnetic field. First, the Hamiltonian commutes with these operators provided that the vector potential satisfies the following condition

$$\left[\left(\vec{\mathbf{p}} + \frac{e}{c}\vec{A}\right)_i, \left(\vec{\mathbf{p}} - \frac{e}{c}\vec{A}\right)_k\right] = i\hbar\frac{e}{c}\left(\frac{\partial A_k}{\partial x_i} + \frac{\partial A_i}{\partial x_k}\right) = 0. \tag{10.29}$$

For the symmetric gauge $\vec{A} = \frac{B}{2}(y, -x, 0)$ with $\vec{B} = B\hat{z}$, this condition is always satisfied. In the symmetric gauge, the translational operators are

$$\begin{aligned}
T_x &= e^{-i\frac{d_x}{\ell^2}[y - \frac{\ell^2}{\hbar}(p_x + \frac{\hbar}{\ell^2}\frac{y}{2})]} = e^{-i\frac{d_x}{\ell^2}[y - \frac{\ell^2}{\hbar}\Pi_x]} = e^{id_x\frac{\mathbf{Y}}{\ell^2}}, \\
T_y &= e^{-i\frac{d_y}{\ell^2}[x + \frac{\ell^2}{\hbar}(p_y - \frac{\hbar}{\ell^2}\frac{x}{2})]} = e^{-i\frac{d_y}{\ell^2}[x + \frac{\ell^2}{\hbar}\Pi_y]} = e^{-id_y\frac{\mathbf{X}}{\ell^2}}.
\end{aligned} \tag{10.30}$$

Note that T_x and T_y are expressed in terms of the guiding center operators X and Y. We assume that these results hold any gauge. Then the translation operator $T_{\vec{d}}$ should have the following form

$$T_{\vec{d}} = \left(e^{id_x \frac{Y}{\ell^2}}, e^{-id_y \frac{X}{\ell^2}}\right). \tag{10.31}$$

One can show that the commutation relation of these translational operators is

$$T_y T_x = e^{-i2\pi\Phi/\Phi_0} T_x T_y, \tag{10.32}$$

which follows from

$$\left[-\frac{id_x}{\hbar}\left(\mathbf{p_x} - \frac{eBy}{2c}\right), -\frac{id_y}{\hbar}\left(\mathbf{p_y} + \frac{eBx}{2c}\right)\right] = 2\pi i\Phi/\Phi_0 \tag{10.33}$$

and

$$e^A e^B = e^B e^A e^{[A,B]}. \tag{10.34}$$

(See exercise in Appendix E.) Here Φ is the flux through the area $d_x d_y$, see Fig. 10.6. Two translational operators for any vectors $\vec{d_1}$ and $\vec{d_2}$ satisfy

$$T_{\vec{d_1}} T_{\vec{d_2}} = T_{\vec{d_2}} T_{\vec{d_1}} e^{i\frac{\hat{z}\cdot(\vec{d_1}\times\vec{d_2})}{\ell^2}}. \tag{10.35}$$

When do these operators commute with each other? These magnetic translational operators **commute** with each other in any gauge **provided** $\frac{\Phi}{\Phi_0} \in \mathbb{Z}$, where Φ is the flux through the area $\hat{z}\cdot\vec{d_1}\times\vec{d_2}$ (see Fig. 10.6). The

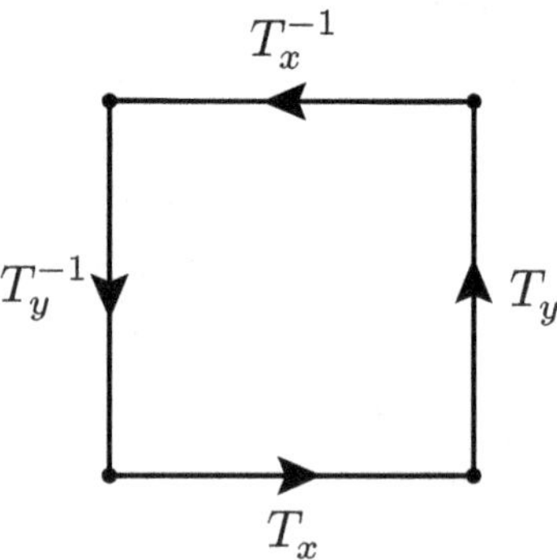

Fig. 10.6.　An electron follows a closed path during successive translations $T_y^{-1} T_x^{-1} T_y T_x$. In the process it acquires the Aharonov–Bohm phase.

translational operators also commute with the Hamiltonian. This is easy to show since the guiding center operators can be written in terms of the intra Landau level operators b and $b^\dagger$ while the Hamiltonian is given in terms of inter Landau operators a and $a^\dagger$. We may summarize all these results by saying that the following operator corresponds to the pseudomomentum of translation operator of a lattice in the presence of a magnetic field

$$\vec{K} = \vec{\Pi} + \frac{e}{c}\vec{B} \times \vec{r}. \tag{10.36}$$

What is interesting about this result is that $\vec{\Pi}$ (or $\vec{K}$) changes under a gauge transformation even if $\vec{B} = 0$. In other words, guiding centers shift under a gauge transformation. An example of this effect is the shift of the electron wave function of a ring threaded by a flux. We will see that this property has deep consequences for both integer and fractional quantum physics.

Exercise 10.2. Suppose you deform a rectangular system with side lengths L_x and L_y into a torus. Show

$$\frac{\Phi}{\Phi_0} \in \mathbb{Z}, \tag{10.37}$$

where Φ is the flux through the area $L_x L_y$. Hint: Use the commutation relation between T_x and T_y with $d_x = L_x$ and $d_y = L_y$, see Eq. (10.32). Note that on a torus, this translational operator $T_x T_y$ is the identity operator.

Exercise 10.3. Suppose you have a rectangular and periodic system consisting of unit cells, each with lengths a_x and a_y. Let $\Phi_{\text{cell}} = a_x a_y B$ the flux through each unit cell. Show

$$\frac{\Phi_{\text{cell}}}{\Phi_0} = \frac{p}{q}, \tag{10.38}$$

where p and q are integers. Hint: Use the result of Exercise 10.2.

10.4. Gauge Invariance

Before we move on, let us try to understand gauge invariance of quantum Hall systems [1]. It plays a central role in the explanation of both integer and fractional quantum Hall effects. The effect of disorder will be ignored for simplicity.

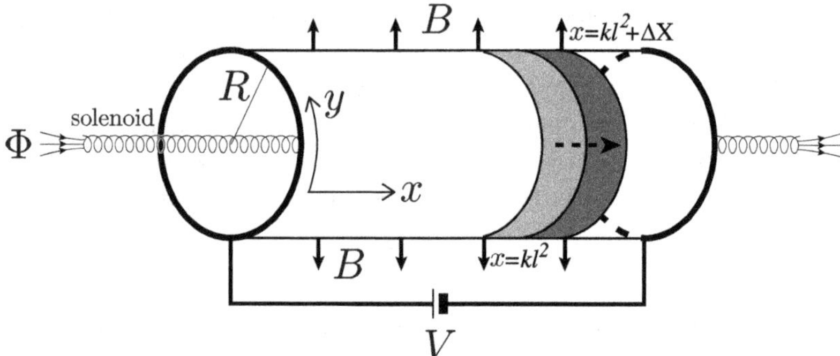

Fig. 10.7. Square Hall bar system is folded into a cylinder. Note that the circumference is $2\pi R = L$ and the length of cylinder is also L. Electrons reside on the surface of the cylinder. The vertical stripes represent probability densities of Landau level states. The x-position of each wave function shifts by amount ΔX after the adiabatic insertion of Φ_0.

- Shifting of Guiding Centers

First we need to understand shifting of guiding centers of Landau level states as B changes. This result will be used in Laughlin's thought experiment to explain the integer quantum Hall effect.

Consider a cylinder adiabatically threaded by a thin magnetic flux of a solenoid (see Fig. 10.7). The magnetic field is confined inside the solenoid and does not directly affect the electrons on the surface of the cylinder. In addition, imagine that a magnetic field is present normal to the surface of the cylinder. The total vector potential in the Landau gauge is

$$\vec{A} + \vec{A}_\phi = \left(0, Bx - \frac{\Phi}{2\pi R}, 0\right) = \left(0, B\left(x - \frac{\Phi}{2\pi RB}\right), 0\right). \quad (10.39)$$

The first vector potential $\vec{A}$ gives the magnetic field applied normal to the surface of the cylinder. The second vector potential $\vec{A}_\phi$ originates from the solenoid. Suppose the flux of the solenoid changes by the elementary **quantum flux Φ_0** : $\Phi \to \Phi + \Phi_0$. This is accompanied by a change in the vector potential. The translational operator is also changed by this operation, see Eq. (10.30). The transformed wave function is

$$\psi_{nk}(x, y) = \frac{1}{\sqrt{L}} e^{iky} \phi_n(x - \Delta X - k\ell^2) \quad (10.40)$$

with the x-position of the wave function shifted by amount (see Eq. (10.39))

$$\Delta X = \frac{\Phi_0}{2\pi RB} = \frac{\Phi_0}{LB} = \frac{2\pi \ell^2}{L}. \quad (10.41)$$

(Note that B does not originate from the solenoid.) The shift ΔX is precisely the distance between neighboring Landau wave functions, see Eq. (10.25). **This means that each state in a Landau level is equally shifted by amount ΔX**, see Fig. 10.7. As a result, a charge e^- leaves the system through the right boundary of the cylinder and, at the same time, the same amount of charge enters from the left boundary (the voltmeter serves as the charge reservoir, see Fig. 10.7). The system has returned to its original state before the application of the gauge transformation.

- Large Gauge Transformation

When a flux threads the cylinder only certain local gauge transformations leave the system invariant (for other values it is not possible). The system is not gauge invariant under flux change that differs from an integer multiple of Φ_0. Let us show this [9].

The change in the vector potential is facilitated by the scalar function $\lambda(x,y) = \frac{\Phi}{2\pi R}y$. The periodic boundary condition requires that the transformed wave function satisfies

$$\psi'(x, y + 2\pi R) = \psi'(x, y). \tag{10.42}$$

The left-hand side of this equation is

$$\psi'(x, y + 2\pi R) = \psi(x, y + 2\pi R)e^{-i\frac{e}{\hbar c}\lambda(x,y+2\pi R)}$$
$$= \psi(x, y)e^{-i\frac{e}{\hbar c}\lambda(x,y)-i\frac{e}{\hbar c}\Delta\Phi}. \tag{10.43}$$

This result must be equal to the right-hand side of Eq. (10.42)

$$\psi(x, y)e^{-i\frac{e}{\hbar c}\lambda(x,y)}. \tag{10.44}$$

From this, it follows that when $\Delta\Phi = \frac{hc}{e} \times$ (integer) the system is invariant under the transformation. This discrete transformation is called a **large gauge transformation**.

10.5. Laughlin's Thought Experiment

To explain the exact quantized values of the transverse conductivity, Laughlin proposed a thought experiment. This argument based on gauge transformation [1]. Again, the effect of disorder will be ignored for simplicity. The argument applies at arbitrary strength of the magnetic field as long as the Fermi energy E_F is in a gap of the density of states.

Let us map the Hall bar system onto an equivalent system of a cylinder, see Fig. 10.7. To produce a steady flow of charge carriers, you need a charge pump, i.e., an electromotive force (emf) is needed. For this purpose, the Faraday law comes in handy: an emf is induced when the solenoid magnetic flux that threads the loop changes. Suppose the flux changes by amount Φ_0 during the time period $\Delta t = T$. From the Faraday law, the emf produces an electric field along the y-axis

$$E_y L = -\frac{1}{c}\frac{\Delta \Phi}{\Delta t}. \tag{10.45}$$

This electric field E_y will generate the Lorentz force, which in turn will induce an electric field E_x that gives rise to a transverse **Hall current** $\langle I_x \rangle$. We recover the original state every time Φ_I increases by Φ_0 in a time interval $\Delta t = T$: according to gauge invariance, when the Fermi energy E_{F} is in a gap of the density of states or in the energy region of localized states, the state of the ring returns to the previous state after addition of an elementary flux, see Sec. 3.3. The only change, then, should be **a transfer of an electron per occupied Landau level** from one electrode to another, which we assume to be attached to either edge as Hall probes. From the Faraday law, Eq.(10.45), we find that the inverse time is given by

$$\frac{1}{\Delta t} = -\frac{cE_y L}{\Phi_0}. \tag{10.46}$$

Since an electron has transferred during the period T this result gives the current density

$$\langle j_x \rangle = \frac{e}{L^2}\frac{L}{\Delta t} = -\frac{e^2}{h}E_y. \tag{10.47}$$

This result gives the following quantized transverse conductivity

$$\sigma_{xy} = -\frac{e^2}{h}. \tag{10.48}$$

We have ignored the localization effect of disorder (more about it in Sec. 10.10). Localized states are not expected to change much during the flux insertion. For above gauge argument to work, one must assume that there must be at least one **delocalized** state which evolves in a complicated manner during the flux insertion. The effect of a change in the magnetic flux in the solenoid is to carry extended wave functions in the direction parallel to the axis of the cylinder like a conveyor belt, see Fig. 10.8. This process is crucial in Laughlin's thought experiment.

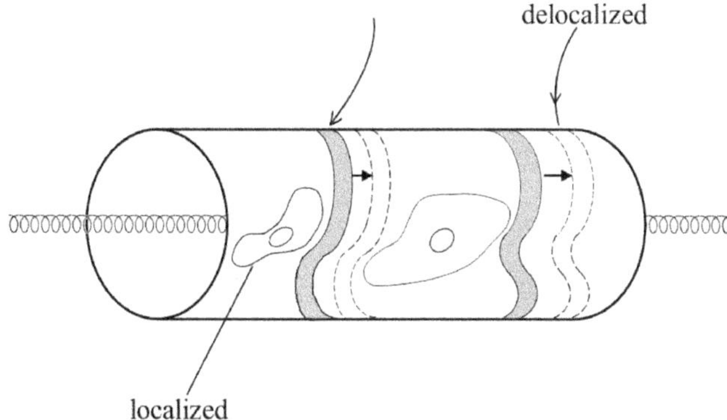

Fig. 10.8. Localized states are not expected to change much during the flux insertion. However, the delocalized states shift as shown in the figure and an electron is transported from one edge to the other [9].

10.6. Integer Quantum Hall Effect and Chern Number of Torus

Laughlin's thought experiment is based on single electron picture. In reality, the number of transferred electrons may vary each time the flux is inserted (as long as it is an integer gauge invariance is satisfied). Only the average value is expected to be precise [12], which may be obtained after repeated flux insertions. As we will explain in this section, the reason that the average value is quantized to a very high precision is because it is a topological Chern number. This insight was provided by Thouless *et al.*[g] Weak disorder will also not change a topological Chern number, which explains the robustness of the integer quantum Hall effect [13].

• Charge Pumping

Consider a Hall bar with a magnetic field pointing in the perpendicular direction. First, roll the Hall bar into a long cylinder. Then make the cylinder into a torus with a large radius, see Fig. 10.9. Consider inserting two long but thin solenoids, such that one is pointing along the x-axis with flux Φ_I and the other along the axis of the y-axis with flux Φ_V. If Φ_I changes the longitudinal current is also changed (again it is the Faraday law applied backwards).

---

[g]David Thouless won a Nobel Prize partly for this work.

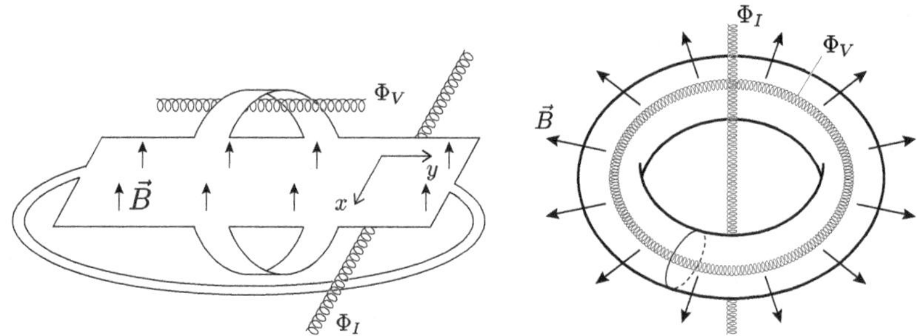

Fig. 10.9. **Left**: Hall bar system of Fig. 10.2 imagined by Thouless [13]. **Right**: Two solenoids thread two holes of the torus, one hole is inside and the other is outside of the torus. This set up is topologically equivalent to the set up shown in the left figure: both manifolds have two holes.

Let us first compute the off-diagonal conductivity $\sigma_{xy}(\Phi_I, \Phi_V)$ at some given values of Φ_I and Φ_V. The Kubo formula gives the transverse conductivity of the sample (see Appendix F.2). In the static and zero temperature limits $E \to 0$ and $T \to 0$, the conductivity is

$$\sigma_{xy}(E) = \frac{i\hbar}{EL^2} \sum_n \left[\frac{\langle \Psi_0 | I_x | \Psi_n \rangle \langle \Psi_n | I_y | \Psi_0 \rangle}{E + i\eta - [E_n - E_0]} - \frac{\langle \Psi_0 | I_y | \Psi_n \rangle \langle \Psi_n | I_x | \Psi_0 \rangle}{E + i\eta + [E_n - E_0]} \right],$$

$$(10.49)$$

where $|\Psi_n\rangle$ is a **disordered many-body state** and E_n is its energy ($|\Psi_0\rangle$ and E_0 are the ground state and its energy). Excited states are $|\Psi_n\rangle$. In this expression, we use current operator I_i instead of current density operator j_i (note $I_i = j_i L^{d-1}$, where d is spatial dimension).

We use the expansion

$$\frac{1}{E + E_n - E_m} \approx \frac{1}{E_n - E_m} - \frac{E}{(E_n - E_m)^2}.$$

$$(10.50)$$

The contribution from the first term in the expansion is zero, see Exercise 10.4. We find from the second term that

$$\sigma_{xy} = -\frac{i\hbar}{L^2} \sum_{n \neq 0} \frac{\langle \Psi_0 | I_x | \Psi_n \rangle \langle n | I_y | \Psi_0 \rangle - \langle \Psi_0 | I_y | \Psi_n \rangle \langle \Psi_n | I_x | \Psi_0 \rangle}{(E_n - E_0)^2}.$$

$$(10.51)$$

When Φ_i changes, the Hamiltonian H also changes. Using the expression for the change $\Delta H = I_i \Delta \Phi_i / cL$ ($I = jL^{d-1}$; see Exercises 10.5 and 10.6), we find that the first order change of wave function is

$$|\Delta \Psi_0\rangle = \sum_{n \neq 0} \frac{\langle \Psi_n | \Delta H | 0 \rangle}{E_n - E_0} |\Psi_n\rangle = \frac{\Delta \Phi_i}{cL} \sum_{n \neq 0} \frac{\langle \Psi_n | I_i | \Psi_0 \rangle}{E_n - E_0} |\Psi_n\rangle. \quad (10.52)$$

This gives

$$\frac{1}{L} \sum_{n \neq 0} \frac{\langle \Psi_n | I_x | \Psi_0 \rangle}{E_n - E_0} |n\rangle = c |\frac{\partial \Psi_0}{\partial \Phi_i}\rangle. \quad (10.53)$$

From this we find the off-diagonal conductivity

$$\sigma_{xy} = i\hbar c^2 \left[\left\langle \frac{\partial \Psi_0}{\partial \Phi_I} \bigg| \frac{\partial \Psi_0}{\partial \Phi_V} \right\rangle - \left\langle \frac{\partial \Psi_0}{\partial \Phi_V} \bigg| \frac{\partial \Psi_0}{\partial \Phi_I} \right\rangle \right]$$

$$= i\hbar c^2 \left[\frac{\partial}{\partial \Phi_I} \left\langle \Psi_0 \bigg| \frac{\partial \Psi_0}{\partial \Phi_V} \right\rangle - \frac{\partial}{\partial \Phi_V} \left\langle \Psi_0 \bigg| \frac{\partial \Psi_0}{\partial \Phi_I} \right\rangle \right]$$

$$= \hbar c^2 \left[\frac{\partial A_{\Phi_V}}{\partial \Phi_I} - \frac{\partial A_{\Phi_I}}{\partial \Phi_V} \right]. \quad (10.54)$$

We change the variables to angular variables $\theta_i = \frac{\Phi_i}{\Phi_0}$. Then the Berry connections are

$$A_{\Phi_V} = \frac{A_{\theta_1}}{\Phi_0}, \quad A_{\Phi_I} = \frac{A_{\theta_2}}{\Phi_0}, \quad (10.55)$$

where

$$A_{\theta_1} = i \left\langle \Psi_0 \bigg| \frac{\partial \Psi_0}{\partial \theta_1} \right\rangle$$

$$A_{\theta_2} = i \left\langle \Psi_0 \bigg| \frac{\partial \Psi_0}{\partial \theta_2} \right\rangle. \quad (10.56)$$

The conductivity can now be written as

$$\sigma_{xy} = \frac{e^2}{h} \frac{1}{2\pi} B(\theta_1, \theta_2), \quad (10.57)$$

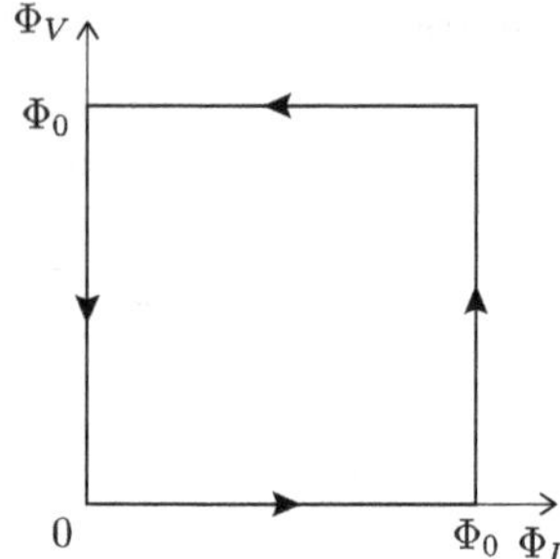

Fig. 10.10.   Applied fluxes form a closed loop around the Brillouin zone in the parameter space of (Φ_I, Φ_V).

where the Berry curvature

$$B(\theta_1, \theta_2) = \left[\frac{\partial A_{\theta_1}}{\partial \theta_2} - \frac{\partial A_{\theta_2}}{\partial \theta_1} \right]. \tag{10.58}$$

The system returns to the initial non-degenerate ground state[h] after following a closed loop along the edges of the Brillouin zone[i] in the (Φ_1, Φ_2) parameter space, see Fig. 10.10. We will compute the value of the off-diagonal conductivity when the cycle is completed. The line integral is equivalent to integration over a closed surface and leads to an integer Chern number: One can compute the **average value** of $B(\theta_1, \theta_2)$ over one adiabatic cycle using Stokes' theorem. We find that

$$\frac{1}{2\pi} \int_0^1 d\theta_1 \int_0^1 d\theta_2 B(\theta_1, \theta_2) = \text{integer}, \tag{10.59}$$

because the integral of $B(\theta_1, \theta_2)$ over the Brillouin zone is a Chern number (in the usual definition of the Chern number the factor $\frac{1}{2\pi}$ is included). From this result, we conclude that the averaged value of the off-diagonal conductivity is $\frac{e^2}{h}$. This effect is an example of **charge pumping** [12] in which electrons are transferred from one charge reservoir (voltmeter) to another during a cyclic process.

---

[h] Fractional quantum Hall states can be degenerate on torus. This property is intimately connected to the existence of anyons, see Sec. 11.10.

[i] A Brillouin zone appears because the dependency on Φ_1 and Φ_2 is periodic with the period Φ_0. This is a consequence of gauge invariance under insertion of Φ_0 through a hole, see Sec. 3.2. Note that the Brillouin zone can be deformed into a closed surface.

Exercise 10.4. Show that the first term in Eq.(10.50) does not contribute to the conductivity. Answer: Since $I_{x,y} \sim v_{x,y}$ it is sufficient to show that the following holds for single particle states $\{|\psi_n\rangle\}$ and energies $\{\epsilon_n\}$

$$\sum_{n\neq 0} \frac{\langle\psi_0|v_x|\psi_n\rangle\langle\psi_n|v_y|\psi_0\rangle + \langle\psi_0|v_y|\psi_n\rangle\langle\psi_n|v_x|\psi_0\rangle}{\epsilon_0 - \epsilon_n} = 0.$$

(10.60)

The left-hand side of this equation is the contribution of the first term to σ_{xy}. Since $v_x = [H, x]$ we have

$$\langle\psi_0|v_x|\psi_n\rangle = \langle\psi_0|Hx - xH|\psi_n\rangle = (\epsilon_0 - \epsilon_n)\langle\psi_0|x|\psi_n\rangle.$$

(10.61)

Plugging this result into the left-hand side of Eq. (10.60) we find

$$\sum_{n}[\langle\psi_0|x|\psi_n\rangle\langle\psi_n|v_y|\psi_0\rangle - \langle\psi_0|v_y|\psi_n\rangle\langle\psi_n|x|\psi_0\rangle]$$

$$= \langle\psi_0|xv_y - v_yx|\psi_0\rangle = \langle\psi_0|[x, v_y]|\psi_0\rangle = 0. \qquad (10.62)$$

Exercise 10.5. Show that the one-dimensional current operator of an electron in the presence of a vector potential is given by

$$j_x = -\frac{c}{L}\frac{\partial H}{\partial A_x} = \frac{e}{m^*L}\left(\frac{\hbar}{i}\frac{\partial}{\partial x} - \frac{e}{c}A_x\right). \qquad (10.63)$$

Here H is the single-particle Hamiltonian in a magnetic field. The current density is $j = nev$, where n is the electron density. The relation between the current and current density is $I = jL^{d-1}$.

Exercise 10.6. Consider an electron. In the Landau gauge, the magnetic field is related to the vector potential as follows

$$A_x = By. \qquad (10.64)$$

The Faraday law gives $E_x L = -\frac{1}{c}\frac{\partial\Phi_y}{\partial t}$ and $E_y L = -\frac{1}{c}\frac{\partial\Phi_I}{\partial t}$. Using

$$\mathbf{E} = \nabla\varphi + \frac{1}{c}\frac{\partial\mathbf{A}}{\partial t} \qquad (10.65)$$

with $\nabla\varphi = 0$ we find

$$A_x = \frac{\Phi_V}{L}, \ A_y = \frac{\Phi_I}{L}. \tag{10.66}$$

The current density operator is (see Exercise 10.5)

$$j_i = \frac{c}{L}\frac{\Delta H}{\Delta A_i}. \tag{10.67}$$

From this and Eq. (10.66) show that the change in the Hamiltonian

$$\Delta H = I_i \Delta\Phi_i/cL. \tag{10.68}$$

10.7. Magnetic Bands/Hofstadter Butterfly

Consider an electron of a disorder-free two-dimensional system in a **periodic potential** in the presence of a magnetic field. The electron band structure has an intriguing structure: as a function of applied flux it looks like a butterfly with a fractal structure and is called Hofstadter's butterfly [14]. Moreover, at certain values of flux, the band structure exhibits quantized values Hall conductivity expressed by the Chern number. Hofstadter's butterfly is observed in Moiré superlattices [15].

Consider a square lattice. Its tight-binding Hamiltonian is

$$H|i,j\rangle = E_0|i,j\rangle - t\left(|i+1,j\rangle + |i-1,j\rangle + |i,j+1\rangle + |i,j-1\rangle\right), \tag{10.69}$$

where $|i,j\rangle$ denotes a state localized at a lattice point represented by $\vec{r} = a(i,j)$ (a is the lattice constant). In the presence of a magnetic field, the nearest-neighbor tunneling coefficient from $\vec{r}_1$ to $\vec{r}_2$ changes to

$$t \to te^{-i\frac{q}{\hbar c}\vec{A}\cdot(\vec{r}_1 - \vec{r}_2)}, \tag{10.70}$$

where $\vec{A}$ is evaluated at a point halfway between sites 1 and 2 [16]. In the Landau gauge $\vec{A} = (0, Bx, 0)$, we find

$$\epsilon|i,j\rangle = E_0|i,j\rangle$$

$$-t\left(|i+1,j\rangle + |i-1,j\rangle + |i,j+1\rangle e^{-iqBxa/\hbar c} + |i,j-1\rangle e^{iqBxa/\hbar c}\right). \tag{10.71}$$

This equation may be transformed into Harper's equation using $|\psi\rangle = \sum_{ij} e^{ikaj} \psi_i |ij\rangle$, where $x = ai$ and $y = aj$ (compare this with the continuum version of the wave function in the Landau gauge, see Eq. (10.17)),

$$\epsilon \psi_i = \psi_{i+1} + \psi_{i-1} + 2\cos\left(2\pi i \frac{\Phi}{\Phi_0} - ka\right)\psi_i. \tag{10.72}$$

Here we set $t = 1$ and $E_0 = 0$. We leave it as an exercise to derive this equation (see Exercise 10.7). This equation is better suited for numerical computation because it is a one-dimensional equation while Eq. (10.71) is a two-dimensional equation. Energy bands exist when the flux $\Phi = Ba^2$ through a unit cell satisfies (see Exercise 10.3)

$$\frac{\Phi}{\Phi_0} = \frac{p}{q}, \quad p \text{ and } q \text{ are integers.} \tag{10.73}$$

At these values, a Brillouin zone can be defined. For example, when $\Phi/\Phi_0 = 1$, $1/2$, $1/3$, and $1/4$ there are 1, 2, 3, and 4 bands, respectively. Otherwise energies form a fractal structure, as shown in Fig. 10.11.

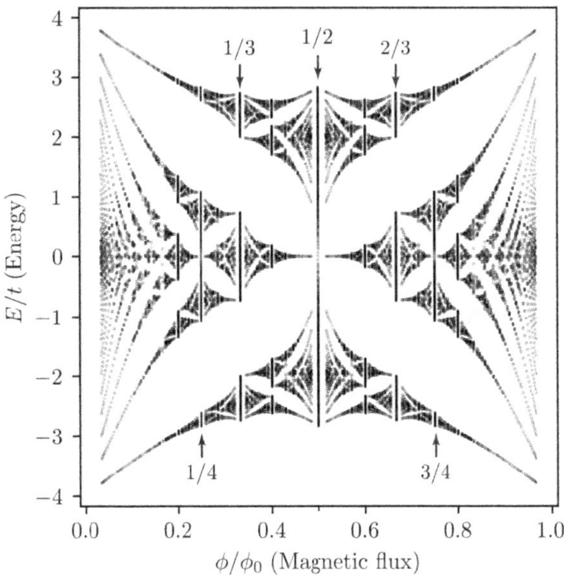

Fig. 10.11. Hofstadter's butterfly is displayed.

> **Exercise 10.7.** Drive Eq.(10.72). Solve this equation numerically and verify the result in Fig. 10.11. Note $\Phi = Ba^2$ and $\Phi_0 = hc/e$.

When energy bands are well-defined, the Chern number may be evaluated. In single-electron approximation, the transverse conductivity is given by the Kubo formula (see Eq. (F.19))

$$\sigma_{xy}(E) = i\hbar \sum_{n,m} \frac{f(E_m) - f(E_n)}{E} \frac{\langle \psi_m | j_x | \psi_n \rangle \langle \psi_n | j_y | \psi_m \rangle}{E + i\eta - [E_n - E_m]}, \quad (10.74)$$

where ψ_n are the **single-electron states**, $n = (\alpha, \vec{k})$ stands for the band index and wave vector, and $f(E)$ is the Fermi distribution function at $T = 0$. The ith component of the current density operator is

$$j_i = \frac{e}{L^d} \sum_{\vec{k}} c_{\vec{k}}^{\dagger} \frac{1}{\hbar} \frac{\partial H_{\vec{k}}}{\partial k_i} c_{\vec{k}}, \quad (10.75)$$

where $c_{\vec{k}}^{\dagger}$ is the electron creation operator (see Exercise 10.5). Note that the current density $\langle \psi | j_x | \psi \rangle$ should have the same unit as nev, where n is the electron density and v is the electron velocity. Here the Hamiltonian $H_{\vec{k}}$ is

$$H_{\vec{k}} u_{n\vec{k}} = \left[\frac{1}{2m^*} \left(-i\hbar\nabla + \hbar\vec{k} - \frac{e}{c}\vec{A}(\vec{r}) + V(\vec{r}) \right) \right] u_{n\vec{k}} = E_{n\vec{k}} u_{n\vec{k}}. \quad (10.76)$$

(We also used a similar Hamiltonian in Eq. (5.10).) Applying the expansion given in Eq. (10.50) we find

$$\sigma_{xy} = -i\hbar \sum_{m,n} [f(E_m) - f(E_n)] \frac{\langle \psi_m | j_x | \Psi_n \rangle \langle \psi_n | j_y | \psi_m \rangle}{(E_n - E_m)^2}. \quad (10.77)$$

In the static limit $E \to 0$, we have $f(E_m) - f(E_n) = 1$ for $E_m < E_F$ and $E_n > E_F$, and $f(E_m) - f(E_n) = -1$ for $E_m > E_F$ and $E_n < E_F$. This result gives

$$\sigma_{xy} = -i\hbar \sum_{E_m < E_F < E_n} \frac{\langle \psi_m | j_x | \psi_n \rangle \langle \psi_n | j_y | \psi_m \rangle - \langle \psi_m | j_y | \psi_n \rangle \langle \psi_n | j_x | \psi_m \rangle}{(E_n - E_m)^2}. \quad (10.78)$$

Using the result of Exercise 10.8 and replacing m with $(\alpha, \vec{k})$, we can rewrite the conductivity as

$$\sigma_{xy} = \frac{ie^2}{\hbar} \int_{\vec{k} \in \mathrm{BZ}} \frac{d^2 k}{(2\pi)^2}$$

$$\times \sum_{E_{\beta,\vec{k}} < E_\mathrm{F} < E_{\alpha,\vec{k}}} \frac{\langle u_{\alpha,\vec{k}} | \nabla_{\vec{k}} H(\vec{k}) | u_{\beta,\vec{k}} \rangle \times \langle u_{\beta,\vec{k}} | \nabla_{\vec{k}} H(\vec{k}) | u_{\alpha,\vec{k}} \rangle \cdot \hat{z}}{(E_\beta - E_\alpha)^2}.$$

$$(10.79)$$

The usual volume factor L^d in front of k-space integral is cancelled with $1/L^d$ appearing in j_i, see Eq. (10.77).

> **Exercise 10.8.** Show that when $\alpha \neq \beta$ we have
>
> $$\langle \psi_{\alpha,\vec{k}} | j_x | \psi_{\beta,\vec{k}} \rangle = \langle u_{\alpha,\vec{k}} | j_x | u_{\beta,\vec{k}} \rangle. \qquad (10.80)$$
>
> Here u_α is the periodic part of the Bloch wave function, determined by the Hamiltonian given by Eq. (10.78). Its normalization is $\langle u_{\alpha,\vec{k}} | u_{\beta,\vec{k}'} \rangle = \delta_{\alpha,\beta} \delta_{\vec{k},\vec{k}'}$ over a unit cell.

Utilizing the result of Exercise 10.9, we write the off-diagonal conductivity as

$$\sigma_{xy} = \frac{e^2}{h} \left[\frac{i}{2\pi} \int_{\vec{k} \in \mathrm{BZ}} d^2 k \sum_{E_{\beta,\vec{k}} < E_\mathrm{F}} \langle \nabla_{\vec{k}} u_{\beta,\vec{k}} | \times | \nabla_{\vec{k}} u_{\beta,\vec{k}} \rangle \cdot \hat{z} \right]. \quad (10.81)$$

The quantity in the bracket is the Berry curvature. Its integral over the Brillouin zone[j] is the Chern number (see Eq. (2.63)) and is thus equal to an integer.

---

[j] It forms a closed surface because it is periodic in k_x and k_y.

> **Exercise 10.9.** Derive the off-diagonal conductivity given in Eq. (10.81) from the expression (10.79). Hint: Use
>
> $$\langle u_{\beta,\vec{k}}|\nabla_{k_i}H_{\vec{k}}|u_{\alpha,\vec{k}}\rangle = \langle u_{\beta,\vec{k}}|\nabla_{k_i}\left[H_{\vec{k}}|u_{\alpha,\vec{k}}\rangle\right] - \langle u_{\beta,\vec{k}}|H_{\vec{k}}|\nabla_{k_i}u_{\alpha,\vec{k}}\rangle$$
> $$= -(E_{\alpha,\vec{k}} - E_{\beta,\vec{k}})\langle\nabla_{k_i}u_{\beta,\vec{k}}|u_{\alpha,\vec{k}}\rangle, \qquad (10.82)$$
>
> where $H_{\vec{k}}$ is defined in Eq. (10.78). Note that $\langle\nabla_{k_i}u_{\beta,\vec{k}}|u_{\alpha,\vec{k}}\rangle = -\langle u_{\alpha,\vec{k}}|\nabla_{k_i}u_{\beta,\vec{k}}\rangle$. Then use the following: the sum over the unfilled bands can be replaced by the sum over filled bands, $\sum_{\alpha,\vec{k}}|u_{\alpha,\vec{k}}\rangle\langle u_{\alpha,\vec{k}}| = 1 - \sum_{\beta,\vec{k}}|u_{\beta,\vec{k}}\rangle\langle u_{\beta,\vec{k}}|$.

10.8. Edge States

The quantization of σ_{xy} can be also explained evoking edge states. In the presence of disorder, the bulk states in the gap between two Landau levels are localized and cannot carry a current. Current must thus flow only at the edges via delocalized edge states [17], see Figs. 10.12 and 10.13. (Edge states must exist because of topological reasons, see Fig. 10.1.) This suggests that the quantized values of σ_{xy} may be explained in terms of edge states. On one side of the edge, the current flows to the right, for example. Then on the opposite edge, the current flows to the left. If there is a voltage difference ΔV between the opposite edges, the current will flow more in one direction. The resulting current is related to the voltage difference

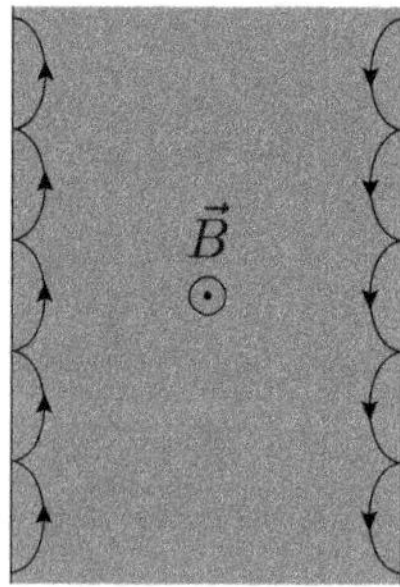

Fig. 10.12. In classical physics, an edge state represents a skipping orbit moving along the boundary of the Hall bar system. An edge state is chiral, i.e., it moves only in one direction, but opposite edges have opposite velocities. Weak disorder cannot couple the left and right edge states.

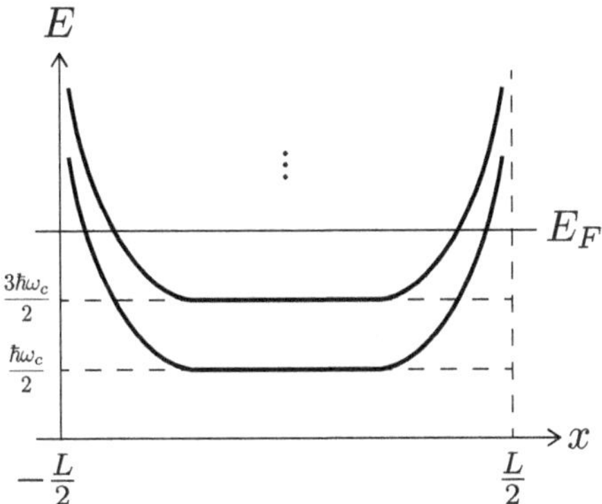

Fig. 10.13. Energy of edge states is plotted. Spin splitting is ignored (spin–up electrons have lower energy). The boundary of a Hall bar system can be modeled by a confining potential. The Fermi wave vectors give two opposite velocities of the edge states. The Fermi velocity at the guiding center $X = k\ell^2$ is given by $v_F = \frac{1}{\hbar}\frac{dE_k}{dk}$. There are two Fermi wave vectors k_F and $-k_F$. Directions of these two Fermi velocities are pointing perpendicular to the page and they have opposite directions: they are given the drift velocity $v_F = c\frac{\vec{E}\times\vec{B}}{B^2}$, where $\vec{E}$ is the electric field due to the confinement potential. Note that the electric field increases as the electron approaches the Hall bar boundaries, i.e., its energy increases. Classically edge states represent velocity of skipping orbits, see Fig. 10.12.

ΔV as follows

$$I = e \int \frac{dk}{2\pi}\langle v\rangle = e \int \frac{dk}{2\pi}\frac{1}{\hbar}\frac{dE}{dk} = \frac{e}{h}\int_{\mu}^{\mu+eV} dE = \frac{e^2}{h}\Delta V. \quad (10.83)$$

From this result, it follows that the conductance has the quantized value of $\frac{e^2}{h}$, provided that the edges are well separated.

The quantization of conductance is a general feature of one-dimensional systems. This effect is present even in the absence of a magnetic field. This is because in one dimension, the DOS goes as $\frac{1}{\sqrt{E}}$ and velocity as $\sqrt{E}$ and they cancel each other in a product, see Exercise 10.10.

Exercise 10.10. Consider one-dimensional electrons in the absence of a magnetic field. Show that the current is given by Eq. (10.83). Hint: Electrons in a small energy interval $\Delta E = e\Delta V$ contribute to the current (ΔV is the voltage difference):

$$\Delta I = e \times \frac{1}{2}D(E) \times \Delta E \times v(E), \quad (10.84)$$

where the DOS per spin is

$$D(E) = \frac{1}{\hbar\pi}\sqrt{\frac{m^*}{2E}}.$$

(10.85)

The factor $\frac{1}{2}$ is present in front of the DOS because only states with positive k contribute to the current. Note that the current and current density are identical in one-dimensional systems: the current density $j = env$ leads to $[j] = [I] = \text{Charge/Time}$, where n is the one-dimensional electron density.

10.9. Interplay Between Disorder and Electron Interactions: Coulomb Gap

So far we have ignored electron–electron interactions. Now suppose that a Landau is partially filled and that the disorder potential dominates over the Coulomb repulsion of electrons. We use the following effective many-electron Hamiltonian

$$H_N = \sum_i^N \left[\frac{1}{2m^*}\left[\frac{\hbar}{i}\vec{\nabla}_i + \frac{e}{c}\vec{A}(\vec{r}_i)\right]^2 + V_D(\vec{r}_i) \right] + \sum_{i<j}^N V(\vec{r}_i - \vec{r}_j),$$

(10.86)

where $\vec{A}(\vec{r})$ is a vector potential, m^* the electron effective mass (see Appendix A), $\vec{r} = (x, y)$ is an electron position vector, $V(\vec{r}) = e^2/\epsilon r$ is the Coulomb interaction, N is the number of electrons, and $V_D(\vec{r})$ is the disorder potential due to donors (ϵ is the background dielectric constant). The entire system consisting of electrons and donors is charge neutral. We assume that the electron correlation effects leading to fractional quantum effect may be ignored and treat electron interactions using the Hartree–Fock approximation. In addition, we assume that electrons are spin-polarized. Our model disorder consists of randomly located δ function impurities with random strength uniformly distributed from $-\lambda$ to $+\lambda$. For this model the energy scale which characterizes the Landau level energy width is $\Gamma = (\lambda^2 N_I/\ell^2 L^2)^{1/2}$, where N_I is the number of impurities and L^2 is the area of the sample.

- Hartree–Fock Hamiltonian

The relative strength of Coulomb interactions and disorder is specified by the parameter $\gamma = (e^2/\epsilon\ell)/\Gamma$. The self-consistent Hartree–Fock equations are

$$\sum_j [E_i\delta_{ij} + \langle i|V_{\mathrm{D}}|j\rangle + \langle i|V_{\mathrm{H}}|j\rangle + \langle i|V_{\mathrm{X}}|j\rangle]\langle j|\alpha\rangle = E\langle i|\alpha\rangle, \quad (10.87)$$

where E, E_i, and $|\alpha\rangle$ are the Hartree–Fock eigenenergy, the ith Landau level energy, and the Hartree–Fock eigenstates, respectively. The Hartree and exchange matrix elements of Coulomb interactions are

$$\begin{aligned}\langle i|V_{\mathrm{H}}|j\rangle &= \sum_\beta f_\beta\langle i\beta|V|j\beta\rangle, \\ \langle i|V_{\mathrm{X}}|j\rangle &= -\sum_\beta f_\beta\langle i\beta|V|\beta j\rangle,\end{aligned} \qquad (10.88)$$

where f_β is the occupation number of a self-consistent Hartree–Fock eigenstate $|\beta\rangle$.

- Hartree–Fock Approximation is Exact in Strongly Localized Regime

Consider the limit where spin-polarized electrons are extremely localized. The Hartree–Fock single-particle energy is

$$E_i = \sum_j \langle ij|V|ij\rangle n_j - \sum_j \langle ij|V|ji\rangle n_j + V_i, \qquad (10.89)$$

where V_i is the random site energy due to a disorder potential. The exchange self-energy is

$$-\sum_j \langle ij|V|ji\rangle n_j = -\langle ii|V|ii\rangle. \qquad (10.90)$$

(Since states i and j are very well localized one finds $\langle ij|V|ji\rangle = 0$ if $i \neq j$.) This term will cancel the Hartree self energy, $\langle ii|V|ii\rangle$. The **electron self-interaction terms vanish!** The renormalized site energy is

$$E_i = \sum_{j\neq i} \langle ij|V|ij\rangle n_j + V_i. \qquad (10.91)$$

Since electrons are well localized, this result is equal to

$$E_i = V_i + \sum_{i\neq j} \frac{e^2}{\epsilon r_{ij}} n_j. \qquad (10.92)$$

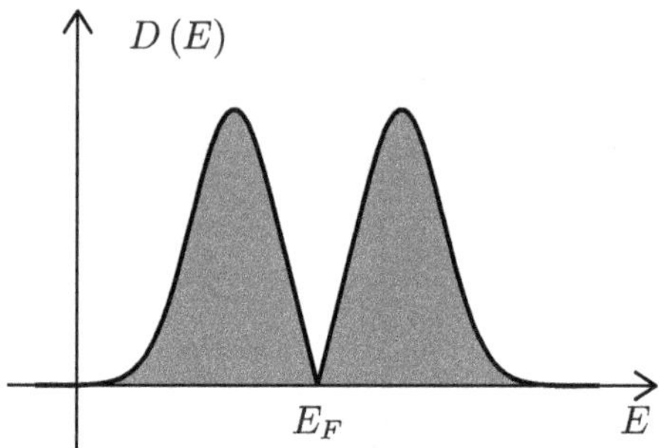

Fig. 10.14. In the presence of electron–electron interactions, the DOS of a Landau level is zero at the Fermi energy E_F (this effect appears even when $E_\mathrm{F} \neq E_c$, where E_c denotes the center of the broadened Landau level). This is a Coulomb gap and its physical origin is explained in the text.

- Tunneling Density of States

Let us numerically compute the DOS in the presence of long-range Coulomb interactions using the Hartree–Fock approximation. The presence of electron–electron interactions leads to a soft gap in the Gaussian DOS[k]: the **tunneling** DOS (see Sec. 6.8) increases linearly with energy E near the Fermi energy [18, 19]

$$D(E) \propto |E - E_\mathrm{F}|. \tag{10.93}$$

A plot of the DOS is shown in Fig. 10.14. When E_F coincides with the critical energy E_c, equal to the bare Landau level energy, a metal insulator transition takes place (see Sec. 10.10).

Let us give a simple argument that explains how a soft gap may arise near E_F [20]. Consider strongly localized electrons. A possible electron transfer from sites i to j (see Fig. 10.15) must have an excitation energy E that satisfies the condition

$$E = E_j - E_i - \frac{e^2}{\epsilon r_{ij}} > 0. \tag{10.94}$$

(This is a stability condition.) The attractive energy $-\frac{e^2}{\epsilon r_{ij}}$ represents the interaction between a hole at i and an electron at j. This condition means

$$r_{ij} > \frac{e^2}{\epsilon E} = r_{\min}. \tag{10.95}$$

---

[k]Disorder will broaden a DOS that has the shape of a delta-function (see Fig. 10.5) into a Gaussian form.

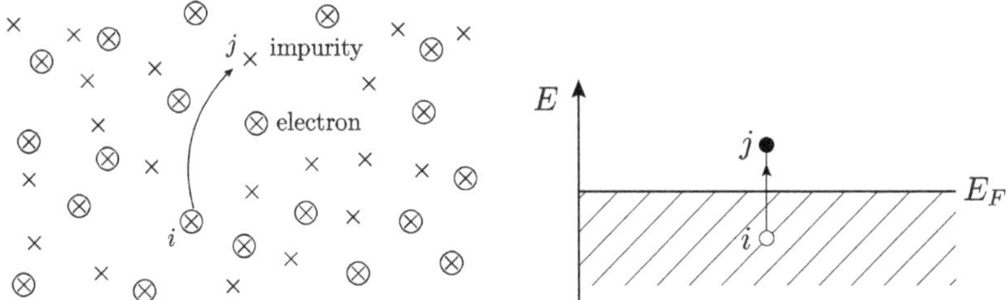

Fig. 10.15. Crosses indicate localized states and circles represent electrons. An electron moves from a localized state to an empty state (left panel). This process is equivalent to an excitation across the Fermi energy (right panel).

The distance $r_{\min}$ is thus large for adjacent energy states. This has an important significance: electrons near the Fermi energy are **spatially separated.**[1] The number of excitations with energy less than E per area is

$$n(E) < \frac{1}{\pi r_{\min}^2} = \frac{E^2 \epsilon^2}{\pi e^4}. \tag{10.96}$$

This implies that the DOS satisfies

$$D(E) = \frac{dn}{dE} < \frac{2\epsilon^2}{\pi e^4} E. \tag{10.97}$$

The upper bound of DOS is thus proportional to E.

What is the physical origin of a soft gap in the regime where electrons are less localized? The essential physics is as follows: it is difficult for the tunneling electron to avoid other electrons since it takes a long time for interacting electrons to diffuse away from each other (see Girvin and Yang, Ref. [6], pp. 290–645). Electrons try to stay away from each other as much as possible.

10.10. Localization–Delocalization Transition

Let us try to understand qualitatively the observed B-dependence of "plateau" transitions, see Fig. 10.3. First, we discuss non-interacting electrons. If all the states are localized one cannot apply Laughlin's gauge argument since localized states do not shift under gauge transformations, see Sec. 10.5. Thus there must be some extended bulk states below the

---

[1]Even in the absence electron interactions, electron states with nearly same energy are spatially separated in the localized phase. If not, tunneling occurs between them and delocalization may happen [6].

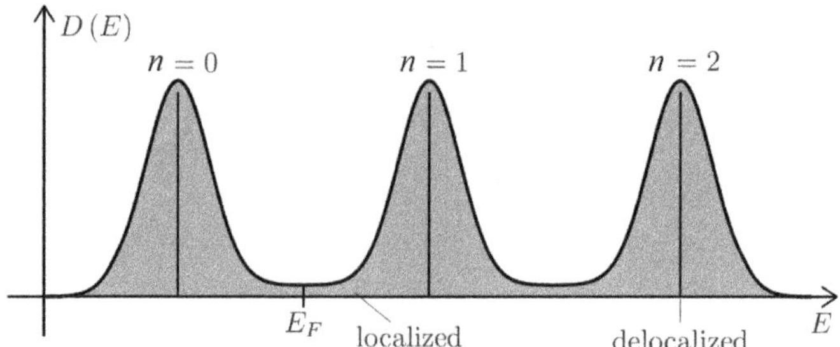

Fig. 10.16.　In the limit $L \to \infty$ only states at the center of disorder-broadened Landau levels are delocalized. Here delocalized edge states are not shown.

Fermi energy. When the flux changes, a delocalized state carries an electron from one edge to the opposite edge [9]. Surprisingly, it turns out that the bulk electron states in the middle of disorder broadened Landau levels are delocalized [21], see Fig. 10.16. All other states are localized except for edge states. (Before reading this section, the reader is encouraged to study Anderson localization from Ref. [6].)

A "plateau" transition may be understood as follows. As B increases the Landau level energy, $\hbar\omega_c(\frac{1}{2} + n)$ and the energy separation between the Landau levels, $\hbar\omega_c$, increase (ω_c is the cyclotron frequency and $n = 0, 1, \ldots$ labels the Landau levels). This effect is illustrated in Fig. 10.17. As magnetic field changes, a delocalized state at the center of disordered Landau level "floats" above the Fermi energy and the number of delocalized states below the Fermi energy decreases by one (the gap states are all localized). When the Fermi level is in the gap between two Landau levels, the **edge states** carry current and give rise to the quantized off-diagonal conductivity (the role of edge states is explained in Sec. 10.8). On the other hand, when the center of disordered Landau level and the Fermi energy coincides, a bulk current can flow and the diagonal conductivity is finite. (The reader should reread this section after studying Secs. 10.10 and 10.8.)

• Localized and Delocalized Bulk States

For the non-interacting electron, picture the electron wave function near the middle of disorder broadened Landau levels behaves as

$$\psi(\vec{r}) \propto e^{-\vec{r}/\xi}, \qquad (10.98)$$

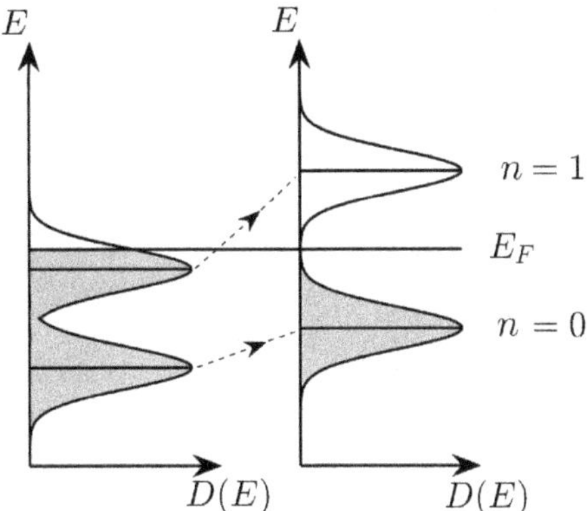

Fig. 10.17. Change of the DOS $D(E)$ as magnetic field increases. Note that disorder broadens discrete Landau levels in contrast to the discrete DOS in Fig. 10.5. This can be understood by invoking Fermi's golden rule for the probability of a transition per unit time from one Landau level state to a group of other Landau level states in a continuum. Edge states are not shown here.

where the localization length diverges at the center of a broadened Landau level E_c

$$\xi \propto |E - E_c|^{-\alpha} \tag{10.99}$$

with the exponent $\alpha = 2.33 \approx 7/3$, **independent** of types of impurities. The critical energy E_c corresponds to the metal-insulator point. In classical **percolation** model of metal-insulator transition, the value of the localization length exponent is $\alpha = 4/3$. Classical percolation is analogous to the following situation: Imagine isolated lakes, which represent puddles of electrons. If the water level (the Fermi energy) of the lakes rises continuously, some lakes will be connected to the nearby lakes. And at some point, lakes far away will also be connected, which represents a metal-insulator transition. In quantum mechanical treatment of the metal-insulator transition, the electron wave functions at the critical energy exhibit a **fractal** like behavior. The quantum corrections change the value of α to $7/3$. For a review of scaling theory of the integer quantum Hall effect, see Ref. [22].

- Participation Ratio, Localization Exponent, and Multifractal

Is the localization exponent ν unaffected by electron–electron interactions? Localization properties can be investigated by evaluating participation ratio for the self-consistent Hartree–Fock eigenfunctions ϕ_α

$$P_\alpha = \frac{[\int d\vec{r}|\phi_\alpha(\vec{r})|^2]^2}{L^2 \int d\vec{r}|\phi_\alpha(\vec{r})|^4}. \tag{10.100}$$

(Remember that L is the length of a square.) One expects the following single parameter scaling Ansatz of the participation ratio

$$P_\alpha(E, L) = L^{D(2)-2}\Pi(L^{1/\nu}|E - E_c|), \tag{10.101}$$

where $\Pi(x)$ is a scaling function. At the critical energy $E = E_c$, i.e., at the middle of the disorder-broadened Landau level, this gives

$$P_\alpha \sim L^{D(2)-2}. \tag{10.102}$$

The participation ratio is computed as a function of the dimensionless quantity $L^{1/\nu}|E - E_c|/\Gamma$. The values of $D(2)$ and ν are determined by data collapse, which gives the universal values $D(2) \approx 1.62^{\mathrm{m}}$ and $\nu = 2.3$, both with and without interactions [23]. The value of $D(2)$ is related to the multifractal[n] character of the extended states responsible for the anomalous diffusion which occurs at E_c. (For a non-critical state, a simple dimensional analysis gives $P_\alpha \sim L^{-2}$ since $[\phi_\alpha(\vec{r})] = 1/\text{Length}$.)

10.11. Critical Conductivity and Thouless Number⋆

At the critical energy, the wave function is delocalized and the diagonal conductivity has a universal value of

$$\sigma_{yy} \sim \frac{1}{2}\frac{e^2}{h}. \tag{10.103}$$

It should be independent of the strength of disorder and electron interactions.

But how can the conductivity be finite when DOS is zero at the Fermi energy? The answer is that the tunneling DOS should not be used to compute the conductivity. The reason is as follows. According to the Einstein

---

[m] In a metal $D(2) = 2$ and in an insulator $D(2) = 0$.
[n] There is a difference between a fractal and a multifractal. A multifractal consists of many fractals with different fractal dimensions. There is a distribution of fractal dimensions.

relation, the conductivity is given by

$$\sigma = e^2 \frac{dn}{d\mu} D_f, \tag{10.104}$$

where D_f is the diffusion constant,[∘] μ is the chemical potential, and n is the electron density. Here

$$\frac{dn}{d\mu} = 2\rho(E_F), \tag{10.105}$$

where the density of states $\rho(E)$ is the **thermodynamic** DOS (factor 2 is due to spin). The thermodynamic DOS is non-zero at E_F. In the presence of electron interactions, the thermodynamic $\rho(E)$ and tunneling DOS $D(E)$ are not the same. $\rho(E_F)$ and $D(E)$ appear, respectively, in the calculation of the specific heat and tunneling current. Since the integer quantum Hall system is compressible, the thermodynamic density of states is finite, as shown in Ref. [24].

- **Thouless Number**

Is the critical conductivity of interacting systems really identical to that of non-interacting systems? Below we argue that the value of the critical conductivity is universal. The Thouless number is related to the geometric mean of the eigenvalue differences $\langle \delta E \rangle$ between periodic and antiperiodic boundary conditions (the bracket stands for disorder averaging). The boundary conditions are changed by applying a flux in the cylindrical geometry. It is defined as the ratio between $\langle \delta E \rangle$ and the mean level spacing ΔE

$$g_T(E) = \frac{\langle \delta E \rangle}{\Delta E}, \tag{10.106}$$

where ΔE is given by

$$\frac{1}{\Delta E} = L^2 D(E). \tag{10.107}$$

[∘]Classically the distance traveled by a random walker is determined by the diffusion constant $\langle r^2 \rangle = D_f t$. But for a quantum particle it may happen that $\langle r^2 \rangle \to$ constant as $t \to \infty$. When this happens one has Anderson localization, where quantum interference is important.

How is the Thouless number related to the conductivity? The Kubo–Greenwood formula [25] gives

$$\sigma \sim D^2(E_{\mathrm{F}})\langle v^2 \rangle. \tag{10.108}$$

Using perturbation theory, one can estimate the energy shift induced by the change of the boundary conditions:

$$\langle \delta E \rangle \sim \frac{\langle v^2 \rangle}{\Delta E} \sim \langle v^2 \rangle D(E_{\mathrm{F}}). \tag{10.109}$$

(It is a second-order effect in the perturbation $\vec{v}\vec{A}$, where $\vec{v}$ is the velocity operator and $\vec{A}$ is the vector potential inducing the change of the boundary conditions; read about the twist boundary condition in Sec. 3.2 and Ref. [26]. The first-order term vanishes after disorder averaging because there is equal chance for any level to go up or down in energy.) From this result, we find $\langle v^2 \rangle \sim \frac{\langle \delta E \rangle}{D(E_{\mathrm{F}})}$, and plugging it into the expression for the conductivity Eq. (10.108), we find that the conductance is proportional to the Thouless number

$$G(L) \sim \langle \delta E \rangle \times D(E_{\mathrm{F}}) \sim \frac{\delta E}{\Delta E} = g_{\mathrm{T}}(E). \tag{10.110}$$

A different argument also suggests that the conductivity is related to the Thouless number: the conductance of a sample with the linear size L is

$$G(L) = L^{d-2}\sigma = e^2 \frac{D_f}{L^2}\rho(E_{\mathrm{F}}), \tag{10.111}$$

where $L^2\rho(E_{\mathrm{F}}) = 1/\Delta E$. The diffusion time to traverse the sample is $\Delta t = L^2/D_{\mathrm{f}}$. One expects that $\Delta t \sim 1/\delta E$, where the energy uncertainty δE originates from the coupling to the leads. Thus

$$G(L) \sim \frac{\langle \delta E \rangle}{\Delta E} = g_{\mathrm{T}}(E). \tag{10.112}$$

The Thouless number is expected to be proportional to the longitudinal conductivity. We will choose the proportionality factor such that the conductivity is given by

$$\sigma_{xx} = \frac{e^2}{h}\frac{\pi g_{\mathrm{T}}(E)}{2}. \tag{10.113}$$

> **Exercise 10.11.** Derive the Kubo–Greenwood formula (10.108) from the Kubo formula. It is a simplified version of the Kubo formula and is well-suited for study of localization and delocalization of one-electron states. Hint: Consult Ref. [25].

- Hartree–Fock Result for Thouless Number

Yang *et al.* [23] numerically investigated the Thouless number $g_T(E)$ in the presence of long-range Coulomb interactions [27]. The results thus obtained are expected to contain effects of the time-dependent Hartree–Fock approximation (see Sec. 6.14): the vertex corrections are included indirectly in the changes of the energy levels as a result of the change in the boundary conditions.[P] The self-consistent Hartree and exchange fields are **different** under different boundary conditions.

The disorder-broadened width of the Landau level is $\sim \Gamma$ and the mean energy spacing is $\Delta E \approx \frac{\Gamma}{N_\phi}$. From this, we find

$$D_{\Delta E}(E) = D(E)\Delta E = D(E)\frac{\Gamma}{N_\phi}, \qquad (10.114)$$

which leads to

$$\frac{1}{\Delta E} \times \Gamma = L^2 D(E) \times \Gamma = L^2 D_{\Delta E}(E)N_\phi. \qquad (10.115)$$

This quantity is plotted in the first row of Fig. 10.18. The geometric mean is used to compute $\delta\langle E\rangle$ (see Exercise 10.12). Numerical results suggest that the Thouless number at the critical energy is independent of the strength of disorder and electron interactions, see Fig. 10.18. There is a cancellation between the changes in $\langle \delta E\rangle$ and ΔE so that their ratio remains unchanged.

[P]One should compute the longitudinal conductivity within the time-dependent Hartree–Fock approximation (see Sec. 6.14) that takes into account self-consistently self-energy and vertex corrections. However, due to the large dimension of the resulting Hamiltonian matrix this is not yet feasible.

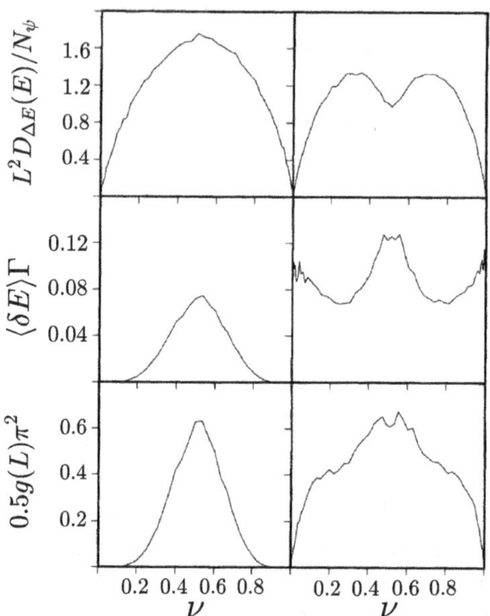

Fig. 10.18. Results for $L^2 D_{\Delta E}(E) N_\phi$, $\delta E/\Gamma$, and $g_T(E)$ are displayed. The ratio between strengths of electron interactions and disorder is denoted by $\gamma = \frac{e^2/(\epsilon \ell)}{\Gamma}$. Here Γ determines the range of strength $[-\Gamma, \Gamma]$ of delta function impurities. Left panel: $\gamma = 0$. Right panel: $\gamma = 0.4$. Note that the critical value at $\nu = 1/2$ is **independent** of γ. Reprinted with permission from Ref. [23].

Exercise 10.12. What is a geometric mean? Answer: It is a useful quantity to compute when we want to compare things with **very different** properties. A geometric mean of a set of numbers $\{a_i\}$ is defined as

$$\left(\prod_{i=1}^{n} a_i \right)^{\frac{1}{n}} = \exp\left[\frac{1}{n} \sum_{i=1}^{n} \ln a_i \right] = \sqrt[n]{a_1 a_2 \cdots a_n}. \qquad (10.116)$$

If negative values of the a_i are allowed,

$$\left(\prod_{i=1}^{n} a_i \right)^{\frac{1}{n}} = ((-1)^m)^{\frac{1}{n}} \exp\left[\frac{1}{n} \sum_{i=1}^{n} \ln |a_i| \right], \qquad (10.117)$$

where m is the number of negative numbers.

10.12. Dynamical Critical Exponent

Experimental data are consistent with the dynamic critical exponent $z \approx 1$ [28]. There is a discrepancy between non-interacting electron theory and experimental findings of the value of the dynamical scaling exponent z [29]. For non-interacting electrons, z equals the space dimension $d = 2$. This remains true also for short-ranged interactions since they are irrelevant at the non-interacting fixed point.

A naive scaling argument incorporating the linearly vanishing DOS appears to give the correct exponent. (For a nice introduction of dimensional analysis, see Ref. [30].) The argument goes as follows: the timescale for critical slowing down near a phase transition is related to the localization length ξ

$$t \sim \frac{1}{\omega} \sim \xi^z, \tag{10.118}$$

where the electron energy is $E = \hbar\omega$. This means

$$\xi \sim \omega^{-1/z}. \tag{10.119}$$

The dimension of the DOS is

$$[D(E)] = \frac{1}{\text{Energy} \times \text{Length}^2}, \tag{10.120}$$

which suggests the dimension of the localization length as follows

$$\xi \sim \frac{1}{\sqrt{D(E)E}}. \tag{10.121}$$

For long-range Coulomb interactions, the density of states $D(E) \propto E$. This result and $\xi \sim \omega^{-1/z}$ imply $z = 1$, in apparent agreement with the experimental result. But, as we mentioned below Eq. (10.105), the tunneling $D(E) \propto E$ must **not** be used here [24], and this naive scaling argument is incorrect.

It still remains an open question whether the implementation of the time-dependent Hartree–Fock approximation, that takes into account self-consistently self-energy and vertex corrections of long-range Coulomb interactions, can give rise to the observed dynamic critical exponent for the integer quantum Hall transition.

Bibliography

[1] R. B. Laughlin, Quantized Hall conductivity in two dimensions, *Phys. Rev. B* **23**, 5632 (1981).

[2] K. v. Klitzing, G. Dorda, and M. Pepper, New method for high-accuracy determination of the fine-structure constant based on quantized hall resistance, *Phys. Rev. Lett.* **45**, 494 (1980).

[3] B. I. Halperin, The quantized Hall effect, *Sci. American* **254**, 4, 52 (1986). •

[4] H. C. Lee and S.-R. Eric Yang, Spin–charge separation in quantum Hall edge liquids, *Phys. Rev. B* **56**, R15529 (1997).

[5] E. Abrahams, P. W. Anderson, D. C. Licciardello, and T. V. Ramakrishnan, Scaling Theory of Localization: Absence of Quantum Diffusion in Two Dimensions, *Phys. Rev. Lett.* **42**, 673 (1979).

[6] S. M. Girvin and K. Yang, *Modern Condensed Matter Physics* (Cambridge University Press, Cambridge, 2019). •

[7] A. Lagendijk, B. van Tiggelen, and D. S. Wiersma, Fifty years of Anderson localization, *Phys. Today* **62**, 8, 24 (2009). •

[8] D. Tong, Lectures on the quantum Hall effect, arXiv:1606.06687v2.

[9] D. Yoshioka, *The Quantum Hall Effect* (Springer, Berlin, 2002).

[10] E. Brown, *Bull. Amer. Phys. Soc.* 8, 256 (1963); Bloch electrons in a uniform magnetic field, *Phys. Rev.* **133**, A1038 (1964).

[11] J. Zak, Magnetic translation group, *Phys. Rev.* **134**, A1602 (1964).

[12] J. E. Averon, D. Osadchy, and R. Seiler, A topological look at the quantum Hall effect, *Phys. Today* **56**, 8, 38 (2003). •

[13] D. Thouless, M. Kohmoto, M. Nightingale, and M. den Nijs, Quantized Hall conductance in a two-dimensional periodic potential, *Phys. Rev. Lett.* **49**, 405 (1982).

[14] D. R. Hofstadter, Energy levels and wave functions of Bloch electrons in rational and irrational magnetic fields, *Phys. Rev. B* **14**, 2239 (1976).

[15] C. R. Dean, L. Wang, P. Maher, C. Forsythe, F. Ghahari, Y. Gao, J. Katoch, M. Ishigami, P. Moon, M. Koshino, T. Taniguchi, K. Watanabe, K. L. Shepard, J. Hone, P. Kim, Hofstadter's butterfly and the fractal quantum Hall effect in Moiré superlattices, *Nature* **497**, 598 (2013).

[16] R. P. Feynman, R. B. Leighton, and M. Sands, *The Feynman Lectures On Physics Vol. III: Quantum Mechanics* (Addison Wesley, Boston, 1977). •

[17] B. I. Halperin, Quantized Hall conductance, current-carrying edge states and the existence of extended states in a two-dimensional disordered potential, *Phys. Rev. B* **25**, 2185 (1992). •

[18] R. C. Ashoori, J. A. Lebens, N. P. Bigelow, and R. H. Silsbee, Equilibrium tunneling from the two–dimensional electron gas in GaAs: evidence for a magnetic-field-induced energy gap, *Phys. Rev. Lett.* **64**, 681 (1990).; J. P. Eisenstein, L. N. Pfeiffer, and K. W. West, Coulomb Barrier to tunneling between parallel two–dimensional electron systems, *Phys. Rev. Lett.* **69**, 3804 (1992).

[19] S.-R. Eric Yang and A. H. MacDonald, Coulomb gaps in a strong magnetic field, *Phys. Rev. Lett.* **70**, 4110 (1993).

[20] A. L. Efros and B. I. Shklovskii, Coulomb gap and low temperature conductivity of disordered systems, *J. Phys. C: Solid State Phys.* **8**, L49 (1975).

[21] H. Levine, S. B. Libby, and A. M. M. Pruisken, Electron delocalization by a magnetic field in two dimensions, *Phys. Rev. Lett.* **51**, 1915 (1983).

[22] B. Huckenstein, Scaling theory of the integer quantum Hall effect, *Rev. Mod. Phys.* **67**, 357 (1995).

[23] S.-R. Eric Yang, A. H. MacDonald, and B. Huckenstein, Interactions, localization, and the integer quantum Hall effect, *Phys. Rev. Lett.* **74**, 3229 (1995).

[24] S.-R. Eric Yang, Z. Wang, and A. H. MacDonald, Thermodynamic and tunneling density of states of the integer quantum Hall critical state, *Phys. Rev. B* **65**, 041302 (2001).

[25] O. Madelung, *Introduction to Solid State Physics* (Springer, Berlin, 1978). •

[26] P. A. Lee, *8.512 Theory of Solids II*, Spring 2009. Massachusetts Institute of Technology: MIT OpenCourseWare, https://ocw.mit.edu.

[27] D. Thouless, Electrons in disordered systems and the theory of localization, *Phys. Rep.* **13**, 93 (1974).

[28] L.W. Engel, D. Shahar, C. Kurdak, and D.C. Tsui, Microwave frequency dependence of integer quantum Hall effect: evidence for finite-frequency scaling, *Phys. Rev. Lett.* **71**, 2638 (1993).

[29] S. L. Sondhi, S. M. Girvin, J.C. Carini, and Shahar, Continuous quantum phase transitions, *Rev. Mod. Phys.* **69**, 315 (1997). •

[30] G. B. West, "Scale and dimension — from animals to quarks," in *Los Alamos Science Vol. XI Summer/Fall* (Los Alamos National Laboratory, Los Alamos, 1984).•

Chapter 11

Fractional Quantum Hall Effect

"Just as the beauty of a woman's breasts or the delicious curve of her hips is actually concerned with childbearing, and isn't merely for the delight of painters and photographers, so a math idea's beauty also has something to do with its "fertility," with the extent to which it enlightens us, illuminates us, and inspires us with their ideas and suggests unsuspected connections and new viewpoints."

Gregory Chaitin

The phenomenology of fractional quantum Hall effect is similar to that of the integral quantum Hall effect. One of the main differences is the value of the quantized Hall conductance associated with fractionally charged excitations [1].

In the integer quantum Hall case, the relevant Landau level are completely filled. There is thus only one possible Slater determinant ground state, and the Hartree–Fock approach becomes exact in the strong magnetic field limit.[a] However, it is difficult to solve the many-particle problem when the filling factor deviates from integer values, see Fig. 11.1. Since each Landau level is highly degenerate, there are many possible degenerate Slater determinant states and the Hartree–Fock approximation fails. If one applies degenerate perturbation theory, the resulting Hamiltonian matrix

---

[a] In this limit the energy separation between Landau levels is much larger than the interaction energy scale $e^2/\epsilon\ell$, where ϵ is the dielectric constant and ℓ is the magnetic length. The coupling between Landau levels may be ignored.

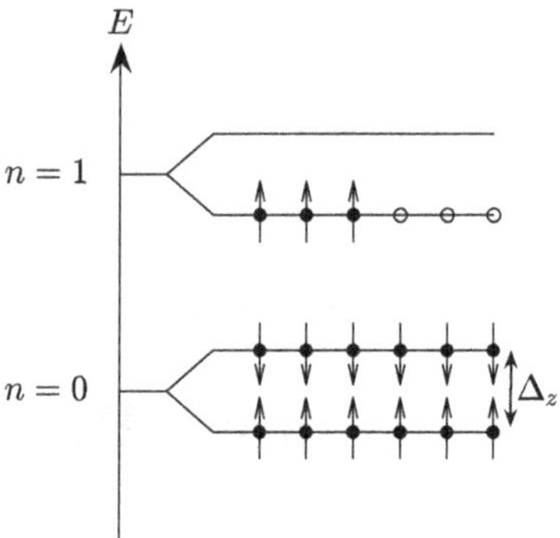

Fig. 11.1. An example of a non-integer filling factor. A system with the total filling factor $\nu = 5/2$ is displayed. Spin-up level of the $n = 1$ Landau level is only partially filled. The $n = 0$ Landau levels are completely filled. Δ_z is the Zeeman energy (whose magnitude is exaggerated in the figure).

will have a prohibitively large dimension to handle analytically. This is the crux of the difficulty.[b]

Instead Laughlin guessed, in a Nobel Prize winning work,[c] a good trial wave function for the ground state with correct electron correlations. Quasiparticles and collective modes may be constructed from the ground state trial wave function, as in the cases of superconductivity and toric code (see Chapters 8 and 13). The special correlations at certain filling factors[d] lead to fractional charges [2], non-trivial edge excitations, and ground state degeneracy, which are some of the hallmark properties of topological ordered phases. The fractional quantum Hall states can be detected in clean samples. But this does not mean disorder is unimportant. Disorder plays the important role of localizing quasiparticles in the bulk, which gives rise to fractional values of the off-diagonal conductivity.

There are several excellent articles and books explaining fractional quantum Hall effect [3–10]. In this chapter our goal is modest. We only explain some basic concepts that are relevant to topologically ordered phases, such as ground state degeneracy, fractional charge, and edge states.

---

[b]However, one can numerically use an exact diagonalization method and solve the problem when the electron number is ~ 10. Many useful insights can be gained from this method.

[c]The prize was shared with Horst L. Störmer and Daniel C. Tsui for their explanation of the fractional quantum Hall effect.

[d]The **filling factor** is defined as $\nu = N/N_\phi$, where N_ϕ is the Landau level degeneracy and N is the number of electrons.

11.1. Phenomenology of Fractional Charge

The measured transverse conductivity of the lowest Landau level at filling factor $\nu = 1/3$ is

$$\sigma_{xy} = \frac{1}{3}\frac{e^2}{h}. \qquad (11.1)$$

This result indicates that something very unusual is going on. The following thought experiment is revealing. Let us adiabatically thread flux Φ_0, over a time period T, through a point in the sample, see Fig. 11.2. Then, as $\sigma_{xy} \neq 0$, an azimuthal electric field is generated, which in turn produces a radial current. Integrating this current over time period T gives

$$q = \sigma_{xy}\int_0^T dt \oint \vec{E}\cdot\vec{dl} = \sigma_{xy}\int_0^T dt\frac{1}{c}\frac{d\Phi}{dt} = \frac{\sigma_{xy}\Phi_0}{c} = \frac{e}{3}. \qquad (11.2)$$

So a quasi-hole thus created has charge $\frac{e}{3}$ and flux Φ_0, and it is an anyon [11,12]! When an extra electron is inserted into a Laughlin state, it divides into three quasiparticles each with charge $e^-/3$. Measurement of the tunneling DOS gives evidence of fractional charges [13]. Quantum shot noise directly measures [14] the tunneling fractional charge. Resonant tunneling measurement through a quantum dot structure also provides evidence for the presence of a fractional charge [15].

11.2. Single-Electron Wave Functions of Landau levels

Before we investigate the many-body ground state and its excitations, let us first study single-particle wave functions. The ground state trial wave

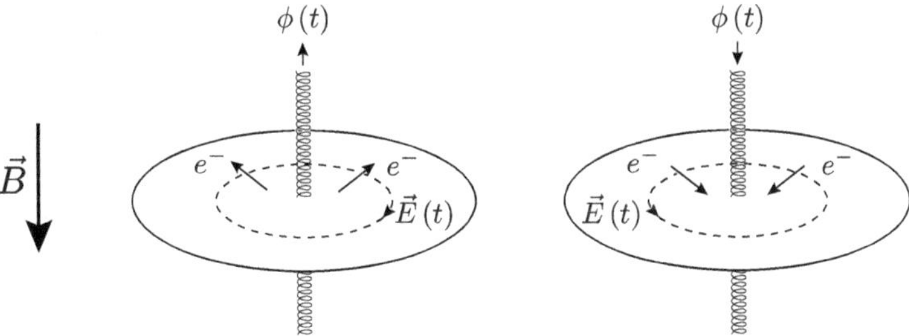

Fig. 11.2. Threading adiabatically flux from 0 to Φ_0 through a point in the sample accumulates a charge. An azimuthal electric field is generated, which in turn produces radial current as the off-diagonal conductivity $\sigma_{xy} \neq 0$. Depending on the direction of the magnetic field inside the solenoid, the sign of the accumulated charge changes. In the left (right) figure a quasihole (quasielectron) is formed.

function can be built from these single-particle wave functions. The single particle Hamiltonian is

$$H = \frac{\Pi^2}{2m^*} = \frac{1}{2m^*}\left[\frac{\hbar}{i}\vec{\nabla} + \frac{e}{c}\vec{A}(\vec{r})\right]^2.$$

(11.3)

- Eigenstates in Symmetric Gauge

The eigenstates in the symmetric gauge $\vec{A} = -\frac{1}{2}\vec{r} \times \vec{B}$ with $\vec{B} = -B\hat{z}$ are

$$\psi_{n,m}(\vec{r}) = A_{nm}e^{i(m-n)\theta - \frac{r^2}{4\ell^2}}\left(\frac{r}{\ell}\right)^{|m-n|}L^{|m-n|}_{(n+m-|m-n|)/2}\left(\frac{r^2}{2\ell^2}\right),$$

(11.4)

where (r, θ) are polar coordinates and the normalization constant is

$$A_{nm} = \frac{1}{\ell}[2\pi 2^\alpha \Gamma(\alpha + \beta + 1)/\beta!]^{-1/2},$$
$$\alpha = |m - n|, \quad \beta = (n + m - \alpha)/2.$$

(11.5)

Here $L^\alpha_\beta(x)$ is the Laguerre polynomial. The angular momentum is $L_z = \hbar(m - n)$ with $m \geq 0$. The guiding center operators are

$$(\mathbf{X}, \mathbf{Y}) = \left(\frac{1}{2}x + i\ell^2\frac{\partial}{\partial y}, \frac{1}{2}y - i\ell^2\frac{\partial}{\partial x}\right).$$

(11.6)

When magnetic field is applied in the opposite direction ($\vec{A} = \frac{1}{2}\vec{r} \times \vec{B}$ with $\vec{B} = B\hat{z}$) the wave functions are

$$\psi_{n,m}(\vec{r}) = A_{n,m}e^{i(n-m)\theta - \frac{r^2}{4\ell^2}}\left(\frac{r}{\ell}\right)^{|m-n|}L^{|m-n|}_{(n+m-|m-n|)/2}\left(\frac{r^2}{2\ell^2}\right).$$

(11.7)

In this case $L_z = \hbar(n - m)$.

Exercise 11.1. Show in the symmetric gauge $\vec{A} = -\frac{1}{2}\vec{r} \times \vec{B}$ with $\vec{B} = -B\hat{z}$ that

$$a = -\frac{i}{\sqrt{2}}\left(\frac{z}{2} + 2\frac{\partial}{\partial z^*}\right),$$
$$a^+ = \frac{i}{\sqrt{2}}\left(\frac{z^*}{2} - 2\frac{\partial}{\partial z}\right).$$

(11.8)

Here z is measured in units of the magnetic length ℓ.

- Analytical Property of Single-Particle Wave Functions of the Lowest Landau level

The following analytical properties of single-electron wave functions of the lowest Landau level are useful in constructing many-electron wave functions.

In the symmetric gauge with $\vec{B} = -B\hat{z}$, the lowest Landau level eigenstates with the inter Landau level index $n = 0$ are

$$\psi_m(z) = \frac{1}{\sqrt{2\pi\ell^2 2^m m!}} (z/\ell)^m e^{-\frac{1}{4\ell^2}|z|^2}, \tag{11.9}$$

where $z = x + iy$ and the intra Landau level index $m = 0, 1, 2, \ldots$. These wave functions are peaked at the radius $R = \sqrt{2m}\ell$ with a width $\Delta R \sim \ell$ (but $\langle r^2 \rangle = (2m+1)\ell^2$). These states are all degenerate and have the same Landau level energy $\hbar\omega_c/2$.

Formally, a function f is analytic in z if it has a Taylor series expansion

$$f(z) = \sum_{p=0}^{\infty} a_p (z - z_0)^n. \tag{11.10}$$

Any analytic function $f(z)$ can be written as a linear combination of degenerate eigenfunctions of Eq. (11.9), omitting the exponential function $e^{-\frac{1}{4\ell^2}|z|^2}$ [16]. This means $f(z)$ is an eigenstate of the lowest Landau level. Note that the absolute value $|z| = \sqrt{x^2 + y^2}$ is non-analytic, i.e., it is not differentiable. In other words, a function of $|z|$ does not satisfy the Cauchy–Riemann equations.

11.3. Laughlin Trial Wave Function

The many-body Hamiltonian in the presence of disorder is given in Eq. (10.86). The fractional quantum Hall effect is only observed in relatively clean samples.[e] So let us ignore for a moment the effect of disorder. Before we explain the Laughlin trial wave function, let us talk about the many-body Hilbert space. Many-body interactions are invariant under rotation about the z-axis. The total z-component of angular momentum[f]

---

[e] In the fractional quantum Hall regime electron–electron interaction effects dominate over disorder effects. However, disorder is still relevant for the observed Hall plateaus.

[f] There are two different angular momenta. One is the canonical angular momentum $\vec{r} \times \vec{p}$ and the other is mechanical angular momentum $\vec{r} \times \vec{\Pi}$. Here the canonical angular momentum is called the angular momentum.

L_z thus commutes with the Hamiltonian

$$[L_z, H] = 0. \tag{11.11}$$

For the lowest Landau level, the total z-component of angular momentum is

$$L_z = \sum_m n_m m, \tag{11.12}$$

where n_m is the occupation number. Note also that the total z-component of spin S_z is conserved. So each Hilbert subspace can be labeled by two quantum numbers (L_z, S_z). One performs calculations in each Hilbert subspace. This division labor is very useful in performing exact diagonalization calculations as it reduces the dimension of the Hamiltonian matrix.

Suppose there are N electrons in the lowest Landau level and consider the ground state at the filling factor $\nu = 1$. Assume a model where electrons are confined in a weak parabolic potential which prevents electrons from repelling each other [17]. These electrons are all spin polarized due to the strong magnetic field. Their wave functions are analytic functions of their position coordinates $z_1, \ldots, z_N$. The ground state is a Slater determinant state where the single particle states from $m = 0, 1, \ldots, N-1$ are occupied

$$\Psi[z] = \begin{vmatrix} z_1^0 & z_1^1 & z_1^2 & \cdots \\ z_2^0 & z_2^1 & z_2^2 & \cdots \\ \vdots & \vdots & \vdots & \vdots \end{vmatrix} e^{-\sum_i^N |z_i|^2/4\ell^2} = \prod_{i<j}(z_i - z_j)e^{-\sum_i^N |z_i|^2/4\ell^2}.$$

$$\tag{11.13}$$

The highest power of each coordinate z_i that appears in this state is $N-1$. The occupation number of each single particle states is $\langle n_m \rangle = 1$ for $m \leq m_c = N - 1$, see Fig. 11.3. This state is a circular droplet with radius $R = \sqrt{2m_c}\ell$. Its total z-component of angular momentum is

$$L_z = \frac{N(N-1)}{2}. \tag{11.14}$$

It has the lowest possible value of L_z and is the only state that has this value of L_z. If the coupling between Landau levels is negligible, this is the exact ground state.

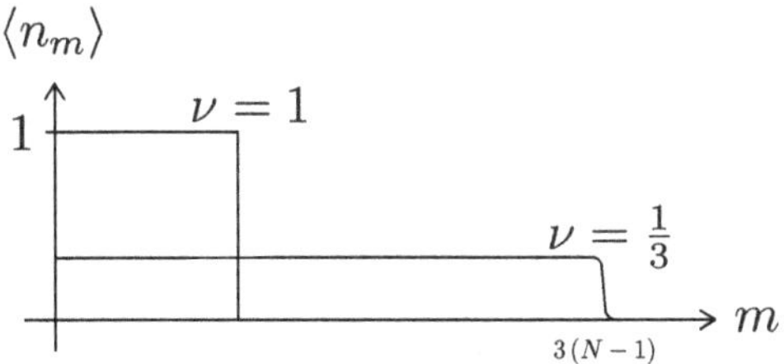

Fig. 11.3. The occupation numbers of quantum Hall droplets for filling factors $\nu = 1$ and $\nu = 1/3$ ($N \gg 1$ is assumed). The occupation number of the $\nu = 1$ state has a discontinuity at "Fermi vector" $k_F = R/\ell^2$ (R is the radius of the droplet) and is analogous to the Fermi–Dirac distribution function of a Fermi liquid. For $\nu = 1/3$, in contrast to $\nu = 1$, the discontinuity at k_F no longer exists and a chiral Luttinger liquid is present (the behavior near k_F will be investigated in Sec. 12.4).

Laughlin proposed the following trial wave function at filling factor $\nu = 1/p$[g]

$$\Psi_p[z] = \prod_{i<j}(z_i - z_j)^p e^{-\sum_i^N |z_i|^2/4\ell^2}. \tag{11.15}$$

The Pauli exclusion principle and antisymmetry under particle exchange are clearly satisfied: when two electrons approach each other, i.e., $z_i = z_j$ the wave function vanishes; when two electrons are exchanged, i.e., $z_i \to z_j$ and $z_j \to z_i$, the wave function changes sign. The highest power of each coordinate z_i is $m_c = p(N-1)$. This state is thus a circular droplet with radius $\sqrt{2 \times p(N-1)}\ell$. The total angular momentum is

$$L_z = \frac{N(N-1)}{2}p. \tag{11.16}$$

There are numerous Slater determinant states with this value of angular momentum. This means $\Psi_p[z]$ is rather complicated when written as a linear combination of Slater determinant states. In the Laughlin state at $1/p$, filling the essential correlation is that no two electrons can have relative angular momentum less than p (read about the Haldane pseudopotentials in [19]). This trial wave function describes a large quantum Hall droplet at filling factor $\nu = 1/p$ with **one** edge. The occupation number of each single particle states is $\langle n_m \rangle = 1/p$ for $m \lesssim m_c = 3(N-1)$, see Fig. 11.3. Notice that the occupation numbers do not drop discontinuously at the edge, in

---

[g]Note that Laughlin solved the three electron problem first before proposing the full many-body trial wave function [18]. The reader should read this work before studying the Laughlin trial wave function.

contrast to the case of $\nu = 1$. This effect is common in Luttinger liquids and will be studied in Sec. 12.4.

The ultimate justification of this trial wave function comes from exact diagonalization studies of small systems [19, 20]. These numerical works utilize the method that any many-body states can be written as a linear combination of numerous Slater determinant wave functions.[h] These works also demonstrate that there is a **finite excitation gap**, which is an essential ingredient of a topologically ordered phase.

11.4. Plasma Analogy and Fractional Charge

There is a very useful "plasma analogy", which provides many useful insights. It is a mapping from a quantum problem in a magnetic field to a two-dimensional problem of a classical plasma without magnetic field. Let us rewrite the probability density of the Laughlin wave function as

$$\left|\Psi_p[z]\right|^2 = e^{-\beta U[z]}, \tag{11.17}$$

where

$$U[z] = -p^2 \sum_{i<j} \ln(|z_i - z_j|/\ell) + \frac{p}{4} \sum_{i=1}^{N} |z_i|^2/\ell^2. \tag{11.18}$$

This $U[z]$ can be interpreted as the potential energy for a classical plasma of charged particles moving in two-dimensions at an artificial inverse temperature $\beta = 2/p$. Each particle has a charge $q = p$. Suppose you have a particle with a charge q at position z_0, its two-dimensional potential is

$$\phi(r) = -q\ln(r/\ell). \tag{11.19}$$

So the first term in Eq. (11.18) represents charge–charge interaction energy. The second term in Eq. (11.18) represents the potential energy due the positive background charge (a plasma is charge neutral). From the Poisson equation,

$$-\nabla^2(r^2/4) = -1/\ell^2 = 2\pi\rho_b. \tag{11.20}$$

From this, it follows that the background density is $\rho_b = -\frac{1}{2\pi\ell^2}$. The number of positive ions is thus equal to the degeneracy of the lowest Landau

---

[h]Slater determinant wave functions form a complete set of basis states.

level ($N_D = A|\rho_b| = A/2\pi\ell^2$). The condition for charge neutral plasma is

$$p\rho + \rho_b = 0 \rightarrow \rho = \frac{1}{p}\frac{1}{2\pi\ell^2}, \tag{11.21}$$

where ρ is the electron density.

> **Exercise 11.2.** Show the result of Eq. (11.19). Hint: Use the two-dimensional Poisson equation
>
> $$-\nabla^2\phi = 2\pi q\delta^2(r). \tag{11.22}$$

Using this plasma analogy, let us show that a quasihole has a fractional charge $1/p$. The detailed argument goes as follows. Consider **one** quasihole at Z in the Laughlin liquid

$$\Psi_Z[z] = \left[\prod_{j=1}^{N}(z_j - Z)\right]\Psi_p[z]. \tag{11.23}$$

One can write the inner product

$$\langle \Psi_Z|\Psi_Z\rangle = e^{-\beta U}, \tag{11.24}$$

where

$$U = -p^2\sum_{i<j}\ln|z_i - z_j|/\ell + \frac{p}{4}\sum_i |z_i|^2/\ell^2 - p\sum_i \ln|z_i - Z|/\ell. \tag{11.25}$$

The third term stands for the interaction between a quasihole with **unit charge** at Z and the plasma particles with charge p at z_i (For sake of argument assume $p = 3$.) The properties of the classical problem is very well understood: To minimize the total energy, charged particles will be repelled from the quasihole at Z, i.e., the quasihole will be **completely screened**, see Fig. 11.4.[i] The screened charge relative to the electron charge is precisely $1/p$. So the effective charge that is missing is $1/p$, i.e., it is **fractional**, as we saw in Sec. 11.4.

---

[i]Here screening is treated as a classical effect. At zero temperature quantum effects must be included and the screening length is given by the Thomas–Fermi screening length $\lambda_{TF} = \frac{1}{2}(\frac{\pi}{3\rho})^{1/3}$. But we do not consider quantum effects here since the temperature is $2/p$.

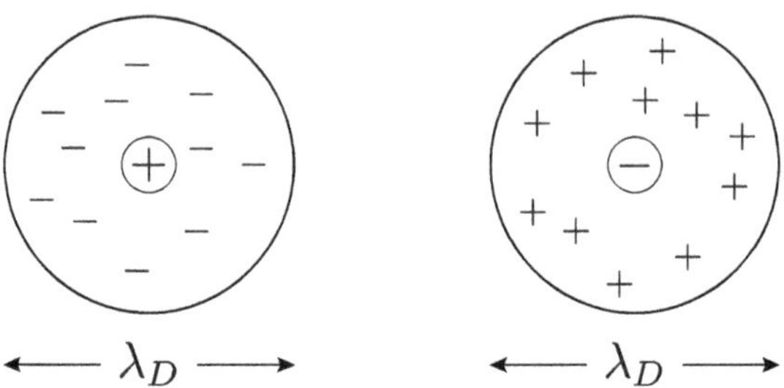

Fig. 11.4. One of the most important features of a plasma is its ability to screen a charge. A cloud of opposite charges within the Debye length $\lambda_D = \sqrt{\frac{k_B T}{4\pi\rho q^2}}$ effectively makes the charge neutral (here ρ is particle density and T is plasma temperature). Note that this is a classical effect.

11.5. Trial Wave Function of Quasihole and Aharonov–Bohm Phase

Suppose you create a quasihole at position $Z = 0$, the many-body wave function of a such a state is

$$\Psi_Z[z] = e^{-\frac{|Z|^2}{4p\ell^2}} \prod_{j=1}^{N}(z_j - Z)\Psi_p[z], \qquad (11.26)$$

where $\Psi_p[z]$ is the Laughlin wave function at inverse filling factor $p = 3$. Note the exponential function $e^{-\frac{|Z|^2}{4p\ell^2}}$ is now attached to the many-electron wave function. This insures its Z-dependent normalization. (Wave functions must all be normalized in computing the Berry connection.) The norm $\langle\Psi_Z|\Psi_Z\rangle$ is independent of Z because of the complete screening property of plasma [6]. The quasihole wave function $\Psi_Z[z]$ has a vortex at Z with vorticity 1. In this many-body state, a fractional charge $-e/3$ is removed from the $m = 0$ to $m_c = 3(N-1)+1$ single particle states. The resulting single-particle occupation number is $\langle n_0 \rangle = 0$ and $\langle n_{m_c} \rangle = 1/3$ while other occupation numbers are unchanged. So a quasihole with a fractional charge is created at $r = 0$. This process corresponds to piercing the droplet with an elementary quantum flux Φ_0 during which **a fractional charge is effectively transferred to the outer edge**, see Fig. 11.5.

Let us compute the Aharonov–Bohm phase of a quasihole using the trial wave function given in Eq. (11.26). Imagine a quasihole adiabatically encircling the origin of the coordinate system. The Berry connections are

$$\mathcal{A}_{Z^*} = i\langle\Psi_Z|\partial_{Z^*}\Psi_Z\rangle = -\frac{i}{4p}Z \qquad (11.27)$$

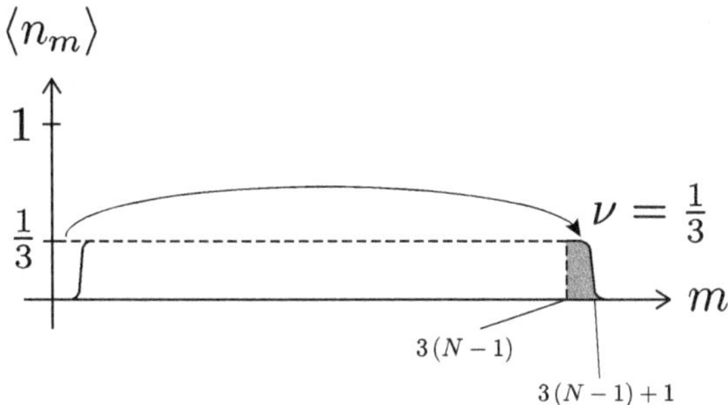

Fig. 11.5. Occupation number of the $\nu = 1/3$ state is shown for $N \gg 1$ (the dashed line). Piercing the droplet with a solenoid with an elementary quantum flux Φ_0 effectively transfers a fractional charge from the center of the droplet to its outer edge. The magnetic flux pushes out the liquid and forms a hole at the center of the liquid and an excited state with a fractional charge is created (as mentioned in Sec. 3.2, the effect of a unit of flux through an empty hole can always be removed by a singular gauge transformation).

and

$$\mathcal{A}_Z = i\langle \Psi_Z | \partial_Z \Psi_Z \rangle = \frac{i}{4p} Z^*. \tag{11.28}$$

To compute the curvature, let us convert these into

$$\mathcal{A}_X = \mathcal{A}_Z + \mathcal{A}_{Z^*} = Y/2p, \quad \mathcal{A}_Y = i(\mathcal{A}_Z - \mathcal{A}_{Z^*}) = -X/2p. \tag{11.29}$$

The Berry phase is computed from the Berry curvature $\nabla \times \vec{A}$ by multiplying it with area A enclosed by the quasihole adiabatic loop

$$\gamma = -\frac{1}{p} \frac{A}{\ell^2}. \tag{11.30}$$

The Berry phase is negative because the external magnetic field is along the $-\hat{z}$. The magnetic length ℓ is restored to make the result dimensionless.

> **Exercise 11.3.** Show Eq. (11.27).

> **Exercise 11.4.** Show Eq. (11.28). Hint: Use $i\partial_Z \langle \Psi_Z | \Psi_Z \rangle = 0$

Exercise 11.5. The quasihole wave function, (11.26) represents a vortex and not a flux tube. A quasihole is, strictly speaking, not a composite of a flux tube and a fractional charge. This is a subtle point. Explain this [6]. Answer: When a flux quantum is actually inserted adiabatically, a charge depletion takes place. The wave function of the state containing the flux tube must thus have the following form

$$\tilde{\Psi}_Z[z] = e^{-\frac{|Z|^2}{4p\ell^2}} \prod_{j=1}^{N} |z_j - Z| \Psi_p[z], \tag{11.31}$$

which differs by the absolute value $|z_j - Z|$ in comparison to the state with a quasihole. The flux tube may be removed without changing the quasihole density profile by performing a singular gauge transformation at the location where the tube is. This singular gauge transformation also generates a phase factor $e^{i\arg(z_j - Z)}$ in the wave function $\tilde{\Psi}_Z[z]$. The resulting wave function is identical to the quasihole wave function $\Psi_Z[z]$.

11.6. Anyon Statistics

When an anyon is exchanged with another anyon, a statistical phase is generated, see Figs. 11.6 and 11.7. Fractional statistics of quasiholes were shown first in Refs. [21, 22]. Here we study this effect following Ref. [6]. The normalized many-body wave function of the state containing two holes at Z_1 and Z_2 is

$$\Psi_{Z_1 Z_2}[z] = [(Z_1 - Z_2)^*(Z_1 - Z_2)]^{\frac{1}{2p}} e^{-\frac{|Z_1|^2 + |Z_2|^2}{4p\ell^2}}$$

$$\times \prod_{j=1}^{N} (z_j - Z_1)(z_j - Z_2)\Psi_p[z]. \tag{11.32}$$

In this wave function, a multiplying factor $e^{-\frac{|Z_i|^2}{4p\ell^2}}$ will not make $\langle \Psi_{Z_1 Z_2} | \Psi_{Z_1 Z_2} \rangle$ depend on Z_i. This can be seen easily in the plasma analogy, which shows that an impurity at Z_i will be **completely screened**, independent of where Z_i is located. The resulting energy will be independent of where Z_1 is located.

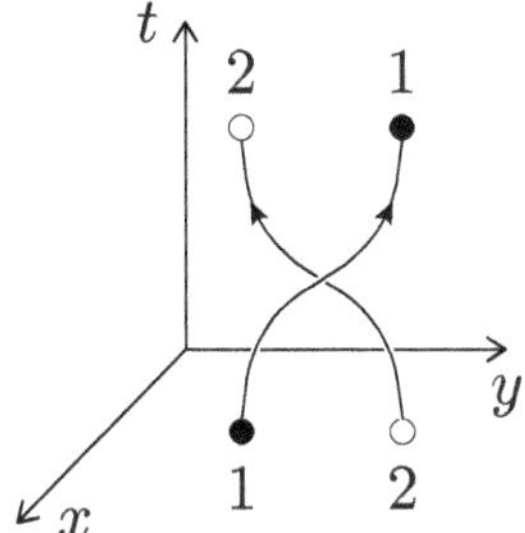

Fig. 11.6.   Exchange between two quasiparticles are visualized braiding two worldlines.

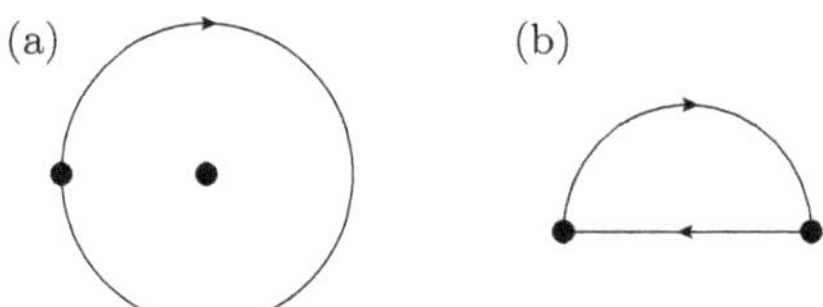

Fig. 11.7.   (a) Suppose there are two quasiparticles on a plane. Let a quasiparticle encircle another quasiparticle. The distance vector from the left particle to the right particle rotates by 360 degrees. However, in three dimensions the loop can be **lifted out of the plane** so that the first quasiparticle does not encircle the second one. This means that no anyons can exist in three dimensions. (b) An exchange is shown. The left particle moves to the position where the right particle is located. Then the right particle moves to the position where the left particle was. During the entire process, the distance vector from the left particle to the right particle rotates by 180 degrees in (b). A quasiparticle encircling another quasiparticle is equivalent to **two** exchanges.

Place a quasihole at $Z_1 = Z$ and the second quasihole at $Z_2 = 0$. The wave function containing these two quasiholes is

$$\Psi_Z[z] = (Z^*Z)^{\frac{1}{2p\ell^2}} e^{-\frac{|Z|^2}{4p}} \prod_{j=1}^{N}(z_j - Z)z_j \Psi_m[z]. \qquad (11.33)$$

Let a hole at $Z_1 = Z$ encircle another hole at $Z_2 = 0$ **counterclockwise**, see Fig. 11.8 (a magnetic field is pointing along minus z-axis).

Two contributions of the Berry phases arise: the Aharonov–Bohm phase[j] due to the external magnetic field and the statistical phase. The

______

[j]Remember that the Aharonov–Bohm phase is a Berry phase, see Sec. 3.1

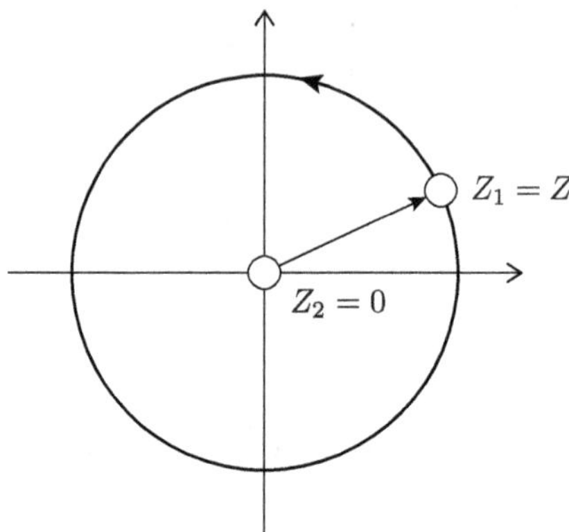

Fig. 11.8. A quasihole at $Z_1 = Z$ encircles **counterclockwise** another quasihole at $Z_2 = 0$.

total Berry connections are (we follow the derivations given in Ref. [6])

$$\mathcal{A}_{Z^*} + a_{Z^*}, \quad \mathcal{A}_Z + a_Z. \tag{11.34}$$

The connections $\mathcal{A}_{Z^*}$ and $\mathcal{A}_Z$ arise from the Aharonov–Bohm effect and are already computed in Eqs. (11.27) and (11.28). The other connections arising from particle exchange can also be computed using the wave function (11.33). We find

$$a_{Z^*} = \frac{i}{2p} \frac{1}{Z^*} \tag{11.35}$$

and

$$a_Z = \frac{-i}{2p} \frac{1}{Z}. \tag{11.36}$$

Exercise 11.6. Show Eq. (11.35). Hint: Note that Z^* appears as $[Z^* Z]^{\frac{1}{2p}}$ only in the exponent of the wave function (11.33).

Exercise 11.7. Show Eq. (11.36). Hint: Use $i\partial_Z \langle \Psi_Z | \Psi_Z \rangle = 0$.

The statistics angle is determined by a_X and a_Y: Since

$$a_X = a_Z + a_{Z^*}, \quad a_Y = i(a_Z - a_{Z^*}) \tag{11.37}$$

we find

$$a_X = -\frac{1}{p}\frac{Y}{|Z|^2}, \quad a_Y = \frac{1}{p}\frac{X}{|Z|^2},\tag{11.38}$$

where $(X, Y) = \vec{Z}$ (X and Y should not be confused with the guiding center coordinate operators). This result can be written as

$$\vec{a}(\vec{r}) = -\frac{\vec{Z} \times \hat{z}}{p|Z|^2} = \frac{1}{p}\frac{1}{|Z|}\hat{\phi},\tag{11.39}$$

where $\hat{z}$ is a unit vector pointing along the z-axis and $\hat{\phi} = -\frac{\vec{Z}}{|Z|} \times \hat{z}$. These Berry connections are precisely the vector potential of a flux tube. Remember that the Berry connection rising from an adiabatic transport can be thought of as a vector potential (see Sec. 3.1). From the vector potential the statistics angle may be computed

$$\theta = \frac{1}{2} \oint d\vec{R} \cdot \vec{a}(\vec{r}) = \frac{\pi}{p}.\tag{11.40}$$

The total phase is the sum of the Aharonov–Bohm and the statistical phases

$$\gamma + 2\theta = \frac{2\pi}{p}\left[-\frac{A}{2\pi\ell^2} + 1\right],\tag{11.41}$$

where the Aharonov–Bohm phase γ was computed in Eq. (11.30) and 2θ appears because encircling is equivalent to two exchanges, see Fig. 3.11. Anticlockwise exchange of an anyon is defined to give a positive statistical phase (see Sec. 3.7). Note that the flux of the quasihole points along $-z$ (see Fig. 11.2). Anyon statistics have been recently confirmed experimentally [23, 24].

11.7. Composite Fermions

Let us briefly explain composite fermion theory of fractional quantum Hall effect. As we learned in the previous section, bare electrons are not a good starting point for the description of the system, but anyons are. We learnt in Sec. 3.7 that an anyon may be generated by attaching a flux tube to an electron. This is a rather counter-intuitive description of the system and is a consequence of strong electron correlations. Composite fermion approach is based on the similar idea of flux attachment. A composite fermion consists of an electron with an attached flux tube with an even integer value of Φ_0 [25]. (This flux field is thought to be dynamically generated, i.e., it is created by the system itself.) In this theory, fractional quantum Hall effect is viewed as the integer quantum Hall effect of a **filled** Landau levels of composite fermions. For experimental investigation of composite fermions, see Ref. [26].

The key idea is as follows. Consider a Laughlin state at filling factor $\nu = \frac{1}{2p+1}$. Its wave function can be rewritten as

$$\Psi[z] = \left[\prod_{i'<j'}^{N} (z_{i'} - z_{j'})^{2p} \right] \left[\prod_{i<j}^{N} (z_i - z_j) \right]. \tag{11.42}$$

(Here the exponential functions are suppressed.) This suggests that each electron has a vortex with vorticity $2p$. In other words, electrons are in the $\nu = 1$ integer quantum Hall state

$$\prod_{i<j}^{N} (z_i - z_j) \tag{11.43}$$

with a flux tube with $2p\Phi_0$ attached to each of them and pointing along $+\hat{z}$ (the external field points along the opposite direction $-\hat{z}$, as shown in Fig. 11.9). When the composite fermions are at uniform density, they cancel part of the external magnetic field. The resulting effective magnetic field generates new Landau levels for composite fermions. At the observed filling factors, the composite fermions are assumed to completely fill these Landau levels. There is ample numerical studies that support this scenario.

How does one modify the Hamiltonian such that the attachment of flux is included properly? Such an effective Hamiltonian may be obtained as follows. One attaches a flux tube to each electron by performing a singular gauge transformation, which introduces phase factors in the many-body wave function

$$\Psi[z] \rightarrow \tilde{\Psi}[z] = e^{i2p \sum_{i<j} \arg(z_i - z_j)} \Psi[z]. \tag{11.44}$$

Note that it does not represent an orthodox gauge transformation as the magnetic field at the location of a flux tube is finite (see Sec. 3.2). In the corresponding effective Hamiltonian, the new vector potential is

$$\vec{A}(\vec{r}_i) \rightarrow \vec{A}(\vec{r}_i) + \vec{a}(\vec{r}_i) \quad \text{with} \quad \nabla_i \vec{a}(\vec{r}_i) = 0. \tag{11.45}$$

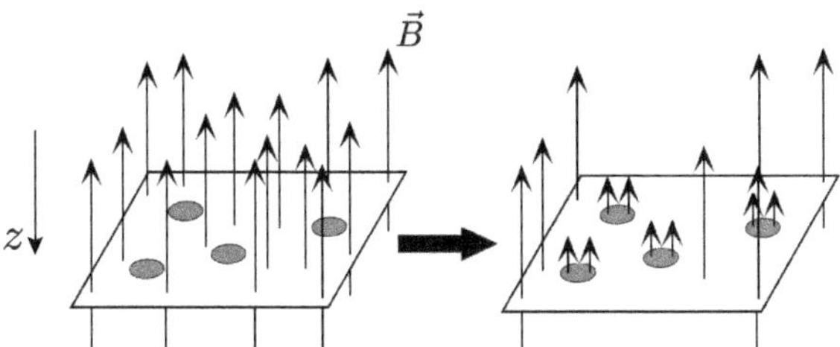

Fig. 11.9. To each electron, an infinitely thin solenoid carrying $2p$ flux quanta is attached pointing antiparallel to B (note that the external field points along $-\hat{z}$). If each discrete flux quantum is smeared an electron experiences a uniform magnetic field $B_{\text{tube}} = -2p\rho\Phi_0$ (this follows from $B_{\text{tube}}A = N2p\Phi_0$ and $\rho = N/A$). Consequently, the total effective magnetic field B^* acting on an electron is reduced.

This statistical gauge field generates the following magnetic field at the positions of electrons

$$\nabla \times \vec{a}(\vec{r}_i) = \vec{b}(r_i) = -\hat{z}2p\Phi_0 \sum_{i \neq j} \delta(\vec{r}_i - \vec{r}_j). \tag{11.46}$$

In other words, flux tubes with flux $2p\Phi_0$ are attached to the electrons. A composite fermion thus moves in statistical and external (magnetic) gauge fields. The extra magnetic field can be rewritten as

$$\nabla \times \vec{a}(\vec{r}) = -\hat{z}2p\Phi_0\rho(\vec{r}), \tag{11.47}$$

where $\rho(\vec{r})$ is the two-dimensional electron density

$$\rho(\vec{r}) = \sum_j \delta(\vec{r} - \vec{r}_j). \tag{11.48}$$

Exercise 11.8. Explain why electron depletion around each z_i is not included in the wave function (11.44). Answer: There is no magnetic field outside each flux tube and hence no real effect on the electron density. In contrast, a quasihole is generated by adiabatically inserting a flux and this process leads to charge flow from or to the flux. See also Exercise 11.5.

Now we make a drastic approximation. We smear out each discrete flux quantum. This approximation constitutes a mean field approach. (The ultimate justification of this method comes from numerical simulations.)

Consequently, composite electrons move in a weaker magnetic field

$$B^* = B - 2p\rho\Phi_0, \tag{11.49}$$

where the second term is the average value of the total smeared field. The filling factor of the composite fermions is

$$\nu^* = \rho\Phi_0/B^* = \frac{\nu}{1 + 2p\nu}, \tag{11.50}$$

where the electron filling factor is

$$\nu = \rho\Phi_0/B = \frac{\nu^*}{2p\nu^* \pm 1}. \tag{11.51}$$

The minus sign corresponds to negative values of B^*. The analogous integer quantum Hall effect for composite fermions occurs at integer values of ν^*.

11.8. Magnetorotons⋆

So far we have investigated the properties of quasi-particle excitations. Another way to justify the trial wave function is to compute the collective excitation mode called magnetorotons [27].

- Two-Particle Interactions: Haldane Pseudopotentials

The collective magnetoroton mode can be built from the Laughlin trail wave function. In order to do this, we need to express many-body interactions in terms of two-body interactions terms. Consider a two-electron problem with the Hamiltonian

$$H = \frac{1}{2m^*}(\Pi_1^2 + \Pi_2^2) = \frac{1}{4m^*}\Pi_R^2 + \frac{1}{m^*}\Pi_r^2, \tag{11.52}$$

where the center of mass and relative momenta are

$$\Pi_R = \Pi_1 + \Pi_2, \quad \Pi_r = (\Pi_1 - \Pi_2)/2. \tag{11.53}$$

Note that the magnetic lengths in $\frac{1}{4m^*}\Pi_R^2$ and $\frac{1}{m^*}\Pi_r^2$ are, respectively,

$$\ell_R = \ell/\sqrt{2}, \quad \text{and} \quad \ell_r = \ell\sqrt{2}. \tag{11.54}$$

(This is explained in Exercise 11.9.)

Exercise 11.9. The center of mass and relative coordinates are

$$Z = \frac{z_1 + z_2}{2}, \ z = z_1 - z_2. \tag{11.55}$$

Show that the Hamiltonians in the center of mass and relative coordinates are

$$H_R = \tfrac{1}{4m^*}(\vec{P} + 2\tfrac{e}{c}\vec{A}(Z))^2,$$
$$H_r = \tfrac{1}{m^*}(\vec{p} + \tfrac{1}{2}\tfrac{e}{c}\vec{A}(z))^2, \tag{11.56}$$

where

$$\vec{P} = \vec{p}_1 + \vec{p}_2 \ \vec{p} = (\vec{p}_1 - \vec{p}_2)/2. \tag{11.57}$$

The center-of-mass motion has the total mass $2m^*$ and the relative motion has the reduced mass $m^*/2$. Note the factors in front of the vector potentials; they give different effective magnetic lengths than ℓ. Even though two electrons repel each other, they form a bound state because electrons execute cyclotron motion in a magnetic field.

In the strong magnetic field limit, Landau level mixing induced by electron interactions may be ignored. For the electrons in the **lowest** Landau level, the eigenstates of H are

$$\langle z_1 z_2 | mM \rangle = A(z_1 + z_2)^M (z_1 - z_2)^m e^{-\frac{1}{4\ell^2}(|z_1|^2 + |z_2|^2)}, \tag{11.58}$$

where A is the normalization constant. Let us consider the two-particle interaction energy. It depends on the intra Landau level indices m and M of the relative and center-of-mass angular momenta, respectively. The mth Haldane pseudopotential is defined as the two–particle interaction energy

$$V_m = \frac{\langle mM|V|mM \rangle}{\langle mM|mM \rangle}, \tag{11.59}$$

where $V(z_1 - z_2)$ is the Coulomb potential that depends only on the relative coordinate. Note that this interaction energy does not depend on the center of mass coordinate. The pseudopotential decreases monotonously with increasing m. (For higher Landau levels the Haldane pseudopotential values are dependent on the Landau level index n, see the next section.)

- Many-electron density operators

Let us express the many-body electron–electron interaction $V = \sum_{i<j} V(\vec{r}_i - \vec{r}_j)$ using the Haldane pseudopotentials.

We first need to introduce single-electron density operators projected onto the nth Landau level. Various physical quantities can be expressed in terms of the electron field operator

$$\psi(\vec{r}) = \sum_{n,m} \psi_{n,m}(\vec{r}) c_{n,m}. \tag{11.60}$$

It contains a linear combination of the destruction operators of various Landau level states. However, in the strong magnetic field limit, the coupling between Landau levels induced by electron–electron interactions may be ignored (the energy separation between the Landau levels is much larger than the characteristic Coulomb energy scale $\hbar\omega_c \gg e^2/\epsilon\ell$). It is not difficult to project various quantities onto a given Landau level using guiding center density operators, as we show in the following.

We start with the following commutation relation of an electron

$$[\mathbf{R}, H] = 0, \tag{11.61}$$

where the electron guiding center is

$$\mathbf{R} = (\mathbf{X}, \mathbf{Y}) = \vec{r} - \ell^2(\hat{z} \times \vec{\Pi})/\hbar. \tag{11.62}$$

Note that $\mathbf{R}$ can be written in terms of the intra Landau level operators b_m and $b_m^\dagger$, which implies

$$\langle \psi_{n_1,m_1} | \mathbf{R} | \psi_{n_2,m_2} \rangle \sim \delta_{n_1,n_2}. \tag{11.63}$$

Since Π_i can be written in terms of inter Landau level operators a_n and $a_n^\dagger$, the second term of $\mathbf{R}$ in Eq. (11.62) satisfies

$$\langle \psi_{n,m} | \hat{z} \times \vec{\Pi} | \psi_{n,m'} \rangle = 0. \tag{11.64}$$

This implies that $\vec{R}$ can be expressed as the projection of the first term in Eq. (11.62)

$$\vec{R} = P_n \vec{r} P_n, \tag{11.65}$$

where $P_n = |n\rangle\langle n|$ is the projection operator onto the nth Landau level.

The single-electron density operator is defined as

$$\rho_{\vec{k}} = e^{-i\vec{k}\cdot\vec{r}}. \tag{11.66}$$

One can show (see Exercise 11.10) that its projection onto the nth Landau level is

$$\bar{\rho}_{\vec{k}}^{\,n} = P_n \rho_{\vec{k}} P_n = e^{-i\vec{k}\cdot\mathbf{R}} P_n e^{i\vec{k}\cdot(\mathbf{R}-\vec{r})} P_n = e^{-i\vec{k}\cdot\mathbf{R}} P_n e^{-i\vec{k}\cdot(\hat{z}\times\vec{\Pi})\ell^2/\hbar} P_n$$

$$= e^{-i\vec{k}\cdot\mathbf{R}} \langle n|e^{-i\vec{k}\cdot(\hat{z}\times\vec{\Pi})\ell^2/\hbar}|n\rangle = \bar{\rho}_{\vec{k}} e^{-k^2\ell^2/4} L_n(k^2\ell^2/2), \tag{11.67}$$

where the **guiding center density operator** is

$$\bar{\rho}_{\vec{k}} = e^{-i\vec{k}\cdot\mathbf{R}}. \tag{11.68}$$

Here we have used the quantity $\bar{\rho}_{\vec{k}}$ to be independent of the Landau level index n! (This is because $\vec{R}$ can be written in terms of the intra Landau level operators, see Eq. (10.10).) One of the fundamental results of guiding center density operators is the Girvin–MacDonald–Platzman commutation relation

$$[\bar{\rho}_{\vec{k}}, \bar{\rho}_{\vec{q}}] = 2i \sin[(\vec{k}\wedge\vec{q})\ell^2/2]\bar{\rho}_{\vec{k}+\vec{q}}, \tag{11.69}$$

where

$$\vec{k}\wedge\vec{q} = \hat{z}\cdot(\vec{k}\times\vec{q}). \tag{11.70}$$

We will see the power of this method in Sec. 11.8 by computing the dispersion relation of magnetoroton collective modes.

Exercise 11.10. Show Eq. (11.67). Hint: The guiding center operator $e^{-i\vec{k}\cdot\mathbf{R}}$ is **independent** of the Landau level index n. Use also

$$\langle n|e^{-i\vec{k}\cdot(\hat{z}\times\vec{\Pi})\ell^2}|n\rangle = e^{-k^2\ell^2/4} L_n(k^2\ell^2/2), \tag{11.71}$$

where $L_n(x)$ is the Laguerre polynomial. This result may be proven using the following: [1] $\Pi_x = \frac{\hbar}{\sqrt{2\ell}}(a + a^\dagger)$ and $\Pi_y = \frac{\hbar}{\sqrt{2\ell}}(a - a^\dagger)$. [2] $e^{-ua}|n\rangle = \sum_{l=0}^{n} \frac{(-u)^l}{l!} [\frac{n!}{(n-l)!}]^{1/2}|n-l\rangle$ and $\langle n|e^{u^*a^\dagger} = \sum_{m=0}^{n} \frac{(u^*)^m}{m!} [\frac{n!}{(n-m)!}]^{1/2}\langle n-m|$. [3] $L_n(x) = \sum_{k=0}^{n} \binom{n}{k} \frac{(-1)^k}{k!} x^k$.

One can show that the many-body electron–electron interaction $V = \sum_{i<j} V(\vec{r}_i - \vec{r}_j)$ can be written in terms of many-electron density operators

projected onto the nth Landau level

$$V^n = \frac{1}{2A} \sum_{\vec{k}} V^n_{\text{eff}}(\vec{k}) \bar{\rho}_{\vec{k}} \bar{\rho}_{-\vec{k}} + \text{constant}, \tag{11.72}$$

where the effective Coulomb interaction is

$$V^n_{\text{eff}}(\vec{k}) = V_{\vec{k}} e^{-k^2 \ell^2/2} [L_n(k^2 \ell^2/2)]^2. \tag{11.73}$$

(A is the area of the system and $V_{\vec{k}}$ is the Fourier transform of the Coulomb potential.) Here the many-body guiding center density operators are

$$\bar{\rho}_{\vec{k}} = \sum_i e^{-i\vec{k}\cdot\vec{R}_i}. \tag{11.74}$$

This quantity should be contrasted with the many electron density operator defined as

$$\rho_{\vec{k}} = \sum_i e^{-i\vec{k}\cdot\vec{r}_i}. \tag{11.75}$$

(It should not be confused with the single-electron density operator defined previously.) The Haldane's pseudopotentials may be obtained by expanding $V^n_{\text{eff}}(\vec{k})$ in terms of Laguerre polynomials and pseudopotential parameters

$$V^n_{\text{eff}}(\vec{k}) = 2\ell^2 \sum_m V_m L_m(k^2 \ell^2/2) e^{-k^2 \ell^2/2}, \tag{11.76}$$

which gives the Haldane pseudopotentials

$$V_m = \int_0^\infty dk\, k V_{\vec{k}} [L_n(k^2 \ell^2/2)]^2 L_m(k^2 \ell^2) e^{-k^2 \ell^2}. \tag{11.77}$$

- Magnetoroton Dispersion

The basic idea in the calculation of the magnetoron energy is the single mode approximation which assumes that all the oscillator strength is exhausted by the magnetorotons: the dynamic structure factor is related to the structure factor as follows

$$S(k, \omega) = \bar{s}(k) \delta(\omega - \Delta(k)), \tag{11.78}$$

where $\Delta(k)$ is the energy dispersion of magnetorotons.

The magnetoroton density excitation is created by acting upon the lowest Landau level ground state $|\Psi_p\rangle$ at filling factor $1/p$ with a density operator of definite momentum

$$|\Psi_k\rangle = \bar{\rho}_k^0 |\Psi_p\rangle, \tag{11.79}$$

where

$$\bar{\rho}_k^0 = \sum_j e^{-ik\frac{\partial}{\partial z_j}} e^{-ik^* z_j/2}. \tag{11.80}$$

is the projection of the density operator

$$\rho_k = \sum_j e^{-ikz_j^*/2 - ik^* z_j/2} \tag{11.81}$$

onto the lowest Landau level. Here the overbar and the superscript 0 indicate that the density operator has been projected into the lowest Landau level. ($\bar{\rho}_k^0$ is **not** the projection operator of the guiding center density operator $\bar{\rho}_k$.) We have used $k = k_x + ik_y$. **After projection onto the lowest Landau level, the z_j^*s replaced by the derivatives.** For an explanation of this operation, see Ref. [16]. (The reader should study this paper first before reading this section.) Note that $|\Psi_p\rangle$ is an analytic function of $z_1, \ldots, z_N$, except for the exponential factors, as we explained in Sec. 11.2.

The magnetoroton dispersion energy is given by

$$\Delta(k) = \frac{\langle \Psi_k | \bar{V} | \Psi_k \rangle}{\langle \Psi_k | \Psi_k \rangle}. \tag{11.82}$$

(The kinetic energy is irrelevant here.) Here the lowest Landau level projection of the interacting term is

$$\bar{V} = \frac{1}{2} \int \frac{d^2 q}{(2\pi)^2} V_q (\bar{\rho}_q^0 \bar{\rho}_q^0 - \rho e^{-q^2 \ell^2/2}), \tag{11.83}$$

where $(\bar{\rho}_{\vec{q}}^0)^\dagger = \bar{\rho}_{-\vec{q}}^0$ and ρ is the average density. This result is almost identical to the interaction term (11.72) given by the guiding center density operators. Using the roton state (11.79) we find

$$= \frac{\bar{f}(k)}{\bar{s}(k)}, \tag{11.84}$$

where the form factor is

$$
\begin{aligned}
\bar{f}(k) &= \langle \Psi_k | (\bar{V} - E_0) | \Psi_k \rangle \\
&= \frac{1}{N} \langle \Psi_p | (\bar{\rho}_k^0)^\dagger (\bar{V} - E_0) \bar{\rho}_k^0 | \Psi_p \rangle \\
&= \frac{1}{N} \langle \Psi_p | (\bar{\rho}_k^0)^\dagger [\bar{V}, \bar{\rho}_k^0] | \Psi_p \rangle \\
&= \frac{1}{2N} \langle \Psi_p | (\bar{\rho}_k^0)^\dagger, [\bar{V}, \bar{\rho}_k^0] | \Psi_p \rangle
\end{aligned}
\tag{11.85}
$$

Here E_0 is the ground state energy. The structure factor is

$$\bar{s}(k) = \frac{1}{N} \langle \Psi_p | (\bar{\rho}_k^0)^\dagger \bar{\rho}_k^0 | \Psi_p \rangle. \tag{11.86}$$

Note that the Girvin–MacDonald–Platzman commutation relation (11.69) is satisfied

$$[\bar{\rho}_{\vec{k}}^0, \bar{\rho}_{\vec{q}}^0] = (e^{k^* q/2} - e^{kq^*/2}) \bar{\rho}_{\vec{k}+\vec{q}}^0, \tag{11.87}$$

where $\ell = 1$.

As k increases, the dispersion energy of the magnetorotons increases from a finite value $\sim 0.02 - 0.15 e^2/\epsilon\ell$ and then decreases, developing a roton minimum at a finite value of $k\ell \sim 1$, just like in ^{4}He (but in the case of ^{4}He, the energy dispersion goes to zero as $k \to 0$). The magnetorotons are intra-Landau excitations.[k] Magnetorotons have been observed in Raman scattering [28].

> **Exercise 11.11.** Show Eqs. (11.84), (11.85), and (11.86). Hint: Consult Refs. [6, 27].

---

[k]These modes are not the inter-Landau level excitation called magnetoplasmon whose energy coincides with the cyclotron energy in the limit $k \to 0$. The magnetoplasmons can be described by the time-dependent Hartree–Fock method, explained in Sec. 6.14. They also develop a roton minimum.

Exercise 11.12. Show that

$$\bar{s}(k) = S(k) - [1 - e^{-|k|^2\ell^2/2}[L_n(|k|^2\ell^2/2)]^2], \qquad (11.88)$$

where $L_0(x) = 1$ for the lowest Landau level and the static structure factor for the Laughlin state is

$$S(k) = \frac{1}{N}\langle P_n\rho_{-\vec{k}}\rho_{\vec{k}}P_n\rangle = \frac{1}{2}|k|^2\ell^2 \text{ for } k\ell \ll 1, \qquad (11.89)$$

where $\rho_{\vec{k}}$ is defined in Eq. (11.75). Show also that

$$\bar{f}(k) = \frac{1}{2}\sum_q v(q)(e^{q^*k/2} - e^{qk^*/2})$$

$$\times [\bar{s}(q)e^{-k^2/2}(e^{-k^*q/2} - e^{-kq^*/2}) + \bar{s}(k+q)(e^{k^*q/2} - e^{kq^*/2})].$$

$$(11.90)$$

Hint: Use Eq. (11.87) and consult Refs. [6, 27].

11.9. Center-of-Mass Magnetic Translations in Torus Geometry

We will show that a Laughlin state in a torus geometry is degenerate. To do this, we need to build the center-of-mass translation operator from guiding center translational operators treated in Sec. 10.3. These are many-body magnetic translational operators [29].

Consider a square with length L ($L_x = L_y = L$) (the periodic boundary conditions along x and y directions transform the rectangle into a torus). We first consider the gauge independent **guiding center** $\vec{\mathbf{R}} = (\mathbf{X}, \mathbf{Y})$ translational operators of an electron (Eq. (10.7)). These translate an **electron** by a distance $d = L/N_\phi$ along $\vec{L}_x = (L, 0)$ and $\vec{L}_y = (0, L)$. They may be written as

$$T_x = e^{i\frac{L}{N_\phi}\frac{\mathbf{Y}}{\ell^2}}; \ T_y = e^{-i\frac{L}{N_\phi}\frac{\mathbf{X}}{\ell^2}}, \qquad (11.91)$$

where the guiding center operators $\mathbf{X}$ and $\mathbf{Y}$ are given in Eq. (10.7). These translational operators commute with the Hamiltonian (see Exercise 11.13)

$$[T_i, H] = 0, \qquad (11.92)$$

where $i = x, y$. This means that eigenstates of T_i are also eigenstates of H.

> **Exercise 11.13.** Show the commutation relation $[T_i, H] = 0$, where
> the Hamiltonian H is given in Eq. (10.4). Hint: Show the commu-
> tation relations $[H, \mathbf{X}] = 0$. and $[H, \mathbf{Y}] = 0$ using the commutation
> relation between Π_x and Π_y, Eq. (10.5).

Translational operators of N electrons can be constructed from T_x and
T_y introduced in Eq. (11.91):

$$\mathbf{T}_x = \prod_{\alpha=1}^{N} e^{i\frac{L}{N_\phi}\frac{\mathbf{Y}_\alpha}{\ell^2}}; \quad \mathbf{T}_y = \prod_{j=\alpha}^{N} e^{-i\frac{L}{N_\phi}\frac{\mathbf{X}_\alpha}{\ell^2}}, \tag{11.93}$$

where $N = N_\phi \nu$. These operators act on the many-body states and
translate the **center of mass** of electrons a distance L/N_ϕ. (Many-body
translational operators are denoted by bold letters.) Using the result of
Exercise 11.13, one can show the commutation relations

$$[H, \mathbf{T}_x] = 0, \quad [H, \mathbf{T}_y] = 0. \tag{11.94}$$

11.10. Ground State Degeneracy on Torus

The translational operator mixes states in the degenerate manifold, and it
generates a matrix Berry phase, see Sec. 2.6. It maps one degenerate ground
state to another degenerate ground state. Using this operator, we will show
that a fraction quantum Hall state exhibits ground state degeneracy in a
torus geometry [29].[1] Reference [6] gives an accessible account of this effect
in fractional quantum Hall systems and we follow it here.

Let us show that many-body translational operators do not commute.
Consider translation operators that shift **one** electron position **by distance**
$d = \frac{L}{N_\phi}$ **along the** x **and** y **directions.** Apply these operators to two
electrons labeled α and β. We consider the following combinations $T_{x\alpha}T_{x\beta}$
and $T_{y\alpha}T_{y\beta}$, where the translational operators are

$$T_{x\gamma} = e^{i\frac{L}{N_\phi}\frac{\mathbf{Y}_\gamma}{\ell^2}}; \quad T_{y\gamma} = e^{-i\frac{L}{N_\phi}\frac{\mathbf{X}_\gamma}{\ell^2}}, \tag{11.95}$$

---

[1]Ground state degeneracy is absent in a disk geometry.

for $\gamma = \alpha$ or β. Exchanging these operators we find

$$(T_{x\alpha}T_{x\beta})(T_{y\alpha}T_{y\beta}) = T_{x\alpha}(T_{y\alpha}T_{y\beta})T_{x\beta}e^{i\theta} = (T_{y\alpha}T_{y\beta})T_{x\alpha}T_{x\beta}e^{i2\theta},$$

$$(11.96)$$

where θ is a phase angle. Here we have used the following result for one electron

$$T_{y\alpha}T_{x\alpha} = T_{x\alpha}T_{y\alpha}e^{i2\pi\frac{\Phi'}{\Phi_0}} = T_{x\alpha}T_{y\alpha}e^{i\theta},$$

$$T_{y\alpha}T_{x\beta} = T_{x\beta}T_{y\alpha} \quad \text{for} \quad \alpha \neq \beta.$$

$$(11.97)$$

where Φ' is the flux through d^2. The phase angle is

$$\theta = 2\pi\frac{\Phi'}{\Phi_0} = 2\pi\frac{d^2 B}{\Phi_0} = 2\pi\left(\frac{L}{N_\phi}\right)^2\frac{B}{\Phi_0} = 2\pi\frac{\Phi}{\Phi_0}\frac{1}{N_\phi^2} = 2\pi\frac{\nu}{N}, \quad (11.98)$$

which follows from the results

$$\frac{\Phi}{\Phi_0} = N_\phi, \quad \frac{1}{N_\phi} = \frac{\nu}{N}, \quad \Phi = L^2 B, \quad (11.99)$$

where ν is the Landau level filling factor. By extending this result for two electrons to N electrons, we find the phase factor

$$e^{i2\theta} \rightarrow e^{iN\theta} = e^{i2\pi\nu}. \quad (11.100)$$

This implies that the total translational operators for N electrons satisfy

$$\mathbf{T}_y\mathbf{T}_x = \mathbf{T}_x\mathbf{T}_y e^{i2\pi\nu}, \quad (11.101)$$

or

$$(\mathbf{T}_y^{-1}\mathbf{T}_x^{-1})(\mathbf{T}_y\mathbf{T}_x) = e^{i2\pi\nu}. \quad (11.102)$$

Thus many-body translational operators do not commute when $\nu \neq 1$.

What is the physical significance of this non-commutation? As depicted in Fig. 11.10, the translation $(\mathbf{T}_y^{-1}\mathbf{T}_x^{-1})(\mathbf{T}_y\mathbf{T}_x)$ can be simulated by changing flux values Φ_1 and Φ_2 of the solenoids (the magnetic fields that are confined inside the solenoids do not directly affect electrons on the surface of the torus). An increase in Φ_1 displaces the center of mass of electrons a distance d along the larger circle of the torus (longitudinal direction). An increase in Φ_2 displaces the center of mass of electrons a distance d along the smaller circle of the torus (transverse direction). Decreases in Φ_1 and Φ_2 returns the center of mass to the starting position. (Remember that gauge invariance under insertion of an elementary quantum flux through

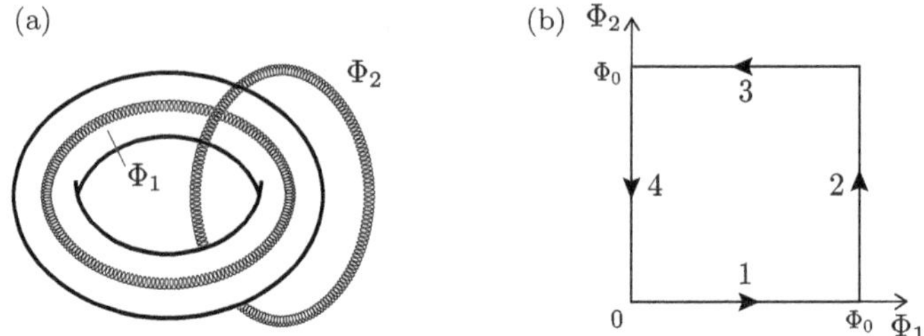

Fig. 11.10. Consider a cylinder in the presence of a perpendicular magnetic field. Simulation of the operation of $(\mathbf{T}_y^{-1}\mathbf{T}_x^{-1})(\mathbf{T}_y\mathbf{T}_x)$ on many-body states. Two fluxes Φ_1 and Φ_2 adiabatically thread two holes of the torus. (a) One solenoid is in the inner of hole and the other solenoid goes through the outer hole. (b) The fluxes are varied as shown in the closed loop.

the holes of a multiply connected domain leads to charge transfer across the system, see Sec. 10.5.) The effect of this translation operation is equivalent to the center of mass of N electrons encircling a square with area d^2, see Eq. (11.93). The resulting Aharonov–Bohm phase of the **center of mass** is

$$\theta = 2\pi\nu = 2\pi N d^2 B/\Phi_0. \tag{11.103}$$

For $\nu = 1/3$ the angle θ is non-zero, i.e., the system does not return to the original state after completing an adiabatic loop, i.e., the ground state must be degenerate. The ground state degeneracy is $1/\nu$. This effect is analogous to the presence of a non-Abelian matrix Berry phase, which results from a rotation of a state in the degenerate subspace, see Sec. 2.6. Translational operators act on the degenerate ground state manifold. From the commutation relation Eq. (11.101) one sees that for a completely filled Landau level with $\nu = 1$, the translational operators commute with each other.

11.11. Exchange of Two Quasiparticles and Statistical Phase

Piercing the quantum Hall droplet with a solenoid with an elementary quantum flux creates a quasiparticle. Suppose a flux tube is adiabatically pushed into torus as in Fig. 11.11(a) so that it threads the larger hole of the torus (let us call this operation $\mathbf{T}_x$). Suppose a second flux tube is adiabatically laid down in the torus as in Fig. 11.11(b) so that it threads the smaller hole inside the torus (let us call this operation $\mathbf{T}_y$). During each of these

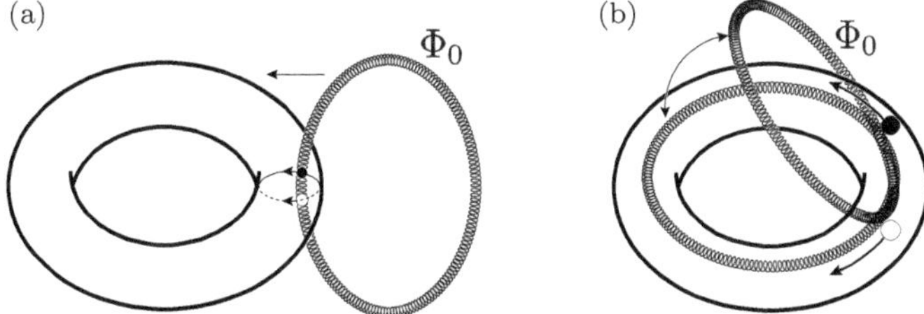

Fig. 11.11.   (a) A solenoid with flux Φ_0 is pushed into torus from outside. A quasielectron (filled circle) and quasihole (open circle) are created at positions where the solenoid pierces the torus (see Fig. 11.2). A quasihole moving in one direction is equivalent to a quasiparticle moving in the opposite direction. So as the solenoid is inserted from outside to the inner hole of the torus, a quasiparticle is transported once around the transverse direction. (b) A solenoid with flux Φ_0 is laid down in the torus from outside. The result is that a quasiparticle is transported once around the longitudinal direction.

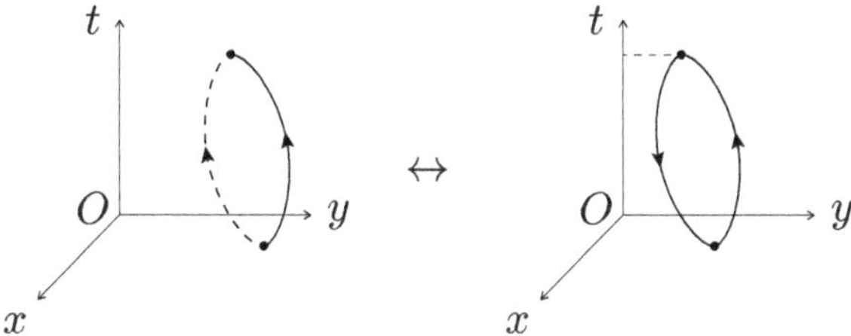

Fig. 11.12.   A quasihole moving clockwise is equivalent to a quasiparticle moving counterclockwise backwards in time. The backwards-in-time point of view is nowadays accepted as completely equivalent to other pictures [30].

processes, a quasielectron move along one of the half circles while a quasihole move along the other half circle as shown in Fig. 11.11(b). A quasihole moving counterclockwise is equivalent to a quasiparticle moving clockwise backwards in time [30], see Fig. 11.12. Thus, during the insertion of a flux tube, a quasiparticle goes around a **full** circle with length L. Under simultaneous insertion of these two fluxes, a quasiparticle encircles another quasiparticle, see Fig. 11.13.

Inserting and removing the two flux tubes (see Fig. 11.11) is equivalent to the translation operation $(\mathbf{T}_y^{-1}\mathbf{T}_x^{-1})(\mathbf{T}_y\mathbf{T}_x)$, which gives the Aharonov–Bohm phase of $2\pi\nu$. The resulting two worldlines of the quasiparticles are shown in Fig. 11.13: one path encircles the other path. Since the phase change after an exchange is half of the value for one full encircling, two

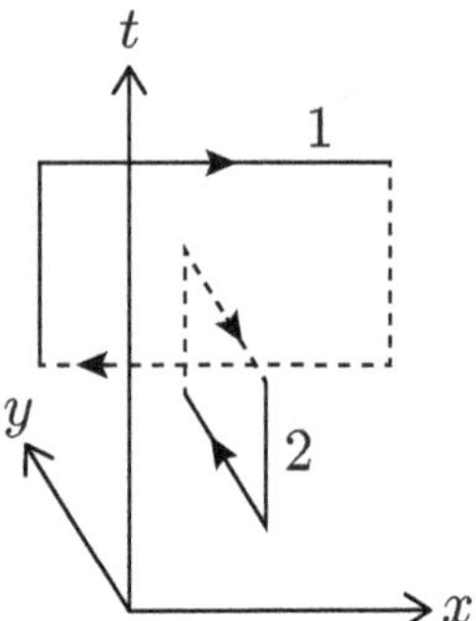

Fig. 11.13.   Consider inserting and removing the two flux tubes. The two processes combined are equivalent to $(\mathbf{T}_y^{-1}\mathbf{T}_x^{-1})(\mathbf{T}_y\mathbf{T}_x)$. World line 1 corresponds to the trajectory along the larger radius of the torus while world line 2 corresponds to the trajectory along t the smaller radius.

quasiparticles acquire the statistical phase that is half of the Aharonov–Bohm phase: $\theta = \pi\nu$. Thus quasiparticles must be **anyons**.

Further reading. An example of a hierarchical fractional quantized Hall state will be discussed in Sec. 12.1. Also, fractional quantum Hall states may be described by different topological Chern–Simons field theories. By attaching odd number of flux quanta, electrons can be transformed into bosons [31], and by attaching even number of flux quanta, they remain fermions [32] (we got a glimpse of this approach in Sec. 11.7). An introduction to Chern–Simons theory may be found in Refs. [7, 33–35]. The Chern–Simons theory is a topological field theory which does not depend on the metric. Thus only topological properties, such as braiding, matter. A fractional quantum Hall state having quasiparticles with **non-Abelian statistics** occurs at the filling factor 5/2. The state is called the Moore–Read state [36, 37]. It is suggested that an attractive interaction leads to a type of superconducting state of composite fermions. It describes a p-wave quasiparticle pairing with zero-energy Majorana excitation modes. The quasiparticle excitations are believed to have $e/4$ charges.

Bibliography

[1] D. C. Tsui, H. L. Stormer, and A. C. Gossard, Two-dimensional magneto-transport in the extreme quantum limit, *Phys. Rev. Lett.* **48**, 1559 (1982).

[2] R. B. Laughlin, Anomalous quantum Hall effect: an incompressible quantum fluid with fractionally charged excitations, *Phys. Rev. Lett.* **50**, 1395 (1983).

[3] R. B. Laughlin, *Nobel Lectures in Physics 1996–2000*, edited by Gösta Ekspong (World Scientific, Singapore, 2002). •

[4] S. Kivelson, D.-H. Lee, and S.-C. Zhang, Electrons in flatland, *Sci. American* **274**, 3, 86 (1996). •

[5] D. Yoshioka, *The Quantum Hall Effect* (Springer, Berlin, 2002).

[6] S. M. Girvin and K. Yang, *Modern Condensed Matter Physics* (Cambridge University Press, Cambridge, 2019). •

[7] X.-G. Wen, *Quantum Field Theory of Many-Body Systems: From the Origin of Sound to an Origin of Light and Electrons* (Oxford University Press, Oxford, 2004).

[8] J. K. Jain, *Composite Fermions* (Cambridge University Press, Cambridge, 2007).

[9] Z. F. Ezawa, *Quantum Hall Effects: Recent Theoretical and Experimental Developments* (World Scientific, Singapore, 2013).

[10] D. Tong, Lectures on the Quantum Hall effect, arXiv:1606.06687v2.

[11] F. Wilczek, Anyons for anyone, *Phys. World* **4**, 1, 40 (1991). •

[12] F. Wilczek, Anyons, *Sci. American* **264**, 5, 58 (1991). •

[13] M. Grayson, D. C. Tsui, L. N. Pfeiffer, K. W. West, and A. M. Chang, Continuum of chiral Luttinger liquids at the fractional quantum Hall edge, *Phys. Rev. Lett.* **80**, 1062 (1998).

[14] R. de-Picciotto, M. Reznikov, M. Heiblum, V. Umansky, G. Bunin, and D. Mahalu, Direct observation of a fractional charge, *Nature* **389**, 162 (1997).; L. Saminadayar, D. C. Glattli, Y. Jin, and B. Etienne, Observation of the $e/3$ fractionally charged Laughlin quasiparticle, *Phys. Rev. Lett.* **79**, 2526 (1997).

[15] V. J. Goldman and B. Su, Resonant Tunneling in the quantum Hall regime: Measurement of fractional charge, *Science* **267**, 1010 (1995).

[16] S. M. Girvin and T. Jach, Formalism for the quantum Hall effect: Hilbert space of analytic functions, *Phys. Rev. B* **29**, 5617 (1984).

[17] S.-R. Eric Yang, A. H. MacDonald, and M. D. Johnson, Addition spectra of quantum dots in strong magnetic fields, *Phys. Rev. Lett.* **71**, 3194 (1993).

[18] R. B. Laughlin, Quantized motion of three two-dimensional electrons in a strong magnetic field, *Phys. Rev. B* **27**, 3383 (1983).

[19] F. D. M. Haldane and E. H. Rezayi, Finite-size studies of the incompressible state of the fractionally quantized Hall effect and its excitations, *Phys. Rev. Lett.* **54**, 237 (1985).

[20] D. Yoshioka, B. I. Halperin, and P. A. Lee, Ground state of two-dimensional electrons in strong magnetic fields and 1/3 quantized Hall effect, *Phys. Rev. Lett.* **50**, 1219 (1983).

[21] B. I. Halperin, Statistics of Quasiparticles and the hierarchy of fractional quantized Hall states, *Phys. Rev. Lett.* **52**, 1583 (1984).

[22] D. Arovas, J. R. Schrieffer and F. Wilczek, Fractional statistics and the quantum Hall effect, *Phys. Rev. Lett.* **53**, 722 (1984).

[23] H. Bartolomei, M. Kumar, R. Bisognin, A. Marguerite, J.-M. Berroir, E. Bocquillon, B. Plaçais, A. Cavanna, Q. Dong, U. Gennser, Y. Jin, and G. Fève, Fractional statistics in anyon collisions, *Science* **368**, 173 (2020).

[24] J. Nakamura, S. Liang, G. C. Gardner, and M. J. Manfra, Direct observation of anyonic braiding statistics, *Nature Physics* **16**, 931 (2020).

[25] J. K. Jain, The composite fermion: a quantum particle and its quantum fluids, *Phys. Today* **53**, 4, 39 (2000). •

[26] W. Kang, H. L. Stormer, L. N. Pfeiffer, K. W. Baldwin, and K. W. West, How real are composite fermions? *Phys. Rev. Lett.* **71**, 3850 (1993).

[27] S. M. Girvin, A. H. MacDonald, and P. M. Platzman, Magneto-roton theory of collective excitations in the fractional quantum Hall effect, *Phys. Rev. B* **33**, 2481 (1986).

[28] A. Pinczuk, B. S. Dennis, L. N. Pfeiffer, and K. West, Observation of collective excitations in the fractional quantum Hall effect, *Phys. Rev. Lett.* **86**, 3983 (1993).

[29] X.-G. Wen and Q. Niu, Ground-state degeneracy of the fractional quantum Hall states in the presence of a random potential and on high-genus Riemann surfaces, *Phys. Rev. B* **41**, 9377 (1990).

[30] R. P. Feynman, *QED: The Strange Theory of Light and Matter* (Princeton University Press, Princeton, 1985). •

[31] S.-C. Zhang, T. H. Hansson, and S. Kivelson, Effective-field-theory model for the fractional quantum Hall effect. *Phys. Rev. Lett.* **62**, 82 (1989).

[32] A. Lopez and E. Fradkin, Fractional quantum Hall effect and Chern–Simons gauge theories, *Phys. Rev. B* **44**, 5246 (1991).

[33] T. Lancaster and S. J. Blundell, *Quantum Field Theory for the Gifted Amateur* (Oxford University Press, Oxford, 2014). •

[34] A. Zee, *Quantum Field Theory in a Nutshell* (Princeton University Press, Princeton, 2010).

[35] A. Altland and B. Simons, *Condensed Matter Field Theory* (Cambridge University Press, Cambridge, 2006).

[36] G. Moore and N. Read, Nonabelions in the fractional quantum Hall effect, *Nucl. Phys. B* **360**, 362 (1991).

[37] N. Read, Topological phases and quasiparticle braiding, *Phys. Today* **65**, 7, 38 (2012). •

Chapter 12

Edge Excitations of Quantum Hall States*

"You must come to Copenhagen to work with us. We like people who can actually perform thought experiments!"

Niels Bohr

A quantum Hall state is an excellent example of display of the bulk edge correspondence [1]. The bulk properties determine the edge properties, such as number of modes and edge structures. Quantum Hall states are incompressible away from the boundary but the edges support low-energy edge excitations. An edge density wave of a quantum Hall droplet is chiral, i.e., it moves only in one direction. Edge electrons constitute a chiral Luttinger liquid. (In a Hall bar system, edge excitations of the opposite edge move in the opposite direction.) Edges of fractional quantum Hall states have excitations with fractional charges. In addition, edges of integer quantum Hall state display spin–charge separation [2]. Edge properties are reviewed in detail in [1, 3, 4]. In this chapter, we examine only some salient features relevant to topological insulators. Before reading this chapter, the reader is encouraged to read Chapter 7 on bosonization.

12.1. Chiral Edges of Fractional Quantum Hall Droplet

• Neutral Edge Excitations of Laughlin States

In the presence of electric and magnetic fields, the deformation of edge propagates (see Fig. 12.1). It is described by the wave equation [1]

$$\frac{\partial \rho}{\partial t} + v \frac{\partial \rho}{\partial x} = 0, \qquad (12.1)$$

329

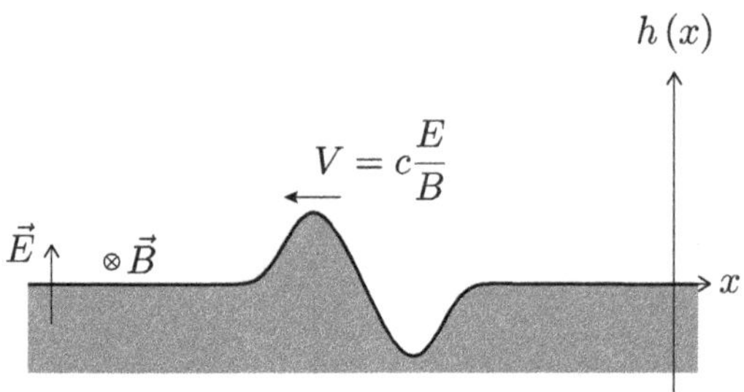

Fig. 12.1. A density wave propagating along an edge of a quantum Hall bar at filling factor ν. Here, electric and magnetic fields are present. The velocity of the wave is given by the drift velocity. Electron system is assumed to be spin-polarized.

where v is the drift velocity (see Eq. (10.1)) and x is the coordinate along the edge. The edge wave $\rho(x)$ is related to the displacement of edge $h(x)$, $\rho(x) = nh(x)$, with the droplet density $n = \frac{\nu}{2\pi\ell^2}$. The edge wave propagates only in one direction, i.e., $k > 0$. The Fourier transform of this equation is given by

$$\frac{\partial \rho_k}{\partial t} + ikv\rho_k = 0. \tag{12.2}$$

The Hamiltonian is

$$H = \frac{1}{2} \int dx \rho(eE)h = \int dx \pi \frac{v}{\nu} \rho^2, \tag{12.3}$$

where E is the electric field (note that $\hbar = 1$ and the drift velocity $v \sim \frac{cE}{B}$). This implies that the Hamiltonian is given by

$$H = 2\pi \frac{v}{\nu} \sum_{k>0} \rho_{-k}\rho_k, \tag{12.4}$$

where $\rho_k = \frac{1}{\sqrt{L}} e^{ikx} \rho(x)$. From the canonical quantization scheme, we have

$$[\rho_k, \pi_{k'}] = i\delta_{k,k'}. \tag{12.5}$$

From this follows

$$[\rho_k, \rho_{k'}] = \frac{\nu}{2\pi} k\delta_{k,-k'}. \tag{12.6}$$

Note that, in comparison with the usual commutation relation between density fields given in Eq. (7.9), this commutation relation **differs by a factor of** ν. At $\nu = 1$, the commutation relations are identical.

- Neutral Edge Excitations of the 2/5 State

The $\nu = 2/5$ hierarchical state [5, 6] is generated by the condensation of quasiparticles on the top of the $\nu = 1/3$ state. The quasiparticles form a $1/15$ state such that the total filling factor is the sum $\nu = 2/5 = 1/3 + 1/15$. From the two wave equations for $\nu = 1/3$ and $\nu = 1/15$, we infer that the Hamiltonian is the sum of the Hamiltonians for the two filling factors

$$H = 2\pi \sum_{k>0} \sum_i \frac{v_i}{\nu_i} \rho_{i,-k} \rho_{i,k}, \tag{12.7}$$

where $i = 1$ and 2 correspond to filling factors $1/3$ and $1/15$ and the density operators satisfy the commutation relation

$$[\rho_{i,k}, \rho_{j,k'}] = \frac{\nu_i}{2\pi} k \delta_{i,j} \delta_{k,-k'}. \tag{12.8}$$

If the edges of these two incompressible states are close to each other, their interaction V_{ij} must be taken into account. The total Hamiltonian has the following form:

$$H = 2\pi \sum_{k>0} \sum_{ij} V_{ij} \rho_{i,-k} \rho_{j,k}, \tag{12.9}$$

where the diagonal elements are $V_{ii} = \frac{v_i}{\nu_i}$. This 2×2 Hamiltonian can be easily diagonalized [1]

$$H = 2\pi \sum_{k>0} \sum_i \mathrm{sgn}(\tilde{\nu}_i) \tilde{v}_i \tilde{\rho}_{i,-k} \tilde{\rho}_{j,k}, \tag{12.10}$$

where $\tilde{\rho}_{j,k}$ describes the new density excitations

$$\tilde{\rho}_{1k} = \frac{\cos(\theta)}{\sqrt{|\nu_1|}} \rho_{1k} + \frac{\sin(\theta)}{\sqrt{|\nu_2|}} \rho_{2k},$$

$$\tilde{\rho}_{2k} = \frac{\cos(\theta)}{\sqrt{|\nu_2|}} \rho_{2k} - \frac{\sin(\theta)}{\sqrt{|\nu_1|}} \rho_{1k}, \tag{12.11}$$

$$\tan(\theta) = 2 \frac{\sqrt{|\nu_1 \nu_2|} V_{12}}{|\nu_1| V_{11} - |\nu_2| V_{22}}$$

and

$$\mathrm{sgn}(\nu_1)\tilde{v}_1 = \frac{\cos^2(\theta)}{\cos(2\theta)} |\nu_1| V_{11} - \frac{\sin^2\theta}{\cos(2\theta)} |\nu_2| V_{22},$$

$$\mathrm{sgn}(\nu_2)\tilde{v}_2 = \frac{\cos^2(\theta)}{\cos(2\theta)} |\nu_2| V_{22} - \frac{\sin^2\theta}{\cos(2\theta)} |\nu_1| V_{11}. \tag{12.12}$$

These coupled modes are analogous to normal phonon modes of diatomic chains [7]. In bosonization of one-dimensional electron systems, a similar coupled Hamiltonian appears (see Chapter 7).

12.2. Electron and Anyon Field Operators

Anyons with a fractional charge may be created when an electron is added to an edge. How can we express the anyon field operator in terms of the density operator? First, we express a fermion field with a boson field using **bosonization**. One of the fundamental results of bosonization approach is the following relation between the electron field operator $\Psi^\dagger(x)$ and density operator $\rho_\alpha(x)$ of an edge denoted by α

$$[\rho_\alpha(x), \Psi^\dagger(x')] = \delta(x - x')\Psi^\dagger(x'). \tag{12.13}$$

(Students should review Chapter 7 on bosonization before reading further.) Second, the fermion field can be written in terms of the phase field $\phi_\alpha(x)/\nu$ in the bosonization approach (see Eqs. (7.30) and (7.26))

$$\Psi(x) \propto e^{\frac{i}{\nu}\phi_\alpha(x)}. \tag{12.14}$$

Note that $\phi_\alpha(x)/\nu$ corresponds to the **electron** phase fields $\phi_{L,R}$ defined in the bosonization theory. The density operator of the Laughlin states obeys a slightly different commutation relation than the one used in the usual bosonization scheme, given by Eq. (12.6). Here $\phi_\alpha(x)$ is the anyon phase field and is related to the charge density field $\rho_\alpha(x)$ (see also Eq. (7.27))

$$\rho_\alpha(x) = \frac{\partial \phi_\alpha}{\partial x},$$
$$\int_{-\infty}^{x} dx' \rho(x') = \phi_\alpha(x) - \phi_\alpha(\infty). \tag{12.15}$$

The relation between the edge charge density $\rho_\alpha(x)$ and the **anyon** phase field $\phi(x)$ is similar to the one depicted in Fig. 7.7. The electron field $\Psi(x)$ at $\nu = 1/3$ may be regarded as consisting of three **anyon fields** [8]

$$\Psi_{\text{any}}(x) \propto e^{i\phi_\alpha(x)}. \tag{12.16}$$

We can verify that this result gives the correct complex phase factor for the exchange two anyons

$$\Psi_{\text{any}}(x)\Psi_{\text{any}}(y) = \Psi_{\text{any}}(y)\Psi_{\text{any}}(x)e^{i\pi\nu},$$
$$\Psi_{\text{any}}^\dagger(x)\Psi_{\text{any}}(y) - e^{i\pi\nu}\Psi_{\text{any}}(x)\Psi_{\text{any}}^\dagger(y) = \delta(x - y). \tag{12.17}$$

> **Exercise 12.1.** Derive Eq. (12.17). Hint: Use the Hausdorff formulas, see Appendix E. Consult Ref. [8].

12.3. Tunneling Density of States and Green's Function

The tunneling rate into a Luttinger liquid at $E = E_F$ is suppressed to zero and displays a power law at low voltages and temperatures. To compute the DOS of a strongly correlated system, it is useful to employ Green's functions (it is instructive to contrast this method to that of the Hartree–Fock approximation given in Sec. 6.8, which describes independent electrons). We will see in this and the next Sections that the Green's function contains a lot of information about one-particle properties, such as the occupation numbers and density of states. The tunneling density of states is related to the spectral density of the retarded electron Green's function.[a] In terms of field operators, the Green's function is defined as

$$iG(x_2, t_2, x_1, t_1) = \langle T(\Psi^\dagger(x_2, t_2)\Psi(x_1, t_1))\rangle. \tag{12.18}$$

The expectation value is with respect to the ground state. The time ordering operator T is defined as

$$T\{\Psi^\dagger(x_2, t_2)\Psi(x_1, t_1)\} = \begin{cases} \Psi^\dagger(x_2, t_2)\Psi(x_1, t_1), & \text{if } t_2 > t_1, \\ -\Psi(x_1, t_1)\Psi^\dagger(x_2, t_2), & \text{if } t_2 < t_1. \end{cases} \tag{12.19}$$

Let us try to understand the physics behind these mathematical expressions. The Green's function gives the response at any point due to an excitation at any other. The electron field operator can be written in terms of creation operators: $\Psi^\dagger(x, t) = \sum_k \psi_k^*(x, t)c_k^\dagger$. In the presence of translational invariance, the Green's function of interacting electrons for $t_1 > t_2$ is

$$iG(k_2, t_2; k_1, t_1) = iG(k_1, t_1 - t_2)\delta_{k_1, k_2} = -\langle\psi_0|c_{k_1}(t_1)c_{k_2}^\dagger(t_2)|\psi_0\rangle, \tag{12.20}$$

where $c_k^\dagger$ is the electron creation operator and ψ_0 is the many-body ground state of N particles. (We drop vector signs on wave vectors and coordinate positions in this section.) The Green's function gives the probability amplitude for the following. Suppose, at time t_2, we add a particle with the wave

---

[a] For an easy introduction to Green's functions, see [9].

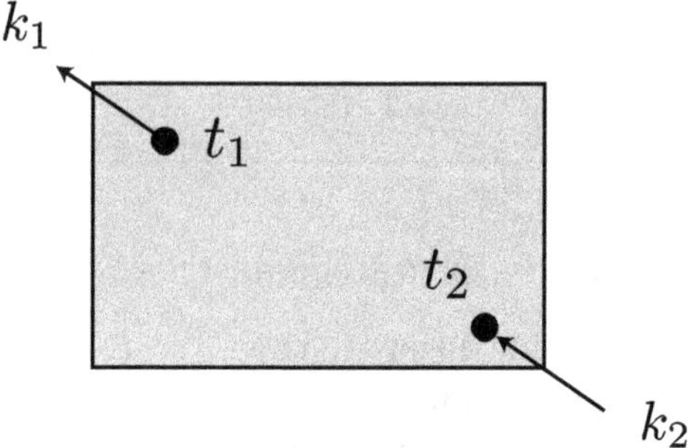

Fig. 12.2.   A particle with momentum k_2 is inserted at time t_2. The particle propagates, and at time t_1 particle has momentum k_1 $(t_2 < t_1)$.

function $\psi_{k_2}(x)$ to the ground state. What is the probability amplitude that, at time t_1, the system will be in the ground state with an added particle in the wave function $\psi_{k_1}(x)$? (See Fig. 12.2.)

Let us explain these ideas by considering the non-interacting case. The retarded Green's function of a non-interacting Fermi gas at zero temperature is given by (see Exercise 12.2)

$$G^R(k,\omega) = \frac{1}{\omega - \epsilon_k + i\eta},\qquad(12.21)$$

where $\eta > 0$ is an infinitely small convergence factor and the electron energy dispersion is ϵ_k. Note that the retarded Green's function has a **pole** at $\omega = \epsilon_k - i\eta$, which represents an exponentially decaying wave function $\sim e^{-\eta t/\hbar}$ (the advanced Green's function behaves in the opposite way). The imaginary part of the retarded Green's function contains information about the **spectral density**

$$A(k,\omega) = -\frac{1}{\pi}G^R(k,\omega),\qquad(12.22)$$

from which one can find the DOS [7]

$$D(\omega) = \int dk\, A(k,\omega) \propto \int dk\, \delta(\hbar\omega - \epsilon_k).\qquad(12.23)$$

(See Exercise 12.2.)

Exercise 12.2. (a) In the case of non-interacting electrons, show that the retarded Green's function is

$$G^R(k,\omega) = \frac{1}{\omega - \epsilon_k + i\eta}. \qquad (12.24)$$

(b) Show that the tunneling DOS is given by Eq. (12.23).
Hint: Use the following results: (1) At t_1, the wave function is $\psi_{k_1}(x)$ and at t_2, it is $e^{-i\epsilon_{k_2}(t_2-t_1)}\psi_{k_2}(x)$. Note that $\int dx \psi_{k_1}^*(x)\psi_{k_2}(x) = \delta_{k_1,k_2}$. (2) $G(k_2,k_1,t_2,t_1) = -i\theta(t_2-t_1)e^{-i(\epsilon_{k_2}-i\eta)(t_2-t_1)}\delta_{k_2,k_1}$. A convergence factor $i\eta$ is inserted. (3) $G(k,\omega) = -i\int d(t_2-t_1)e^{i\omega(t_2-t_1)}G(k,k,t_2,t_1)$.

When mutual electron interactions are present, the pole in the Green's shifts. Note that both the real and imaginary parts of the shift can be finite, $\omega = \epsilon_k - \Delta_k - i\gamma_k$. Consequently, the spectral density consists of a shifted and/or broadened delta function. A broadened delta function means that an electron acquires a finite lifetime as a consequence of interaction with other electrons (this can be anticipated from Fermi's golden rule for the probability of a transition per unit time of a quantum state).

The Green's function of the interacting one-dimensional system is qualitatively different from that of an interacting Fermi gas. This is because perturbation theory breaks down, as we discussed in Chapter 7. Using the bosonization approach of Chapter 7, one can calculate the edge Green's function of the Laughlin state at filling factor ν

$$iG(x,t) = \langle T\psi^\dagger(x,t)\psi(0,0)\rangle = e^{\frac{1}{\nu^2}\langle\phi(x,t)\phi(0,0)\rangle} \propto \frac{1}{(x-vt)^{\frac{1}{\nu}}}, \qquad (12.25)$$

where the boson correlation function is

$$\langle\phi(x,t)\phi(0,0)\rangle = -\nu\ln(x-vt) + \text{constant}. \qquad (12.26)$$

(Do Exercises 12.3 and 12.4.) The Fouirer transform is

$$G(k,\omega) \propto \frac{(vk+\omega)^{\frac{1}{\nu}-1}}{\omega - vk + i\eta\,\text{sgn}(\omega)}. \qquad (12.27)$$

Note how different this result is from the result of the non-interacting electron gas, Eq. (12.21). From this result, the tunneling DOS can be computed

$$D(\omega) \propto \int dk \mathrm{Im} G(k,\omega) \propto |\omega|^{\frac{1}{\nu}-1}. \tag{12.28}$$

Exercise 12.3. Show $\langle e^{a\phi(x)}e^{b\phi(0)}\rangle = C e^{ab\langle\phi(x)\phi(0)\rangle}$, where C is a constant. Hint: Use $\langle e^A e^B\rangle = e^{\langle\frac{1}{2}(A^2+B^2)\rangle}e^{\langle AB\rangle}$. (The commutator $[A,B]$ is a c-number.) The phase field $\phi(x)$ can be expressed in terms of boson operators, see Eqs. (7.17), (7.18) and (7.26).

Exercise 12.4. Show $\langle\phi(x)\phi(0)\rangle = \sum_{q>0}\frac{2}{L_q}e^{iqx} = \ln\frac{L}{2\pi(\eta-ix)}$, where $L_q = qL/2\pi$ and η is a small convergence factor. This gives the result $e^{ab\langle\phi(x)\phi(0)\rangle} = (\frac{L}{2\pi(\eta-ix)})^{ab}$ in Eq. (12.25).

12.4. Momentum Distribution Function of Narrow Hall Bar: Luttinger Liquid

From the k-space Green's function (12.20), one can compute the momentum occupation number $n(k)$

$$n(k) = \langle\psi|c^{\dagger}_{\alpha,k}c_{\alpha,k}|\psi\rangle = iG_{\alpha}(k,t\to 0^+) = \int dx e^{ikx}iG_{\alpha}(x,t\to 0^+), \tag{12.29}$$

where $t = t_1 - t_2$ and $|\psi\rangle$ is the interacting ground state. The result of $n(k)$ is shown in Fig. 12.3. In a Fermi liquid, the Fermi surface exists and quiasiparticles are well-defined. On the other hand, interaction effects are singular in one dimension and the susceptibility diverges at $k = 2k_F$, in contrast to two- and three-dimensional systems, see Fig. 7.1 (k_F is the Fermi wave vector). In one-dimensional systems, single-particle excitations are thus ill-defined and only collective **bosonic** excitations are meaningful. This implies that the sharpness of the Fermi surface is no longer present, as the dashed line in Fig. 12.3 shows.

Next, we consider a narrow Hall bar with two edges. An added electron fractionalizes into anyons (see Eq. (12.16)). Suppose that these anyons are distributed in such a way that n anyons reside on the right edge and

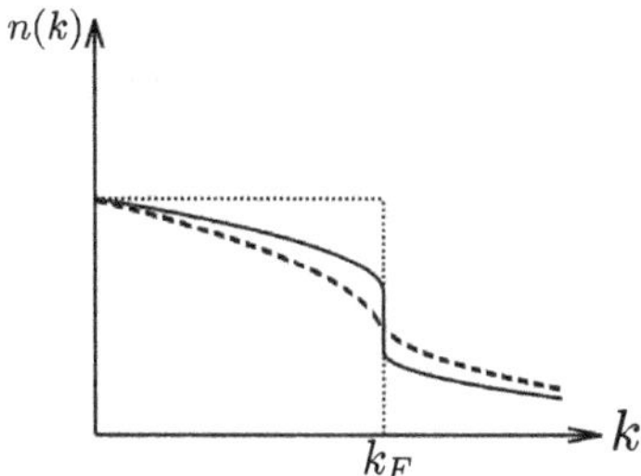

Fig. 12.3.   Momentum occupation number $n(k)$ of a Fermi liquid as a function of wave vector k is shown. The step function is present in $n(k)$ of an electron gas described by the Hartree–Fock approximation (dottted line). For a two- or three-dimensional interacting electron gas $n(k)$ deviates from a step function but a discontinuity is still present(solid line). For an interacting one-dimensional electron system, the discontinuity at k_F may vanish and a Luttinger liquid forms (dashed line), see Sec. 12.4.

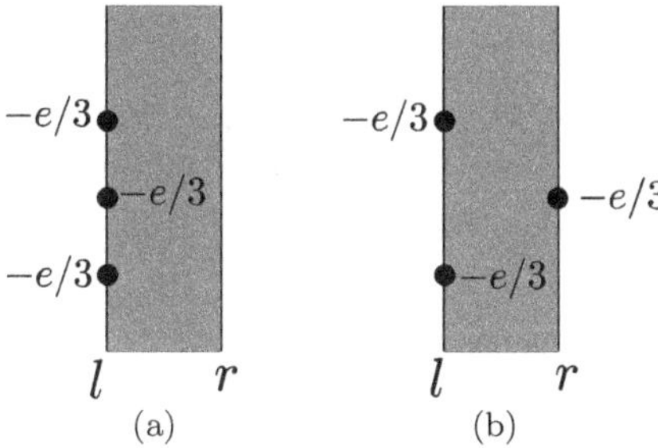

Fig. 12.4.   Division of an electron between the two edges of a narrow Hall bar. In this case, the momentum distribution function may have singularities not only at $k = k_F$ but also at other values of k. The left and right edges may be interchanged.

d anyons reside on the left edge so that $n + d = 1/\nu$ (see Fig. 12.4). The creation operator of such an electron is

$$\Psi^{\dagger}_{n,d}(x) = e^{in\phi_R(x)}e^{id\phi_L(x)}. \tag{12.30}$$

From this and using Eq. (12.29) we find [10]

$$\langle \Psi_{n,d}(x)\Psi^{\dagger}_{n,d}(0)\rangle = \frac{e^{i(n-d)k_F x}}{x^{2\Delta_{n,d}}}, \tag{12.31}$$

where

$$2\Delta_{n,d} = \frac{1}{2}\left(\frac{1}{\nu} + (n-d)^2\nu\right). \tag{12.32}$$

By Fourier transforming $\langle \Psi_{n,d}(x)\Psi^{\dagger}_{n,d}(0)\rangle$, we find that the **momentum distribution function** $n(k)$ has a power-law contribution

$$\delta n(k) \sim |k \pm pk_F|^{\alpha}, \tag{12.33}$$

where the possible values of p are

$$p = n - d = \frac{1}{\nu}, \frac{1}{\nu} - 2, \frac{1}{\nu} - 4, \ldots, 1. \tag{12.34}$$

It has a power-law singularity at $\pm pk_F$ with the exponent $\alpha = 2\Delta_p - 1$. For $\nu = 1/3$, the exponents are $2\Delta_{3,0} - 1 = 2$ and $2\Delta_{2,1} - 1 = 2/3$ at k_F are $3k_F$, respectively. The presence of a power law at $3k_F$ is consistent with the higher harmonics in the Luttinger liquid paradigm [11] (see also the introduction to Chap. 7).

12.5. Spin–Charge Separation of Integer Quantum Hall Edges

The separation of spin and charge in Luttinger liquids is an enormous simplification [12]. When this happens, a Hamiltonian breaks into two parts, namely a charge-density part and a spinwave part

$$H = H_{\rho} + H_{\sigma}. \tag{12.35}$$

Since these two parts of the Hamiltonian are **independent and do not interact,** the Hilbert space is made of two separate spaces: one for spinwaves and the other one for charge-density waves. These two excitations

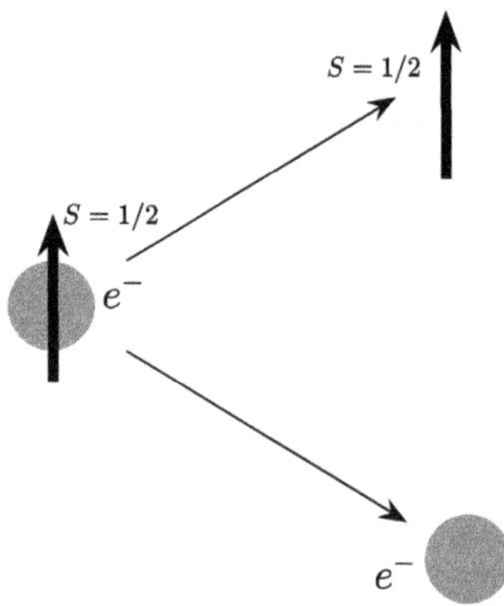

Fig. 12.5. Illustration of spin and charge separation. Spin and charge move independently with different velocities.

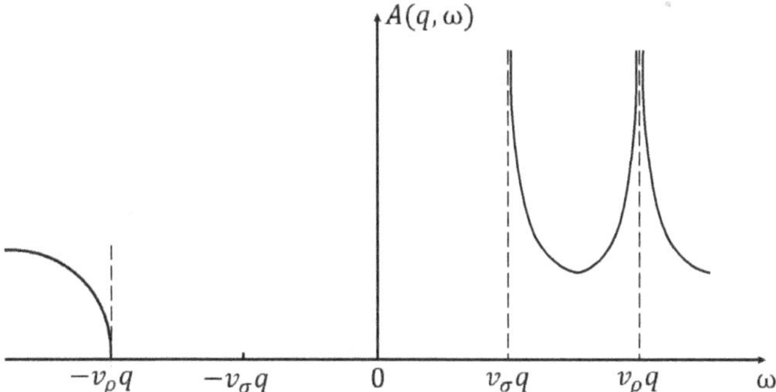

Fig. 12.6. The spectral density displays power-law singularities at $E = \pm\hbar v_\rho q$ and $\pm\hbar v_\sigma q$, where v_ρ and v_σ are the charge and spin mode velocities. See Ref. [13].

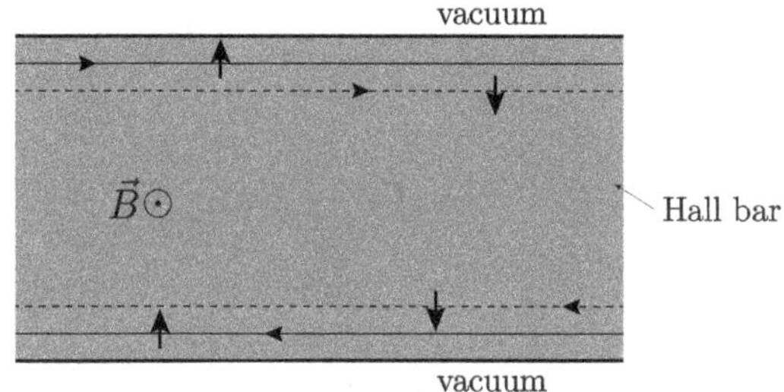

Fig. 12.7. Edge states of a quantum Hall bar at filling factor $\nu = 2$. An edge has both spin-up ($\uparrow$) and -down ($\downarrow$) states. Upper (lower) edge states are left (right) moving.

are therefore totally separate and will move independently of one another (see Fig. 12.5). In this case, the spectral density is no longer of a Lorentzian form. Instead it has power-law singularities at the energies of the spin and charge modes, see Fig. 12.6. The wave functions and space–time correlation functions factorize into products and the spin and charge collective modes also have different velocities. We saw an example of this effect when we studied superconductors in Sec. 8.4: a quasiparticle of a superconductor has spin 1/2 but no charge. Another notable example is a domain wall soliton in polyacetylene, see Fig. 4.12.

Spin–charge separation has received little attention in connection with quantum Hall edge electrons. However, Lee and Yang [2] showed that the tunneling DOS of a quantum Hall bar (see Fig. 12.7) at $\nu = 2$ displays spin–charge separation, provided that the long range of the Coulomb interaction is present. Bosonization is well suited for the investigation of this phenomenon. When the filling factor is $\nu = 2$, the spin-up and -down $n = 0$ lowest Landau levels are completely filled, see Fig. 12.7. Only edge excitations are allowed. One can write down the effective low-energy Hamiltonian for these edge excitations [1] (these excitations consist of electron–hole pairs).

The effective Hamiltonian of edge electrons interacting via the long-range Coulomb interaction has the following form:

$$H = \sum_{s,p>0} 2\pi v_s : \rho_{sp}\rho_{s,-p} : + \sum_{s,p>0} V_{\text{intra}}(p) : \rho_{sp}\rho_{s,-p} :$$
$$+ \sum_{s,p>0} V_{\text{inter}}(p) : \rho_{sp}\rho_{-s,-p} :, \tag{12.36}$$

where the density operators satisfy the commutation relation

$$[\rho_{sp}, \rho_{sp'}] = \frac{p}{2\pi}\delta_{pp'}. \tag{12.37}$$

Here spin and momentum are denoted by s and p. The Fermi velocity of the edge with spin s is v_s (see Fig. 10.13). The intra- and inter-edge interactions are given by $V_{\text{intra}}(p) = \frac{2}{\epsilon}\ln(\frac{2}{|p|a})$ and $V_{\text{inter}}(p) = \frac{2e^2}{\epsilon}\ln(\frac{2e^2}{|p|b})$. The lengths a and b are comparable to the width W of the Hall bar [14]. Note that the dimension of the density field is $[\rho_{sp}] = 1/\sqrt{\text{Length}}$.

The Hamiltonian can be transformed into

$$H = H_\rho + H_\sigma + H_{\rho\sigma}, \tag{12.38}$$

where

$$H_\rho = 2\pi \sum_{p>0} v_{c,p} : \rho_p\rho_{-p} :,$$
$$H_\sigma = 2\pi \sum_{p>0} v_{s,p} : \sigma_p\sigma_{-p} : \tag{12.39}$$

with charge and spin density operators

$$\rho_p = \frac{1}{\sqrt{2}}[\rho_{\uparrow,p} + \rho_{\downarrow,-p}],$$
$$\sigma_p = \frac{1}{\sqrt{2}}[\rho_{\uparrow,p} - \rho_{\downarrow,-p}] \tag{12.40}$$

and the commutation relation

$$[\rho_{s,p}, \rho_{s,-p}] = \left(\frac{p}{2\pi}\right)\delta_{p,p'}. \tag{12.41}$$

The coupling between spin and charge modes is given by

$$H_{\rho\sigma} = 2\pi \sum_{p>0} g\sigma_p \rho_{-p}. \tag{12.42}$$

The coupling constant g is given by the difference between the Fermi velocities: $g = v_\uparrow - v_\downarrow$ (this difference is due to spin-splitting). The total Hamiltonian can be written using a 2×2 matrix

$$H = \frac{2\pi}{e^2} \sum_{p>0} (\rho_p, \sigma_{-p}) \begin{pmatrix} v_{cp} & g/2 \\ g/2 & v_{sp} \end{pmatrix} \begin{bmatrix} \rho_{-p} \\ \sigma_{-p} \end{bmatrix}. \tag{12.43}$$

(Here normal ordering is omitted.) The 2×2 matrix can be easily diagonalized and two independent modes are found.

One-dimensional fermion field can be written in terms of the boson density operators (see Chapter 7). This allows one to compute the electron Green's function, from which one can compute the tunneling/quasi-particle density of states. We find the following result for quasi-particle density of states for each spin in the energy region $|\omega| \ll \omega_0 = \frac{e^2}{\epsilon W}$,

$$D(\omega) \sim \frac{1}{\sqrt{\ln\left|\dfrac{\omega_0}{\omega}\right|}} e^{\frac{sg}{\pi v_0}\frac{1}{\ln\left|\frac{\omega_0}{\omega}\right|}}, \tag{12.44}$$

where $v_0 = \frac{2e^2}{\epsilon\pi}$ and $s = 1(-1)$ for spin-up (-down). The exponential factor is due to the coupling between spin-up and -down edges. However, in the limit $\omega \to 0$, the DOS approaches $D(\omega) \to \frac{1}{\sqrt{\ln\left|\frac{\omega_0}{\omega}\right|}}$. Note that this $D(\omega)$ is precisely square root of the DOS of the spin-polarized case $\nu = 1$

$$D(\omega) \sim \frac{1}{\ln\left|\dfrac{\omega_0}{\omega}\right|}. \tag{12.45}$$

It is also equal to the result for independent spin and charge modes, i.e., $g = 0$. The appearance of the square root in the DOS in the limit $\omega \to 0$ is thus a consequence of spin–charge separation (Fig. 12.5) and depends crucially on the long range of the Coulomb interaction.[b] The ratio between

---

[b]Spin–charge separation is also present for short-term interactions, but in this case the DOS does not possess a pseudogap.

the spin-up and -down DOS is given by

$$\ln\left[\frac{D_\uparrow(\omega)}{D_\downarrow(\omega)}\right] = \frac{2g}{\pi v_0}\,\frac{1}{\ln\left|\dfrac{\omega_0}{\omega}\right|}. \qquad (12.46)$$

The new predicted behavior is expected to show up in the tunneling current between a bulk doped GaAs and the abrupt edge of a quantum Hall fluid.

Bibliography

[1] X.-G. Wen, *Quantum Field Theory of Many-Body Systems: From the Origin of Sound to an Origin of Light and Electrons* (Oxford University Press, Oxford, 2004).

[2] H. C. Lee and S.-R. Eric Yang, Spin–charge separation in quantum Hall edge liquids, *Phys. Rev. B* **56**, R15529 (1997).

[3] D. Yoshioka, *The Quantum Hall Effect* (Springer, Berlin, 2002).

[4] S. M. Girvin and K. Yang, *Modern Condensed Matter Physics* (Cambridge University Press, Cambridge, 2019).

[5] F. D. M. Haldane, Fractional quantization of the Hall effect: A hierarchy of incompressible quantum fluid states, *Phys. Rev. Lett.* **51**, 605 (1983).

[6] B. I. Halperin, Statistics of quasiparticles and the hierarchy of fractional quantized Hall states, *Phys. Rev. Lett.* **52**, 1583 (1984).

[7] N. W. Ascroft and N. David Mermin, *Solid State Physics* (Thomson Learning, London, 1976). •

[8] Z. F. Ezawa, *Quantum Hall Effects: Recent Theoretical and Experimental Developments* (World Scientific, Singapore, 2013).

[9] R. D. Mattuck, *A Guide to Feynman Diagrams in the Many-Body Problem* (Dover Publications, Mineola, 1992). •

[10] S.-R. Eric Yang, S. Mitra, A. H. MacDonald and M. P. A. Fisher, Momentum distribution function of a narrow Hall bar in the FQHE regime, *J. Kor. Phys. Soc.* **29**, 10 (1996).

[11] F. D. M. Haldane, Luttinger Liquid Theory of One-dimensional Quantum Fluids. I. Properties of the Luttinger model and their extension to the general 1D interacting spinless Fermi gas, *J. Phys. C: Solid State Phys.* **14**, 2585 (1981); F. D. M. Haldane, Effective Harmonic-Fluid Approach to Low-Energy Properties of One-Dimensional Quantum Fluids, *Phys. Rev. Lett.* **47**, 1840 (1981).

[12] T. Giamarchi, *Quantum Physics in One Dimension* (Oxford University Press, Oxford, 2003).

[13] J. Voit, Charge–spin separation and the spectral properties of Luttinger liquids, *Phys. Rev. B* **47**, 6740 (1993).

[14] U. Zülicke and A. H. MacDonald, Plasmon modes and correlation functions in quantum wires and Hall bars, *Phys. Rev. B* **54**, 16813 (1996).

Chapter 13

Toric Code for Pedestrians

"There's more than one way to skin a cat."

Richard P. Feynman[a]

We study a quantum spin liquid[b] called Kitaev's toric code [2, 3], which exhibits topological order [4–6]. This system is attractive since many of its properties can be analytically explored. As in the BCS and fractional quantum Hall cases, one first guesses the correct trial wave function for the ground state and then constructs quasiparticle excitations. In the toric code, some of these excitations exhibit Abelian anyon (semion) statistics. The topological entanglement entropy of this phase is finite, equal to $\ln 2$. Moreover, the ground state is four-fold degenerate in a torus topology. (Note that the Laughlin state at filling factor $1/3$ has degeneracy 3 and the superconducting state has degeneracy 4.) Some of these properties of the toric code, especially the appearance of a pair of non-local excitations, are similar to those of interacting disordered zigzag nanoribbons and the reader is encouraged to study this chapter carefully. The presentation of this chapter is somewhat heuristic. For a more refined treatment, see, for example Refs. [2, 3].

---

[a]This was his response to the claim that string theory is unique in explaining the existence gravity (read a very enjoyable book on superstrings, see Ref. [1]).

[b]Quantum spin liquids are generally characterized by their long-range quantum entanglement, fractionalized excitations, and absence of ordinary magnetic order.

343

13.1. Hamiltonian of Toric Code

The Hamiltonian of Kitaev's toric code is

$$H = -\sum_s A_s - \sum_p B_p, \tag{13.1}$$

where, for each "star" (vertex) s and plaquette p, the operators A_s and B_p are defined, respectively, as

$$A_s = \sigma_1^z \sigma_2^z \sigma_3^z \sigma_4^z, \quad B_p = \sigma_1^x \sigma_2^x \sigma_3^x \sigma_4^x. \tag{13.2}$$

Here σ^x and σ^z are the Pauli spin matrices. Figure 13.1 shows how a star and a plaquette are defined.

The operators A_s and B_p commute with each other (see Exercise 13.1):

$$[A_s, A_{s'}] = [B_p, B_{p'}] = [A_s, B_p] = 0. \tag{13.3}$$

They also commute with the Hamiltonian

$$[H, A_s] = [H, B_p] = 0. \tag{13.4}$$

Moreover, one can also show that

$$A_s^2 = B_p^2 = 1. \tag{13.5}$$

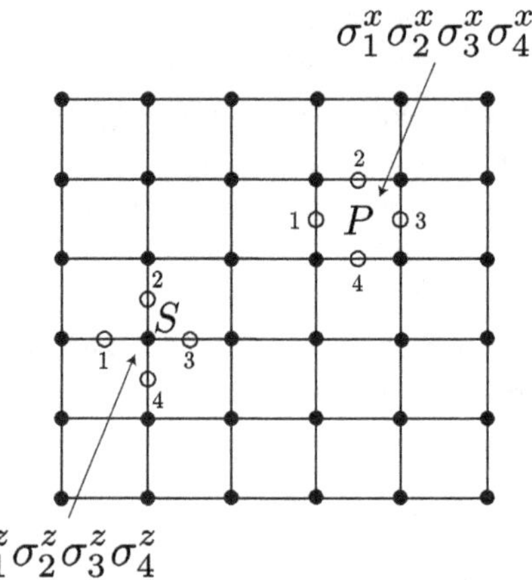

Fig. 13.1. Each link represents a spin-$\frac{1}{2}$ qubit. Spin interaction terms in the Hamiltonian can be divided into two groups of "stars" and plaquettes. A star and a plaquette each has four links $1, 2, 3, 4$. A star operator consists of four σ^z and a plaquette operator of four σ^x. Imagine that each star and plaquette are labeled.

Exercise 13.1. Show the commutation relations Eqs. (13.3) and (13.4). Demonstrate also Eq. (13.5). Hint: Use $(\sigma^z)^2 = (\sigma^x)^2 = 1$, which follows from the Pauli spin matrices satisfying the following anticommutation relations

$$\{\sigma^a, \sigma^b\} = 2\delta_{ab}\, I. \tag{13.6}$$

The following commutation relations between Pauli spin matrices are useful

$$\begin{aligned}
[\sigma^x, \sigma^y] &= 2i\sigma^z, \\
[\sigma^y, \sigma^z] &= 2i\sigma^x, \\
[\sigma^z, \sigma^x] &= 2i\sigma^y, \\
[\sigma^x, \sigma^x] &= 0.
\end{aligned} \tag{13.7}$$

Exercise 13.2. (a) Show that $|\uparrow\uparrow\uparrow\uparrow\rangle$ and $|\downarrow\downarrow\downarrow\downarrow\rangle$ are eigenstates of a star operator $A_s = \prod_{i \in s} \sigma_i^z$ with an eigenvalue of $+1$. (b) Show that $|\rightarrow\rightarrow\rightarrow\rightarrow\rangle$ and $|\leftarrow\leftarrow\leftarrow\leftarrow\rangle$ are eigenstates of a plaquette operator $B_p = \prod_{i \in p} \sigma_i^x$ with an eigenvalue of $+1$. (c) Using a tensor product of matrices write $\sigma_z \otimes \sigma_z \otimes \sigma_z \otimes \sigma_z$ and $\sigma_x \otimes \sigma_x \otimes \sigma_x \otimes \sigma_x$ as 16×16 matrices. Find all the eigenvectors and their eigenvalues (± 1) using Mathematica.

13.2. Ground State of Square Lattice with Open Boundary Condition

The ground state Ansatz is

$$|\Psi\rangle = c\sum_{\alpha} |\psi_\alpha\rangle = \prod_p \frac{1}{2}(1 + B_p)|0\rangle, \tag{13.8}$$

where $|0\rangle$ is the **vacuum** state with all the links in the **spin-up** state (one can also choose them to be in the spin-down state).[c] The ground

---

[c] From the previous exercise we learned that operator $\frac{1}{2}(1 + A_s)$ is projection operator onto the $+1$ eigenspace of A_s. This means one can also write the ground state in terms of $\frac{1}{2}(1 + A_s)$

$$|\Psi\rangle = \prod_s \frac{1}{2}(1 + A_s)|0'\rangle, \tag{13.9}$$

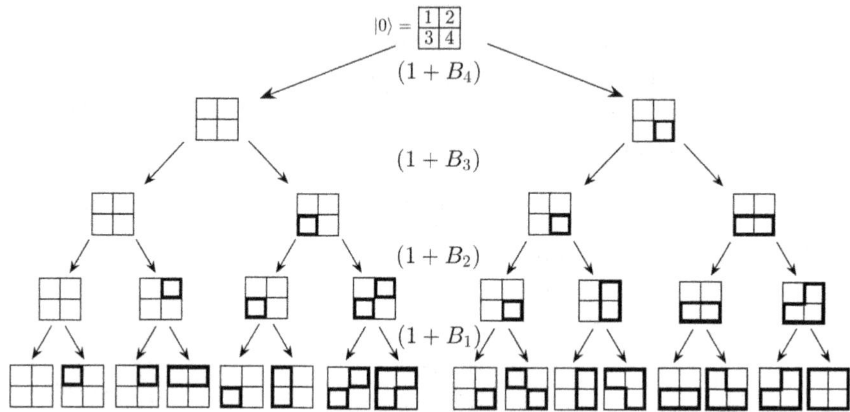

Fig. 13.2.   The ground state $|\Psi_1\rangle = \frac{1}{16}(1+B_1)(1+B_2)(1+B_3)(1+B_4)|0\rangle$ is displayed. Four squares are labeled $1, 2, 3, 4$. Note that it consists of **all possible loop configurations with equal weight**. There are in total $(2N)^2$ number of configurations, where N is the length of the square. **Thick** (thin) lines represent **spin-downs** (-ups).

state Ansatz $|\Psi\rangle$ is given in terms of plaquette operators B_p. Acting on $|0\rangle$ with B_p will flip all the spins in the plaquette p. This can be seen by remembering the following relations:

$$\sigma^x|\uparrow\rangle = |\downarrow\rangle, \quad \sigma^x|\downarrow\rangle = |\uparrow\rangle. \tag{13.12}$$

The result of the operation $\prod_p \frac{1}{2}(1+B_p)$ on the vacuum state $|0\rangle$ is displayed in Figs. 13.2 and 13.3, (see Exercise 13.4). The ground state Ansatz is a superposition of different states with **equal weight** c

$$|\Psi\rangle = c\sum_\alpha |\psi_\alpha\rangle, \tag{13.13}$$

where each $|\psi_\alpha\rangle$ contains closed paths.

Let us show that the Ansatz $|\Psi\rangle$ is an eigenstate of H. All the eigenstates have eigenvalues ± 1 when acted on by A_s or B_p, see Eqs. (13.4) and

---

where

$$|0'\rangle = |\rightarrow\rangle \otimes \cdots \otimes |\rightarrow\rangle \tag{13.10}$$

with

$$|\rightarrow\rangle = \frac{|\uparrow\rangle + |\downarrow\rangle}{\sqrt{2}}. \tag{13.11}$$

$|0'\rangle$ is the product state in which each link is in the state $|\rightarrow\rangle$. Note that A_s will flip all the spins belonging to a star s, for example, $|\rightarrow\rangle$ flips into $|\leftarrow\rangle$.

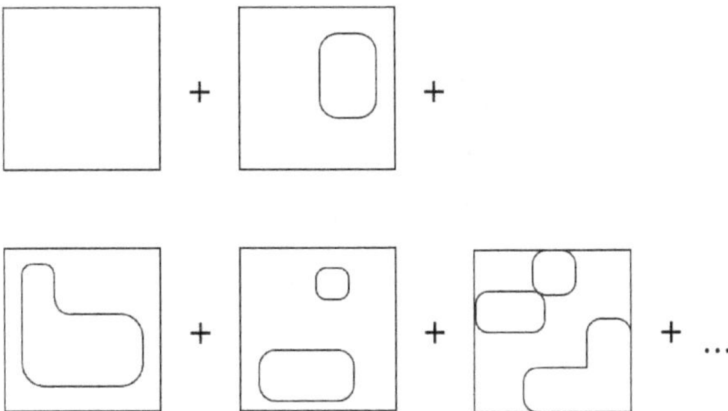

Fig. 13.3. Bird's-eye view of the ground state is shown as a sum of loop states. Only states represented by closed loops are included in the ground state.

(13.5). However, those states with the eigenvalue -1 of B_p are projected out and only states with the eigenvalue $+1$ are kept in the Ansatz. Since A_s and B_p commute, we have

$$A_s|\Psi\rangle = B_p|\Psi\rangle = +|\Psi\rangle \tag{13.14}$$

for all s and p. From this, we infer that $|\Psi\rangle$ is also an eigenstate of H, which consists of a sum of A_ss and B_ps. This can be also easily verified graphically by inspecting Fig. 13.4. We have

$$A_s|\psi_\alpha\rangle = |\psi_\alpha\rangle, \tag{13.15}$$

as shown in the second row of Fig. 13.4. This follows from (a) $\sigma_z|\uparrow\rangle = |\uparrow\rangle$ and $\sigma_z|\downarrow\rangle = -|\downarrow\rangle$ and (b) the presence of an even number of spin-downs in any loop. Thus $A_s|\Psi\rangle = |\Psi\rangle$. However, B_p maps $|\psi_\alpha\rangle$ to a different state $|\psi_\gamma\rangle$

$$B_p|\psi_\alpha\rangle = |\psi_\gamma\rangle \quad \text{with } \alpha \neq \gamma. \tag{13.16}$$

(See Fig.13.4.) Nonetheless, we have $B_p|\Psi\rangle = |\Psi\rangle$ because (a) the linear coefficients in $|\Psi\rangle$ Eq. (13.13) are all equal and (b) B_p map the states $\{\psi_\alpha\}$ into each other and is **bijective**.[d]

We have just shown that the Ansatz is an eigenstate of H. Next we argue that the Ansatz has the minimum energy. The operators A_s's and B_p's are mutually independent and among themselves, see Exercise 13.1.

---

[d]A bijective function has the following property: each element of one set is paired with exactly one element of the other set, and each element of the other set is paired with exactly one element of the first set.

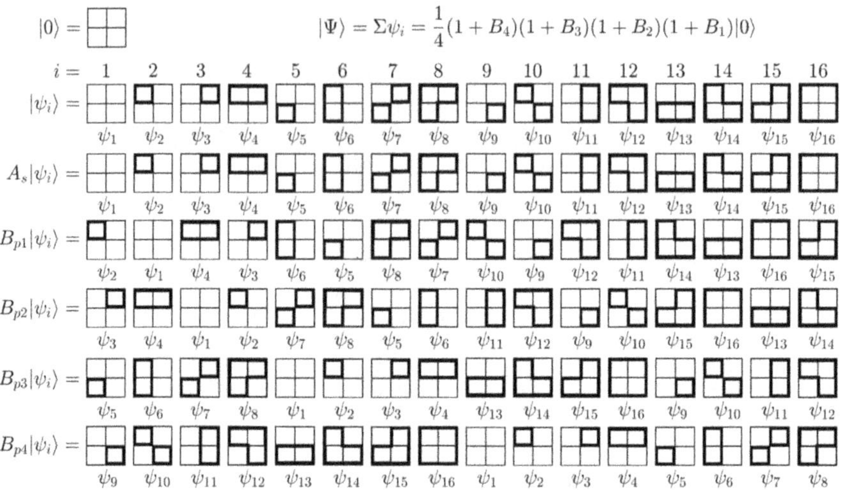

Fig. 13.4. The first row displays the states $\{|\psi_i\rangle\}$ appearing in the ground state linear combination. The second row illustrates $A_s|\psi_i\rangle = |\psi_i\rangle$ and the rest of the rows illustrate $B_p|\psi_i\rangle = |\psi_k\rangle$.

The Hamiltonian can be minimized by maximizing each term of A_s and B_p separately (see the Hamiltonian Eq. (13.1)). A state different from the Ansatz state has two possibilities. First, a loop is cut and the two resulting end vertices have an odd number of $\downarrow$. At each of these vertices the eigenvalue of A_s is -1. This will increase the energy of the state in comparison to that of the Ansatz state. Second, suppose that a different Ansatz state is not a linear combination of states with equal weight. For example, when the plaquette state is in an antisymmetric combination[e]

$$| \uparrow\uparrow\downarrow\uparrow \rangle - | \downarrow\downarrow\uparrow\downarrow \rangle, \tag{13.18}$$

the value of B_p is -1. This will again increase the energy of the state in comparison to that of the Ansatz state. All these results suggest that the Ansatz has the minimum energy and is indeed the ground state.

---

[e]In contrast, when the plaquette state is in a symmetric combination, for example,

$$| \uparrow\uparrow\downarrow\uparrow \rangle + | \downarrow\downarrow\uparrow\downarrow \rangle. \tag{13.17}$$

the value of B_p is $+1$.

Exercise 13.3. Show that the operator $\frac{1}{2}(1 + A_s) = \frac{1}{2}(1 + \sigma_1^z \sigma_2^z \sigma_3^z \sigma_4^z)$ is the projection operator onto the eigenspace of A_s with the eigenvalue $+1$ and $\frac{1}{2}(1 + B_p)$ the projection operator onto eigenspace of B_p with the same eigenvalue $+1$. Hint: Since $A_s^2 = B_p^2 = 1$ (from Eq. (13.5)) all states have eigenvalues ± 1 of A_s and B_p.

Exercise 13.4. Write down the ground state $|\Psi\rangle$ given by Eq. (13.13) for a system consisting of four small squares. Answer is given in Fig. 13.2.

13.3. Ground States in Torus Geometry

When periodic boundary conditions are used in both directions of a square, a torus is formed. A torus has $2N^2$ independent links (N is the length of the square). In torus geometry, the constraints given in Eqs. (13.15) and (13.16) are not all independent since they are related by two additional conditions

$$\prod_s A_s |\Psi\rangle = \prod_p B_p |\Psi\rangle = |\Psi\rangle. \tag{13.19}$$

(See Exercise 13.5.) There are thus $2N^2 - 2$ independent constraints, two fewer than the number of spins. The ground state **degeneracy** is thus $D = 2^2 = 4$.

Exercise 13.5. Show that the ground state $|\Psi\rangle$ fulfills Eq. (13.19) under periodic boundary conditions. Hint: Multiplying all A_s operators together on the torus involves squaring of Pauli matrices, and likewise for multiplying all B_p operators. Note that $\sigma_k^2 = 1$ for $k = x, y, z$.

To construct degenerate ground states that satisfy periodic boundary conditions, we define the states $|0\rangle$, $|Y\rangle$, $|X\rangle$, and $|XY\rangle$, depicted in Figs. 13.5 (a)–(d). The first, second, third, and fourth degenerate ground

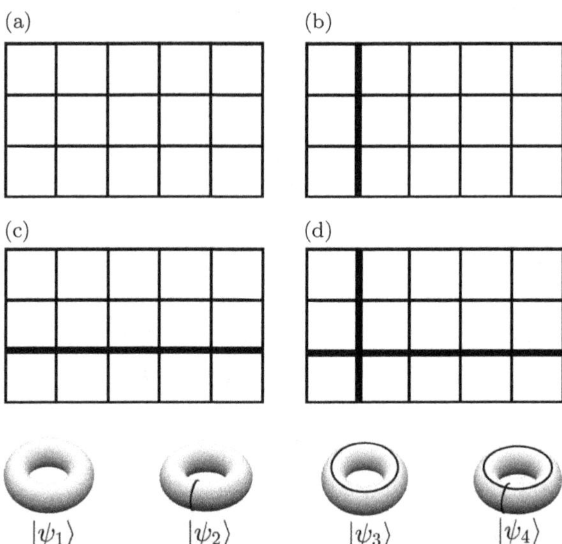

Fig. 13.5. **Upper**: (a) $|0\rangle$ is the vacuum state. (b) The state $|Y\rangle = L_Y|0\rangle$ is obtained from $|0\rangle$ by flipping spins along a thick line parallel to the y direction. (c) The state $|X\rangle = L_X|0\rangle$ is obtained from $|0\rangle$ by flipping spins along a thick line parallel to the x direction. (d) The state $|X,Y\rangle = L_Y L_X|0\rangle$ is obtained from $|0\rangle$ by flipping spins along two thick lines parallel to the x and y directions, respectively. **Lower**: Schematic representation of the four ground states of Fig. 13.5. There are four different types of ground states of a torus (periodic boundary conditions are imposed along x and y directions). States making up $|\psi_1\rangle$ consist of closed loops (see Eq. (13.22) but they do not loop around the radii of the torus. States in $|\psi_2\rangle$ loop odd number of times around the smaller radius of the torus (they loop even number of times around the larger radius of the torus). States in $|\psi_3\rangle$ loop odd number of times around the larger radius of the torus (they loop even number of times around the smaller radius of the torus). States in $|\psi_4\rangle$ loop odd number of times around both the larger and smaller radii of the torus.

states may be constructed as follows:

$$|\Psi_1\rangle = \prod_p \tfrac{1}{2}(1 + B_p)|0\rangle,$$
$$|\Psi_2\rangle = \prod_p \tfrac{1}{2}(1 + B_p)|Y\rangle,$$
$$|\Psi_3\rangle = \prod_p \tfrac{1}{2}(1 + B_p)|X\rangle,$$
$$|\Psi_4\rangle = \prod_p \tfrac{1}{2}(1 + B_p)|XY\rangle.$$
$$(13.20)$$

It is easy to verify that the degenerate ground states are related to each other

$$|\Psi_2\rangle = L_Y|\Psi_1\rangle,$$
$$|\Psi_3\rangle = L_X|\Psi_1\rangle,$$
$$|\Psi_4\rangle = L_Y L_X|\Psi_1\rangle.$$
$$(13.21)$$

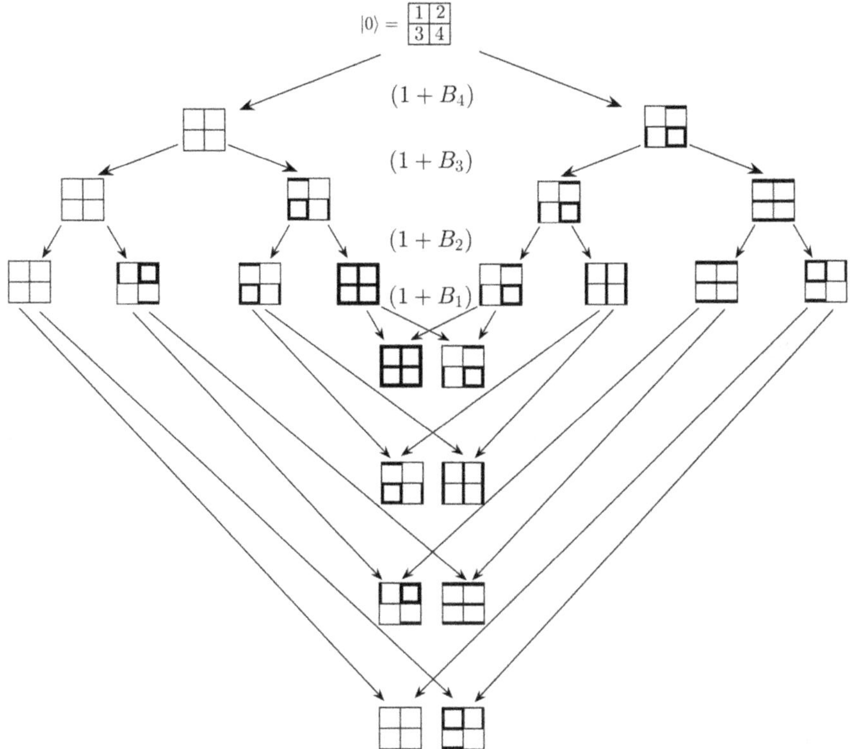

Fig. 13.6. The ground state $|\Psi_1\rangle = \frac{1}{16}(1+B_1)(1+B_2)(1+B_3)(1+B_4)|0\rangle$ is displayed. Periodic boundary conditions are used. Four squares are labeled $1, 2, 3, 4$. Note that each state appears twice in the bottom line.

The meaning of the operators L_X, L_Y, and $L_Y L_X$ are explained in Fig. 13.5. Examples of $|\Psi_1\rangle$, $|\Psi_2\rangle$, and $|\Psi_4\rangle$ are shown in Figs. 13.6, 13.7, and 13.8, respectively. These ground states are represented as a sum of loop states.

The proposed states $|\Psi_i\rangle$ are eigenstates of H. Each state can be written as a superposition of the basis states with **equal weight**

$$|\Psi_i\rangle = c \sum_\alpha |\psi_\alpha\rangle. \tag{13.22}$$

This result follows from the following properties

$$A_s|\psi_\alpha\rangle = |\psi_\beta\rangle, \quad B_p|\psi_\alpha\rangle = |\psi_\gamma\rangle \text{ with } \alpha \neq \beta. \tag{13.23}$$

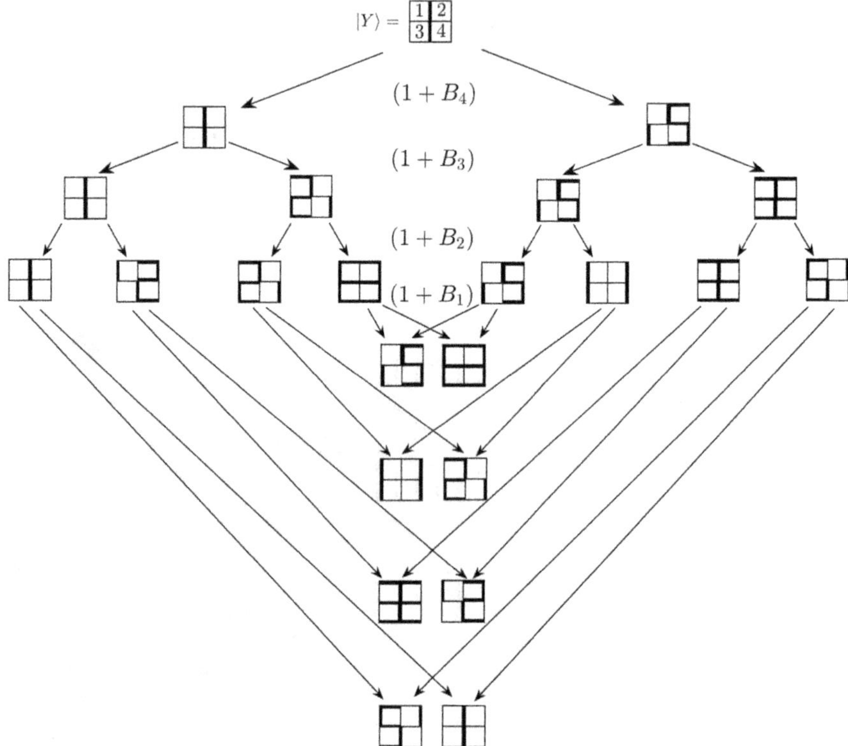

Fig. 13.7.   The ground state $|\Psi_2\rangle = \frac{1}{16}(1+B_1)(1+B_2)(1+B_3)(1+B_4)|Y\rangle$ is displayed.

These mappings A_s and B_p are illustrated in Fig. 13.9 for the state $|\Psi_4\rangle$. Note that, as we mentioned previously, both A_s and B_p are bijective operators.

These states all have the same minimum energy

$$H\left(\sum_\alpha |\psi_\alpha\rangle\right) = E_G \sum_\beta |\psi_\beta\rangle, \tag{13.24}$$

where E_G is the ground state energy

$$E_G = -2N^2. \tag{13.25}$$

(Remember that N is the length of the square.) Let us check this result for the ground states for $N = 2$. There are four vertices and four plaquettes (N^2 vertices and N^2 plaquettes). Each of them contribute -1 to the ground state energy. So the total energy is $E_G = -2N^2 = -8$.

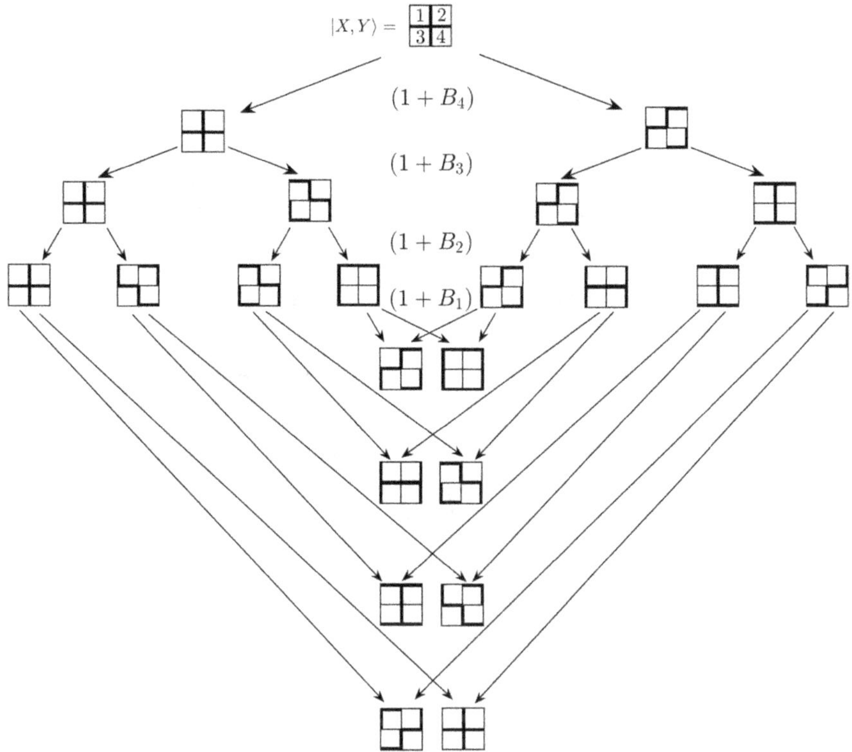

Fig. 13.8. The ground state $|\Psi_4\rangle = \frac{1}{16}(1+B_1)(1+B_2)(1+B_3)(1+B_4)|X,Y\rangle$ is displayed. Periodic boundary conditions are used. Note that each state appears twice in the bottom line.

13.4. Excitations in Torus Geometry

There are three types of excitations. The first type is constructed as follows. When an open string C of σ_i^x (Fig. 13.10(a)) is applied to the ground state $|\Psi\rangle$,

$$X(C)|\Psi\rangle = \prod_{i\in C} \sigma_i^x |\Psi\rangle, \tag{13.26}$$

some spins flip and excitations are created at the endpoints of C, see Fig. 13.10(b). (Here i labels links on C and $|\Psi\rangle$ is one of the four degenerate ground states.) This can be understood as follows. **When the vertex has an even number of $\uparrow$ the value of A_s is $+1$; when the vertex has an odd number of $\downarrow$ it is -1.** The operator induces an energy change $-1 \rightarrow 1$ at each endpoint (a vertex always has an even number

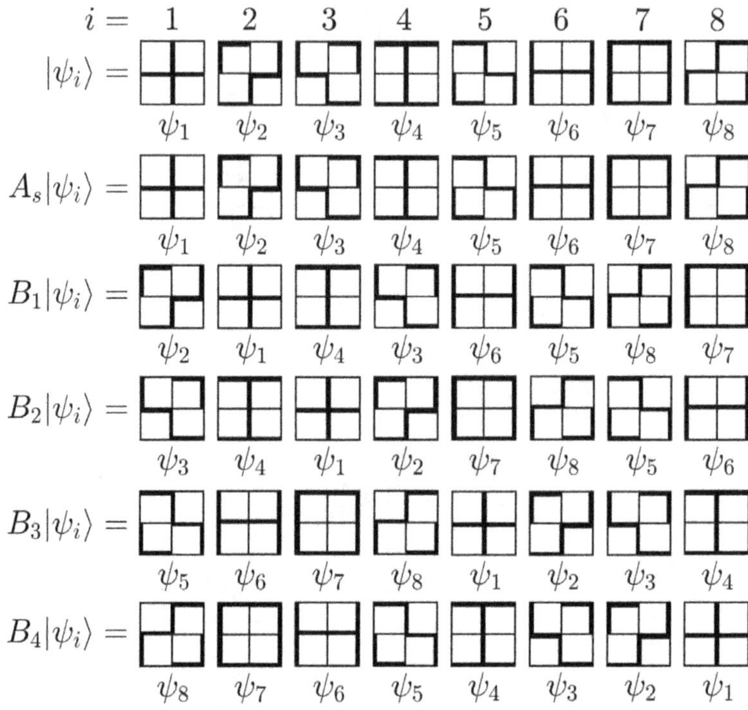

Fig. 13.9. The first row displays the states $\{|\psi_i\rangle\}$ appearing in the ground state linear combination of $|\Psi_4\rangle$. The second row illustrates $A_s|\psi_i\rangle = |\psi_i\rangle$ and the rest of the rows illustrate $B_p|\psi_i\rangle = |\psi_k\rangle$.

of $|\uparrow\rangle$ in the ground state). An excitation consists of two endpoints and, consequently, its energy is $2 \times 2 = 4$. Such an endpoint excitation acts like a charge in an Aharonov–Bohm-like effect (see below) and called a **vertex charge** (however, it does not have a real electric charge). We call such a particle an X-type anyon. Consider a long string C with an excitation at each endpoints of the string, see Fig. 13.10(a). No matter how long the string is, it only costs the end excitations energy to create such a string. These excitations can feel the presence of each other no matter how far away they are from each other, i.e., they are **non-local**. They can only be created using a non-local operators.

The second type excitation is constructed as follows using plaquette operators. An open string C' of σ_i^z (see Fig. 13.10(a)) is applied to the ground state $|\Psi\rangle$. It also creates excitations at its end links (an end link is

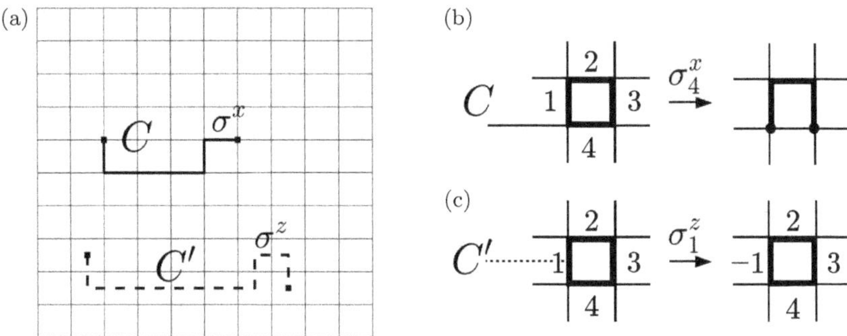

Fig. 13.10. (a) A string C of links is displayed. σ_i^x is operated on $|\Psi\rangle$ at link i. Also shown is a string C' of σ_i^z. String C' **crosses** different links perpendicularly. (b) An end link of C, labeled by 4, is shown. When σ_4^x operates on $|\Psi\rangle$ at link 4 two charges are created (filled circles). The excitation energy is thus 4: two A_s change sign and each sign gives rise an energy change of 2. (c) An end plaquette of C' is shown. Operation of $\sigma_1^z|\downarrow\rangle = -|\downarrow\rangle$ (the link is symbolically denoted as -1 in the figure). This gives an energy change of 2. Since there is another end plaquette at the other end of C' the total energy change is 4.

shown in Fig. 13.10(c))

$$Z(C')|\Psi\rangle = \prod_{i\in C'} \sigma_i^z |\Psi\rangle, \tag{13.27}$$

where i labels a link on C'. Since there are two end plaquettes the **excitation** energy is again 4. The ends of open loop of plaquettes create "magnetic vortices". We call such a particle a Z-type anyon.

It is clear that moving one X-quasiparticle (the end of a σ_x-string) around a second X-quasiparticle, i.e., does not change the state. The same is also true when a Z-quasiparticle goes around a Z-quasiparticle since all the operators involved commute. Also, moving a pair of X-quasiparticles around a pair of Z-quasiparticle has no effect since all spin operators occur either zero or 2 times. Charges and fluxes are (hardcore-) bosons. But what about moving a single Z-quasiparticle around a single X-quasiparticle (see Fig. 13.11)? The resulting state has the opposite sign

$$|\Psi_{\text{fin}}\rangle = L_z(C'')|\Psi_{\text{ini}}\rangle = L_z(C'')\left[X(C)Z(C')\right]|\Psi\rangle$$
$$= -X(C)L_z(C'')Z(C')|\Psi\rangle = -|\Psi_{\text{ini}}\rangle. \tag{13.28}$$

Here $L_z(C'')$ consists of a series of σ_zs forming a loop C'' shown in Fig. 13.11. (To understand the sign change, do Exercise 13.6.) Hence,

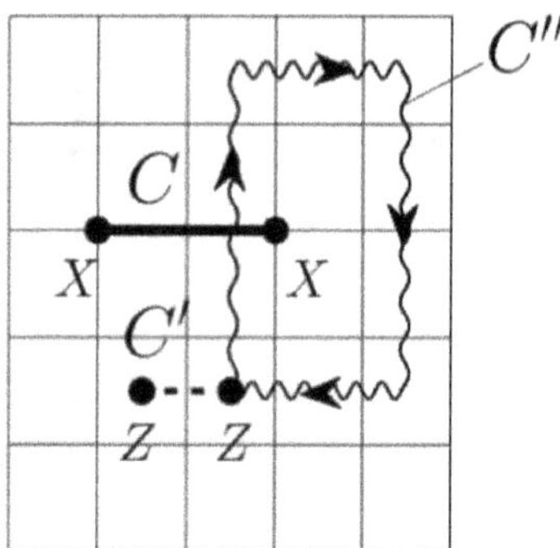

Fig. 13.11. A Z-quasiparticle encircles an X-quasiparticle. The encircling is achieved by applying σ_i^z repeatedly from the position of Z-quasiparticle and along the wavy path shown in the figure.

when a Z-quasiparticle encircles an X-quasiparticle there is a sign change $-1 = e^{i\pi}$. When an X-quasiparticle and a Z-quasiparticle are **interchanged** we get a factor $e^{i\pi/2}$ (remember that an encircling is equivalent to two exchanges). X and Z are thus relative semions, see Fig. 13.11. Charges and fluxes are mutual **semions**.[f] These are Abelian anyons despite the presence of ground state degeneracy. (This is also the case with, for example, the $1/3$ fractional quantum Hall state. But non-Abelian anyons may exist when ground state degeneracy is present. A good example is quasiparticles of the $5/2$ fractional quantum Hall state.)

Exercise 13.6. Show that $L_z(C'')X(C) = -X(C)L_z(C'')$. Hint: An X-string crosses a Z-string at a single link. This means that both σ_j^x and σ_j^z operate on a single link j. Use the anticommutation relation $\{\sigma_j^x, \sigma_j^z\} = 0$.

There is a third type of particle. The combination of both X and Z anyons (see Fig. 13.12) creates a composite quasiparticle ϵ, which can be shown to be a fermion [3]. As shown in Fig. 13.12, if the X particle in this composite object is rotated counterclockwise around the Z particle, the resulting phase change is -1 (see Exercise 13.6). This encircling should be viewed as a particle encircling itself, see Sec. 3.7. The spin of the composite

---

[f] Two exchanges of semions give a phase of -1.

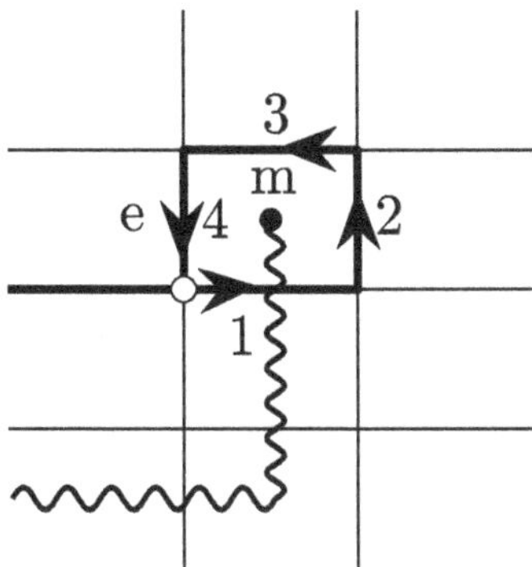

Fig. 13.12. X particle is rotated counterclockwise around Z particle. This operation is facilitated by the operator $\sigma_1^x \sigma_2^x \sigma_3^x \sigma_4^x$, where $1, 2, 3, 4$ label links in the loop.

particle ϵ is thus $1/2$ (this is analogous to the situation described in Fig. 3.15 in connection with the spin of an anyon).

In Sec. 20.3 we will show that the topological entanglement entropy of the toric code is finite. The toric code is thus a topologically ordered insulator.

Bibliography

[1] P. C. W. Davies and J. R. Brown (Eds), *Superstrings: A Theory of Every-thing?* (Cambridge University Press, Cambridge, 1998). •

[2] A. Y. Kitaev, Fault-tolerant quantum computation by anyons, *Ann. Phys.* **303**, 2 (2003).

[3] J. K. Pachos, *Introduction to Topological Quantum Computation* (Cambridge University Press, Cambridge, 2012).

[4] S. M. Girvin and K. Yang, *Modern Condensed Matter Physics* (Cambridge University Press, Cambridge, 2019). •

[5] L. Balents, Spin liquids in frustrated magnets, *Nature* **464**, 199 (2010). •

[6] J. Knolle and R. Moessner, A field guide to spin liquids, *Annu. Rev. Condens. Matter Phys.* **10**, 451 (2019).

PART 3

Topologically Ordered Zigzag Nanoribbons

Chapter 14

Graphene

"Pure analysis puts at our disposal a multitude of procedures whose infallibility it guarantees; it opens to us a thousand different ways on which we can embark in all confidence; we are assured of meeting there no obstacles; but of all these ways, which will lead us most promptly to our goal? Who shall tell us which to choose? We need a faculty which makes us see the end from afar, and intuition is this faculty. It is necessary to the explorer for choosing his route; it is not less so to the one following his trail who wants to know why he chose it."

Henri Poincaré

"You took something that had a meaning; you converted it into a formula, and you got out the answer. In that process you do not need to think any more about what the different stages in the algebra correspond to in the geometry. You lose the insights, and this can be important at different stages. You must not give up the insight altogether! You might want to come back to it later on."

Michael Atiyah

Graphene[a] has numerous remarkable properties [1]. It has great potential for spintronic applications and also provides numerous interesting issues in fundamental physics, such as fractional quantum Hall effect [2].

The band structure of graphene has interesting topological properties. There are two pairs of Dirac cones and each pair touch each other and

---

[a]Andre Geim and Konstantin Novoselov won a Nobel Prize for their discovery of graphene.

form a diabolo. We saw in Sec. 2.4 that a diabolo displays an interesting Berry phase associated with a magnetic monopole sitting at the center of the diabolo. The presence of two independent diabolos leads to fermion doubling and the Nielson–Ninomiya topological theorem.

In this chapter, we do not cover the entire field of graphene; we only present very basic properties of graphene that are relevant for later study of topological properties of zigzag nanoribbons. For other interesting topics of graphene, see excellent reviews given in Refs. [3–8].

14.1. Tight-Binding Hamiltonian of Graphene and Chiral Symmetry

Consider two-dimensional graphene sheet made of carbon atoms. Each atom has four electrons in the orbitals $2s$ and $2p$ (see Fig. 14.1). Three electrons participate in bonding process. In the sp^2 hybridization s, p_x, and p_y wave functions combine to produce three wave functions in the xy-plane, with their maxima pointing in directions making angles of $120°$ [9]. These orbitals in the xy-plane form occupied bonding (low energy) and unoccupied antibonding (high-energy) orbitals with the nearest neighbor carbon atoms, see Fig. 14.2. The remaining orbital $2p_z$ is non-bonding and form the π bands. An electron in the π orbital can hop between carbon atoms and form a linear energy dispersion near zero energy. This linear energy dispersion is of main interest here. It gives rise to an effective low-energy Hamiltonian given by the Dirac Hamiltonian.

Each unit cell of a honeycomb lattice contains one A and one B carbon atoms, as shown in Fig. 14.3. The tight-binding Hamiltonian of π electrons

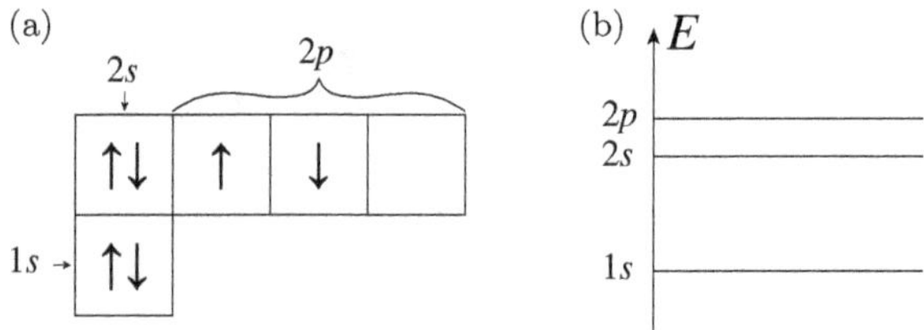

Fig. 14.1. (a) Atomic shell structure of a carbon atom. The $1s$ level is unimportant here. (b) Energy level spacing is shown. Energy levels of $2s$ and $2p$ are close to each other and their orbitals fuse to form newly hybridized orbitals.

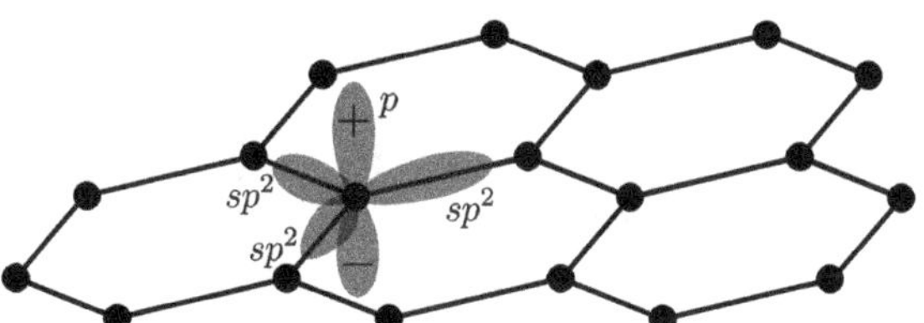

Fig. 14.2. In the sp^2 hybridization s, p_x, and p_y wave functions combine to produce three wave functions in the xy-plane, with their maxima pointing in directions making angles of $120°$.

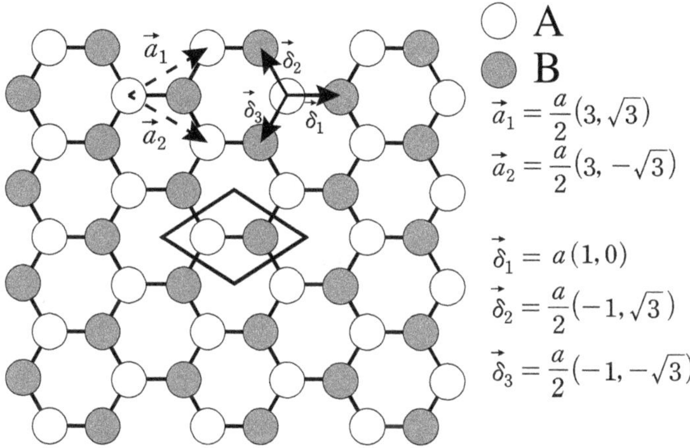

$$\vec{a}_1 = \frac{a}{2}\left(3, \sqrt{3}\right)$$

$$\vec{a}_2 = \frac{a}{2}\left(3, -\sqrt{3}\right)$$

$$\vec{\delta}_1 = a\left(1, 0\right)$$

$$\vec{\delta}_2 = \frac{a}{2}\left(-1, \sqrt{3}\right)$$

$$\vec{\delta}_3 = \frac{a}{2}\left(-1, -\sqrt{3}\right)$$

Fig. 14.3. Honeycomb lattice is shown. It is a triangular Bravais lattice with a two point basis. A unit cell is shown in the middle of the figure. The carbon–carbon distance is $a = 1.42\text{Å}$. The primitive lattice vectors of triangular lattice are $\vec{a}_1$ and $\vec{a}_2$.

of graphene is

$$H = -t \sum_{\vec{R}, \vec{\delta}} a^{\dagger}_{\vec{R}} b_{\vec{R}+\vec{\delta}} + h.c., \tag{14.1}$$

where $\vec{R}$ denotes the position of each unit cell, see Fig. 14.3. Here $a^{\dagger}_{\vec{R}}$ creates an electron on carbon atom A and $b_{\vec{R}}$ destroys an electron on carbon atom B.

The nearest-neighbor tight-binding Hamiltonian may be written as

$$H = -\begin{pmatrix} \mathbf{O} & \mathbf{T} \\ \mathbf{T}^{\dagger} & \mathbf{O} \end{pmatrix}, \tag{14.2}$$

where

$$\mathbf{T} = \begin{pmatrix} t' & 0 & 0 & 0 & \cdots \\ t & t' & 0 & 0 & \cdots \\ 0 & t & t' & 0 & \cdots \\ 0 & 0 & t & t' & \cdots \\ \vdots & \vdots & \vdots & \vdots & \ddots \end{pmatrix} \tag{14.3}$$

and $\mathbf{O}$ is a zero matrix. In writing this matrix, the basis states are ordered as $(a_{\vec{R}_1}, a_{\vec{R}_2}, \ldots, b_{\vec{R}_1}, b_{\vec{R}_2}, \ldots)$. As in the case of polyacetylene, the Hamiltonian anticommutes with the chiral operator $\mathbf{C}$ (see Eq. (14.3))

$$\{\mathbf{C}, H\} = 0. \tag{14.4}$$

> **Exercise 14.1.** Derive the nearest-neighbor tight-binding Hamiltonian Eq. (14.2) and show the anticommutation relation Eq. (14.4).

14.2. Band Structure of Graphene

Using the Fourier transform

$$\begin{bmatrix} a_{\vec{R}}^\dagger \\ b_{\vec{R}}^\dagger \end{bmatrix} = \frac{1}{\sqrt{N_{\mathrm{unit}}}} \sum_{\vec{k}} e^{-i\vec{k}\cdot\vec{R}} \begin{bmatrix} a_{\vec{k}}^\dagger \\ b_{\vec{k}}^\dagger \end{bmatrix}, \tag{14.5}$$

we can rewrite the tight-binding Hamiltonian in k-space. Here N_{unit} is the number of unit cells. Using $\frac{1}{N_{\mathrm{unit}}} \sum_{\vec{R}} e^{i\vec{R}\cdot(\vec{p}-\vec{k})} = \delta_{\vec{p},\vec{k}}$ we find

$$H = -t \sum_{\vec{k},\vec{\delta}} a_{\vec{k}}^\dagger b_{\vec{k}} e^{i\vec{k}\cdot\vec{\delta}} + h.c. \tag{14.6}$$

The k-space Hamiltonian can be written in a more compact form

$$H = \sum_{\vec{k}} \left(a_{\vec{k}}^\dagger, b_{\vec{k}}^\dagger \right) h(\vec{k}) \begin{bmatrix} a_{\vec{k}} \\ b_{\vec{k}} \end{bmatrix}. \tag{14.7}$$

Note that $h(\vec{k})$ is a 2×2 off-diagonal matrix. Its off-diagonal elements are

$$h_{12}(\vec{k}) = -tf(\vec{k}) = -t \sum_i e^{i\vec{k}\cdot\vec{\delta}_i} = -t \left(e^{-ik_x a} + 2e^{ik_x a/2} \cos\left(\frac{\sqrt{3}k_y a}{2}\right) \right) \tag{14.8}$$

and $h_{21}(\vec{k}) = -tf(\vec{k})^*$, where a is the carbon–carbon distance.

The Hamiltonian matrix $h(\vec{k})$ can be written as

$$h(k) = -t|f(k)| \begin{bmatrix} 0 & e^{-i\phi(k)} \\ e^{i\phi(k)} & 0, \end{bmatrix}, \tag{14.9}$$

where the angle $\phi(k)$ is defined by $f(\vec{k}) = |f(\vec{k})|e^{-i\phi(k)}$. The eigenenergy is

$$\epsilon(\vec{k}) = \pm t|f(\vec{k})|. \tag{14.10}$$

Exercise 14.2. Equation (14.8) does not satisfy the Bloch theorem $f(\vec{k}) = f(\vec{k} + \vec{b})$, where $\vec{b} = m\vec{b}_1 + n\vec{b}_2$ are reciprocal vectors and m and n are integers. Equation (14.8) must be multiplied with a phase factor $ie^{-\vec{k}\cdot\vec{\delta}_1}$ (see Ref. [6]). Then the vectors $\vec{\delta}_i$ enter the new $f(\vec{k})$ in combinations of $\vec{\delta}_2 - \vec{\delta}_1$ and $\vec{\delta}_3 - \vec{\delta}_1$, which are lattice vectors. Show that $\epsilon(\vec{k}) = \pm|f((\vec{k})|$. Note that the phase factor does not affect the energy dispersion since it depends on the absolute value of $f(\vec{k})$.

Exercise 14.3. Show that the 2×2 Hamiltonian matrix, (14.9) can be written in the standard form

$$h(\vec{k}) = h_x(\vec{k})\sigma_x + h_y(\vec{k})\sigma_y + h_z(\vec{k})\sigma_z, \tag{14.11}$$

where the terms originating from the nearest-neighbor hopping are

$$\begin{aligned} h_x(\vec{k}) &= t[1 + \cos(\vec{k}\cdot\vec{a}_1) + \cos(\vec{k}\cdot\vec{a}_2)], \\ h_y(\vec{k}) &= t[\sin(\vec{k}\cdot\vec{a}_1) + \sin(\vec{k}\cdot\vec{a}_2)], \\ h_z(\vec{k}) &= m. \end{aligned} \tag{14.12}$$

We have added a mass term, $m\sigma_z$, to the Hamiltonian equation (14.7). Remember the relevant vectors are $\vec{a_1} = \frac{a_0}{2}(\sqrt{3}, 1)$, $\vec{a_2} = \frac{a_0}{2}(\sqrt{3}, -1)$, and $\vec{a_3} = \vec{a}_2 - \vec{a}_1$, where $a_0 = \sqrt{3}a$.

Figure 14.4 displays the first Brillouin zone with six valley points. The energy dispersion is zero at six valley points (the reader should verify this). However, there are only two independent valleys $\vec{K}$ and $\vec{K}'$ (this is an example of fermion doubling), see Fig. 14.5. Given a pair of $\vec{K}$ and $\vec{K}'$ other pairs $\vec{K}$ and $\vec{K}'$ can be obtained by adding the reciprocal lattice

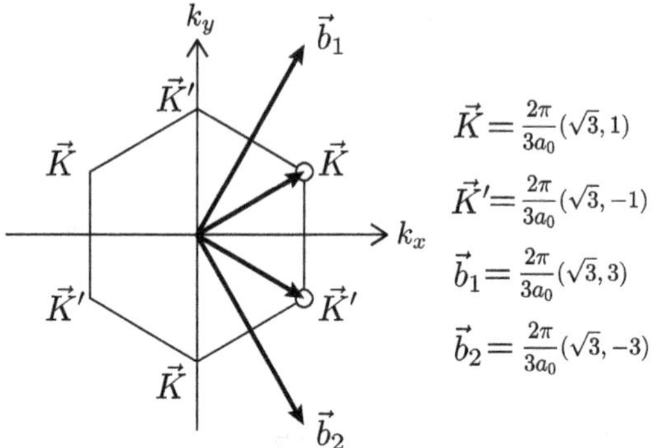

Fig. 14.4. First Brillouin zone with the reciprocal lattice vectors $\vec{b}_1$ and $\vec{b}_2$. Six Dirac cones are at $\vec{K}$ and $\vec{K'}$ points. Here $a_0 = \sqrt{3}a$ is the unit cell length. Naming of these points is arbitrary and their positions may be switched.

vectors (the reader should also verify this). What is noteworthy about the graphene band structure is the presence of a linear dispersion near $\vec{K}$ and $\vec{K'}$ points. Electrons with the linear energy dispersion represent massless Dirac electrons and are expected to display Klein tunneling with perfect transmission through a potential barrier [7] (this effect is well known in quantum electrodynamics). The DOS is semimetallic, as shown in Fig. 14.5.

14.3. Peierls Substitution and Effective Dirac Hamiltonians

Near the valley points $\vec{K}$ and $\vec{K'}$ an effective Dirac Hamiltonian can be derived from $h(\vec{k})$ given in Eq. (14.9). The Taylor expansion of the Hamiltonian matrix $h(\vec{k})$ in $\vec{k}$ gives the following 2×2 Hamiltonian matrix

$$h_{\vec{K}}(\vec{q}) = v_F \vec{\sigma} \cdot \vec{q}, \tag{14.13}$$

where $\vec{q}$ is measured from $\vec{K}$, i.e., $\vec{k} = \vec{q} + \vec{K}$. Here $\vec{\sigma} = (\sigma_x, \sigma_y)$, where σ_x and σ_y are the Pauli spin matrices. The Hamiltonian describes states near the K valley (see Exercise 4.6). For the K' valley

$$h_{\vec{K'}}(\vec{q}) = v_F \vec{\sigma'} \cdot \vec{q} = v_F \vec{\sigma} \cdot \vec{q'}, \tag{14.14}$$

where $\vec{q}$ is measured from $\vec{K'}$, $\vec{\sigma'} = (-\sigma_x, \sigma_y)$, and $\vec{q'} = (-q_x, q_y)$. The energy surface of $h_{\vec{K}}(\vec{q})$ or $h_{\vec{K'}}(\vec{q})$ has the shape of a double cone (diabolo), as shown in Fig. 2.6. The energy dispersion is linear in $\vec{q}$ with $v_F \approx 10^6 m/s$.

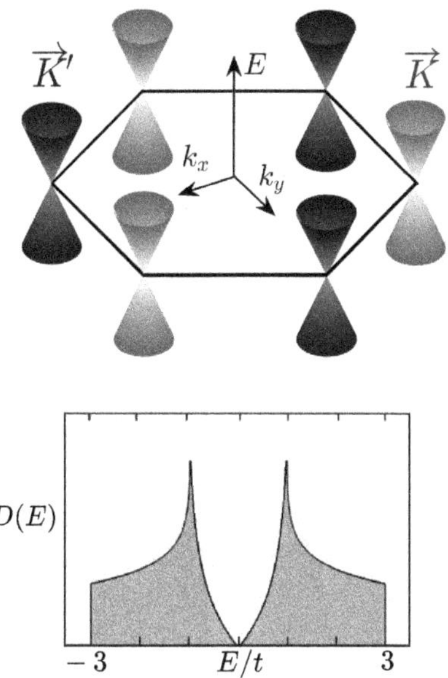

Fig. 14.5. **Upper**: Six Dirac cones. Among them only one $\vec{K}$ and one $\vec{K}'$ are independent. **Lower**: DOS of graphene. The Fermi energy is at $E = 0$. DOS is **linear near** $E = 0$.

An electron with such a dispersion is called a **massless Dirac electron**.

> **Exercise 14.4.** Compare the relativistic Dirac and Weyl Hamiltonians. Also compare the equations that electrons K and K' valleys satisfy. Answer: The 4×4 Dirac Hamiltonian is
>
> $$\begin{pmatrix} -c\vec{\sigma} \cdot \vec{p} & mc^2 \\ mc^2 & c\vec{\sigma} \cdot \vec{p} \end{pmatrix}, \tag{14.15}$$
>
> where $\vec{p}$ is the momentum and c is the velocity of light and m is the electron mass. The Weyl Hamiltonian is obtained from the Dirac equation by setting $m = 0$
>
> $$\begin{pmatrix} -c\vec{\sigma} \cdot \vec{p} & 0 \\ 0 & c\vec{\sigma} \cdot \vec{p} \end{pmatrix}. \tag{14.16}$$

Now suppose a smooth varying external potential is present $V(\vec{r})$. One can include the effect of such a potential in the effective-mass approximation. In this approach, the following wave function Ansatz is used. For the

$\vec{K}$ valley, the eigenstate has the following form

$$\Psi(\vec{r}) = e^{i\vec{K}\cdot\vec{r}} \begin{bmatrix} F_A(\vec{r}) \\ F_B(\vec{r}) \end{bmatrix} \tag{14.17}$$

and for the $\vec{K}'$ valley

$$\Psi(\vec{r}) = e^{i\vec{K}'\cdot\vec{r}} \begin{bmatrix} F'_A(\vec{r}) \\ F'_B(\vec{r}) \end{bmatrix}. \tag{14.18}$$

Unprimed wave functions are the $\vec{K}$ valley wave functions while primed ones are those of the $\vec{K}'$ valley. The envelope wave functions $F_{A,B}$ and $F'_{A,B}$ (see Appendix A) describe the slowly varying envelope part of the wave function in the presence of $V(\vec{r})$. The first and second components of the wave functions give the probability amplitudes of finding an electron on A and B atoms, respectively. (One-dimensional Dirac equation was studied in Exercise 4.4.)

Let us explain how the effective Dirac Hamiltonian is obtained. The goal of the effective mass approximation is to find **partial differential equations for the envelope wave functions**. The basic idea of this approach is to separate slow and fast degrees of freedom. For justification of this envelope wave function procedure in graphene, see [10] and Appendix A. Partial differential equations can be derived from Eqs. (14.13) and (14.14) by replacing wave vectors with their corresponding quantum operators (the Peierls substitution)

$$q_x \to \frac{1}{i}\frac{\partial}{\partial x}, \quad q_y \to \frac{1}{i}\frac{\partial}{\partial y}, \tag{14.19}$$

which can be justified within effective mass approximation (see Exercise 14.5 and Appendix A). One of the advantages of this approach is that perturbations and boundary conditions can be added to these equations. In the presence of a slowly varying external potential $V(\vec{r})$, the resulting effective two-dimensional Dirac Hamiltonians are

$$H_{\vec{K}} = \hbar v_F \left[\sigma_x \frac{1}{i}\frac{\partial}{\partial x} + \sigma_y \frac{1}{i}\frac{\partial}{\partial y} \right] + V(\vec{r})I,$$
$$H_{\vec{K}'} = \hbar v_F \left[-\sigma_x \frac{1}{i}\frac{\partial}{\partial x} + \sigma_y \frac{1}{i}\frac{\partial}{\partial y} \right] + V(\vec{r})I. \tag{14.20}$$

Exercise 14.5. Derive the Dirac Hamiltonian equation (14.20) from the tight-binding description of graphene without using the replacement $\vec{k} \to \frac{1}{i}\nabla$. Consult Ref. [10], which justifies the use of this replacement. See also Appendix A.

The conduction band envelope wave functions of $H_{\vec{K}}$ are

$$\begin{bmatrix} F_A(\vec{r}) \\ F_B(\vec{r}) \end{bmatrix} = \frac{e^{i\vec{k}\cdot\vec{r}}}{\sqrt{2A}} \begin{bmatrix} 1 \\ e^{i\theta_{\vec{k}}} \end{bmatrix} \tag{14.21}$$

and the valence band envelope wave functions are

$$\begin{bmatrix} F'_A(\vec{r}) \\ F'_B(\vec{r}) \end{bmatrix} = \frac{e^{i\vec{k}\cdot\vec{r}}}{\sqrt{2A}} \begin{bmatrix} 1 \\ -e^{i\theta_{\vec{k}}} \end{bmatrix}, \tag{14.22}$$

where $\theta_{\vec{k}}$ is the **polar angle** of $\vec{k}$. These are simplified wave functions that are valid near the $\vec{K}$ and $\vec{K}'$ valley points. They can be used to study various properties of graphene.

14.4. Integer Quantum Hall Effect in Graphene

Let us find the eigenstates and eigenenergies of a Dirac electron in a magnetic field by including the vector potential in the electron momentum. The Dirac Hamiltonian of the $\vec{K}$ valley is

$$H_K = v_F \vec{\sigma} \cdot \vec{\Pi} = E_c \begin{bmatrix} 0 & a^\dagger \\ a & 0 \end{bmatrix}, \tag{14.23}$$

where $E_c = \sqrt{2}\hbar v_F/\ell$ and $a^\dagger$ and a are the ladder operators, see Eq. (10.11). The square of the Hamiltonian is diagonal

$$H_K^2 = E_c^2 \begin{bmatrix} a^\dagger a & 0 \\ 0 & a^\dagger a + 1 \end{bmatrix}. \tag{14.24}$$

The eigenenergy of H_K is

$$E_p = \text{sgn}(p) E_c \sqrt{|p|}, \tag{14.25}$$

where $a^\dagger|\psi_p\rangle = \sqrt{p+1}|\psi_{p+1}\rangle$ and $p = 0, \pm1, \pm2, \ldots$, see Fig. 14.6. (Note that $H_K^2|\psi_p\rangle = E_c^2 p|\psi_p\rangle$, which implies that $E_p = \pm E_c\sqrt{|p|}$.) The

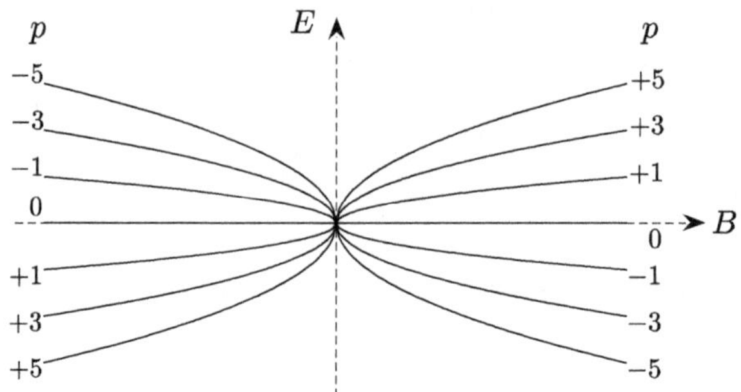

Fig. 14.6. Schematic Landau level energy spectrum of the K valley as a function of an external magnetic field B. Note that the $p = 0$ Landau levels is only half-filled when $E_F = 0$.

eigenstate wave function in the Landau gauge $\vec{A} = (0, -xB, 0)$ is (see Eq. (10.17))

$$\psi_{p,k}(x, y) = \frac{C_p}{\sqrt{L}} e^{iky} \begin{bmatrix} \phi_{|p|,k}(x, y) \\ \phi_{|p|-1,k}(x, y) \end{bmatrix}. \tag{14.26}$$

with

$$\phi_{p,k}(x, y) = M_p e^{\frac{(y - k\ell^2)^2}{2\ell^2}} H_p(x - k\ell^2), \tag{14.27}$$

where $p = 0, \pm 1, \dots$ and H_p are Hermite polynomials. Note that $C_p = \frac{1}{\sqrt{2}}$ for $p \neq 0$ and 1 for $p = 0$. The $p = 0$ the wave function is **chiral** with only one non-zero component

$$\psi_{0,k}(x, y) = \frac{C_0}{\sqrt{L}} e^{ikx} \begin{bmatrix} \phi_{0,k}(x, y) \\ 0 \end{bmatrix}. \tag{14.28}$$

Remarkably its energy is zero, **independent** of B, as shown in Fig. 14.6.

Including spin and valley degrees of freedom the possible values of the quantized Hall conductivity are

$$\sigma_{xy} = 4(p + 1/2)e^2/h. \tag{14.29}$$

The contribution of the $p = 0$ Landau level to the quantized Hall conductance is

$$\sigma_{xy} = 2e^2/h. \tag{14.30}$$

Let us explain this result. In order to understand the integer quantum Hall effect of graphene we need to investigate the structure of edge Landau levels. A simple electrostatic potential barrier cannot confine electrons because of the Klein tunneling. Instead one needs a position-dependent mass potential of the form $M(y)\sigma_z$ acting differently on A- and B-components of the electron wave function [11] (the confinement potential is in the y-direction). The energy spectrum of edge Landau levels may be computed in the presence of a smooth edge potential. As we learned in Chapter 10, each occupied electron Landau level contributes e^2/h to the integer quantum Hall effect. Thus the contribution of the $p = 0$ electron (hole) Landau level of the K (K') valley to the off-diagonal conductivity is $e^2/h(-e^2/h)$. (Note that an electron and a hole have counter-propagating edge modes, see Fig. 14.7.) Spin degrees of freedom gives the final result $2e^2/h$.

Further reading. Fractional quantum Hall effect in graphene was predicted [12] and observed in graphene [2]. For further theoretical and experimental developments, see Ref. [11, 13].

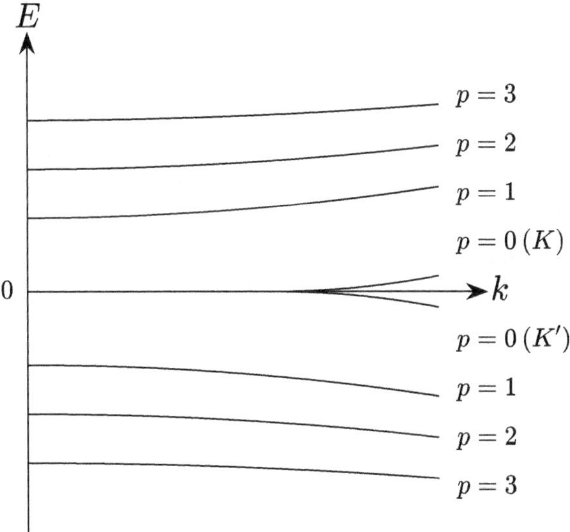

Fig. 14.7. Schematic drawing of edge Landau level energy spectrum of a graphene as a functions of wave vector k. The spectra of K and K' valleys are identical except for the $p = 0$ Landau level. It is instructive to compare this that of an ordinary Hall bar, shown in Fig. 10.13. Note that the Fermi edge velocity is $v_F = \frac{1}{\hbar}\frac{dE_k}{dk}$.

14.5. Berry Phase, Pseudospins, and Winding Number

The conduction and valence band eigenstates of the $\vec{K}$ Hamiltonian equation (14.13) are

$$u_c(\vec{k}) = \frac{1}{\sqrt{2}} \begin{bmatrix} 1 \\ e^{i\theta_{\vec{k}}} \end{bmatrix} \quad \text{and} \quad u_v(\vec{k}) = \frac{1}{\sqrt{2}} \begin{bmatrix} 1 \\ -e^{i\theta_{\vec{k}}} \end{bmatrix}, \qquad (14.31)$$

where $\theta_{\vec{k}}$ is the polar angle of $\vec{k}$. The eigenstates of the $\vec{K}'$ Hamiltonian equation (14.14) are

$$u_c(\vec{k}) = \frac{1}{\sqrt{2}} \begin{bmatrix} 1 \\ -e^{i\theta_{\vec{k}}} \end{bmatrix} \quad \text{and} \quad u_v(\vec{k}) = \frac{1}{\sqrt{2}} \begin{bmatrix} 1 \\ e^{i\theta_{\vec{k}}} \end{bmatrix}. \qquad (14.32)$$

(See Eqs. (5.57) and (5.58) for why the plane wave part is not present.) The states with $\vec{k} = 0$ represent the zero modes of the $\vec{K}$ and $\vec{K}'$ points. The presence of these states leads to the Berry phase. The Berry phase of the conduction band states is

$$\gamma = i \oint_C \langle u_{\vec{k}} | \nabla_{\vec{k}} u_{\vec{k}} \rangle \, d\vec{k} = i \int_0^{2\pi} d\theta_{\vec{k}} u_{c\vec{k}}^*(\vec{r}) \frac{du_{c\vec{k}}(\vec{r})}{d\theta_{\vec{k}}} = -\pi, \qquad (14.33)$$

where C is a closed curve encircling $\vec{K}$ or $\vec{K}'$ point. This result is consistent with the presence of **magnetic monopoles** at the $\vec{K}$ and $\vec{K}'$ points. Note that the Berry phase of a closed curve in k-space is equivalent to **half of the winding number**, see Eq. (2.54) [10].

Let us multiply the wave functions of Eqs. (14.31) and (14.32) with the phase factor $e^{-i\theta_{\vec{k}}/2}$ (note that when $\vec{k}$ rotates 2π the quantity $\theta_{\vec{k}}/2$ does not return to the original value; it is thus **not** a gauge invariant transformation, see Eq. (2.21)). The resulting conduction and valence band $\vec{K}$ valley wave functions are

$$\frac{1}{\sqrt{2}} \begin{bmatrix} e^{-i\theta_{\vec{k}}/2} \\ e^{i\theta_{\vec{k}}/2} \end{bmatrix}, \quad \frac{1}{\sqrt{2}} \begin{bmatrix} e^{-i\theta_{\vec{k}}/2} \\ -e^{i\theta_{\vec{k}}/2} \end{bmatrix}. \qquad (14.34)$$

For the $\vec{K}'$ valley the conduction and valence band wave functions are

$$\frac{1}{\sqrt{2}} \begin{bmatrix} e^{i\theta_{\vec{k}}/2} \\ -e^{-i\theta_{\vec{k}}/2} \end{bmatrix}, \quad \frac{1}{\sqrt{2}} \begin{bmatrix} e^{i\theta_{\vec{k}}/2} \\ e^{-i\theta_{\vec{k}}/2} \end{bmatrix}. \qquad (14.35)$$

The resulting Berry phase is zero! However, these wave functions change sign under a 2π rotation in the k-space, $\theta_{\vec{k}} \to \theta_{\vec{k}} + 2\pi$. This phase change is called **monodromy** phase as opposed to Berry phase. We will use the wave functions given in Eqs. (14.31) and (14.32) since they contain the effect of the Berry phase.

Exercise 14.6. Show Eqs. (14.31) and (14.32).

We further explore topological consequences of the Dirac point where two Dirac cones meet and where the electron energy is zero. It is convenient to introduce the concept of a pseudospin. We have seen that the solutions of the Dirac Hamiltonian can be written as a two-component wave function, i.e., they can be represented as a spinor $\binom{c_1}{c_2}$. The probability amplitudes of finding an electron on A and B atoms are c_1 and c_2, respectively. This feature can be used to define a pseudospin representation: the components of a pseudospin are defined as the expectation values of the Pauli spin matrices $\sigma_x, \sigma_y, \sigma_z$:

$$\langle \vec{\sigma} \rangle_c = \langle u_c(\vec{k})|\vec{\sigma}|u_c(\vec{k})\rangle, \quad \langle \vec{\sigma} \rangle_v = \langle u_v(\vec{k})|\vec{\sigma}|u_v(\vec{k})\rangle. \tag{14.36}$$

We compute the pseudospin expectation values of the conduction and valence bands of $\vec{K}$ and $\vec{K}'$ valleys (see Eq. (2.50)): as shown in Fig. 14.8, for the $\vec{K}$ valley $\langle \vec{\sigma} \rangle_c$ is parallel to $\vec{k}$ and $\langle \vec{\sigma} \rangle_v$ is antiparallel while for the $\vec{K}'$ valley $\langle \vec{\sigma} \rangle_c$ rotates relative to $\vec{k}$. Let us compute the one-dimensional

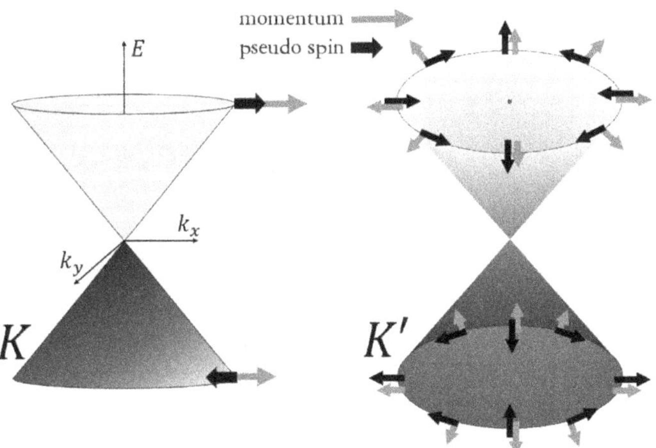

Fig. 14.8. Expectation value of pseudospin at $|\vec{k}| = $ constant for the $\vec{K}$ and $\vec{K}'$ valleys. For the $\vec{K}$ valley the pseudospin is either parallel or antiparallel to $\vec{k}$ while for the $\vec{K}'$ valley the pseudospin is either parallel or antiparallel to $(-k_x, k_y)$. The pseudospins of the conduction and valence bands have the opposite direction.

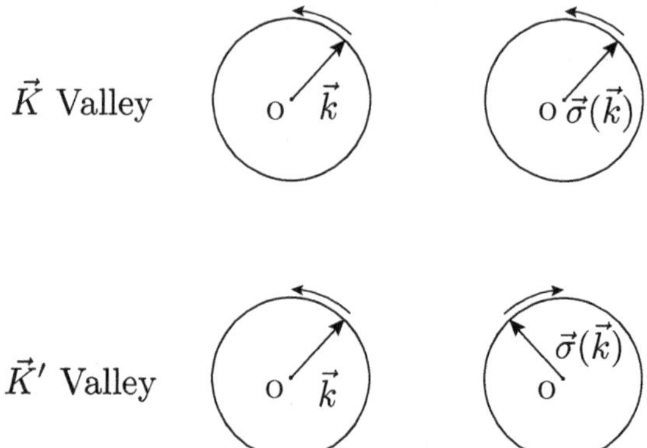

Fig. 14.9. **Upper**: As $\vec{k}$ rotates 2π the winding number of $\vec{\sigma}(\vec{k})$ is 1 for the $\vec{K}$-valley. Origin O represents the $\vec{K}$ valley point. **Lower**: Winding number of $\vec{\sigma}(\vec{k})$ is -1 for $\vec{K}'$-valley. Origin O represents the $\vec{K}'$ valley point.

winding number defined previously in Eq. (2.68): when $\vec{k}$ rotates on a circle the winding number of $\langle \vec{\sigma}(\vec{k}) \rangle$ is $+1(-1)$ for $\vec{K}(\vec{K}')$, see Fig. 14.9. **The sum of winding numbers of $\vec{K}$ and $\vec{K}'$ valleys is zero** (this is the Nielson–Ninomiya theorem [14]).

Consider a short-ranged impurity potential

$$V_{\text{imp}}(\vec{r}) = \begin{pmatrix} V(\vec{r}) & 0 \\ 0 & V(\vec{r}) \end{pmatrix}. \tag{14.37}$$

For a backscattering process $\vec{k} \to -\vec{k}$, the matrix element between two solutions of the Dirac equation is

$$\langle -\vec{k}c | V_{\text{imp}} | \vec{k}c \rangle = 0 \tag{14.38}$$

because the pseudospins have **opposite** values for $\vec{k}$ and $-\vec{k}$ [10]: $\theta_{-\vec{k}} = \theta_{\vec{k}} \pm \pi$ (see Exercise 14.7). So for massless Dirac electrons, there is no backscattering and the transmission coefficient is one. This is analogous to the absence of backscattering in the edges of a quantum Hall bar, see Fig. 10.12 (read about Klein tunneling in Ref. [7]). This means electron localization is non-trivial in graphene related materials. A similar result also holds for the valence band states. For oblique scattering, the transmission can deviate from one, i.e., there can be some backscattering [8].

> **Exercise 14.7.** Show that the impurity matrix element given in Eq. (14.38) is zero. Hint: Use the wave function given in Eq. (14.35). The **interference** between the two components of the pseudospin state cancels out.

Further reading. The envelope wave functions of the Dirac equations satisfy the elliptic second-order partial differential equation

$$A\psi_{xx} + 2B\psi_{xy} + C\psi_{yy} + D\psi_x + E\psi_y + F\psi + G = 0 \qquad (14.39)$$

with

$$B^2 - AC < 0. \qquad (14.40)$$

Here A, B, C, D, E, F, and G are functions of x and y and $\psi_x = \frac{\partial \psi}{\partial x}$ and $\psi_{xy} = \frac{\partial \psi}{\partial x \partial y}$. Other subscripted ψ's are similarly defined. **The Atiyah–Singer index theorem** applies to elliptic partial differential equations and it states that the topological charge of a compact manifold is related to the number of zero modes of this equation. (This theorem connects analysis and topology in a surprising but a beautiful way.) For application of this theorem to graphene, see Ref. [15].

Bibliography

[1] K. S. Novoselov, A. K. Geim, S. V. Morozov, D. Jiang, M. I. Katsnelson, I. V. Grigorieva, S. V. Dubonos, and A. A. Firsov, Two-dimensional gas of massless Dirac fermions in graphene, *Nature* **438**, 197 (2005).

[2] Y. Zhang, Y.-W. Tan, H. L. Stormer, and P. Kim, Experimental observation of the quantum Hall effect and Berry's phase in graphene, *Nature* **438**, 201 (2005).

[3] A. H. Castro Neto, F. Guinea, N. M. R. Peres, K. S. Novoselov, and A. K. Geim, The electronic properties of graphene, *Rev. Mod. Phys.* **81**, 109 (2009). •

[4] M. I. Katsnelson, *Graphene: Carbon in Two Dimensions* (Cambridge University Press, Cambridge, 2012).

[5] A. K. Geim and A. H. MacDonald, Graphene: Exploring carbon flatland, *Phys. Today* **60**, 8, 35 (2007). •

[6] S. M. Girvin and K. Yang, *Modern Condensed Matter Physics* (Cambridge University Press, Cambridge, 2019). •

[7] C. W. J. Beenakker, Colloquium: Andreev reflection and Klein tunneling in graphene, *Rev. Mod. Phys.* **80**, 1337 (2008). •

[8] P. Kim, "Graphene and relativistic quantum physics", in *Poincaré Seminar XVII: Dirac Matters*,(http://www.bourbaphy.fr/, 2014); V. V. Cheianov

and V. I. Fal'ko, Selective transmission of Dirac electrons and ballistic magnetoresistance of n-p junctions in graphene, *Phys. Rev. B* **74**, 041403 (2006). •

[9] M. Alonso and E. J. Finn, "Molecules" in *Fundamental University Physics Vol III: Quantum and Statistical Physics*, (Addison Wesley, Boston, 1976).

[10] T. Ando and T. Nakanishi, Impurity scattering in carbon nanotubes — absence of back scattering, *J. Phys. Soc. Jpn.* **67**, 1704 (1998).

[11] M. O. Goerbig, The quantum Hall effect in graphene — a theoretical perspective, *C. R. Physique* **12**, 369 (2011). •

[12] Y. Zheng and T. Ando, Hall conductivity of a two-dimensional graphite system, *Phys. Rev. B* **65**, 245420 (2002).

[13] C. Dean, P. Kim, J. Li, and A. Young, "Fractional quantum Hall effects in graphene," in *Fractional Quantum Hall Effects: New Developments*, edited by B. I. Halperin and J. K. Jain (World Scientific, Singapore, 2020).

[14] E. Witten, Three Lectures on topological phases of matter, *Riv. Nuovo Cimento*, **39**, 313 (2016). •

[15] J. K. Pachos, Manifestations of topological effects in graphene, *Contemporary Phys.* **50**, 375 (2009). •

Chapter 15

Quantum Anomalous Hall Effect

"The creation of something new is not accomplished by the intellect but by the play instinct acting from inner necessity. The creative mind plays with the objects it loves."

Carl Gustav Jung

The quantum anomalous Hall effect in the **absence** of a net magnetic field was proposed by Haldane [1] (he received a Nobel Prize partly for this contribution). He used a graphene toy model with different A and B site energies and the nearest-and next-nearest-neighbor hopping parameters. In addition, a magnetic field is present such that the average flux through each unit cell is zero. In such a system, both inversion and time-reversal symmetries are broken, and the Hall conductivity acquires quantized values proportional to integer multiple of the conductance quantum (e^2/h). These integers can be identified as a Chern number (see Sec. 2.4), which arises out of topological properties of the underlying band structure. This effect was experimentally confirmed in a magnetic system [2]. Such a system is called a Chern insulator and is **topologically equivalent** to the integer quantum Hall system.

15.1. Main Physics

As we will see later in this chapter, the quantized Hall conductivity value is intimately related to the Chern number. In the presence of time-reversal symmetry the $\vec{K}$ and $\vec{K}'$ valleys have the **same** sign of the regular gap m

(see the gap term appearing in the Dirac Hamiltonian matrix, given in Exercise 14.3). When the mass term is much smaller than the tunneling term one can show that the Dirac Hamiltonians can be written as [3]

$$H_K = v_F(\sigma_x p_x + \sigma_y py) + m\sigma_z$$

$$H_{K'} = v_F(-\sigma_x p_x + \sigma_y p_y) + m\sigma_z. \tag{15.1}$$

In the presence of the Haldane mass m', which breaks time reversal symmetry, the Dirac Hamiltonians are

$$H_K = v_F(\sigma_x p_x + \sigma_y p_y) + m'\sigma_z$$

$$H_{K'} = v_F(-\sigma_x p_x + \sigma_y p_y) - m'\sigma_z. \tag{15.2}$$

(These Dirac Hamiltonians may be derived from the Hamiltonian of the Haldane model, see Sec. 15.2) Note that the mass terms have different signs in the absence of time reversal symmetry. This leads to the same sign in the Berry curvatures (do Exercise 15.1), resulting in a finite value of the Chern number. The presence of a magnetic field is not a necessary condition for a quantum Hall effect, but broken time-reversal symmetry is.

Exercise 15.1. Consider the Dirac Hamiltonian matrix given in Eqs. (15.1) and (15.2). They can be written as

$$h(\vec{k}, m) = \vec{d} \cdot \vec{\sigma}. \tag{15.3}$$

Compute the Berry phases by plotting the pseudospin $\langle u_{\vec{k}} | \vec{\sigma} | u_{\vec{k}} \rangle$ as a function of wave vector $\vec{k} \in C$ (unit circle) for (a) $(d_x, d_y, d_z) = (k_x, k_y, m)$, (b) $(d_x, d_y, d_z) = (-k_x, k_y, m)$, and (c) $(d_x, d_y, d_z) = (-k_x, k_y, -m)$ (the ordinary and Haldane masses are both denoted by m). Compute the sign of the Berry curvatures when the masses of $\vec{K}$ and $\vec{K}'$ valleys are different and when they are the same. Answer: A non-zero m implies a finite gap. When the masses of $\vec{K}$ and $\vec{K}'$ valleys are equal the Chern numbers have opposite signs and cancel each other (see Fig. 15.1(a) and 15.1(b)) (as we mentioned before, this is the Nielson–Ninomiya theorem [4]). When the masses of $\vec{K}$ and $\vec{K}'$ valleys are opposite the Chern numbers have same sign and do not cancel each other (see Fig. 15.1(a) and (c)).

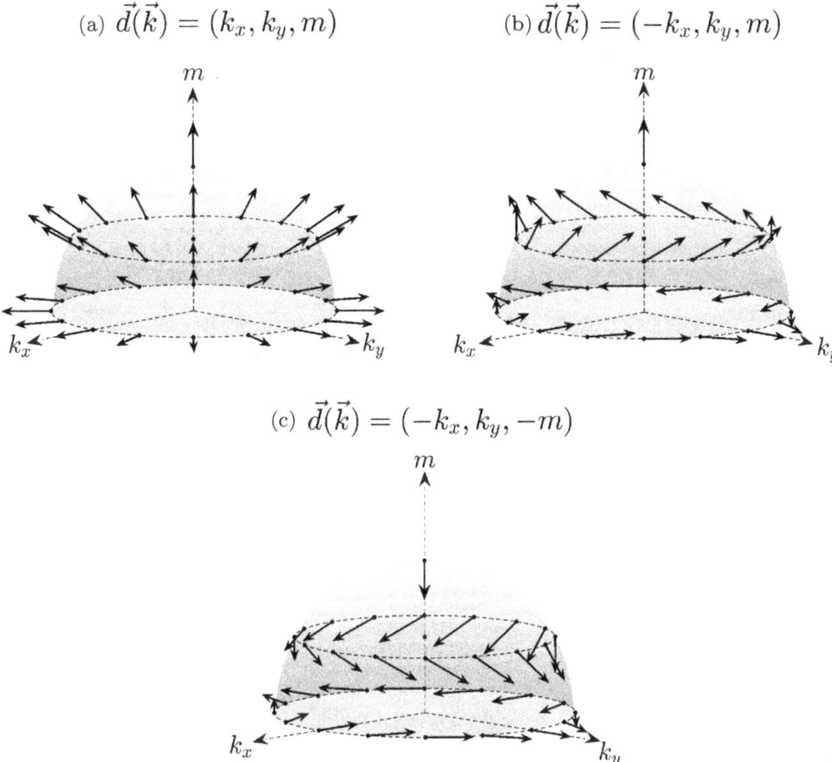

Fig. 15.1. (a) and (b): when the masses of $\vec{K}$ and $\vec{K}'$ valleys are equal their pseudospins point up at the North pole at $(k_x, k_y) = (0,0)$. (a) and (c): when they have different signs, their pseudospins point in opposite directions at the North pole. Note that (a) and (b) depict the pseudospins of the spin-up state rotating in the opposite directions. Thus their Chern numbers have different signs. (a) and (c) depict rotation of the pseudospins of the spin-up and -down states, respectively. They also rotate in the opposite directions but their Berry curvatures have different signs (this follows from Eq. (2.55); the pseudospins point in the opposite directions at the North pole). Thus their Chern numbers have the same sign.

15.2. Hamiltonian of Haldane Model

Let us introduce a graphene model for the quantum anomalous Hall effect.[a] We consider a spinless fermion model without Landau levels. The magnetic field $\vec{B}(\vec{r})$ is in the normal direction to graphene sheet in a such way that

---

[a] Haldane proposed this model before graphene was experimentally discovered.

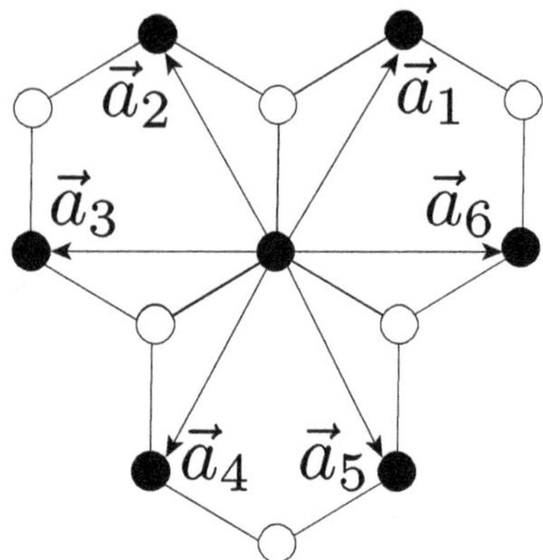

Fig. 15.2. Next-nearest-neighbor sites are displayed. A hexagon can be considered as a primitive unit cell; it has the same area as the one shown in Fig. 14.3. (There is no unique way of choosing a primitive cell.)

the total flux through each primitive unit cell[b] is **zero**.[c] Its Hamiltonian is

$$H = -t \sum_{\text{n.n.}} c_i^\dagger c_j - t' \sum_{\text{n.n.n.}} e^{i\phi_{ij}} c_i^\dagger c_j + m \sum_i s_i c_i^\dagger c_i, \qquad (15.4)$$

where t and t' are, respectively, the nearest-and next-nearest-neighbor hopping parameters (see Fig. 15.2). An electron acquires a phase ϕ_{ij} as it moves between next-nearest-neighbor sites i and j (see Eq. (10.70) for its origin). The second term breaks time-reversal symmetry. For A and B carbon atoms $s_i = 1$ and -1, respectively, and their site energies are $\pm m$ (this breaks inversion symmetry).

The phases can be chosen with any consistent convention such that the total phase accumulated around a closed path adds up to the flux enclosed in units of the flux quantum. Since the total flux through a unit cell is zero, the hopping parameter t does not lead to a complex phase factor. However, t' does lead to a complex phase factor. Consider **three** clockwise next-nearest-neighbor hoppings forming a closed path C (there are two

---

[b]It fills the entire lattice without overlapping when translated through all the vectors in a Bravais lattice.

[c]Imagine a magnetic dipole at the center of each hexagone. The magnetic moments all point normal to the two-dimensional plane.

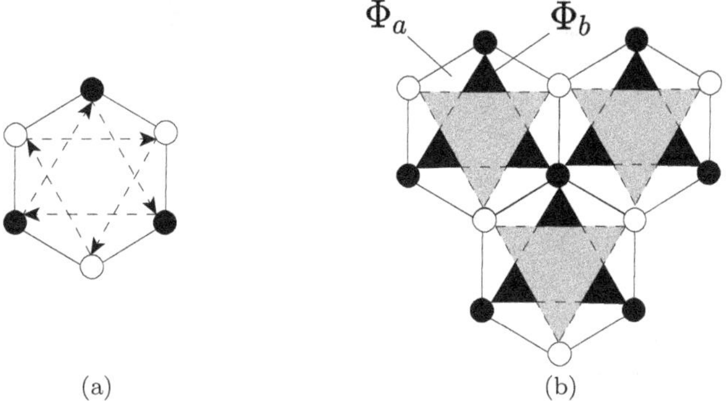

Fig. 15.3.   (a) A loop is formed by the next-nearest-neighbor hopping between A carbon atoms or between B carbon atoms. There are two possible loops. Phase convention in the Hamiltonian of Eq. (15.4) is such that **clockwise** rotation gives **positive** phase. (b) There are six a-type triangles (blank) and three b-type triangles (shaded) outside each gray dotted triangle. The total flux through the unit cell is zero.

such loops, see Fig. 15.3(a)). The total phase accumulated during such a process is

$$3\phi = \frac{e}{\hbar c}\oint_C \vec{A}(\vec{r})\cdot d\vec{r} = -2\pi(6\Phi_a + 3\Phi_b)/\Phi_0, \tag{15.5}$$

where ϕ is the phase acquired during a single hop (it is the phase ϕ_{ij} in the Hamiltonian Eq. (15.4)). $\Phi_a(\Phi_b)$ is the flux through the triangle a (b) (these triangles are displayed in Fig. 15.3(b)). There are six a-type triangles and three b-type triangles outside a triangle formed by the next-nearest-neighbor hopping. Thus the phase acquired during a **single** clockwise next-nearest-neighbor hopping is

$$\phi = -2\pi(2\Phi_a + \Phi_b)/\Phi_0. \tag{15.6}$$

The Hamiltonians for the next-nearest-hopping term are:

$$
\begin{aligned}
H_A &= -t'\sum_{\vec{R}_A}\sum_{j=1,\ldots,6} e^{i\phi_{jA}}\, a^{\dagger}_{\vec{R}_{A,j}}\, a_{\vec{R}_A}, \\
H_B &= -t'\sum_{\vec{R}_B}\sum_{j=1,\ldots,6} e^{i\phi_{jB}}\, b^{\dagger}_{\vec{R}_{B,j}}\, b_{\vec{R}_B}.
\end{aligned}
\tag{15.7}
$$

Here $\vec{R}_A(\vec{R}_B)$ denotes the positions of an A (B) carbon atom and $\vec{R}_{A,j} = \vec{R}_A + \vec{\alpha}_j$ denotes positions of the next-nearest A carbon atoms from $\vec{R}_A$

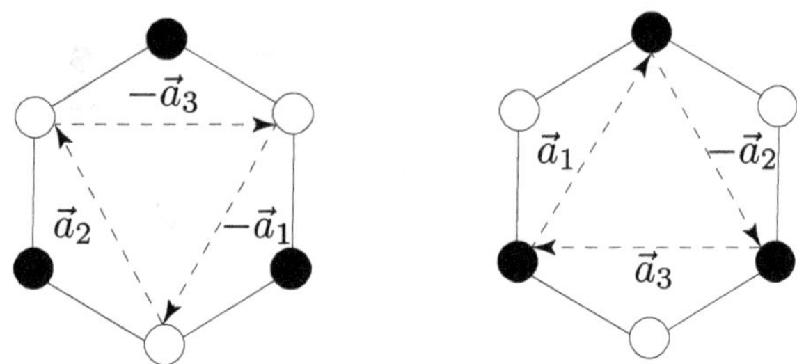

Fig. 15.4. Clockwise hopping between A- (B-)sites are shown in the left (right) figure. The inverted triangle loop consists of the vectors $-\vec{a}_1$, $\vec{a}_2$, and $-\vec{a}_3$. Each hop associated with these vectors gives rise to $+\phi$. Thus the sign of phases ϕ_{jA} associate with the hops represented by $\vec{a}_j$ must be given by the sign in these vectors: ϕ_{1A}, ϕ_{2A}, and ϕ_{3A} are $-\phi$, ϕ, and $-\phi$, respectively. The sign of phases ϕ_{jB} are chosen similarly: ϕ_{1B}, ϕ_{2B}, and ϕ_{3B} are equal to ϕ, $-\phi$, and ϕ, respectively.

($\vec{\alpha}_j$ is one of six vectors shown in Fig. 15.2). $\vec{R}_{B,j}$ is defined similarly: $\vec{R}_{B,j} = \vec{R}_B + \vec{\beta}_j$. The phase ϕ_{jA} is defined as the phase acquired during the hop from $\vec{R}_A$ to $\vec{R}_A + \vec{\alpha}_j$. The phase ϕ_{jB} is defined similarly, see Fig. 15.4.

In $\vec{k}$-space the Hamiltonians are

$$H_A = -t' \sum_{\vec{k}} \sum_{j=1,\ldots,6} e^{i(\phi_{jA} - \vec{k}\cdot\vec{a}_j)} a_{\vec{k}}^\dagger a_{\vec{k}} = \sum_{\vec{k}} g_A(\vec{k}) a_{\vec{k}}^\dagger a_{\vec{k}},$$

$$H_B = -t' \sum_{\vec{k}} \sum_{j=1,\ldots,6} e^{i(\phi_{jB} - \vec{k}\cdot\vec{b}_j)} b_{\vec{k}}^\dagger b_{\vec{k}} = \sum_{\vec{k}} g_B(\vec{k}) b_{\vec{k}}^\dagger b_{\vec{k}}. \tag{15.8}$$

The functions $g_A(\vec{k})$ and $g_B(\vec{k})$ are computed in Exercises 15.2 and 15.3. The total Hamiltonian Eq. (15.4) can be written as

$$H = [h(\vec{k}) + h_1^+(\vec{k})]\mathbf{I} + h_1^-(\vec{k})\sigma_z, \tag{15.9}$$

where $h(\vec{k})$ originates from the nearest-neighbor hopping, see Eq. (14.11). The terms originating from the next-nearest-neighbor hopping are

$$h_1^+(\vec{k}) = -2t' \cos\phi[\cos(\vec{k}\cdot\vec{a}_1) + \cos(\vec{k}\cdot\vec{a}_2) + \cos(\vec{k}\cdot\vec{a}_3)],$$

$$h_1^-(\vec{k}) = -2t' \sin\phi[-\sin(\vec{k}\cdot\vec{a}_1) + \sin(\vec{k}\cdot\vec{a}_2) - \sin(\vec{k}\cdot\vec{a}_3)], \tag{15.10}$$

where $\mathbf{I}$ is a 2×2 identity matrix and $\vec{a}_3 = \vec{a}_2 - \vec{a}_1$.

Exercise 15.2. Derive the Hamiltonian given in Eq. (15.10) from the next-nearest-neighbor Hamiltonian given in Eq. (15.4). Hint: Use the information in Fig. 15.2 and the following

$$g_A(\vec{k}) = -t' \sum_{j=1,\dots,6} e^{i(\phi_{jA}-\vec{k}\cdot\vec{a}_j)} = -2t' \sum_{j=1,2,3} \cos(\phi_{jA} - \vec{k}\cdot\vec{a}_j)$$

$$= -2t' \sum_{j=1,2,3} [\cos\phi_{jA}\cos(\vec{k}\cdot\vec{a}_j) + \sin\phi_{jA}\sin(\vec{k}\cdot\vec{a}_j)].$$

$$(15.11)$$

Here we have used: $\vec{a}_4 = -\vec{a}_1$, $\vec{a}_5 = -\vec{a}_2$, $\vec{a}_6 = -\vec{a}_3$. Then use the results $\sin\phi_{1A} = -\sin\phi$, $\sin\phi_{2A} = \sin\phi$, and $\sin\phi_{3A} = -\sin\phi$ (see Fig. 15.4), which gives

$$g_A(\vec{k}) = -2t'\cos\phi \left[\sum_{j=1,2,3} \cos(\vec{k}\cdot\vec{a}_j) \right]$$
$$- 2t'\sin\phi[-\sin(\vec{k}\cdot\vec{a}_1) + \sin(\vec{k}\cdot\vec{a}_2) - \sin(\vec{k}\cdot\vec{a}_3)].$$

$$(15.12)$$

Exercise 15.3. As in Exercise 15.2, show that the following holds for B carbon atoms

$$g_B(\vec{k}) = -t' \sum_{j=1,\dots,6} e^{i(-\phi_{jB}-\vec{k}\cdot\vec{a}_j)} = -2t' \sum_{j=1,2,3} \cos(-\phi_{jB} - \vec{k}\cdot\vec{a}_j).$$

$$(15.13)$$

Note the sign change in ϕ_{jB} in comparison to ϕ_{jA} in $g_A(\vec{k})$, see Fig.15.4.

$$g_B(\vec{k}) = -2t'\cos\phi \left[\sum_{j=1,2,3} \cos(\vec{k}\cdot\vec{a}_j) \right]$$
$$+ 2t'\sin\phi[-\sin(\vec{k}\cdot\vec{a}_1) + \sin(\vec{k}\cdot\vec{a}_2) - \sin(\vec{k}\cdot\vec{a}_3)].$$

$$(15.14)$$

15.3. Band Structure and Winding Number

The first next-nearest-neighbor hopping term $h_1^+(\vec{k})$ breaks chiral symmetry
(particle–hole symmetry) even when $\phi = 0$. The computed band structure
indeed shows breaking of particle–hole symmetry but no gap is produced,
see Fig. 15.5. However, the second term $h_1^-(\vec{k})$ ($t' \neq 0$ and $\phi \neq 0$) breaks
time symmetry and generates a band gap, as shown in Fig. 15.5.

It is easier to compute the winding number than the off-diagonal conductivity. The Hamiltonian can be written in a more compact form

$$H = E(\vec{k})\mathbf{I} + \vec{h}'(\vec{k}) \cdot \vec{\sigma}, \tag{15.15}$$

where

$$E(\vec{k}) = -2t' \cos \phi \sum_{i=1,2,3} \cos(\vec{k} \cdot \vec{a}_i) \tag{15.16}$$

and $\vec{h}' = (h_x'(\vec{k}), h_y'(\vec{k}), h_z'(\vec{k}))$ with

$$\begin{aligned}
h_x'(\vec{k}) &= t[1 + \cos(\vec{k} \cdot \vec{a}_1) + \cos(\vec{k} \cdot \vec{a}_2)], \\
h_y'(\vec{k}) &= t[\sin(\vec{k} \cdot \vec{a}_1) + \sin(\vec{k} \cdot \vec{a}_2)], \\
h_z'(\vec{k}) &= m - 2t' \sin \phi[-\sin(\vec{k} \cdot \vec{a}_1) + \sin(\vec{k} \cdot \vec{a}_2) - \sin(\vec{k} \cdot \vec{a}_3)].
\end{aligned} \tag{15.17}$$

> **Exercise 15.4.** Derive the Hamiltonian given in Eq. (15.17) from
> the Hamiltonian equation (15.9).

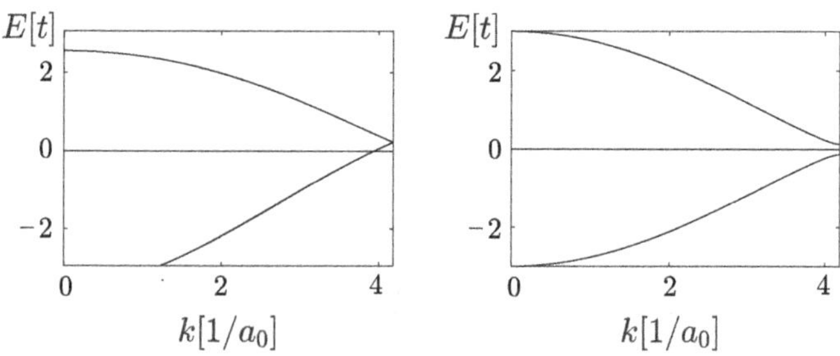

Fig. 15.5. Plot of the energy dispersion from $\vec{k} = (0,0)$ to $\vec{K} = \frac{2\pi}{3a_0}(\sqrt{3}, 1)$ for two
values of $\phi = 0$ (left figure) and $\pi/2$ (right figure). The parameters are $t' = 0.075t$ and
$m = 0$.

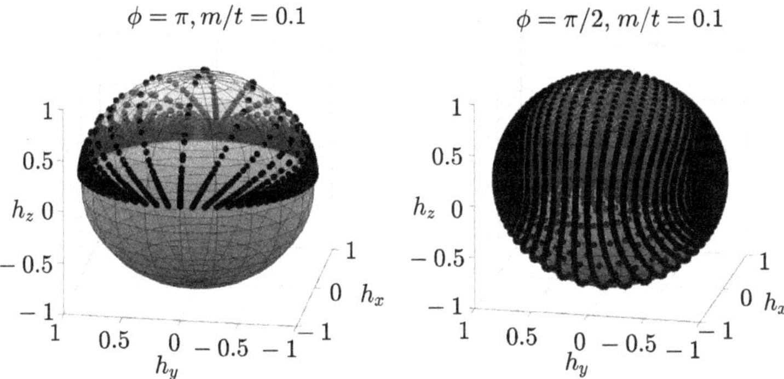

Fig. 15.6. Plot of the unit vector $\hat{h}' = \vec{h}'/|\vec{h}'|$ on the unit sphere as $\vec{k}$ sweeps the entire two-dimensional Brillouin zone. Left (right) figure displays the results for $\phi = \pi$ ($\phi = \pi/2$) (see two points marked in Fig. 15.9). In contrast to the case of $\phi = \pi$, when the grid points become denser at the value of $\phi = \pi/2$ the unit vector covers the entire sphere.

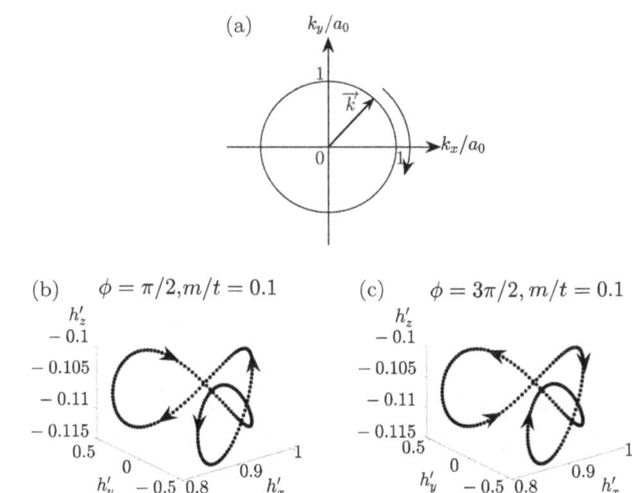

Fig. 15.7. (a) $\vec{k}$ moves on a circle. (b) The unit vector $\hat{h}'(\vec{k})$ moves on the unit sphere. (c) The unit vector $\hat{h}'(\vec{k})$ (measured in units of t) moves on along the same curve but in the opposite direction in comparison to the result in (b).

Figure 15.6 displays a plot of $\vec{h}'(\vec{k})$, which defines the winding number, see Eq. (2.66). (Note the winding and Chern number are identical, see Exercise 9.5.) The results for $\hat{h}'(\vec{k})$ indicate that there are two topological regions in the parameter space. In one region $\hat{h}'(\vec{k})$ rotates clockwise but in the other region it rotates in the opposite direction, see Fig. 15.7.

15.4. Off-Diagonal Conductivity, Chern Number, and Phase Diagram

We have showed in Sec. 10.6 that the off-diagonal conductivity of two-dimensional electron gas in a magnetic field is a Chern number, see Eq. (5.37). Here we used the Streda formula (see Exercise 15.5)

$$\sigma_{xy} = \frac{enc}{B}. \tag{15.18}$$

to show the similar result for the Haldane model in the limit $B \to 0$. Before we do that, let us show the usual quantized off-diagonal conductivity using this formula. The electron density is

$$n = \frac{N_\Phi \nu}{A} = \frac{eB}{hc} \nu. \tag{15.19}$$

where $\nu = N/N_\phi$ **is the filling factor** and $N_\Phi = 2\pi \ell^2 L^2$ is the total number of flux (N is the number of electrons). Plugging this result into the Streda formula, one finds at $\nu = 1$ that the quantized off-diagonal conductivity $\sigma_{xy} = \frac{e^2}{h}$ at finite values of B.

Exercise 15.5. Assume that the Fermi level is in the gap between two Landau levels. Show the Streda formula given in Eq. (15.18). Hint: It follows from the result

$$-e\frac{\partial n}{\partial t} = \sigma_{xy}\frac{1}{c}\frac{\partial B}{\partial t} \tag{15.20}$$

and by replacing the ac conductivity with the static one. This result can be shown using the relation between $E_y(x,t)$ and $A_y(x,t)$ given in Eq. (5.8) and the following result

$$e\frac{\partial n}{\partial t} + \frac{\partial j_x}{\partial x} = 0, \quad j_x = \sigma_{xy}E_y. \tag{15.21}$$

Note that $\vec{A}$ has only the y-component and that $\vec{A}$ and $\vec{E}$ are plane waves $\sim e^{i(kx-wt)}$. Also remember

$$\nabla \times \vec{A} = \frac{1}{c}\frac{\partial \vec{B}}{\partial t}. \tag{15.22}$$

Streda's formula does not apply for localized states at the Fermi energy.

In the limit of **zero** external magnetic field, Haldane used the Streda formula, Eq. (15.18), to show that the off-diagonal conductivity is indeed quantized. Instead of the tight-binding approach of the previous section, he applied the effective mass approximation and expanded near $\vec{K}$ and $\vec{K}'$. The effective Hamiltonians are

$$H_{\vec{K}}(\vec{q}) = 3t' \cos\phi \mathbf{I} + \frac{\sqrt{3}a_0}{2}(q_y\sigma_x - q_x\sigma_y) + M_{\vec{K}}\sigma_z,$$
$$H_{\vec{K}'}(\vec{q}) = 3t' \cos\phi \mathbf{I} - \frac{\sqrt{3}a_0}{2}(q_y\sigma_x + q_x\sigma_y) + M_{\vec{K}'}\sigma_z, \tag{15.23}$$

where $\vec{k} = \vec{K} + \vec{q}$ and $\vec{k} = \vec{K}' + \vec{q}$ and

$$M_{\vec{K}} = m - 3\sqrt{3}t' \sin\phi,$$
$$M_{\vec{K}'} = m + 3\sqrt{3}t' \sin\phi. \tag{15.24}$$

Using the Peierls substitution, Eq. (14.19), one finds the Landau level energies from the Hamiltonians Eq. (15.23)

$$E_p = \pm\sqrt{(M_\alpha v_F^2)^2 + e\hbar v_F^2|pB|}, \quad p \geq 1$$
$$E_0 = \alpha e M_\alpha v_F^2 \mathrm{sgn}(B), \quad p = 0, \tag{15.25}$$

where $\alpha = \pm 1$ stand for the $\vec{K}$ and $\vec{K}'$ valleys. The Landau level energy spectrum in the presence of a gap is displayed in Fig. 15.8.

Using the Streda formula, Eq. (15.18), in the limit $B \to 0$ Haldane computed

$$\sigma_{xy} = \frac{e^2}{h}\nu', \tag{15.26}$$

where the Chern number is

$$\nu' = \frac{1}{2}[\mathrm{sgn}(M_{\vec{K}'}) - \mathrm{sgn}(M_{\vec{K}})]. \tag{15.27}$$

Let us try to understand the implications of this result. One can determine the critical lines in the phase diagram that separate regions with different values of σ_{xy}. The critical lines consist of the parameter points where the gap **closes**. The gap vanishes when $M_{\vec{K}}$ or $M_{\vec{K}'}$ is zero, i.e., when

$$m = 3\sqrt{3}t' \sin\phi \text{ for } \vec{K} \text{ valley,}$$
$$m = -3\sqrt{3}t' \sin\phi \text{ for } \vec{K}' \text{ valley.} \tag{15.28}$$

Note that $a_0 = 1$ is the unit cell length. These critical lines are displayed in Fig. 15.9. The regions inside the critical lines are topologically non-trivial with the Chern numbers $\nu' = -1$ and 1. The origin of these Chern numbers

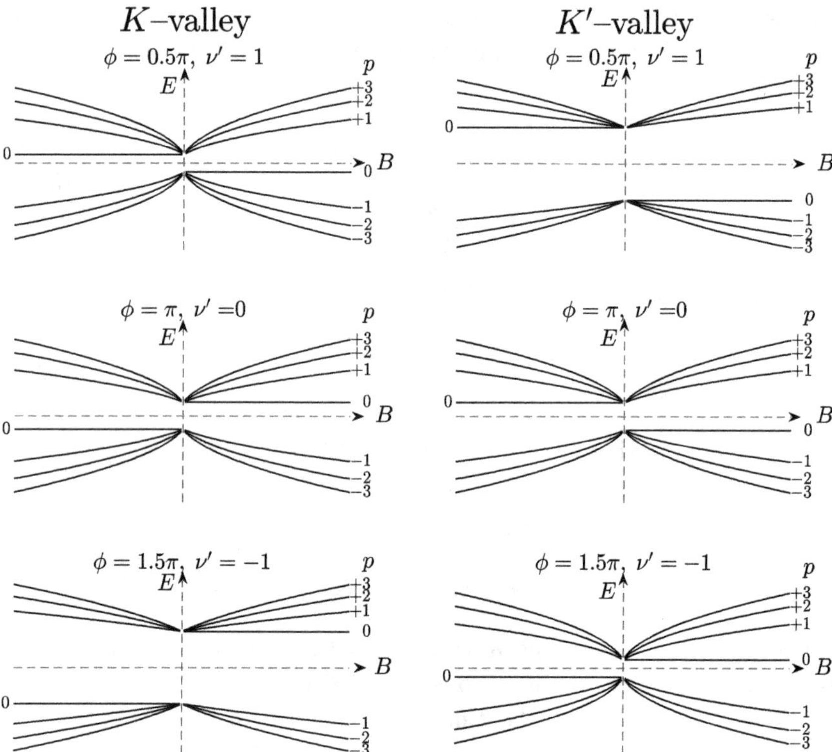

Fig. 15.8. Schematic Landau level energy spectrum as a function of an external magnetic field B. The spectra corresponding to the Chern numbers $\nu' = 1, 0, -1$ are shown for the K-valley (left column) and K′-valley (right column). The mass value is $\frac{m}{3\sqrt{3}t'} = 0.6151$. Note that there is a gap at $E = 0$ and the Fermi level is in the gap.

is as follows. The occupation numbers differ by the complete filling of one Landau level in comparison with those of the time-reversal invariant system, which can be seen by comparing the Landau levels given in Figs. 14.6 and 15.8. This means that an extra field-dependent ground state charge density $e\Delta n = \pm\frac{e^2}{hc}B$ is present, see Eq. (15.19). The Streda formula then gives the Chern numbers $\nu' = -1$ and 1.

These results are consistent with those obtained from the analysis of $\vec{h}'$. In the non-trivial region $\vec{h}'(\vec{k})$ covers the entire unit sphere (see Fig. 15.6) while in the trivial region it does not. These results also agree with those of two different rotations of $\vec{h}'(\vec{k})$ shown in Fig. 15.7 that are computed at values of $\phi = \pi/2$ and $3\pi/2$.

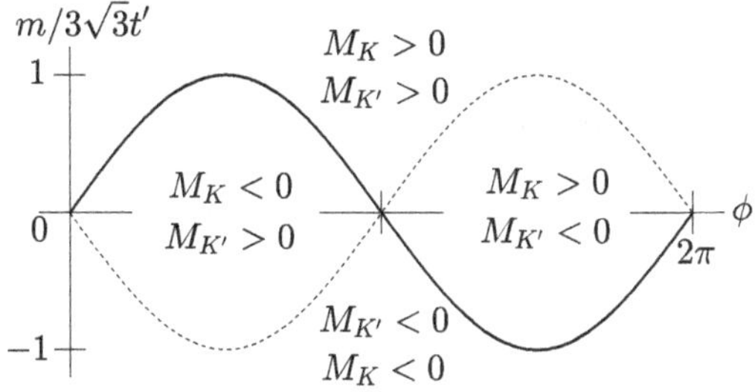

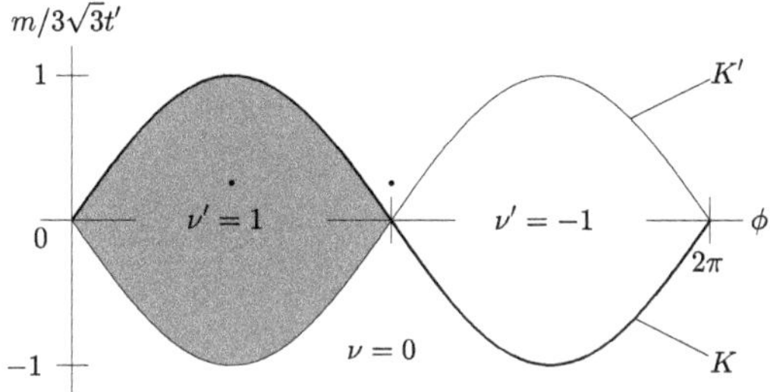

Fig. 15.9. **Upper**: Values of $M_{\vec{K}}$ and $M_{\vec{K}'}$. **Lower**: Display of the quantized off-diagonal-conductivity in the parameter space. The gap at $\vec{K}$ and $\vec{K}'$ closes, respectively, on the critical lines marked $\vec{K}$ and $\vec{K}'$. The vector $\vec{h}'(\vec{k})$ in Fig. 15.6 are plotted for the two marked dots in the figure.

The physical origin of the results shown in Fig. 15.9 is as follows (do Exercise 15.1): When $M_{\vec{K}}$ and $M_{\vec{K}'}$ have opposite signs then $\nu' = \pm 1$, but when they have the same sign then $\nu' = 0$. The mass term is introduced by breaking time-reversal symmetry for the former while not for the latter. In the Haldane model, the $\vec{k} \to -\vec{k}$ symmetry is not present because of broken time-reversal symmetry. Kramers band degeneracy (see Exercise 15.7) is thus not present.

Exercise 15.6. Derive the Hamiltonians given in Eq. (15.23) from the Hamiltonian Eq. (15.17).

Exercise 15.7. Explain Kramers degeneracy in the presence of time-reversal symmetry. Hint: See [5].

Bibliography

[1] F. D. M. Haldane, Model for a quantum Hall effect without Landau levels: condensed-matter realization of the "parity anomaly", *Phys. Rev. Lett.* **61**, 2015 (1988).

[2] C.-Z. Chang, J. Zhang, X. Feng, J. Shen, Z. Zhang, M. Guo, K. Li, Y. Ou, P. Wei, L.-L. Wang, Z.-Q. Ji, Y. Feng, S. Ji, X. Chen, J. Jia, X. Dai, Z. Fang, S.-C. Zhang, K. He, Y. Wang, L. Lu, X.-C. Ma, and Q.-K. Xu, Experimental observation of the quantum anomalous hall effect in a magnetic topological insulator, *Science* **340**, 167 (2013).

[3] S. M. Girvin and K. Yang, *Modern Condensed Matter Physics* (Cambridge University Press, Cambridge, 2019). •

[4] E. Witten, Three lectures on topological phases of matter, *Riv. Nuovo Cimento* **39**, 313 (2016). •

[5] O. Madelung, *Introduction to Solid State Physics* (Springer, Berlin, 1978). •

Chapter 16

Non-interacting Graphene Nanoribbons: Zero-Energy Soliton States of Zigzag Edges

"Our imagination is stretched to the utmost, not as in fiction, to imagine things which are not really there, but just to comprehend those things which are there."

Richard P. Feynman

One prominent feature of graphene is that zigzag edges (see Fig.16.1) can support chiral symmetry protected topological edge states [1–4] displaying an integer charge [4–6]. Chiral symmetry thus plays an important role in zigzag graphene nanoribbons [7]. As in polyacetylene, chiral symmetry suggests that there should be solitonic zero modes in graphene nanoribbons. However, unlike polyacetylene, two different dimerized phases are not needed to form solitons in graphene nanoribbons [6]. Mere presence of a zigzag edge guarantees the existence of a soliton [5]. In non-interacting zigzag graphene nanoribbons, these zero modes localized on the zigzag edges form a singular peak in the DOS (the number of zero modes is proportional to the length of the zigzag edges). The zero-energy states of zigzag graphene nanoribbons are very robust against disorder and magnetic field, as they are a topological property of two interpenetrating triangle sublattices, see Fig. 16.2. An excellent opportunity to observe these topological boundary charges has recently arisen, as rapid progress has been made in the fabrication of atomically precise graphene nanoribbons [8, 9].

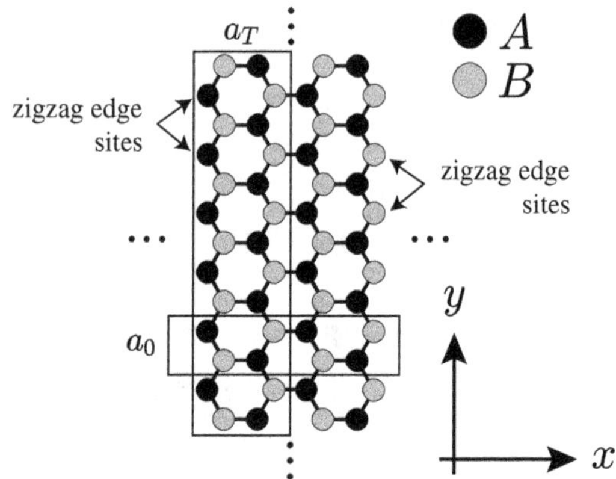

Fig. 16.1. Graphene nanoribbons have a honeycomb lattice structure. The black circles of the left boundary constitute the left zigzag edge sites while the grey circles of the right boundary constitute the right zigzag edge sites. The unit cell length of a zigzag ribbon is $a_0 = \sqrt{3}a$ (here $a = 1.42\text{Å}$ is the carbon–carbon distance of graphene). The zigzag edges have different chiralities A and B and belong to two different triangular lattices, see Fig. 16.2. The bottom and top carbon atoms in the figure form armchair edges. The unit cell length of an armchair ribbon is $a_T = \sqrt{3}a_0$. Note that **the zigzag edge sites consist of either A or B type of carbon atoms and that there is no nearest-neighbor hopping between zigzag edge sites**. This is the key topological property of zigzag ribbons. This allows zero energy edge modes to exist. In contrast, armchair edges consist of A and B types of carbon atoms.

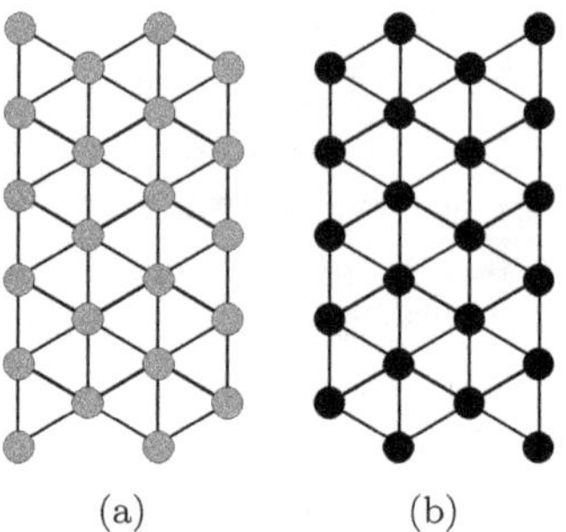

(a)   (b)

Fig. 16.2. A zigzag ribbon consists of two triangular lattices A and B, shown in (a) and (b). When they interpenetrate each other they form a zigzag graphene nanoribbon, see Fig. 16.1. Zigzag edges sites (boundary sites) are only connected to two neighboring sites while interior sites are connected to three sites. This **topological** feature is robust.

16.1. Edge Modes of Graphene Sheet and Zak Phase

Before we study zigzag nanoribbons, we consider non-interacting electrons of a semi-infinite sheet of graphene with a zigzag edge or armchair edge

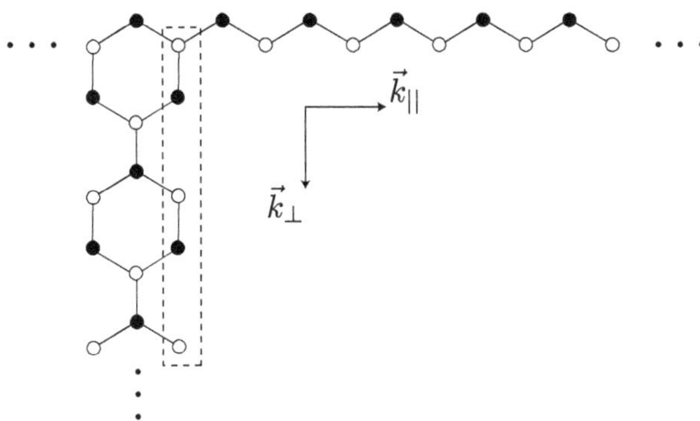

Fig. 16.3. Semi-infinite graphene sheet with a zigzag edge. The value of the Zak phase at $\vec{k}_\parallel$ is obtained by integrating over $\vec{k}_\perp$ in the Brillouin zone.

boundary, see Figs. 16.3 and 16.4. It is possible to gain some insight into the edges by investigating the Zak phase of graphene in torus geometry. Let us investigate the Zak phase [10, 11] $Z_{2D}(k_\parallel)/\pi$ as a function of wave vector $k_\parallel$ parallel to an edge direction

$$Z_{2D}(k_\parallel) = i \oint_{\text{B.Z.}} \langle u_{\vec{k}} | \nabla_{k_\perp} u_{\vec{k}} \rangle \, dk_\perp, \qquad (16.1)$$

where the integration is over the wavevector $k_\perp$ perpendicular to the edge (we have previously studied the Zak phase in Sec. 5.4). Integration over $k_\perp$ is performed as a cut of the 2D Brillouin zone in the direction of $k_\perp$. When $k_\perp$ is perpendicular to the zigzag edge the computed $Z_{2D}(k_\parallel)$ is of π (see Fig. 16.3). On the other hand, when $k_\perp$ is perpendicular to the armchair edge (see Fig. 16.4), $Z_{2D}(k_\parallel)$ is zero. This suggest that a zigzag edge may be topologically non-trivial while an armchair edge may be topologically trivial. This is an example of the bulk-edge correspondence [1]. Tight-binding calculations of graphene also confirm these results.

These results are somewhat similar to the result of the Zak phase of polyacetylene, see Sec. 5.5. Let us try to understand this. Both graphene and polyacetylene have chiral symmetry which guarantees the existence of edge modes [7]. The 2×2 Hamiltonians of polyacetylene and graphene, Eqs. (4.14) and (14.9), have the same structure and their wave functions both have two components. Moreover, as in the case of polyacetylene, the integration is along a chain of dimers (see Figs. 16.3 and 16.4).

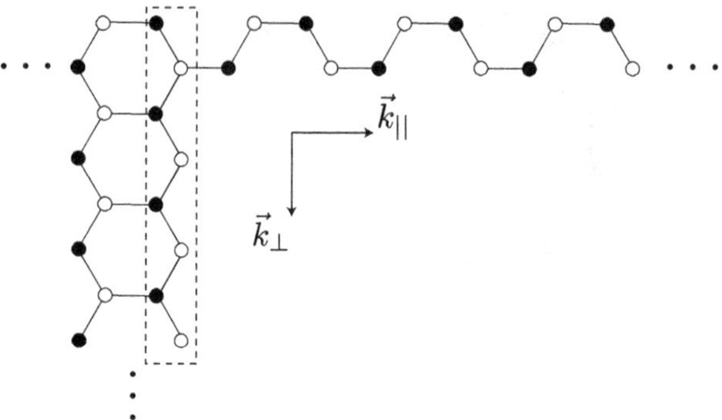

Fig. 16.4. Semi-infinite graphene sheet with an armchair edge. The value of the Zak phase at $\vec{k}_{\parallel}$ is obtained by integrating over $\vec{k}_{\perp}$ in the Brillouin zone.

16.2. Edge States of Zigzag Ribbon

In this section, we find the edge states of graphene zigzag nanoribbons using the Dirac equations combined with the appropriate boundary conditions. The Zak phase of nanoribbons will be computed in Sec. 17.3, where we include the on-site repulsion energy. This effect will split degeneracies at the Fermi level and produce a Mott gap.

- Band Structure

Each carbon atom inside a zigzag ribbon is connected to three other carbon atoms while carbon atoms at the boundary zigzag edges are only connected to two other carbon atoms. The left and right zigzag edges are separated by a distance. It is assumed that all dangling bonds at graphene edges are terminated by hydrogen atoms and thus do not contribute to the electronic states near the Fermi level.

The unit cell length of a zigzag ribbon is $a_0 = \sqrt{3}a$, see Fig. 16.1. Its shape may be guessed from the Brillouin zone of two-dimensional graphene. The zigzag ribbon direction is defined by $\vec{a}_1 - \vec{a}_2$ (see Fig. 14.3), which leads to the reciprocal lattice vector $\vec{b}_3 = \frac{1}{2}(\vec{b}_1 - \vec{b}_2) = \frac{2\pi}{a_0}\hat{y}$. The vectors $\vec{b}_{1,2}$ are shown in Fig. 14.4. The Brillouin zone edge is given by $\frac{\vec{b}_3}{2} = \frac{\pi}{a_0}$ because the first Brillouin zone is defined by bisecting the reciprocal lattice vector $\vec{b}_3$ [12]. The Brillouin zone of graphene zigzag nanoribbons is $-\frac{\pi}{a_0} < k < \frac{\pi}{a_0}$, as shown in Fig. 16.5. $\vec{K}$ and $\vec{K}'$ valley points have $k_y = \pm\frac{2\pi}{3a_0}$. The band structure and the DOS of graphene zigzag nanoribbons in the nearest tight-binding approach are shown in Figs. 16.6 and 16.7. Bands are **flat** in the interval $|k_y| > \frac{2\pi}{3a_0}$ near the Brillouin zone boundary. The states in this

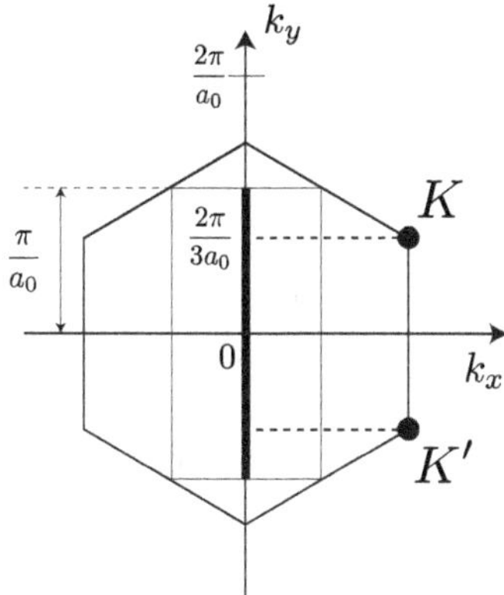

Fig. 16.5. The Brillouin zone of a thin ribbon is shown as the thick line.

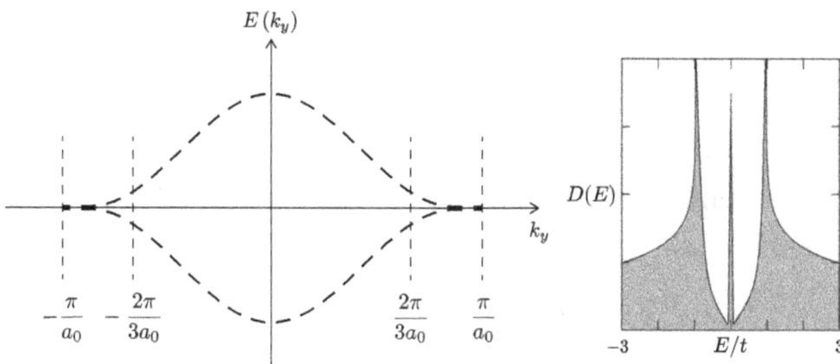

Fig. 16.6. **Left**: Schematic band structure of a graphene zigzag nanoribbon for non-interacting electrons. Zero modes are in the interval $-\frac{\pi}{a_0} < k_y < -\frac{2\pi}{3a_0}$ or $\frac{2\pi}{3a_0} < k_y < \frac{\pi}{a_0}$. Inversion symmetry is present. There are other bands but they are not shown. The number of zero modes is proportional to the ribbon length L, given by $\frac{1}{\sqrt{3}}\frac{L}{a_0}$. **Right**: DOS of a graphene zigzag nanoribbon. The peak in the middle is due to zero energy modes.

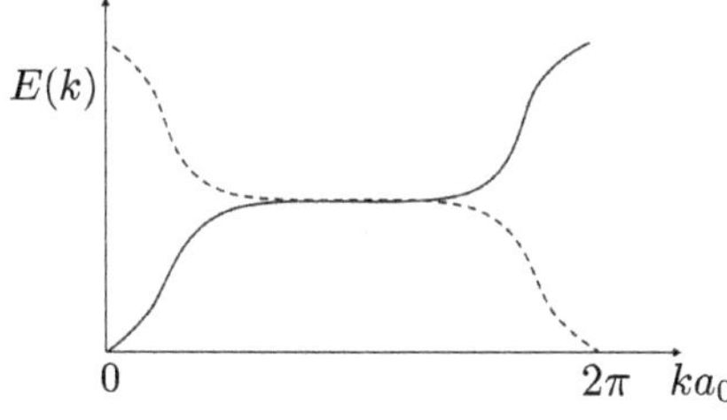

Fig. 16.7. The band structure $E(k)$ displayed in Fig. 16.6 is replotted schematically to display similarity with the band structure of chiral edge modes of a quasi-one-dimensional p-wave superconductor shown in Fig. 9.5. However, note that the edge velocity is nearly zero in zigzag ribbons.

region correspond to **zigzag edge states** and localized on the edges with nearly zero energy [5]. Their number is proportional to the length of the zigzag edges. The DOS is thus sharply peaked at zero energy.

- Zigzag edge states (solitons) without valley mixing

In Sec. 14.3, we found the Dirac equations of graphene that can be applied to numerous problems. From the effective Dirac Hamiltonians, we find the equations that the envelope wave functions satisfy. They are often easier to solve than tight-binding equations and can provide many useful insights. Let us apply these equations and find the zigzag edge states with nearly zero energy displayed in Fig. 16.6.

The envelope functions F_A and F_B of the valley $\vec{K}$ are coupled to each other. In the absence of $V(\vec{r})$, one can show from Eq. (14.20) that

$$(\partial_x^2 + \partial_y^2)F_B + E'^2 F_B = 0, \tag{16.2}$$

where the eigenenergy is $E' = \frac{E}{\hbar v_F}$. From now on, we will exchange x and y for notational convenience (this means $\frac{\partial}{\partial x}$ and $\frac{\partial}{\partial y}$ are also exchanged in Eq. (14.20), but not σ_x and σ_y). Then the longitudinal (y-direction) and transverse (x-direction) parts of the wave function separate as follows

$$F_B(x, y) = e^{ik_y y}\phi_B(x), \tag{16.3}$$

where $\phi_B(x)$ describes the electron motion perpendicular to the ribbon direction. It satisfies the following equation

$$(-\partial_x^2 + k_y^2)\phi_B(x) = E'^2 \phi_B(x). \tag{16.4}$$

From $\phi_B(x)$, one can also determine the A part of the envelope function

$$\phi_A = \frac{1}{E'}(k_y - \partial_x)\phi_B. \tag{16.5}$$

One can find similar differential equations for the electron wave functions of the $\vec{K}'$ valley.

The nature of the edge states can be understood by examining their wave functions obtained solving the Dirac equations [13]. Consider first the $\vec{K}$ valley. Suppose the zigzag edge at $y = 0$ consists of A-type carbons and of B-type carbons at $y = w$, see Fig. 16.1. Then the wave functions are zero at the boundaries: $F_B(x = 0) = 0$ and $F_A(x = w) = 0$. (On the other hand, the wave function along the y-axis, i.e., along the zigzag edges is a plane wave.) Although there are other solutions we are mainly interested in **decaying edge modes**.

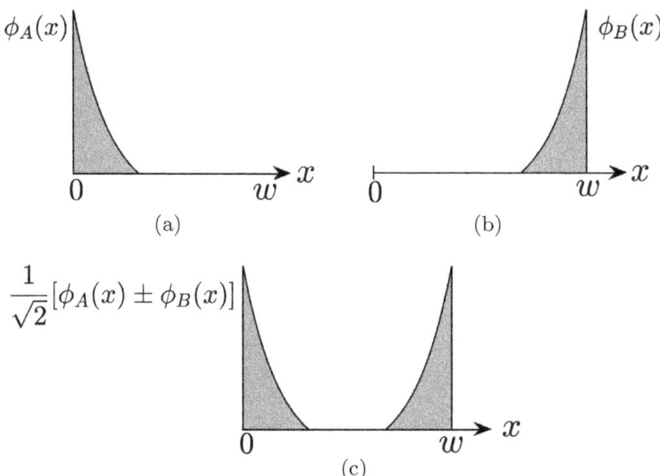

Fig. 16.8. Ambiguity of choosing between possible zigzag edge states in the absence of disorder and electron interactions. Schematic drawing of the wave functions ϕ_l and ϕ_r of the two degenerate edge states is shown. They are localized on the (a) left and (b) right edges, respectively. (c) Another degenerate representation can be formed: antibonding $\frac{1}{\sqrt{2}}(\phi_l - \phi_r)$ and bonding edge states $\frac{1}{\sqrt{2}}(\phi_l + \phi_r)$. These states have opposite chiralities at the opposite zigzag edges. All these edge states are delocalized along the zigzag edge direction. We will see later in Chapter 18 that these states give rise to stable **non-local** fractional charges in the presence of disorder.

Let us find a solution depicted in Fig. 16.8(b). We are looking for solutions of type $F_B(x, y) = e^{ik_y y}\phi_B(x)$ with

$$\phi_B(x) = c_1 e^{\alpha x} + c_2 e^{-\alpha x}, \tag{16.6}$$

where α is a real number (see Fig. 16.8(c)). The boundary conditions on the zigzag edges yield that the coefficients satisfy

$$\phi_B(x = 0) = 0 \rightarrow c_1 + c_2 = 0. \tag{16.7}$$

From this we find

$$\phi_B(x) = A(e^{\alpha x} - e^{-\alpha x}). \tag{16.8}$$

We find its energy from Eq. (16.4)

$$E^2 = k_y^2 - \alpha^2. \tag{16.9}$$

From Eq.(16.5) we find the boundary condition on the A-component

$$\phi_A(x = w) = 0 \rightarrow (k_y - \alpha)c_1 e^{\alpha w} - (k_y + \alpha)c_2 e^{-\alpha w} = 0. \tag{16.10}$$

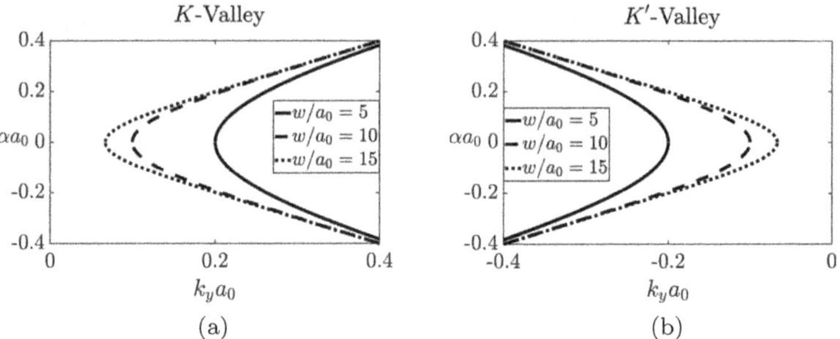

Fig. 16.9. Inverse decay length of a zigzag edge state as a function of wave vector. For K-valley (K'-valley) possible values are $k_y > 0$ ($k_y < 0$), consistent with the result of Fig. 16.6. Note that k_y is measured from K or K' valley. As the width w increases α increases, meaning that the decay length becomes shorter. Solutions with $\alpha > 0$ and $\alpha < 0$ are equivalent.

Only solutions with $k_y > 0$ exist, as shown in Fig. 16.9(a). In fact, for this mode, $\phi_A(x) = 0$ for all x but $\phi_B(x) \neq 0$: it is a chiral mode. From Eq.(16.10) we find that the inverse decay length α satisfies

$$e^{-2\alpha w} = \frac{k_y - \alpha}{k_y + \alpha}. \tag{16.11}$$

The solution of this equation is

$$k_y = \frac{\alpha(1 + e^{-2\alpha w})}{1 - e^{-2\alpha w}}. \tag{16.12}$$

Figure 16.9 shows how the inverse decay length α varies as a function of k_y. Since the ratio $\alpha a_0 \ll \alpha w$ we have $\alpha w \gg 1$. This means that $k_y \approx \alpha$ implying that there are two degenerate states with $E = 0$.

For the $\vec{K}'$ valley we find a similar solution shown in Fig. 16.8(a). It satisfies

$$e^{-2\alpha w} = \frac{k_y + \alpha}{k_y - \alpha}. \tag{16.13}$$

Only solutions with $k_y < 0$ exist, as shown in Fig. 16.9(b). In contrast to the result of the $\vec{K}$ valley, these wave functions $\phi_A(x)$ are decaying exponentially from the A-type zigzag edge, see Fig. 16.8(a). However, they are negligible near the B-type zigzag edge. These are chiral modes: $\phi_B(x) = 0$ for all x while $\phi_A(x) \neq 0$. Note these solutions of $\vec{K}$ and $\vec{K}'$ are independent of each other, i.e., there is no valley mixing. We will call these independent states with nearly zero energy the chiral edge states (do Exercise 16.1).

Exercise 16.1. Argue that chiral edge states have zero energy. Hint: Use the nearest-neighbor tight-binding Hamiltonian.

For a given value k_y, there are two degenerate solutions with $E = \pm E \to 0$, see Eq. (16.9). The probability density of these states is divided between the left and right zigzag edges. However, in the absence of electron–electron interactions and disorder there is some ambiguity in choosing degenerate zigzag edge states: this is because one can generate other degenerate solutions by forming linear combinations of them that are localized only on one of the two zigzag edges, see Figs. 16.8(a) and 16.8(b). If we assume that there is **weak disorder** in the ribbon, bonding and antibonding states are more stable over states that are localized only on one of the two zigzag edges. On the other hand, when electron interactions are included the opposite is true, see Sec. 17.1.

16.3. Undoped Armchair Ribbons

In this section, we find the eigenstates of graphene armchair nanoribbons using the Dirac equations combined with the appropriate boundary conditions.

- Band Structure

The unit cell length of an armchair ribbon is $a_T = \sqrt{3}a_0$, as shown in Fig. 16.1 (here $a_0 = \sqrt{3}a$). The Brillouin zone of graphene zigzag nanoribbons is $-\frac{\pi}{\sqrt{3}a_0} < k < \frac{\pi}{\sqrt{3}a_0}$, see Fig. 16.10. Its shape may be guessed from the Brillouin zone of two-dimensional graphene: the direction of the armchair ribbon is along $\vec{a}_1 + \vec{a}_2$, see Fig. 14.3 (this direction is along $\hat{x}$, as shown in Fig. 16.1). Then the first Brillouin zone edge is given by $\frac{\vec{b}_4}{2} = \frac{1}{2} \times \frac{1}{2}(\vec{b}_1 + \vec{b}_2) = \frac{\pi}{\sqrt{3}a_0}\hat{x}$ with the reciprocal lattice vector $\vec{b}_4 = \frac{1}{2}(\vec{b}_1 + \vec{b}_2)$. The vectors $\vec{b}_1$ and $\vec{b}_2$ are shown in Fig. 14.4.

- Envelope Wave Functions with Valley Mixing

Before we investigate solitons, we examine the effect of the boundary conditions of the armchair edges. As we will see below, unlike zigzag edges, the armchair edges **couple** the two valleys. Consider an infinitely long armchair graphene nanoribbon. Because the armchair edges couple the valleys, the A- and B- components of the **total** wave function are given by a linear

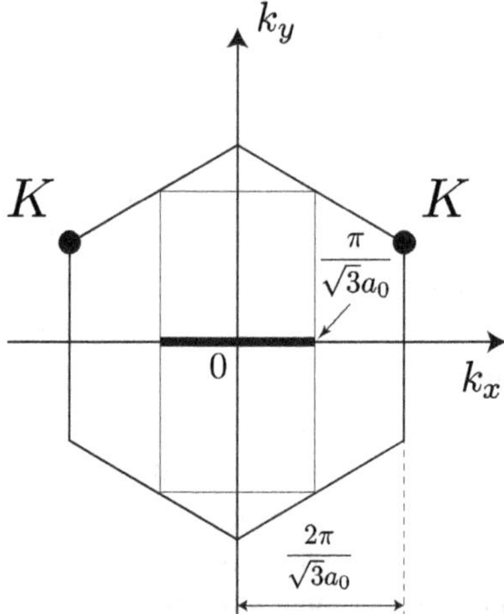

Fig. 16.10.　Black thick line represents the Brillouin zone of an armchair ribbon.

combination of $\vec{K}$ and $\vec{K}'$ valley wave functions

$$\Psi_A(\vec{r}) = e^{i\vec{K}\cdot\vec{r}}F_A(\vec{r}) + e^{i\vec{K}'\cdot\vec{r}}F'_A(\vec{r}),$$
$$\Psi_B(\vec{r}) = e^{i\vec{K}\cdot\vec{r}}F_B(\vec{r}) + e^{i\vec{K}'\cdot\vec{r}}F'_B(\vec{r}). \tag{16.14}$$

(Here the symbol $'$ is used for $\vec{K}'$ valley wave functions). Using degenerate perturbation theory we find that the four envelope wave functions [14] $(F_A(\vec{r}), F_B(\vec{r}), F'_A(\vec{r}), F'_B(\vec{r}))$ are eigenstates of the Hamiltonian

$$H = \begin{pmatrix} H_{\vec{K}} & 0 \\ 0 & H_{\vec{K}'} \end{pmatrix}, \tag{16.15}$$

where $H_{\vec{K}}$ and $H_{\vec{K}'}$ are the Dirac Hamiltonians equation (14.20). The $\vec{K}$ valley components of the envelope wave function are

$$F_{A,B}(\vec{r}) = e^{ik_x x} \begin{bmatrix} \phi_A(y) \\ \phi_B(y) \end{bmatrix}, \tag{16.16}$$

and the $\vec{K}'$ valley components are

$$F'_{A,B}(\vec{r}) = e^{ik_x x} \begin{bmatrix} \phi'_A(y) \\ \phi'_B(y) \end{bmatrix}. \tag{16.17}$$

Here we have assumed that **armchair edges are along the x-axis**, see Fig. 16.1. Note that the x and y parts of the wave function separate.

It should be noted that the wave functions are zero at both A and B sites of an armchair edge. The boundary condition on the A-components of the total wave function is $\Psi_A(y=0) = \Psi_A(y=w) = 0$, which gives

$$\Psi_A(y=0) = F_A(x,0) + F'_A(x,0) = 0,$$
$$\Psi_A(y=w) = e^{iK_y w} F_A(x,0) + e^{iK'_y w} F'_A(x,0) = 0,$$
(16.18)

where K_y and K'_y are the y-components of $\vec{K}$ and $\vec{K}'$. The boundary condition on the B-components is $\Psi_B(y=0) = \Psi_B(y=w) = 0$, which gives

$$\Psi_B(y=0) = F_B(x,0) + F'_B(x,0) = 0,$$
$$\Psi_B(y=w) = e^{iK_y w} F_B(x,0) + e^{iK'_y w} F'_B(x,0) = 0.$$
(16.19)

The wave functions $\phi_B(y)$ and $\phi'_B(y)$, given by Eqs. (16.16) and (16.17), are of type $e^{\pm i k_n y}$ (see Eq. (16.4)). The other components $\phi_A(y)$ and $\phi'_A(y)$ are found from Eq. (16.5). We find solutions satisfying the above boundary conditions

$$\psi_{s,n,k_x}(r) = \begin{pmatrix} F_A(\vec{r}) \\ F_B(\vec{r}) \\ F'_A(\vec{r}) \\ F'_B(\vec{r}) \end{pmatrix} = \frac{e^{ik_x x}}{2\sqrt{Lw}} \begin{pmatrix} e^{ik_n y} \\ e^{ik_n y} \\ -e^{-ik_n y} \\ -e^{-ik_n y} \end{pmatrix},$$
(16.20)

where the allowed wave vectors are (L is the length of the ribbon)

$$k_n = \frac{\pi n}{w} - \frac{2\pi}{3a_0}.$$
(16.21)

The corresponding eigenenergies are

$$\epsilon_{s,n}(k_x) = s\sqrt{k_x^2 + k_n^2},$$
(16.22)

where $s = +1$ and -1 are for the conduction and valence bands, respectively. These four-component wave functions show that the $\vec{K}$ and $\vec{K}'$ valleys are coupled, in contrast to the case of zigzag ribbons with two-component wave functions. It can be shown that no boundary states exist in armchair ribbons and only propagating modes exist.

The computed band structure shows the following results. According to the effective mass approximation [14], an armchair graphene nanoribbon

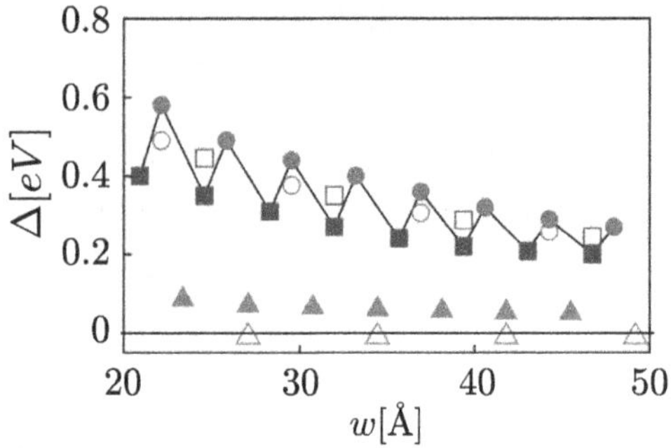

Fig. 16.11. Size of gap Δ of undoped graphene armchair ribbon is shown. Effective mass approximation results for the same widths are open squares, open circles, and open triangles. Local density approximation results [15] for widths $w = 3(M + 1)a_0$ (filled circles), $w = 3Ma_0$ (filled squares), and $w = 3(M + 2)a_0$ (filled triangles). Reprinted with permission from Ref. [16].

has a gap and is semiconducting when the ribbon width is $w = (3M + 1)a_0$ or $3Ma_0$, in qualitative agreement with the result of local density approximation [15] (M is an integer; $a_0 = \sqrt{3}a$ is the unit cell length of the graphene lattice, where $a = 1.42$ Å is the carbon–carbon distance). When $w = (3M + 2)a_0$, the effective mass approximation predicts no gap. However, according to the result of local density approximation, there is gap when $w = (3M+2)a_0$. Electronic properties of undoped graphene armchair ribbons thus display unusual dependence on ribbon **width**, see Fig. 16.11: the size of gap displays a highly oscillatory behavior as a function of width. This behavior is not present in zigzag ribbons, which has a different edge topology.

16.4. Domain Solitons Without Dimerization and Chiral Gauge Theory

A gap-edge soliton of an armchair ribbon may be described by **chiral gauge theory**. Such a soliton state can display an approximate **quantization of fractional charge** (the effects of electron repulsion will be included in Sec. 17.2). Armchair edges do not support edge states. However, a domain-wall soliton can be created inside an armchair ribbon. We describe below a domain-wall soliton of a semiconducting armchair graphene nanoribbon under a local tensile strain [4]. In contrast to polyacetylene, Kekulé-like bond alternation is not required to form such a domain-wall soliton. A soliton mode of a graphene nanoribbon connects sites of two sublattices with opposite chiralities. Some of the salient features of the domain-wall soliton are discussed in this section and the similarities and differences in comparison to those of polyacetylene are delineated.

- Chiral Gauge Theory of Graphene

In the continuum description of graphene, change of the site hopping parameters can be simulated by a chiral gauge field $\vec{A}_f(\vec{r})$ [17–19]. For the K valley, we have

$$H_{\vec{K}} = v_F \vec{\sigma} \cdot \left(\vec{p} - \frac{e}{c} \vec{A}_f(\vec{r}) \right),$$ (16.23)

where the Pauli spin matrices are $\vec{\sigma} = (\sigma_x, \sigma_y)$, $\vec{p}$ is the momentum operator, and v_F is the Fermi velocity of bulk graphene. Similarly, for the $\vec{K}'$ valley, we have

$$H_{\vec{K}'} = v_F \vec{\sigma}' \cdot \left(\vec{p} + \frac{e}{c} \vec{A}_f(\vec{r}) \right),$$ (16.24)

where $\vec{\sigma}' = (-\sigma_x, \sigma_y)$. Note that the signs of the gauge vector fields differ in these equations [20] — hence the name chiral vector potential. A chiral gauge vector potential is not a real vector potential. It describes the change in the hopping parameters in an effective Hamiltonian.

Let us give a short derivation of the chiral vector potential. Consider a two-dimensional graphene system. When site-dependent hopping parameters are present, the tight-binding Hamiltonian equation (14.1) is modified as follows:

$$H = -\sum_{\vec{R},i} t_i a_{\vec{R}}^\dagger b_{\vec{R}+\vec{\delta}_i} + \text{h.c.},$$ (16.25)

where the position vectors $\vec{\delta}_i$ are defined in Fig. 14.3. Here we make an approximation that the external perturbation only changes the hopping parameters and not the atomic positions. We Fourier transform the Hamiltonian using

$$a_{\vec{R}}^\dagger = \frac{1}{\sqrt{N_{\text{cell}}}} \sum_{\vec{k}} e^{-i\vec{k}\vec{R}} a_{\vec{k}}^\dagger,$$ (16.26)

$$b_{\vec{R}+\vec{\delta}_i} = \frac{1}{\sqrt{N_{\text{cell}}}} \sum_{\vec{k}} e^{i\vec{k}\cdot(\vec{R}+\vec{\delta}_i)} b_{\vec{k}},$$ (16.27)

and obtain the following Hamiltonian

$$H = -\sum_{\vec{k},i} t_i e^{i\vec{k}\cdot\vec{\delta}_i} a_{\vec{k}}^{\dagger} b_{\vec{k}}. \tag{16.28}$$

(N_{cell} is the number of unit cells of the ribbon). The off-diagonal element of matrix Hamiltonian $h(k)$ is (see Eq. (14.7))

$$h_{12} = -\sum_{i} t_i e^{i\vec{k}\cdot\vec{\delta}_i} = -\sum_{i} t_i e^{i(\vec{K}+\vec{q})\cdot\vec{\delta}_i}, \tag{16.29}$$

Note that $\vec{k} = \vec{K} + \vec{q}$, where $\vec{q}$ is measured from valley point $\vec{K}$. Since $|\vec{q}| \ll |\vec{K}|$ we expand the matrix element as follows

$$h_{12} = -(t - \delta\gamma_1)e^{i(\vec{K}\cdot\vec{\delta}_1+\vec{q}\cdot\vec{\delta}_1)} - (t - \delta\gamma_2)e^{i(\vec{K}\cdot\vec{\delta}_2+\vec{q}\cdot\vec{\delta}_2)} - (t - \delta\gamma_3)e^{i(\vec{K}\cdot\vec{\delta}_3+\vec{q}\cdot\vec{\delta}_3)}$$

$$= \frac{3t}{2}(q_x - iq_y) - \frac{\sqrt{3}}{2}(\delta\gamma_2 - \delta\gamma_3) + i\left(\delta\gamma_1 - \frac{\delta\gamma_2 + \delta\gamma_3}{2}\right) + \mathcal{O}(\delta\gamma_i q_j). \tag{16.30}$$

The parameter $\delta\gamma_1$ represents the change of the hoping integral of the bonds along the ribbon direction while $\delta\gamma_2$ and $\delta\gamma_3$ represent those of side bonds, see Fig. 16.12.

The first part of h_{12} together with its Hermitian conjugate, gives the matrix Hamiltonian

$$h_{\vec{K}}(\vec{q}) = v_F \vec{\sigma} \cdot \vec{q}, \tag{16.31}$$

with $v_F = 3t/2$. Similarly, the second and the third terms will form the deformation-induced perturbed Hamiltonian

$$h_{\text{def}} = -\frac{\sqrt{3}}{2}(\delta\gamma_2 - \delta\gamma_3)\sigma_x - \left(\delta\gamma_1 - \frac{\delta\gamma_2 + \delta\gamma_3}{2}\right)\sigma_y. \tag{16.32}$$

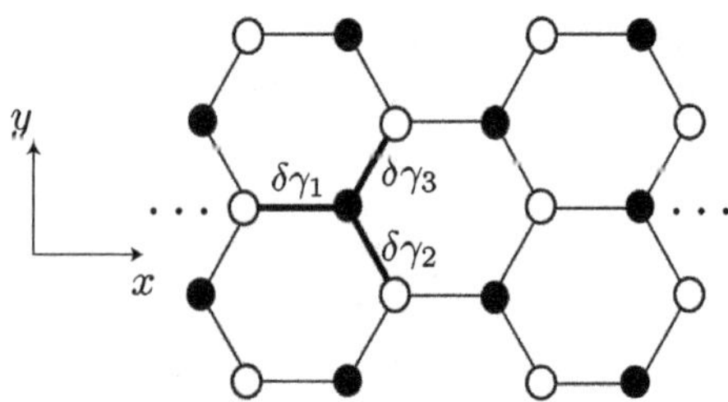

Fig. 16.12.   Three hopping parameters of each site, $(\gamma_1, \gamma_2, \gamma_3)$, are modified.

Similar result may be obtained for the $\vec{K}'$ valley. Comparing thus obtained results with the effective Hamiltonian Eqs. (16.23) and (16.24), we finally obtain the components of chiral vector potential $\vec{A}_f$

$$\frac{ev_F}{c}A_{f,x} = \frac{\sqrt{3}}{2}(\delta\gamma_2 - \delta\gamma_3),$$

$$\frac{ev_F}{c}A_{f,y} = \delta\gamma_1 - (\delta\gamma_2 + \delta\gamma_3)/2. \qquad (16.33)$$

There is no z component of $\vec{A}_f(\vec{r})$.

• Model for Domain Wall Soliton

Sasaki *et al.* [21] explored a graphene nanoribbon with a domain-wall soliton connecting two distinct bonding structures such that each structure has a bond alternation similar to a dimerization in polyacetylene. However, Jeong *et al.* [6] showed that dimerization is **not** required to generate a soliton in graphene-related systems. They applied a local tensile strain perpendicular to the ribbon direction. Such a strain force induces changes in the hopping parameters between the carbon atoms in a rectangular area D, where strain is applied, see Fig. 16.13. They showed that a soliton can be formed on two **zigzag edge sites** near the boundary of region D of carbon atoms with opposite chirality.

Let us explain how this result is obtained. In the effective mass approximation, the strain produces the chiral vector potential $\vec{A}_f(\vec{r}) = (0, A_f\theta(\vec{r}), 0)$, where $\theta(\vec{r}) = 1$ in D and 0 outside it. The chiral vector is a constant in D with its direction along the y-axis. The y-component of $\vec{A}_f(\vec{r})$ is given in Eq. (16.33). For tensile strain $\delta\gamma_2 = \delta\gamma_3$ so the x-component of $\vec{A}_f(\vec{r})$ is also zero

$$\frac{ev_F}{c}A_{f,x} = \frac{\sqrt{3}}{2}(\delta\gamma_2 - \delta\gamma_3) = 0. \qquad (16.34)$$

The z-component of the induced magnetic field is non-zero only at the edges of D located at $x = \pm d/2$

$$B_f = A_f\left[\delta(x + d/2) - \delta(x - d/2)\right]. \qquad (16.35)$$

The values of the magnetic field have different signs at $x = \pm d/2$.

An eigenstate of the total Hamiltonian can be written as a linear combination of the propagating modes $\psi_{s,n,k_x}(r)$ that are solutions of H (see Eqs. (16.20)–(16.22)). The eigenstates are written as a linear combination

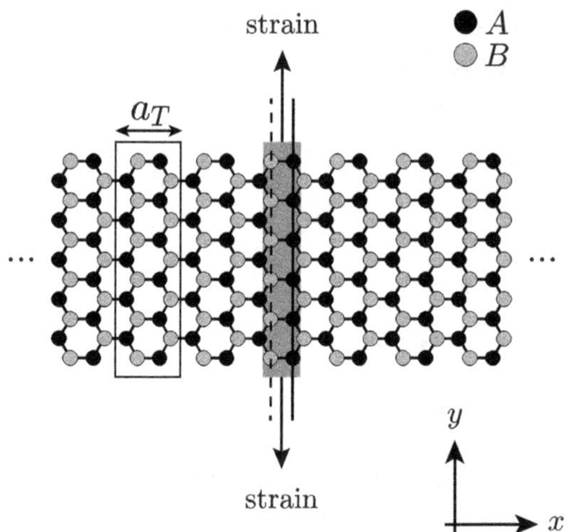

Fig. 16.13. Armchair graphene nanoribbon with local tensile strain applied perpendicular to ribbon direction. A and B zigzag edge sites are shown, respectively, on the solid and dashed lines. The ribbon is periodic along the ribbon direction. The length of the unit cell is $a_T = \sqrt{3}a_0$. Strain is applied only in region D (dark rectangle in the figure). Reprinted with permission from Ref. [7].

of states of the lowest-energy conduction subband ($s = 1$ and $n = 1$) and highest energy valence subband ($s = -1$ and $n = 1$)

$$\psi(\vec{r}) = \sum_{k_x} [C_{1,1,k_x}\psi_{1,1,k_x}(\vec{r}) + C_{-1,1,k_x}\psi_{-1,1,k_x}(\vec{r})], \qquad (16.36)$$

where C_{s,n,k_x} are the expansion coefficients. These expansion coefficients are eigenvectors of the Hamiltonian matrix. The matrix dimension is two times the number of discrete points in the Brillouin zone of k_x. In our approximation, only the lowest (highest) conduction (valence) subband is included.

For a ribbon with the shortest width, there are **two** solutions with $E \approx 0$ (see Fig. 16.14), representing gap states, i.e., a domain-wall soliton and antisoliton solutions with opposite energies [6]. (The number of gap states is proportional to the length of the zigzag edges.) The antisoliton state is unoccupied at half-filling. **Half** of the spectral weight of a soliton/antisoliton arises from the conduction band and **the other half** from the valence band. The solitonic states are **not** eigenstates of the chiral operator: the chirality of the wave function on the left and right hand edges of the domain is of A- and B-type, respectively (see Figs. 1.1 and 16.15).

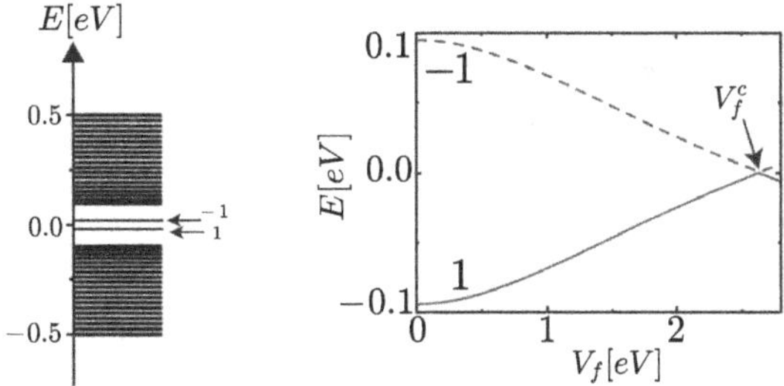

Fig. 16.14. (a) Eigenenergies of a ribbon in the presence of a stripe D with width $d = 5\text{Å}$, the number of $\vec{k}$-space grid points $N_{int} = 41$, and the strength of deformation potential $V_f = 2$ eV (the ribbon width and length are $w = 25a_0$ and $L_{\text{arm}} = 1504$ Å. Here $a_0 = 2.46$ Å is the length of the unit cell). Label $1(-1)$ in the gap denotes kink (antikink) with lower (higher) energy. (b) Kink and antikink energies as a function of deformation potential V_f for $w = 25a_0$. Near the critical value V_c^f they become almost degenerate. For larger (smaller) values of w the energy gap is smaller (larger). Reprinted with permission from Ref. [6].

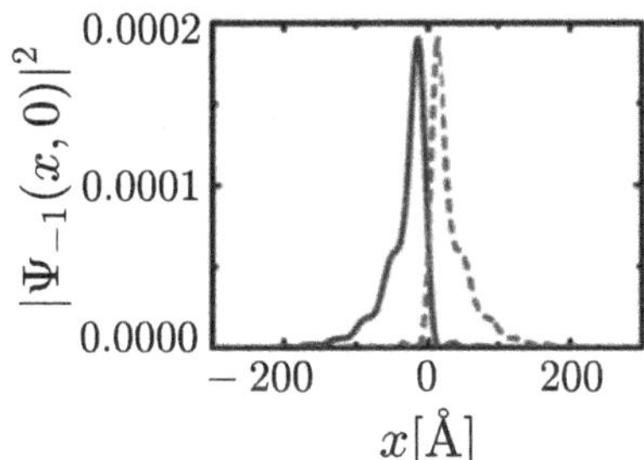

Fig. 16.15. Probability density of an antikink state with energy $E = 0.020\,\text{eV}$. The A-component (solid line) and B-component (dashed line) are shown for $N_{\text{int}} = 81$ and $V_f = 2\,\text{eV}$. Reprinted with permission from Ref. [6].

These states are mixed chiral states that connect A- and B-sublattices. However, a soliton and antisoliton are connected to each other by chiral operation (they have opposite energy). Figure 16.16 displays the pseudospin of a solitonic state. It rotates by π across the domain wall, indicative of a topological kink. The crucial feature of a topological kink is that the total change is π irrespective of the manner in which the phase changes as the coordinate changes.

Additional investigation shows that fractional charges may be present. However, the zigzag edges shown in Fig. 16.13 are spatially close to each other and charge fractionalization will not be precise because fractional charges overlap with each other. For a more precise charge quantization,

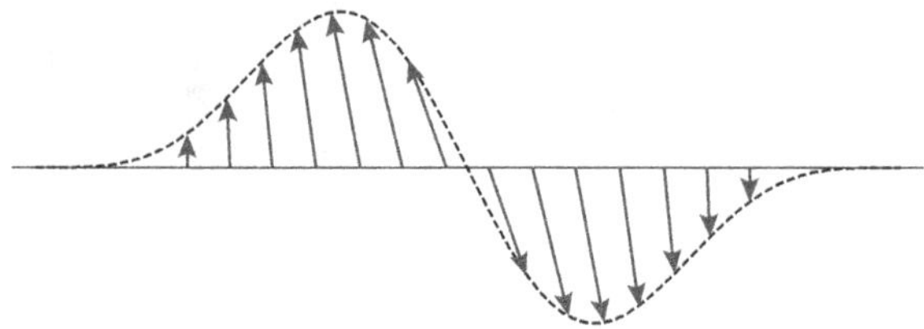

Fig. 16.16. Domain-wall soliton corresponds to a topological kink. Pseudospin vector rotates by an angle of π out of the plane. Reprinted with permission from Ref. [7].

the width of the strain region D should be larger, in addition to a sufficiently large energy separation between the bonding and antibonding states, which will reduce quantum fluctuations. Electron localization induced by a random potential will also reduce quantum fluctuations, see Sec. 3.8. More fractional charges will be present for a longer zigzag edges.

The presence of both a soliton and an antisoliton bound to a domain wall D is different from the result of polyacetylene which has only one soliton or antisoliton. The reason is fermion doubling, i.e., there are two valleys in graphene instead of one. Unlike that of polyacetylene, graphene lattice is two dimensional. Chiral symmetry then leads to a phenomenon called "fermion doubling" [22], meaning that two valleys $\vec{K}$ and $\vec{K}'$ exist instead of one (polyacetylene has only one valley). This has an interesting effect on the number of soliton modes per domain wall. A graphene zigzag nanoribbon with the shortest width has **two** domain-wall states, in contrast to polyacetylene with only one soliton per domain wall (for other values of the width the number of soliton modes is even [6]). Unusual spin and charge relations are not expected in graphene nanoribbons [6], as a **single** domain wall can simultaneously support both a soliton and an antisoliton. Note that, when both the soliton and antisoliton states are unoccupied, the valence band misses the total charge of e^- per spin, which is just the usual spin and charge relation.

Bibliography

[1] S. Ryu and Y. Hatsugai, Topological origin of zero-energy edge states in particle-hole symmetric systems, *Phys. Rev. Lett.* **89**, 077002 (2002).

[2] X.-G. Wen, Colloquium: Zoo of quantum-topological phases of matter, *Rev. Mod. Phys.* **89**, 041004 (2017).

[3] X.-G. Wen, Topological order: From long-range entangled quantum matter to a unified origin of light and electrons, *ISRN Condens. Matter Phys.* **2013**, 198710 (2013). ●

[4] Y. H. Jeong and S.-R. Eric Yang, Topological end states and Zak phase of rectangular armchair ribbon, *Ann. Phys.* **385**, 688 (2017).

[5] M. Fujita, K. Wakabayashi, K. Nakada, and K. Kusakabe, Peculiar localized state at zigzag graphite edge, *J. Phys. Soc. Jpn.* **65**, 1920 (1996).

[6] Y. H. Jeong, S. C. Kim, and S.-R. Eric Yang, Topological gap states of semiconducting armchair graphene ribbons, *Phys. Rev. B* **91**, 205441 (2015).

[7] S.-R. Eric Yang, Soliton fractional charges in graphene nanoribbon and polyacetylene: Similarities and differences, *Nanomaterials* **9**, 885 (2019). •

[8] J. Cai, P. Ruffieux, R. Jaafar, M. Bieri, T. Braun, S. Blankenburg, M. Muoth, A. P. Seitsonen, M. Saleh, X. Feng, K. Müllen, and R. Fasel , Atomically precise bottom-up fabrication of graphene nanoribbons, *Nature* **466**, 470 (2010).

[9] M. Kolmer, A.-K. Steiner, I. Izydorczyk, W. Ko, M. Engelund, M. Szymonski, A.-P. Li, and K. Amsharov, Rational synthesis of atomically precise graphene nanoribbons directly on metal oxide surfaces, *Science* **369**, 571 (2020).

[10] P. Delplace, D. Ullmo, and G. Montambaux, Zak phase and the existence of edge states in graphene, *Phys. Rev. B* **84**, 195452 (2011).

[11] J. Zak, Berry's phase for energy bands in solids, *Phys. Rev. Lett.* **62**, 2747 (1989).

[12] N. W. Ascroft and N. David Mermin, *Solid State Physics* (Thomson Learning, London, 1976). •

[13] A. H. Castro Neto, F. Guinea, N. M. R. Peres, K. S. Novoselov, and A. K. Geim, The electronic properties of graphene, *Rev. Mod. Phys.* **81**, 109 (2009). •

[14] L. Brey and H. A. Fertig, Electronic states of graphene nanoribbons studied with the Dirac equation, *Phys. Rev. B* **73**, 235411 (2006).

[15] L. Yang, C.-H. Park, Y.-W. Son, M. L. Cohen, and S. G. Louie, Quasiparticle energies and band gaps in graphene nanoribbons, *Phys. Rev. Lett.* **99**, 186801 (2007).

[16] J. W. Lee, S. C. Kim and S.-R. Eric Yang, Spintronic properties of one-dimensional electron gas in graphene armchair ribbons, *Solid State Commun.* **152**, 1929 (2012).

[17] K.-I. Sasaki and R. Saito, Pseudospin and deformation-induced gauge field in graphene, *Prog. Theor. Phys. Suppl.* **176**, 253 (2008).

[18] C. L. Kane and E. J. Mele, Size, shape, and low energy electronic structure of carbon nanotubes, *Phys. Rev. Lett.* **78**, 1932 (1997).

[19] M. I. Katsnelson and A. K. Geim, Electron scattering on microscopic corrugations in graphene, *Phil. Trans. R. Soc. A* **366**, 195 (2008).

[20] R. Jackiw and S.-Y. Pi, Chiral gauge theory for graphene, *Phys. Rev. Lett.* **98**, 266402 (2007).

[21] K.-I. Sasaki, R. Saito, M. S. Dresselhaus, K. Wakabayashi, and T. Enoki, Soliton trap in strained graphene nanoribbons, *New J. Phys.* **12**, 103015 (2010).

[22] E. Witten, Three Lectures on topological phases of matter, *Riv. Nuovo Cimento*, **39**, 313 (2016). •

Chapter 17

Interacting Graphene Nanoribbon and Mott Insulator

"It is different with the magicians. They are, to use mathematical jargon, in the orthogonal complement of where we are and the working of their minds is for all intents and purposes incomprehensible....Richard Feynman is a magician of the highest caliber."

Marc Kac

"To discover something new you have to take the time to make every mistake possible along the way."

Richard P. Feynman

"Unfortunately, creative people are at their most creative when writing their autobiographies. Historians have scrutinized their diaries, notebooks, manuscripts, and correspondence looking for signs of the temperamental seer periodically struck by bolts from the unconscious. Alas, they have found that the creative genius is more Salieri than Amadeus."

Steven Pinker

"Knowledge and technology are usually generated by stochastic tinkering rather than by top-down directed research."

Nassim Nicholas Taleb

Generation mechanism of gap and topological objects has become one of the centers of the study of graphene material [1]. In this chapter, we will show that interacting graphene nanoribbons at half-filling are a Mott insulator.[a] The presence of a gap suggests that graphene nanoribbons may be a symmetry protected topological insulator. Indeed, as guaranteed by chiral symmetry, a zigzag edge possesses degenerate chiral edge states that are delocalized along the ribbon direction.

When a gap is present, a topological Zak phase may be present. (Note that the Zak phase is a bulk and not a boundary property, see Sec. 5.4.) A calculation shows that the Zak phase of both periodic armchair and zigzag graphene nanoribbons is 0 mod 2π [3]. When a periodic graphene nanoribbon is transversely cut, a rectangular ribbon with two zigzag edges and two armchair edges is formed. The resulting armchair edges have no edge states, but zigzag edges do. These results are both consistent with the possible values of the Zak phase [4] 0 and 2π. This is an example of bulk-edge correspondence.

In graphene zigzag nanoribbons, a gap accompanies edge magnetism and makes it more interesting: each zigzag edge is ferromagnetic but opposite zigzag edges are coupled antiferromagnetically [5,6]. Note that the opposite zigzag edges have opposite chiralities [7,8].

17.1. Mott Insulator and Edge Antiferromagnetism of Zigzag Graphene Nanoribbon

Among different magnetic states of zigzag ribbons, the antiferromagnetic state has the lowest ground state energy when the ribbon width is less than $\sim$100Å [6]. This result is valid not only in the Hubbard model with on-site repulsion but also in models with other types of electron interactions.

- Model Hamiltonian

On-site repulsive interaction between electrons produce an excitation gap and zigzag graphene nanoribbons become a Mott insulator. There are still zigzag edge states but their energy is non-zero. At half-filling, there is no spin flip scattering that couples spin-up and -down electrons in the Hubbard Hamiltonian. The total interacting Hamiltonian of zigzag graphene

---

[a]For an accessible introduction to Mott insulator, see Ref. [2]. In a Mott insulator electron interactions induce an excitation gap. A well-known example is the antiferromagnetic gap in Hubbard model at half-filling, see Sec. 6.12.

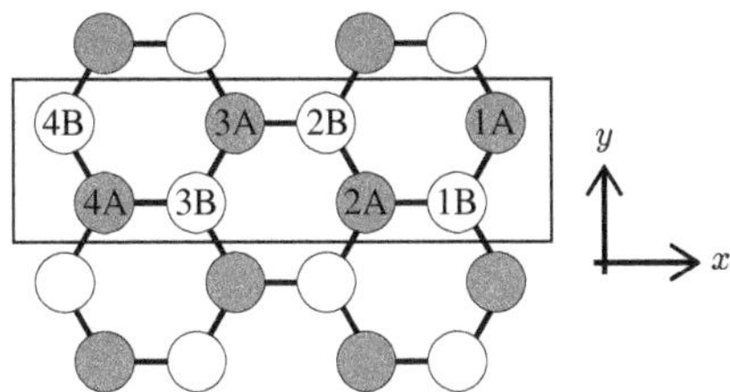

Fig. 17.1. Sites in a unit cell are labeled as shown. Ribbon direction is vertical.

nanoribbon at half filling is

$$H = -t \sum_{l,\sigma} \sum_{m=1}^{M/2} [a^\dagger_{l,m,\sigma} b_{l-1,m,\sigma} + b^\dagger_{l,m,\sigma} a_{l,m,\sigma}]$$

$$-t \sum_{l,\sigma} \sum_{m=1}^{M/2} a^\dagger_{l,m+1,\sigma} b_{l,m,\sigma} + U \sum_{l} \sum_{m=1}^{M/2} n_{l,m,\uparrow} n_{l,m,\downarrow} + \text{h.c.} \tag{17.1}$$

Here l denotes a unit cell, m denotes a site in a unit cell, and M is the number of carbon atoms in a unit cell (the ribbon shown in Fig. 17.1 with eight carbon atoms in a unit cell). The operators $a^\dagger_{l,m}$ and $b^\dagger_{l,m}$ create an electron, respectively, at A type and B type carbon atoms.

The first term represents hopping between neighboring cells. The second term describes an intra cell downward tunneling, see Fig. 17.1. The third denotes an intra cell sideway tunneling. Applying the mean-field approximation we find a bilinear Hamiltonian

$$H = - \sum_{\langle ij \rangle \sigma} t_{ij} c^\dagger_{i\sigma} c_{j\sigma}$$

$$+ U \sum_{i} (n_{i\uparrow} \langle n_{i\downarrow} \rangle + \langle n_{i\uparrow} \rangle n_{i\downarrow}) - \frac{U}{2} \sum_{i} (n_{i\uparrow} + n_{i\downarrow}), \tag{17.2}$$

where each site is represented by $i = (l,m)$. At half-filling, the average site occupation number is $n_f = 1$. In this Hamiltonian, the constant terms are chosen such that the Fermi energy is $E_F = 0$ (see Fig. 17.2 for an explanation). The mean-field Hamiltonian can be split into spin-up and -down parts

$$H \rightarrow H_\uparrow + H_\downarrow. \tag{17.3}$$

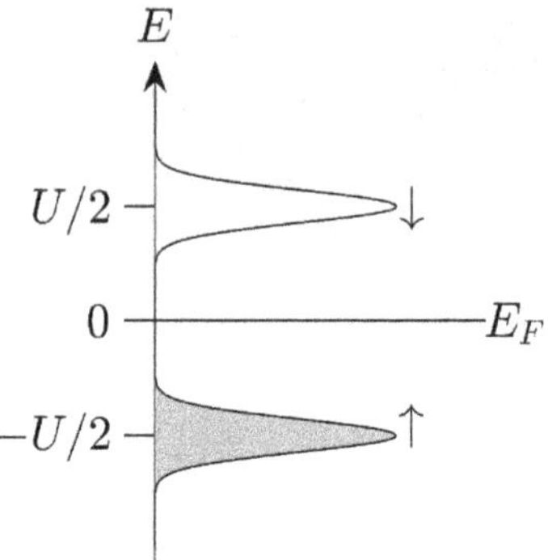

Fig. 17.2.   The DOS of a nearly isolated site i is displayed (the coupling of the impurity with the rest of the system gives a level broadening). Suppose $\langle n_{i\downarrow}\rangle = 0$ and $\langle n_{i\uparrow}\rangle = 1$. Then the on-site terms give the following energy $E_i = U(n_{i\uparrow}\langle n_{i\downarrow}\rangle + \langle n_{i\uparrow}\rangle n_{i\downarrow}) - \frac{U}{2}(n_{i\uparrow} + n_{i\downarrow}) \rightarrow -U/2$, which represents the lower energy peak in the figure. Suppose both spin-up and -down states are occupied so that $\langle n_{i\downarrow}\rangle = \langle n_{i\uparrow}\rangle = 1$. This gives the upper peak $E_i \rightarrow +U/2$ in the figure. Particle–hole symmetry at half-filling suggests that the Fermi level is zero and is in the middle of $-U/2$ and $+U/2$.

Note that there is no spin flip scattering that couples spin-up and -down electrons in this decoupling scheme.[b]

We introduce new destruction operators $A_{km\sigma}$ and $B_{km\sigma}$

$$
\begin{aligned}
a_{lm\sigma} &= \frac{1}{\sqrt{L}} \sum_k e^{ik_y y_{lm}} A_{km\sigma}, \\
b_{lm\sigma} &= \frac{1}{\sqrt{L}} \sum_k e^{ik_y y_{lm}} B_{km\sigma},
\end{aligned}
\tag{17.4}
$$

where L is the length of the ribbon. We find $H_\uparrow$ and $H_\downarrow$ in the k-space (see Exercise 17.1)

$$
H = H_\uparrow + H_\downarrow
\tag{17.5}
$$

---

[b]However, spin-up electrons are still influenced by spin-down electrons because of the on-site repulsion between electrons of opposite spins. The Hamiltonian $H_\uparrow$ contains spin-down occupation numbers $\langle n_{i\downarrow}\rangle$.

with

$$H_\sigma = \sum_{k \in \text{BZ}} [A_{k1\sigma}^\dagger, B_{k1\sigma}^\dagger, A_{k2\sigma}^\dagger, B_{k2\sigma}^\dagger \ldots] H_\sigma(k) \begin{bmatrix} A_{k1\sigma} \\ B_{k1\sigma} \\ A_{k2\sigma} \\ B_{k2\sigma} \\ \vdots \end{bmatrix}. \tag{17.6}$$

The Hamiltonian matrix $H_\uparrow(k)$ is

$$H_\uparrow(k) = \begin{pmatrix} U\langle n_{\downarrow,1A}\rangle & -g_k & 0 & 0 & 0 & \ldots \\ -g_k & U\langle n_{\downarrow,1B}\rangle & -1 & 0 & 0 & \ldots \\ 0 & -1 & U\langle n_{\downarrow,2A}\rangle & -g_k & 0 & \ldots \\ 0 & 0 & -g_k & U\langle n_{\downarrow,2B}\rangle & -1 & \ldots \\ 0 & 0 & 0 & -1 & U\langle n_{\downarrow,3A}\rangle & \ldots \\ \vdots & \vdots & \vdots & \vdots & \vdots & \end{pmatrix}. \tag{17.7}$$

There is another Hamiltonian matrix for spin-down electrons, which has the same structure but with reversed spins. In this Hamiltonian matrix [5]

$$g_k = 2\cos(ka_0/2), \tag{17.8}$$

and sites in a unit cell are ordered as 1A, 1B, 2A, 2B, etc., see Fig. 17.1. The dimension of the matrix M is equal to the number of carbon atoms in a unit cell. Here $t = 1$.

Exercise 17.1. Derive the Hamiltonians $H_\uparrow$ and $H_\downarrow$ of Eq. (17.3). Hint: Use the Hamiltonian equation (17.1) and the decoupling scheme Eq. (17.2).

For a given value of k and σ, the expansion coefficients $\{\alpha_{km\sigma}, \beta_{km\sigma}\}$ for $m = 1, \ldots, M/2$ constitute eigenstates

$$\sum_m (\alpha_{km\sigma} A_{km\sigma}^\dagger + \beta_{km\sigma} B_{km\sigma}^\dagger)|0\rangle \tag{17.9}$$

of the Hamiltonian matrix. The site occupation number can be found by summing over all occupied eigenstates i:

$$n_{km\sigma} = \sum_{i \in occ} (|\alpha_{km\sigma}^i|^2 + |\beta_{km\sigma}^i|^2). \tag{17.10}$$

The initial site occupation numbers are chosen as those of an antiferromagnetic state. The occupation numbers are to be determined self-consistently for each k by the iteration process. At each wave vector k there are $M/2$ **occupied** electrons with spin σ.

Exercise 17.2. Derive the Hamiltonian matrix given in Eq. (17.7). Hint: Use the result of Exercise 17.1.

- Band Structure and Edge Antiferromagnetism

Theoretical calculations [6] of disorder-free zigzag ribbons show that antiferromagnetism is favored over ferromagnetism for ribbon widths smaller than 100 Å. In the rest of book we assume these ribbon width values.

We use the antiferromagnetic Hartree-Fock initial state and compute band structure displayed in Fig. 17.3. Electron–electron interactions play

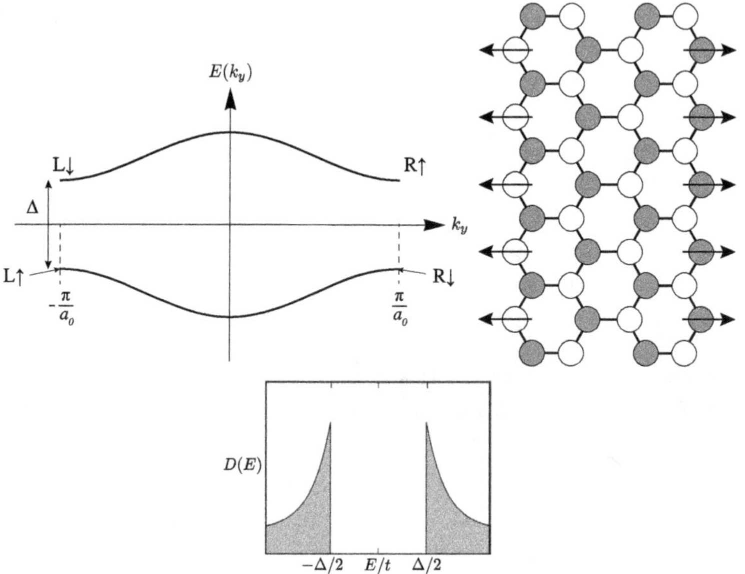

Fig. 17.3. **Upper left**: Schematic band structure of an interacting zigzag graphene nanoribbon. Only one conduction and one valence bands are displayed. States near the wave vectors $k = \pm\pi/a_0$ are shown: R and L represent states confined on the right and left zigzag edges, respectively (a_0 is the unit cell length of the zigzag graphene nanoribbon). The small arrows indicate spins. Another degenerate ground state can be obtained by exchanging $\uparrow$ and $\downarrow$ spins (this will be discussed in Chapter 18). The states near the Brillouin zone edges produce the van Hove singularities in the DOS. When a unit cell has N_{cell} carbon atoms there are N_{cell} bands. **Upper right**: Zigzag edge antiferromagnetism in the presence of electron interaction. In the limit $U \to \infty$ the edge site spin value is $s_{iz} = \frac{1}{2}(\langle n_{i,\uparrow}\rangle - \langle n_{i,\downarrow}\rangle) = \pm 1/2$. **Lower**: DOS of an interacting zigzag graphene nanoribbon.

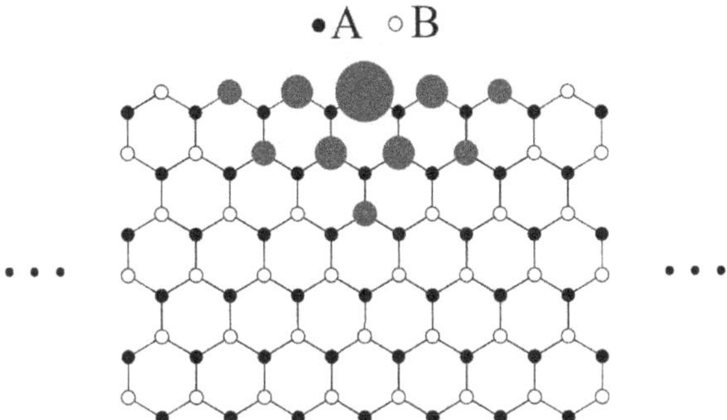

Fig. 17.4. Site probability of a zigzag edge state with nearly **zero** energy is plotted. Such a state is **chiral**: the site probability is non-zero only on A- or B-type carbon atoms and not on both. It is located either on the upper or lower zigzag edge. (These edges have opposite chirality.)

an important role in zigzag graphene nanoribbons: they produce an excitation gap [6,9] near the Fermi energy, see Fig. 17.3. This means a zigzag graphene nanoribbon is a Mott insulator. Electron–electron interactions also induce antiferromagnetism between the left and right zigzag edges while generating ferromagnetism on each zigzag edge, see Fig. 17.3. Degenerate edge states that are extended along the ribbon direction are found. Although edge states in non-interacting ribbons can be either localized on one edge or divided between the opposite edges (see Fig. 16.8), on-site repulsion removes this ambiguity: it will repel electrons with opposite spins to the opposite zigzag edges. Thus bonding and antibonding states are not stable, and the edge states are all localized entirely on the left or right edge, see Fig. 17.4. According to the Hund's rule, degenerate single electron states generate ferromagnetism on each edge.

17.2. Fractional Charge?

We saw in Sec. 16.4 that a domain-wall soliton may be constructed from a linear combination of chiral zigzag edge modes. It is instructive to analyze this problem in the presence of electron-electron interactions [10]. Suppose we consider an undoped periodic armchair graphene nanoribbon (ribbon ring) with length L_{arm} and **short** width w, and assume, for simplicity, that only two bonds are affected by tensile strain, as shown in Fig. 17.5(a). When $t' \neq 0$, a domain soliton mode forms, which can be expressed as the linear combination $\phi_L \pm \phi_R$ of the chiral modes of the left and right zigzag edges near the dashed line shown in Fig. 17.5(a). These linear combinations

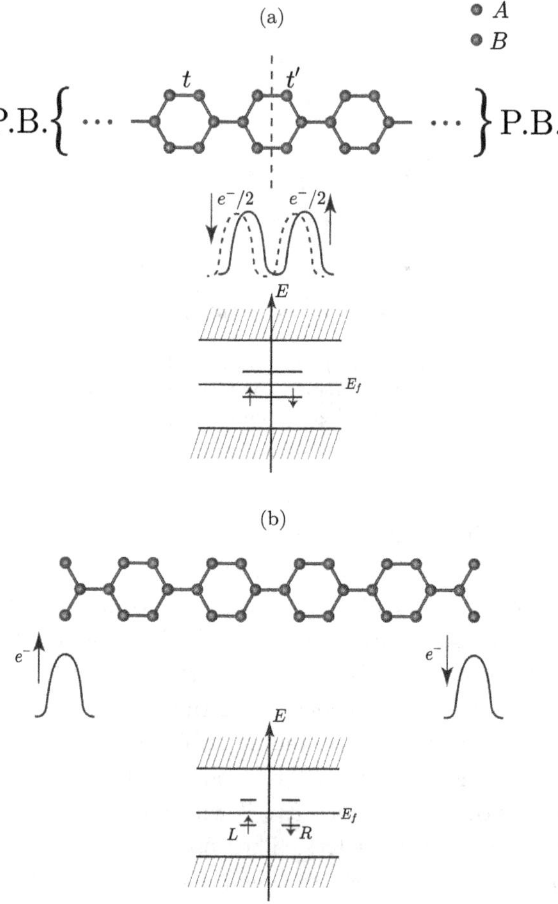

Fig. 17.5. (a) Periodic armchair graphene nanoribbon with shortest width and modified hopping t' between two neighboring zigzag edges near the center. P.B. represents periodic boundary conditions. A schematic illustration of the probability densities of the soliton gap states is also shown. The energy spectrum of the gap states for a finite on-site electron repulsion U is also shown. In the limit of small U, the energy splitting of the solitonic gap states vanishes and the energies approach zero. (b) When $t' = 0$ (corresponding to cutting of the bond), a rectangular armchair graphene nanoribbon is realized. In this case, fractional charges do not form on the left and right zigzag edges, in contrast to the case with $t' \neq 0$. Reprinted with permission from Ref. [11].

correspond to bonding and antibonding states. An electron in a solitonic state resides near these two neighboring zigzag edges with **opposite chirality** [10]. Its tight-binding probability density is divided equally between the left and right zigzag edges (the wavefunction exhibits some overlap between the edges). This domain soliton mode is robust as the edge chiral modes ϕ_L and ϕ_R are topologically protected [12, 13] (they persist as long as the zigzag edges are not destroyed).

In the limit where $t' \to 0$, a rectangular armchair graphene nanoribbon is realized, as shown in Fig. 17.5(b). One might think that bonding and antibonding states persist in a rectangular armchair graphene nanoribbon with the electron probability density is divided between the left and right zigzag edges, producing two well-separated $e^-/2$ fractional charges. However, as we have seen in this chapter, the effect of on-site repulsion is more significant in comparison to the effect of forming bonding and antibonding states. It makes the bonding and antibonding states $\phi_L \pm \phi_R$ no longer stable on the zigzag edges. Instead **chiral** modes ϕ_L and ϕ_R develop on the left and right zigzag edges (see Fig. 17.5(b)). This is because on-site repulsion induces antiferromagnetic coupling between edge charges and mitigates the formation of fractional edge charges.

17.3. Non-quantized Zigzag Edge Charge in Staggered Potential

We learned in Sec. 5.5 that, in one-dimensional systems inversion symmetry guarantees the quantization of the Zak phase. But suppose now that inversion symmetry of a quasi-one-dimensional ribbon is broken. An example of such a system is a periodic armchair ribbon in the presence of a staggered potential defined as follows: the site potential energy is $\epsilon_i = V/2$ on sublattice A and $\epsilon_i = -V/2$ on sublattice B. A staggered potential will break reflection and inversion symmetries of a graphene ribbon, see Fig. 17.6. The band structure of a periodic armchair ribbon in the presence of a staggered potential is shown in Fig. 17.7. There is no spin-splitting. Note that a periodic armchair does not have zigzag edges.

Let us cut the periodic armchair ribbon in such a way that a **rectangular ribbon** with two **short** zigzag edges and two **long** armchair edges are created, see Fig. 17.6. In this case, a boundary edge can **affect** the bulk property and induce spin-splitting, see Ref. [3]. In a staggered potential, the sites on the opposite zigzag edges experience different electric potentials, as shown in Fig. 17.6. Is the charge of a zigzag edge state an integer and does the edge antiferromagnetism survive?

We find that the difference in the zigzag edge occupation numbers of spin-up and -down electrons continuously decreases as the strength of the staggered potential increases, leading to reduced edge antiferromagnetism.

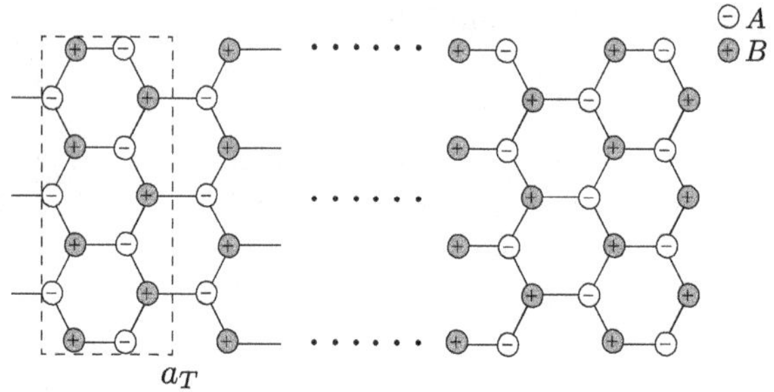

Fig. 17.6. A rectangular armchair graphene ribbon has two long armchair edges and two short zigzag edges. Left (right) end is made of A (B)-type sites. Note that the unit cell length is $a_T = \sqrt{3}a_0$.

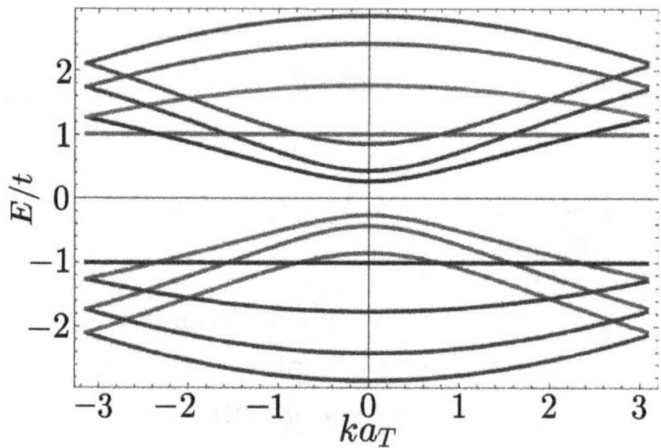

Fig. 17.7. Band structure of an armchair ribbon in the presence of a staggered potential with the parameters $V = 0.3t$ and $U = 0.5t$. The ribbon width is $w = 3a_0$, Reprinted with permission from Ref. [3].

Let us describe this process in more detail. Taking electron–electron interactions into account we find that, as the strength of the staggered potential varies, three types of couplings between the **zigzag edge states** occur: antiferromagnetic without or with spin-splitting, and paramagnetic without spin-splitting. We find that a **spin-splitting** is present only in the staggered potential region $0 < V < V_C$.[c] At $V = 0.03t$, the occupied end states are located on the opposite zigzag edges with antiparallel spins (see Fig. 17.8(a)). However, at a larger value of $V = 0.3t$, they are both located

---

[c]In the initial stage of the Hartree–Fock iteration process, the total site occupation number of each site is chosen to be one, but the spin-up and -down site occupation numbers are chosen to be slightly different. Such a choice will generate a non-zero spin-splitting when the iteration process is completed.

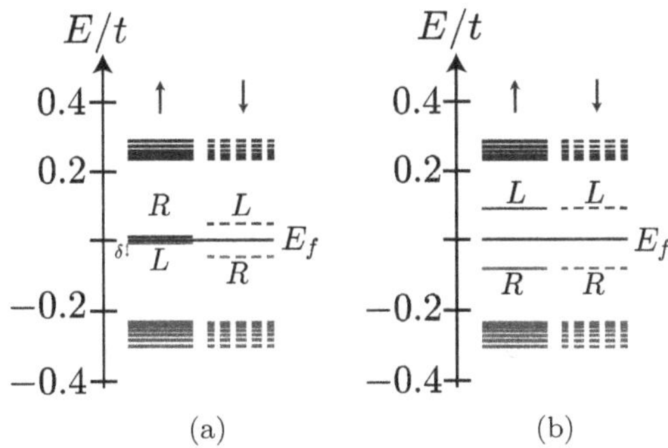

Fig. 17.8. Schematic drawing of energy spectrum for spin-up and -down end/gap states of a rectangular armchair graphene ribbon with width $w = 3a_0 = 3\sqrt{3}a$. Two values of the strength of the staggered potential, $0.03t$ (a) and $0.3t$ (b), are used (on-site $U = 0.5t$). The first value is just below the critical value V_c and the second one is far above it. Spin splitting δ is present in (a) while not in (b). Symbols R and L mean that an end state is localized, respectively, on the right and left zigzag edges. Reprinted with permission from Ref. [3].

on the right zigzag edge with zero total spin (see Fig. 17.8(b)). The transition from the antiferromagnetic state to the paramagnetic state at $V = 0$ thus goes through an intermediate spin-split antiferromagnetic state, and this spin-splitting disappears suddenly at V_c, see Fig. 17.8.

- Zak Phase and Edge Charge

Let us invoke the bulk-edge correspondence and relate the boundary charges to the Zak phase in the presence of a staggered potential. The Zak phase can be computed as follows. Since the valence bands have band crossings, we must use the expressions for the Zak phase given in Eq. (5.74)

$$Z/2\pi = \frac{1}{2\pi} \sum_{s=0}^{M-1} \mathrm{Arg}[\det\langle C_{l,k_s} | C_{l',k_{s+1}}\rangle]. \tag{17.11}$$

The expansion coefficients $C_{l,k}$ are eigenvectors. Wave vector space is divided into M intervals. The edge charge is expected to be related to Z as follows: $Q = eZ/2\pi$. But when is this result correct? Here we will argue that it is correct provided that the edge is insulating.

First let us discuss the Zak of periodic armchair and zigzag ribbons in the **absence** of a staggered potential. The numerical value of the Zak phase of both zigzag and armchair graphene nanoribbons is 0 mod 2π.[d] This means that the edge charge can be zero or an integer. As we saw in Sec. 17.1, an actual calculation of the boundary zigzag charges gives an integer edge charge. This quantization of the edge charges is related to chiral symmetry, see Sec. 5.5. Graphene nanoribbons are thus topologically non-trivial.

Now let us discuss the Zak in the **presence** of a staggered potential. For small and large values of the staggered potential V, the end charge of a rectangular armchair graphene ribbon can be connected to the Zak phase of the periodic armchair graphene ribbon with the same width, see Fig. 17.9. For $V \gtrsim V_c$ (see Fig. 17.9) the edge charges computed from the numerical work and the Zak phase approximately agree with each other. They also agree when $V/t \gg 1$. Note that for small and large values of V the edge charges are, respectively, close to 1 and 0. This agreement is expected because the zigzag edges are almost insulating as the occupation number

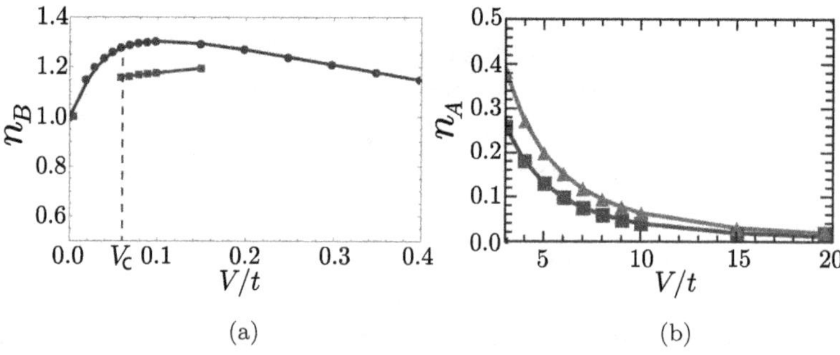

(a)   (b)

Fig. 17.9. (a) Charge per right end site, including both spin-up and -down electrons, computed from the Zak phase (circles), $n_B = q/e = \frac{Z}{2\pi}$, see Eq. (5.59) ($U = 0.5t$). The numerically computed occupation number is also shown (squares) for $V \gtrsim V_c$. (In the staggered potential region $0 < V < V_c$ a spin-splitting is present, and therefore, the comparison with the value computed using the Zak phase is not meaningful.) (b) Charge per left end site computed from the Zak phase (triangles) and from the occupation number (squares) in the region $V/t \gg 1$. Parameters are identical to those of Fig. 17.7. Reprinted with permission from Ref. [3].

---

[d]For an armchair ribbon, this result seems to be in conflict with $Z_{2D}(k_{\parallel}) = \pi$ mod 2π of a **graphene sheet**, obtained by taking the integration variable to be the wave vector along the direction of the armchair ribbon (see Sec. 16.1). However, since the unit cells of the two systems are different there is no reason why these two results should be identical (the cell of a graphene and zigzag nanoribbon are displayed, respectively, in Figs. 14.3 and 17.1). The result obtained from a simple procedure of dimensional reduction appears to be incorrect.

of each edge site is nearly filled or empty (see Sec. 3.8). For other values of V, the computed edge charge does not agree with the one computed from the Zak phase.

17.4. Doped Disorder-Free Armchair Nanoribbon

The properties of doped graphene armchair nanoribbons depend significantly on the width of the ribbon. The main reason for this is because the band gap of undoped armchair nanoribbons varies significantly on the width. When the system is doped, electrons form a one-dimensional electron gas and occupy subbands. Since the dielectric constant is rather small, $\epsilon \sim 1$, the effects of Coulomb interactions are expected to be rather strong. Many-body self-energies should thus be one order of magnitude larger in comparison to the corresponding values of an electron gas of ordinary semiconductors with $\epsilon \sim 10$. We explain this effect below.

We will assume, for simplicity, that only the lowest conduction subband is occupied and disorder is not present. The electron energy gets renormalized due to the mutual electron interactions. The exchange self-energy is (see Eq. (6.76))

$$\Sigma_{\mathrm{ex}}(k) = -\sum_{k'} n_{k'} \langle kk'|V|k'k\rangle. \tag{17.12}$$

Here n_k is the Fermi distribution function and V is the Coulomb potential. The states $|k\rangle$ are given in Eq. (16.16). They are identical to the eigenstates of a **non-interacting** armchair ribbon. This is because, even in the presence of electron interactions, translational invariance is present and $|k\rangle$ are the eigenstates of the Hartree–Fock equations (see Exercise 6.10). There should also be a Hartree self-energy correction but it cancels with the contribution from the background positive charges, as explained in detail in Ref. [14]. (The contribution from the occupied valence band electrons is already included in the value of the gap Δ.) It is assumed that the electron density is such that the exchange self-energy correction is smaller than the Fermi energy. This allows a partially spin-polarized electron gas, see Fig. 17.10. The magnitude of the exchange self-energy depends on k. However, we will assume that the subband shifts rigidly. The subband energy shifts due to the exchange self-energy, $\Sigma_{\mathrm{ex}}(k = 0)$, is shown in Fig. 17.11 (the exchange self-energy is always negative). We find that the magnitude of this self-energy depends significantly on ribbon width. Note that, in contrast to the band gap Δ

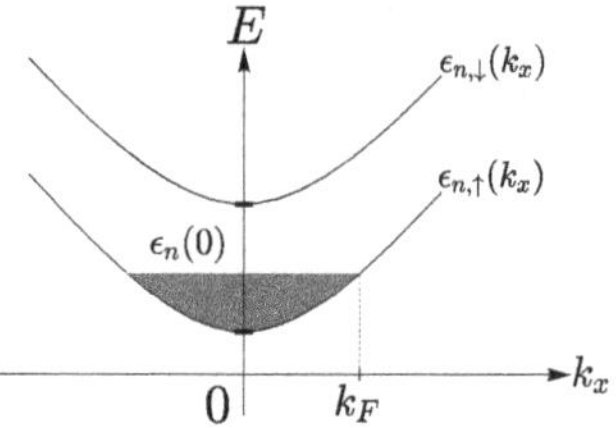

Fig. 17.10.   For a smaller electron density the exchange self-energy can be larger than the Fermi energy and the electron gas is fully spin-polarized (the nth spin-up subband is occupied while the spin-down subband is unoccupied). The Fermi wave vector is $k_{F,\uparrow}$. Reprinted with permission from Ref. [15].

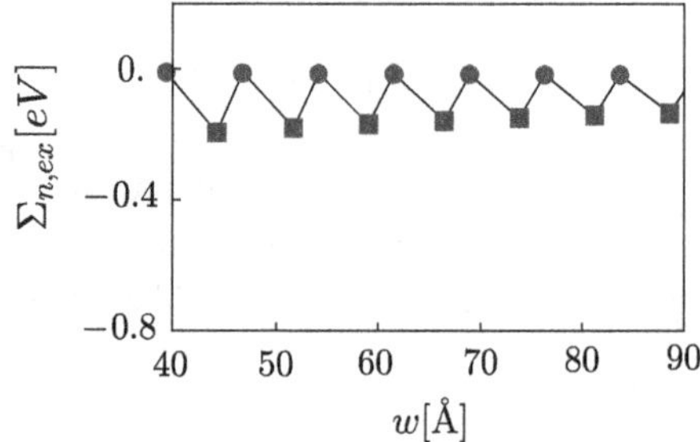

Fig. 17.11.   Exchange self-energy of the $n = 0$ subband of doped graphene for ribbon widths $w = 3(M + 1)a_0$ (circles) and $w = 3Ma_0$ (squares), where $a_0 = 2.46\text{Å}$ and M is an integer. The effective mass approach is used. Here the dielectric constant is $\epsilon = 3$ and the dimensionless Fermi wavevector is $k_F a_0 = 0.07$. Reprinted with permission from Ref. [15].

(Fig. 16.11), the exchange self-energy increases, on average, with ribbon width. Magnetic properties of the one-dimensional electron gas may thus depend sensitively on the **width** of the ribbon. Width-dependent properties are a consequence of eigenstates that have a subtle width-dependent mixture of $\vec{K}$ and $\vec{K}'$ states, and can be understood by examining the wave function overlap that appears in the expression for the many-body exchange self-energy. Ferromagnetic and paramagnetic states may be used for the purposes of spintronics.

Bibliography

[1] J. K. Pachos, Manifestations of topological effects in graphene, *Contemporary Phys.* **50**, 375 (2009). ●

[2] F. Gebhard, *The Mott Metal-Insulator Transition* (Springer, New York, 2013). •

[3] Y. H. Jeong and S.-R. Eric Yang, Topological end states and Zak phase of rectangular armchair ribbon, *Ann. Phys.* **385**, 688 (2017).

[4] J. Zak, Berry's phase for energy bands in solids, *Phys. Rev. Lett.* **62**, 2747 (1989).

[5] K. Wakabayashi, K.-I. Sasaki, T. Nakanishi, and T. Enoki, Electronic states of graphene nanoribbons and analytical solutions, *Sci. Technol. Adv. Mater.* **11**, 054504 (2010).

[6] L. Pisani, J. A. Chan, B. Montanari, and N. M. Harrison, Electronic Structure and Magnetic Properties of Graphitic Ribbons, *Phys. Rev. B* **75**, 064418 (2007); L. Yang, C.-H. Park, Y.-W. Son, M. L. Cohen, and S. G. Louie, Quasiparticle energies and band gaps in graphene nanoribbons, *Phys. Rev. Lett.* **99**, 186801 (2007).

[7] L. Brey and H. A. Fertig, Electronic states of graphene nanoribbons studied with the Dirac equation, *Phys. Rev. B* **73**, 235411 (2006).

[8] T. Stauber, P. Parida, M. Trushin, M.V. Ulybyshev, D.L. Boyda, and J. Schliemann, Interacting electrons in graphene: Fermi velocity renormalization and optical response, *Phys. Rev. Lett.* **118**, 266801 (2017).

[9] M. Fujita, K. Wakabayashi, K. Nakada, and K. Kusakabe, Peculiar localized state at zigzag graphite edge, *J. Phys. Soc. Jpn.* **65**, 1920 (1996).

[10] Y. H. Jeong, S. C. Kim, and S.-R. Eric Yang, Topological gap states of semiconducting armchair graphene ribbons, *Phys. Rev. B* **91**, 205441 (2015).

[11] S.-R. Eric Yang, Soliton fractional charges in graphene nanoribbon and polyacetylene: Similarities and differences, *Nanomaterials* **9**, 885 (2019).

[12] S. Ryu and Y. Hatsugai, Topological Origin of zero-energy edge states in particle–hole symmetric systems, *Phys. Rev. Lett.* **89**, 077002 (2002).

[13] P. Delplace, D. Ullmo, and G. Montambaux, Zak phase and the existence of edge states in graphene, *Phys. Rev. B* **84**, 195452 (2011).

[14] N. W. Ascroft and N. David Mermin, *Solid State Physics* (Thomson Learning, London, 1976). •

[15] J. W. Lee, S. C. Kim and S.-R. Eric Yang, Spintronic properties of one-dimensional electron gas in graphene armchair ribbons, *Solid State Commun.* **152**, 1929 (2012).

Chapter 18

Topologically Ordered Interacting Disordered Ribbon: Mott-Anderson insulator and Soliton Fractional Charge

"I have no special talent. I am only passionately curious."

Albert Einstein

"Asking the right questions of your data and knowing what you are looking to find is a critical component for gaining insights from your data that drive specific actions."

Kevin Hanegan

"That is a good explanation- hard to vary, because all its details play a functional role. ...Some of the resulting ideas have enormous reach: they explain more than what they were originally designed to."

David Deutsch

We saw in Chapter 8 that the electron–phonon interaction acts as a singular perturbation in the presence of a sharp Fermi surface. In graphene zigzag nanoribbons, a disorder potential acts as a singular perturbation on the zigzag edge states. In contrast to what is usually expected of a topological insulator, even a weak disorder potential profoundly affects an interacting zigzag graphene nanoribbon [1]. (Another example of a topologically ordered insulator where disorder plays an essential role is quantum Hall bars displaying the quantized off-diagonal conductivity.) We will learn in this chapter that disorder breaks chiral symmetry of graphene nanoribbons, inducing a transition from a symmetry protected phase to a topologically ordered phase [2]. A topologically ordered zigzag nanoribbon is a Mott–Anderson insulator [3] with local magnetic moments.[a]

This new phase has several fascinating properties, such as double ground state degeneracy, spin–charge separation, and $e^-/2$ soliton fractional charges residing on A- and B-zigzag edges. It is highly non-trivial to understand how randomness can lead to fractional charge quantization. Moreover, these fractional charges are expected to be Abelian anyons called semions. These non-perturbative effects are a result of the subtle interplay between topology of the underlying lattice, electron correlations, and disorder. They can all be derived from the simple Hubbard model using the well-known Hartree–Fock method, provided that one chooses the right Hartree–Fock ground state Ansatz. In this chapter, following spirit of the variational approaches of the toric code, BCS, and Laughlin wave functions, we guess the correct Hartree–Fock ground state and explore properties of interacting disordered zigzag ribbons. (In Chapters 19 and 21, we will study quasiparticle excitations from the ground state.)

18.1. Model of Solitonic State of Zigzag Nanoribbons at Half-Filling

- Singular Effect of Disorder Potential

In interacting zigzag nanoribbons, disorder has a unique effect not present in other systems, where a weak disorder potential behaves similarly to a **singular** perturbation [5] on zigzag edge electronic states, generating drastic changes in the energy spectrum. This effect leads to formation of instantons (see Sec. 3.5) and gives rise to **mixed chiral edge states**. This implies that these edge states occur independent of the disorder potential range, density, and strength.

Let us analyze the scattering of the left and right edge states by a short-ranged disorder potential. Consider a spin-up electron near $k = \frac{\pi}{a_0}$ with the wave function $\phi_{R\uparrow}$ localized on the right zigzag edge. For a short-ranged potential, a significant wave vector transfer in a backscattering occurs for $|k - k'| \sim 1/a_0$ [6], see Fig. 18.1. Such a short-ranged disorder potential couples the **chiral** zigzag edge state $\phi_{R\uparrow}$ to another chiral zigzag edge state $\phi_{L\uparrow}$ on the opposite zigzag edge near $k = -\frac{\pi}{a_0}$, as shown in Fig. 18.1

---

[a]Those who wish to learn more about Mott–Anderson insulators can look up Ref. [4]. It has several good chapters on interacting disordered systems.

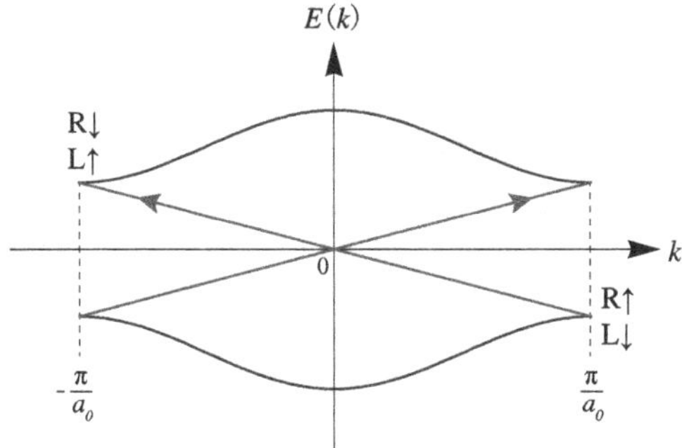

Fig. 18.1. States localized on the right and left zigzag edges are denoted, respectively, by R and L. The long arrows indicate the coupling, induced by a short-ranged disorder potential, between states $R \uparrow$ and $L \uparrow$ or $R \downarrow$ and $L \downarrow$. It is a singular perturbation on zigzag edge states. The resulting spin-split gap states was shown in Fig. 1.4. Reprinted with permission from Ref. [2].

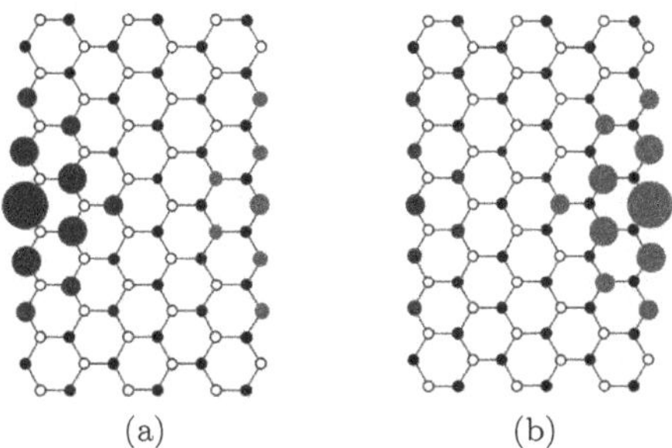

Fig. 18.2. Probability density of two nearly degenerate chiral edge states is displayed (These states are not completely chiral when $k \neq \pm\pi/a_0$). Note that the ribbon is periodic along the vertical direction.

(ϕ_R and ϕ_L are depicted in Figs. 16.8 and 18.2). This process produces the bonding $\frac{1}{\sqrt{2}}(\phi_L + \phi_R)$ or antibonding $\frac{1}{\sqrt{2}}(\phi_L - \phi_R)$ edge state. These bonding and antibonding states are divided nearly equally between the left and right zigzag edges. It suggests that solitonic fractional charges may exist in disordered zigzag graphene nanoribbons. These mixed chiral states are an example of an instanton. An instanton is a non-perturbative effect. When it is applied to a double well, one finds that the energy splitting between the bonding and antibonding states is $\Delta E \propto e^{-1/\lambda}$, which is singular in the coupling constant λ, see Sec. 3.5.

In interacting zigzag ribbons, **van Hove singularities** are present in the DOS, displaying a sharply peaked DOS near the gap edges, see Fig. 17.3. Because of this sharp peak, the singular effect of disorder is amplified: it generates numerous edge states with fractionalized charges between the opposite zigzag edges and these edge states have a broad energy distribution.

- Diagonal and Off-Diagonal Disorder

The properties of the Dirac electrons in zigzag graphene nanoribbons are significantly affected by impurities [6–8]. In our model, we can consider both diagonal and off-diagonal disorder. Below we present results only for diagonal disorder since off-diagonal disorder gives similar results. Both types of disorder do not change the underlying topological structure of the lattice, see Figs. 18.3 and 18.4.

We model disorder by placing randomly N_{imp} defects or impurities at carbon sites $\vec{R}_j$. We take the following simple model for the disorder potential

$$V_i = \sum_j^{N_{\text{imp}}} \epsilon_j e^{-|\vec{r}_i - \vec{R}_j|^2/d^2}, \tag{18.1}$$

where d is the range of the potential. The values of ϵ_j and d depend on the type of charged impurities in the substrate and defects in graphene. Defects have $d \sim a_0$ while impurities have $d \gg a_0$. Note that when $d = 0$, the disorder potential is defined such that it is finite only for $\vec{r}_i = \vec{R}_j$ (in a continuous model this implies that the effective range is equal to the size of a carbon atom $\sim a_0$). When $d \sim a_0\,(d \gg a_0)$ the potential is short

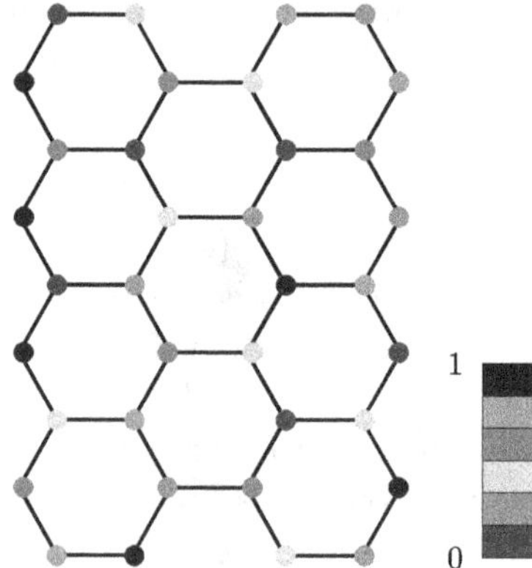

Fig. 18.3. Diagonal disorder: site energies V_i are varied randomly. Colors represent strength of on-site disorder potential. A zigzag edge site is connected to two other carbon atoms while a site away from the edges is connected to three other carbon atoms. But the hopping parameter t is the same for all sites. Reprinted with permission from Ref. [2].

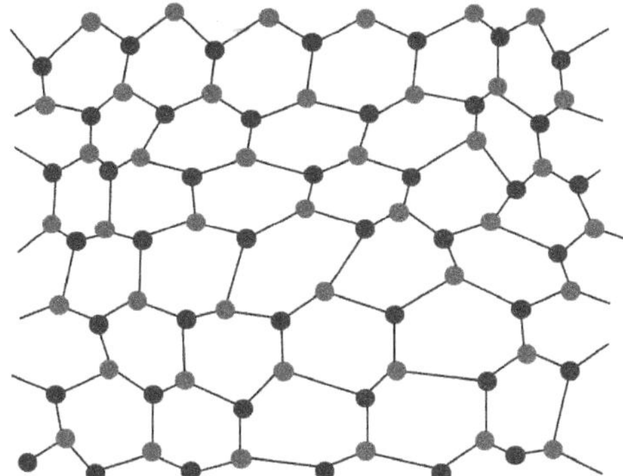

Fig. 18.4. Off-diagonal disorder: random network of hexagons consisting of A and B carbon atoms. Again a zigzag edge site is connected to **two** other carbon atoms while a site away from the edges is connected to **three** other carbon atoms. The hopping parameters t_{ij} are not the same for all sites. The zigzag edges have definite chirality, consisting of A or B carbon atoms. In contrast, armchair edges have mixed chirality. Reprinted with permission from Ref. [2].

(long)-ranged. The strength of the potential ϵ_j is chosen randomly from the energy interval $[-\Gamma, \Gamma]$. In the self-consistent Born approximation, disorder strength is characterized by the parameter $\Gamma \sqrt{n_{\mathrm{imp}}}$, where $n_{\mathrm{imp}} = N_{\mathrm{imp}}/N_C$ is the ratio between the numbers of impurities and total carbon atoms. Thus the number of impurities or defects is also relevant in determining the strength of the disorder potential. Defects with a short-ranged potential are more convenient to use in numerical studies of the quantization of a fractional charge.

- Mean Field Hamiltonian

We include both electron–electron interactions and disorder in a tight-binding model at half-filling. The interplay between on-site repulsion U and disorder can be treated using the self-consistent Hartree–Fock approximation. When $U = 0$ disorder can be treated exactly in this method (see Sec. 10.9) while in the other limit, where disorder is absent, interaction effects can be represented well by the Hartree–Fock approximation, which is widely used in graphene related systems [9]. When both disorder and interactions are present the self-consistency provides an excellent approximation [10]. The results of the Hartree–Fock approximation are consistent with those of density functional theory [11, 12].

At half-filling, electron interactions can be treated using a simple Hubbard Hamiltonian. As we mentioned before, this means that the mean-field

Hamiltonian can be written as a sum of spin-up and -down Hamiltonians with no spin flip scattering, even in the presence of disorder

$$H_{\mathrm{MF}} = H_\uparrow + H_\downarrow = -\sum_{\langle ij\rangle\sigma} t_{ij} c_{i\sigma}^\dagger c_{j\sigma} + \sum_{i\sigma} V_i c_{i\sigma}^\dagger c_{i\sigma}$$

$$+ U\sum_i (n_{i\uparrow}\langle n_{i\downarrow}\rangle + \langle n_{i\uparrow}\rangle n_{i\downarrow}) - \frac{U}{2}\sum_i (n_{i\uparrow} + n_{i\downarrow}). \tag{18.2}$$

Here $c_{i\sigma}^\dagger$ and $n_{i\sigma}$ are the electron creation and occupation operators at site i with spin σ. Since the translational symmetry is broken, the Hamiltonian is written in the site representation. In the hopping term, the summation is over the nearest neighbor sites. The eigenstates and eigenenergies of the resulting tight-binding Hamiltonian matrix of the bilinear Hartree–Fock Hamiltonian are solved self-consistently. The Fermi energy is set to $E_F = 0$ (see Fig. 17.2).

- Hartree–Fock Ground State Ansatz

There are several nearly degenerate Hartree–Fock ground states. The state with the global minimum of the Hartree–Fock total energy does not display solitons, and the Hartree–Fock tunneling excitation spectrum originating from this state does not display the effects of solitons. The solitonic ground state has a local energy minimum. However, we will show in Chapter 21, using the density matrix renormalization approach, that the solitonic Ansatz is the true ground state. In the following, we assume that it is the true ground state.

Solitons in a disordered interacting nanoribbon may be numerically generated as follows. We start the iteration process with the following paramagnetic initial values of the site occupation numbers $n_{i,\sigma}$ with a small spin-splitting:

$$n_{i,\uparrow} + n_{i,\downarrow} = 1,$$

$$n_{i,\uparrow} = \frac{1}{2} + \delta_i,$$

$$\sum_i n_{i,\uparrow} = N_C, \tag{18.3}$$

where δ_i are small random numbers (either positive or negative). These conditions give $N/2$ number of electrons with spin $\uparrow$ or $\downarrow$. However, at each site, although the total site occupation number is one, the spin-up and -down site occupation numbers are slightly different. Such a choice will generate a **non-zero spin-splitting** when the iteration process is completed. This choice of the Hartree–Fock ground state Ansatz is crucial. (With a different initial spin configuration, representing an antiferromagnetic state $n_{i,\sigma} = \pm 1$, a phase without solitons is generated.) This choice of the Hartree–Fock ground state Ansatz is crucial in generating a Mott-Anderson insulator with fractional anyon charges. Zigzag edges of a disordered interacting zigzag ribbon profoundly affect the bulk properties as they induce bulk spin-splitting (armchair edges do not). This feature is different from the case of quantum Hall states where the gapless edge modes are a bulk property. Note that quantum Hall states exist also in spherical geometry that has no boundaries.

The self-consistent occupation numbers $\langle n_{i\sigma} \rangle$ in the Hamiltonian are the sum of the probabilities to find electrons of spin σ at site i

$$\langle n_{i\sigma} \rangle = \sum_{E \leq E_F} |\psi_{i\sigma}(E)|^2. \tag{18.4}$$

The sum is over the occupied eigenstates $\psi_{i\sigma}(E)$ with energy E below the Fermi energy E_F. We will define the weak disorder regime as the regime where the ratio between disorder strength and interaction strength is $\kappa \equiv \Gamma\sqrt{n_{\mathrm{imp}}}/U \ll 1$. For a large Hamiltonian matrix, the problem may be solved using sparse matrix diagonalization technique.[b] We use $U/t = 1$ unless stated otherwise.

18.2. Basic Properties of Interacting Disordered Zigzag Ribbon

In this section, we present numerical results that show how disorder can profoundly affect properties interacting zigzag graphene nanoribbons.

• Nearly Degenerate States

As we mentioned in Sec. 18.1, depending on the initial values of the site occupation numbers, the Hartree–Fock approximation can generate different nearly degenerate many-body states. Our calculation of the total Hartree–Fock energy of each phase, given by Eq. (6.105), shows that the many-body state with solitons is a local minimum of energy and not the global minimum.

Suppose that the Hartree–Fock states with and without solitons $|\Psi_1\rangle$ and $|\Phi_1\rangle$ have nearly degenerate energies, the following argument suggests

---

[b]Mathematica has excellent routines for sparse matrix diagonalization.

that the linear combination $|\Psi\rangle = C|\Psi_1\rangle + D|\Phi_1\rangle$ may give a better variational ground state. From $H|\Psi\rangle = E|\Psi\rangle$, we find that the expansion coefficients satisfy

$$\begin{pmatrix} E_1 & \langle\Psi_1|H|\Phi_1\rangle \\ \langle\Phi_1|H|\Psi_1\rangle & E_2 \end{pmatrix} \begin{pmatrix} C \\ D \end{pmatrix} = E \begin{pmatrix} 1 & \langle\Psi_1|\Phi_1\rangle \\ \langle\Phi_1|\Psi_1\rangle & 1 \end{pmatrix} \begin{pmatrix} C \\ D \end{pmatrix}.$$

$$(18.5)$$

Here the Hartree–Fock total energies are $E_1 = \langle\Psi_1|H|\Psi_1\rangle$ and $E_2 = \langle\Phi_1|H|\Phi_1\rangle$. Note that $|\Psi_1\rangle$ and $|\Phi_1\rangle$ are eigenstates of H_{MF} and not eigenstates of the original Hamiltonian H. The off-diagonal Hamiltonian matrix elements may be estimated using

$$\langle\Psi_1|H|\Phi_1\rangle = \sum_i \langle\Psi_1|H|\Psi_i\rangle\langle\Psi_i|\Phi_1\rangle, \qquad (18.6)$$

where $\{\Psi_i\}$ are the complete set of Hartree–Fock Slater determinant wave functions. One can numerically check that the overlap $\langle\Psi_1|\Phi_1\rangle$ is non-zero. This suggests that $|\Psi_1\rangle$ and $|\Phi_1\rangle$ are coupled by H.

There is a more systematic and better method to go beyond the Hartree–Fock approximation in the presence of nearly degenerate energy states. One can apply a variational scheme called a multi-configuration self-consistent field method [13]. In this approach, both linear expansion coefficients of Slater determinant states and single particle orbitals in each determinant state are chosen optimally. The exact diagonalization method is not applicable to graphene nanoribbons because the dimension of the relevant Hilbert space is too large. However, the matrix product version of the density matrix renormalization group [14] may be applied to investigate site spins in the ground state of a zigzag nanoribbon [15]. Here we are mainly interested in the properties the soliton state, and thus the Hartree–Fock approximation suffices for our purposes.

- Site Spins of Interacting Disordered Zigzag Ribbon

The net site spin values of the soliton state of an interacting disordered zigzag ribbon are shown in Fig. 18.5. We see that the edge magnetization develops magnetic domain walls and that magnetization is nearly zero at some sites. (These reconstructions can be explained by disorder induced fractional charges that accompany solitons, see Sec. 19.4.) The bulk magnetization in the interior region of the ribbon is non-uniform and is disordered. Note that it is **smaller** in comparison to that of the edge magnetization but it displays rather similar variations.

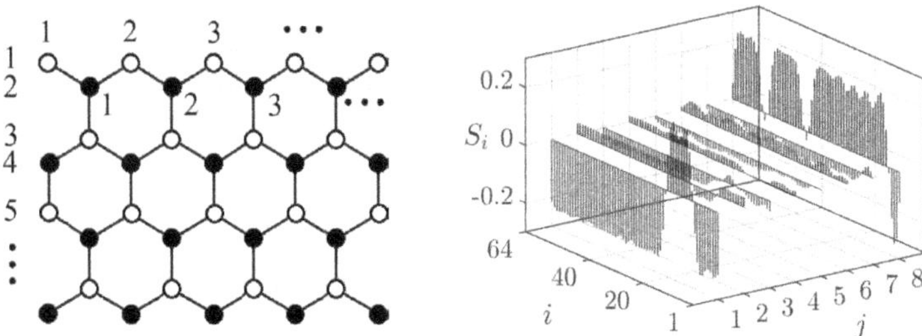

Fig. 18.5. **Left**: Vertical lengths are measured in number of horizontal carbon lines and horizontal lengths in number of vertical carbon lines. **Right**: Net site spin values $S_i = S_{i\uparrow} + S_{i\downarrow} = n_{i\uparrow} - n_{i\downarrow}$ are plotted, where $n_{i\sigma}$ is the site occupation number for spin σ. The index j labels horizontal carbon lines depicted in the upper figure. This result is for one disorder realization with diagonal disorder. Each zigzag edge displays modulated ferromagnetism. The parameters are $U = t$. Reprinted with permission from Ref. [2].

• Doubly Degenerate Soliton States

Since the mean field Hamiltonian splits into spin-up and -down parts the soliton state can be written as a product of spin-up and -down Slater determinants

$$\Psi_1 = \Psi_{L,\uparrow}(\vec{r}_1, \ldots, \vec{r}_{N/2})\Psi_{R,\downarrow}(\vec{r}_{N/2+1}, \ldots, \vec{r}_N), \tag{18.7}$$

where $\Psi_{L,\uparrow}$ ($\Psi_{R,\downarrow}$) describes $N/2$ electrons with spin $\sigma = \uparrow$ ($\downarrow$), including electrons localized on the left (right) zigzag edge (the total number of electrons is N). However, note that each of them also contains mixed chiral edge single-particle states. It is immediately clear that there is another degenerate state with all the spins reversed (see Fig. 18.6)

$$\Psi_2 = \Psi_{R,\uparrow}(\vec{r}_1, \ldots, \vec{r}_{N/2})\Psi_{L,\downarrow}(\vec{r}_{N/2+1}, \ldots, \vec{r}_N). \tag{18.8}$$

One cannot transform between these two degenerate groundstates using a local operator. $e^-/2$ edge fractional charges exist only when these two sublattices have opposite site spin expectation values, which is also why doubly degenerate ground states exist. Note that topologically ordered fractional quantum Hall liquids and toric code spin liquids also have degenerate ground states, see Secs. 11.10 and 13.3, respectively. In the fractional quantum Hall case, the existence of fractional charges is also intimately related to the ground state degeneracy.

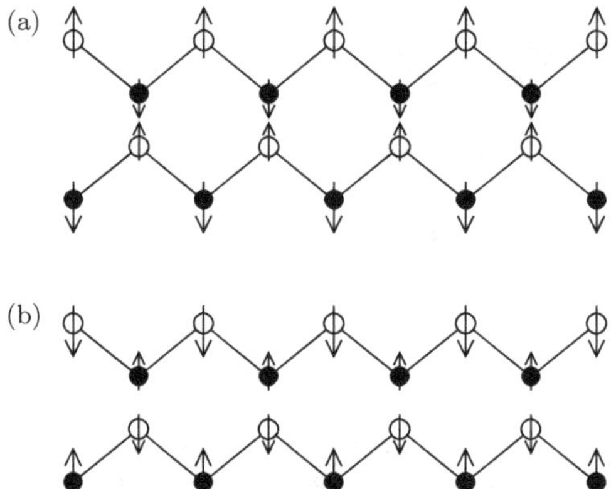

Fig. 18.6. Site spins of the boundary edges of two degenerate groundstates in the absence of disorder (spin directions of A- and B-sublattices are switched between them).

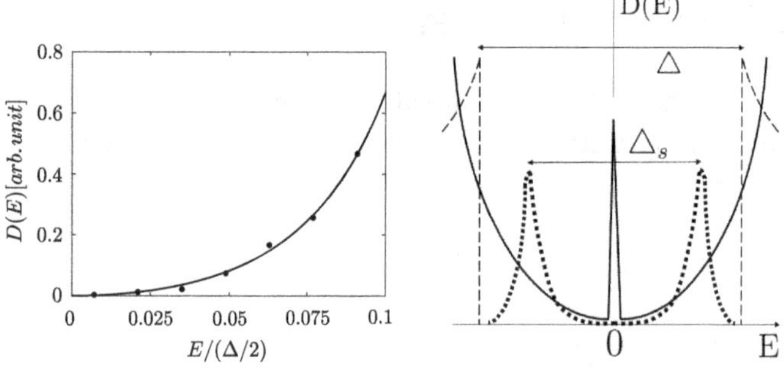

Fig. 18.7. **Left**: The DOS near $E = 0$ of a half-filled interacting disordered zigzag graphene nanoribbon averaged over many disorder realizations. The parameters are $U = t$, $\Gamma = 0.03t$, and $N_D = 2511$. The ribbon length is $L = 1232.5$Å. Histogram interval 0.014. Reprinted with permission from Ref. [2]. **Right**: Schematic drawings of DOS. Solid line: the DOS for non-interacting zigzag ribbon in the limit $\Gamma \to 0$. The states in the peak near $E \approx 0$ represent zigzag edge states delocalized along the zigzag edge direction. Dashed line: the DOS of disorder-free interacting zigzag ribbon. It is zero inside the gap. Dotted line: the tunneling DOS of interacting disordered zigzag ribbon at half-filling in the weak disorder regime. The states in the gap are mixed chiral zigzag edge states that are localized along the zigzag edge direction. The enlarged view near $E = 0$ is shown in the left figure.

- Soft Gap and Localized Zigzag Edge States

Before we show the DOS of interacting disordered zigzag nanoribbons, it is instructive to compare the DOS in the absence and presence of electron interactions, see Fig. 18.7. In the absence of electron interactions, the DOS is peaked near $E = 0$ in the limit small disorder strength. The states in the peak have probability densities that are divided between the opposite zigzag edges. When electron interactions are present these states transform into left and right edge states with $E \approx \pm\Delta/2$, see Fig. 16.8(a). In the absence of disorder, typically, the value of the gap is $\Delta \sim 0.2t$ [12], where $t \sim 3\,\mathrm{eV}$ is the hopping constant.

In the presence of disorder, numerical results show that the gap Δ of an interacting disordered zigzag nanoribbon is filled with edge states. In the thermodynamic limit of large ribbon lengths L and in the weak disorder regime, the gap states of the tunneling DOS form a soft gap (see Sec. 6.8) near $E \approx 0$, as shown in Fig. 18.7. Its physical origin was explained in Sec. 10.9. A fit to the DOS gives $y = A(e^{\alpha x^2} - 1)$ where $x = E/(\Delta/2)$, $A = 0.1029$, and $\alpha = 213$. There are two peaks in the DOS near $|E| = \Delta_s/2$ ($\Delta_s \sim 1/\alpha^{1/2}$), which may be explained as follows. In a random potential and in the absence of on-site repulsion, the energy distribution of localized states will have some mean value near $E = 0$. When two electrons with spin-up and down occupy one of these localized states their site energy will be split by electron–electron interactions, see Fig. 18.8. Since there are numerous localized states, the resulting DOS will be split with two broadened peaks, as shown in Fig. 18.7.

In contrast to massless Dirac electrons, electrons of interacting disordered zigzag ribbon can backscatter and become localized (see the band structure in Fig. 18.1). Localization properties of electron states may be investigated using the typical DOS

$$D_{\mathrm{typ}}(E) = e^{\langle \ln D \rangle} = e^{\frac{1}{N_s} \sum_i^{N_s} \ln D_i(E)}, \tag{18.9}$$

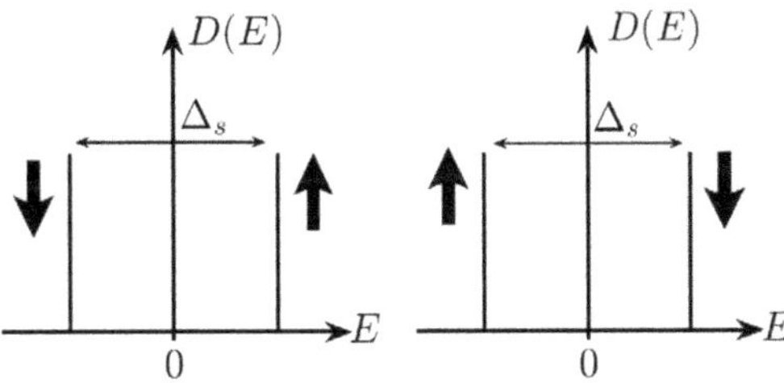

Fig. 18.8. Consider degenerate localized spin-up and -down states. Energy splitting of these spin-up and -down localized states induced by on-site repulsion is shown. When many impurity states are present, these split states form broadened levels, as shown in Fig. 18.7.

It is defined as the geometric average (see Eq. (10.116)) of the local DOS of site i

$$D_i(E) = \sum_n \delta(E - E_n)|\psi_n(i)|^2, \qquad (18.10)$$

where $\psi_n(i)$ is the nth eigenstate with energy E_n. Unlike the DOS, upon localization, the local DOS undergoes a large variation from a continuous to discrete spectrum [3]. (This insight was first provided by P. W. Anderson.[c]) The geometric average is needed to analyze the local DOS because it displays large fluctuations. A small (large) value of the typical DOS at E implies a localized (delocalized) state. A short-ranged disorder potential induces stronger localization of gap-edge states along the zigzag edges than a long-ranged disorder potential [6]. Results of the typical DOS is displayed in Fig. 18.9. It shows that gap-edge states are localized while states outside the gap are delocalized.[d]

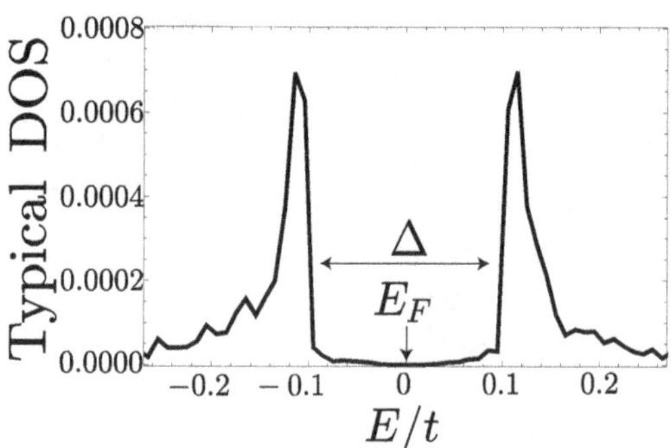

Fig. 18.9. Typical DOS for disorder strength $\Gamma = 0.5t$. The number of disorder realization is $N_D = 800$, impurity concentration $n_{\text{imp}} = 0.01$, and range $d = 0$. The width and length of the ribbon are $w = 7.1\text{Å}$ and $L = 307.4\text{Å}$, respectively. The histogram energy interval is $\Delta E/t = 0.01$. Reprinted with permission from Ref. [1].

---

[c]The Nobel Prize in Physics 1977 was awarded jointly to Philip Warren Anderson, Sir Nevill Francis Mott and John Hasbrouck Van Vleck for their theoretical investigations of the electronic structure of magnetic and disordered systems. It is believed that the TDOS directly determines the conductivity.

[d]In a non-interacting disordered zigzag nanoribbon, disorder induces localization of low-energy electron states. However, not all states are localized [6]. This result differs from the result of the usual one-dimensional localization theory in which all states are localized in one dimension. For massless Dirac electrons, impurity induced backscattering can be absent [7, 16], see Exercise 14.7. Before reading further, the reader is encouraged to study Anderson localization from Ref. [17].

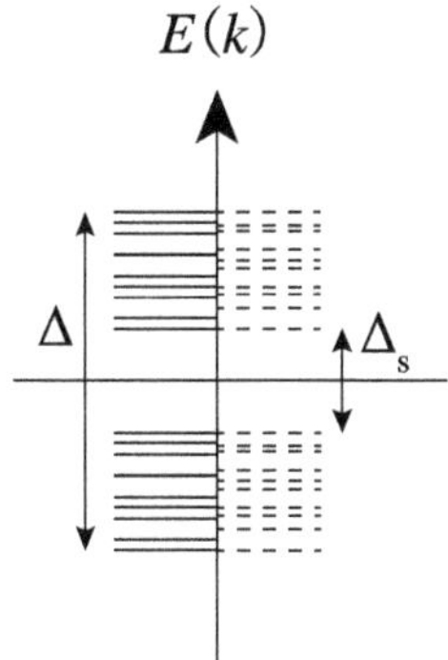

Fig. 18.10. Energy levels of gap-edge states of an interacting zigzag graphene nanoribbon in a disorder potential are shown. Disorder breaks translational invariance. The **spin-split** energy levels of the spin-up (solid lines) and spin-down (dashed lines) **gap-edge** states are shown. These states decay exponentially from the zigzag edges. There are also few states inside Δ_s but are not shown in the figure.

- Mott–Anderson Insulator with Localized Moments

Numerical results show that gap states of an interacting disordered zigzag graphene nanoribbon are **spin-split** [18], as shown in Fig. 18.10. Localized and spin-split states constitute localized magnetic moments. (The formation of a localized moment in the Hartree–Fock approximation is explained in Exercises 18.1 and 18.2). The presence of localized moments strongly suggests that interacting disordered zigzag graphene nanoribbons are a Mott–Anderson insulator, see Fig. 18.11. (The basic properties of a Mott–Anderson insulator are explained in Ref. [3].)

Exercise 18.1. Consider an impurity with a single atomic orbital in a non-interacting electron gas. An electron can hop in and out of the impurity site. Electrons interact with each other only when they simultaneously occupy the impurity. The relevant model is the Anderson model described by the Hamiltonian

$$H = \sum_{k\sigma} \epsilon_{k\sigma} n_{k\sigma} + E(n_{d\uparrow} + n_{d\downarrow}) + U n_{d\uparrow} n_{d\downarrow}$$
$$+ \sum_{k\sigma} V_{dk}(c_{k\sigma}^{\dagger} c_{d\sigma} + c_{d\sigma}^{\dagger} c_{k\sigma}). \qquad (18.11)$$

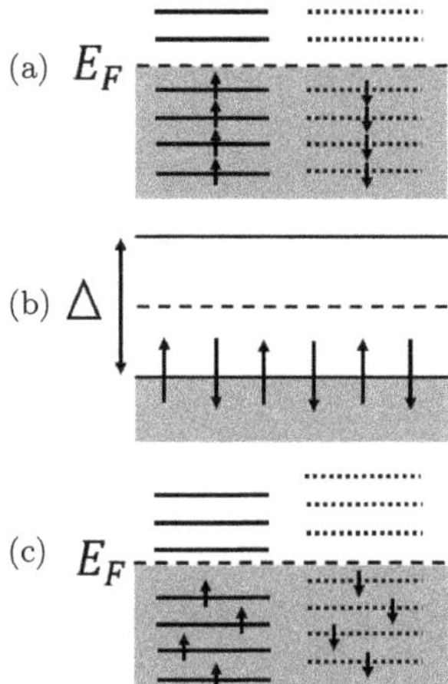

Fig. 18.11. (a) Energy level diagrams of a gapless Anderson insulator, (b) Mott insulator, (c) and Mott–Anderson insulator. In a Mott insulator, on-site repulsion U induces a gap. Such a gap for a zigzag nanoribbon is shown in Fig. 17.3. The states near the gap edge $E = \pm\Delta$ significantly contribute to magnetic effects. An example is antiferromagnetism in the Hubbard model, see Sec. 6.12. A Mott–Anderson insulator has a soft gap and spin-split states with local magnetic moments.

Here $n_{d\sigma}$ is the occupation number of the d-level with spin σ, U is the on-site repulsion, and E is the energy of the atomic orbital. The hybridization V_{dk} is the hopping energy between the orbital and an electron state with wave vector k. The Hartree–Fock Ansatz of this Hamiltonian is $|\Psi\rangle = \prod_{\epsilon_\alpha < \epsilon_F} c_\alpha^\dagger |0\rangle$. Here $c_\alpha^\dagger$ creates a Hartree–Fock eigenstate labeled by α

$$c_\alpha^\dagger = \langle\alpha|d\sigma\rangle c_{d\sigma}^\dagger + \sum_k \langle\alpha|k\sigma\rangle c_{k\sigma}^\dagger. \qquad (18.12)$$

(k is no longer a good quantum number because translational symmetry is broken by the impurity). Find the time derivative $\frac{dc_\alpha^\dagger}{dt}$. Hint: Consult Ref. [19].

Exercise 18.2. This exercise is a continuation of Exercise 18.2. Show that the Hartree–Fock solutions give the following occupation numbers of the orbital:

$$\langle n_{d,\sigma}\rangle = \frac{1}{\pi}\cot^{-1}\frac{E - \epsilon_F + U\langle n_{d,-\sigma}\rangle}{\Delta}, \qquad (18.13)$$

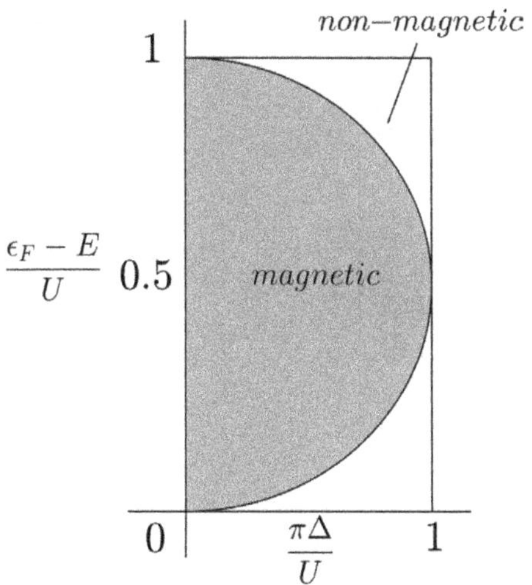

Fig. 18.12. Parameter space for magnetic and non-magnetic behavior.

where the occupation numbers are

$$\langle n_{d\sigma} \rangle = \int_{-\infty}^{\epsilon_F} d\epsilon\, \rho_{d\sigma}(\epsilon) \qquad (18.14)$$

and the DOS of the d-level is

$$\rho_{d\sigma}(\epsilon) = \frac{1}{\pi} \frac{\Delta}{(\epsilon - E_\sigma)^2 + \Delta^2}. \qquad (18.15)$$

Here the renormalized d-level energy is $E_\sigma = E + U\langle n_{d,-\sigma}\rangle$. Show that $\Delta = -\pi \sum_k |V_{dk}|^2 \delta(\epsilon - \epsilon_k)$. The occupation of the impurity site determines the stability of the local magnetic moment: if $n_{d\uparrow} = n_{d\downarrow}$ the solution is **non-magnetic**; otherwise a magnetic moment is stable, see Fig. 18.12. Hint: Consult Ref. [19].

18.3. Mixed Chiral States and Solitons

We have so far explained some basic electronic properties of interacting disordered zigzag nanoribbons. Additional numerical results show that gap-edge states can display fractional end charges of $e^-/2$. These states are localized along the zigzag edges. In this section, we investigate the essential physics leading to charge fractionalization of soliton gap states.

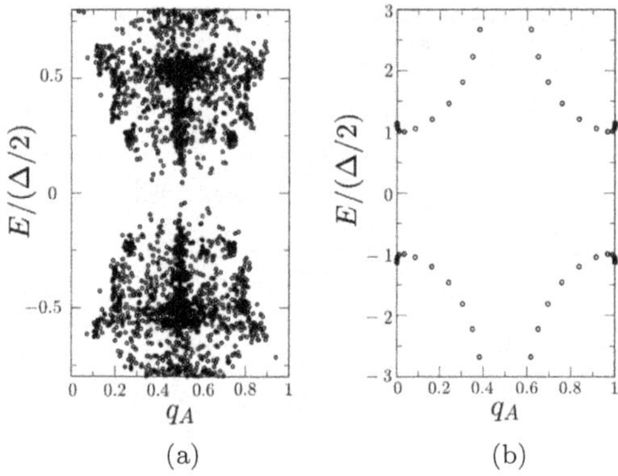

Fig. 18.13. (a) Plot of q_A for disorder potential strength $\Gamma = 0.1t$, where each point represents the probability of finding an electron of a gap-edge state on A carbon sites, q_A. The disorder realization number is $N_D = 300$, and the ribbon length is $L = 196.8\text{Å}$. Here, the impurity/defect to carbon-atom ratio is $n_{\mathrm{imp}} = 0.1$. The gap size is $\Delta_s \approx 0.12\frac{\Delta}{2}$. (b) Plot of q_A for $U = t$ in the absence of disorder, where the on-site electron repulsion and hopping parameter are indicated by U and t, respectively. Reprinted with permission from Ref. [2].

A disorder potential produces **drastic** changes in the energy spectrum when compared to the disorder-free behavior, see Fig. 18.13(a). The probability of finding an electron with energy E on A carbon sites, $q_A = \sum_{i \in A} |\psi_i|^2$, is plotted as a function of E for $\Gamma = 0.1t$. Note that particle–hole symmetry (chiral symmetry) is **broken**. In contrast to the case of $\Gamma = 0$, shown in Fig. 18.13(b), there are numerous states with $q_A \approx 1/2$ in the energy range $|E| < \Delta/2$. If the disorder potential experienced by the left and right edges differs, charge fluctuations will arise. We found that the localization length along the edges decreases as $|E|$ decreases toward $\Delta_s/2$ (Δ_s is defined in Fig. 18.7). A small localization length means that the repulsive energy between an electron in a soliton state and an added electron can be significant, see Fig. 18.8. This effect determines the magnitude of Δ_s. Even a weaker disorder potential with $\Gamma = 0.03t$ produces similar drastic changes in the energy spectrum when compared to the disorder-free behavior, see Fig. 18.14. This figure also shows that disorder-induces **spin-splitting** of energy levels of **gap-edge** states.

The midgap states that $E \approx 0$ are of special interest because they display relatively small disorder-induced charge fluctuations, see Fig. 18.14. Moreover, a self-consistent treatment of disorder and electron interaction within the Hartree–Fock approximation [10, 20] shows that a midgap state

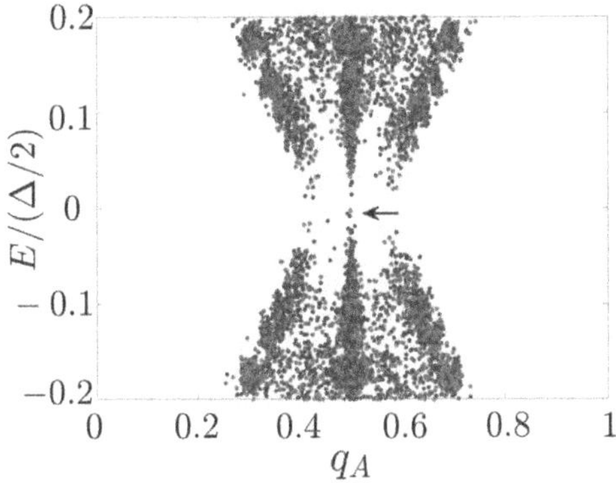

Fig. 18.14. Plot of q_A for $\Gamma = 0.03t$, $U = t$, $L = 1232.5\text{Å}$, and $N_D = 2511$. Dots represent spin-up and spin-down states. Reprinted with permission from Ref. [2].

can fractionalize into two $e^-/2$ fractional boundary charges on the opposite zigzag edges (see Fig. 18.15). This **mixed chiral state** is divided into two equal parts and decays exponentially from the zigzag edges [16]. The wider the distance between the opposite zigzag edges, the better the fractional quantization, as the overlap between the fractional charges on the left and right zigzag edges decreases (these fractional charges are non-local excitations; we also encountered non-local excitations in the toric code, see Sec. 13.4). A **bulk mixed chiral** state with energy larger than Δ is also shown in Fig. 18.15. Mixed chiral states represent a pseudospin kink connecting the left and right zigzag edges of different chiralities, see Fig. 18.16. In other words, a **soliton connects A- and B- sublattices**. This is analogous to a soliton in polyacetylene connecting two different dimerized phases. A zero energy soliton is consistent with the Atiyah-Singer theorem connecting zero modes to a topological charge. It may be thought of as an instanton that connects two vacua, similar to an odd denominator fractional charge that connects degenerate fractional quantum Hall groundstates of a torus, see Sec. 11.11.

Our numerical simulation shows that an interacting disordered zigzag graphene nanoribbon cannot be reached iteratively from a disorder-free chiral symmetry protected topological state. Moreover, an interacting disordered zigzag graphene nanoribbon has doubly degenerate soliton states, $e^-/2$ fractional charges, and broken chiral symmetry. Thus we expect that it is in a topological ordered phase rather than in a symmetry protected topological phase (see [21] for the distinction between them). An

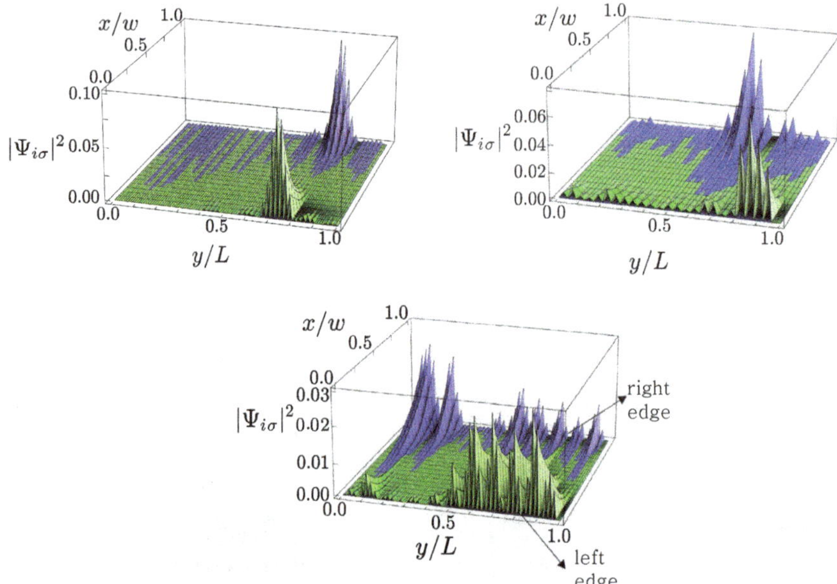

Fig. 18.15. **Left**: Site probability distribution of an electron state displaying charge fractionalization. Two fractional charges $e^-/2$ are located, respectively, on the left (A-type) and right (B-type) zigzag edges of an interacting disordered zigzag graphene nanoribbon. The overlap between A and B parts is practically zero. This state is a good example of a **mixed chiral state**, representing a non-local state. Ribbon width is w, ribbon length is L and i denotes lattice sites. **Right**: Site probability distribution of an electron state displaying charge fractionalization. Although the average y-position of the peaks is the same there are some deviations, as shown in the figure. **Bottom**: Site probability distribution of an electron state with the energy value outside the gap. This is not closely confined to the zigzag edges. Reprinted with permission from Ref. [1].

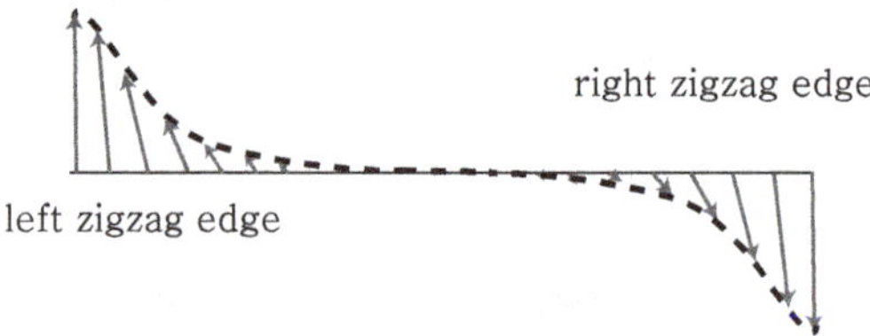

Fig. 18.16. Schematic drawing of the rotation of pseudospin probability of an edge state along a line connecting the two opposite zigzag edges. It rotates out of plane by π and represents a kink. Reprinted with permission from Ref. [1].

interacting disordered zigzag graphene nanoribbon is somewhat analogous to topologically ordered Laughlin states. In both systems, fractional charge and ground state degeneracy are intimately related [17,21]. To establish the topologically ordered nature of the soliton state, we will show in Chapter 20 that the topological entanglement entropy is indeed finite.

18.4. Why Midgap States Exhibit Quantized Charge Fractionalization

It is important that the fractional charge is quantized to a high precision. Several conditions must be met. (1) The fractional charge profile decays fast so that fractional charges do not overlap. Note that, the wider the zigzag graphene nanoribbon, the better the fractional quantization as the overlap between fractional charges on the left and right zigzag edges decreases (this condition is satisfied in our case because the wave functions of edge states decay exponentially). The domain-wall soliton of Fig. 16.15 has two fractional charges that overlap, and it does not represent a true charge fractionalization as the overlap between fractional charges is not small. (2) Fractional boundary charges are well defined only when quantum charge and disorder-induced fluctuations are small.

When a midgap electron is removed, two fractional charges are produced. Let us first discuss the role quantum charge fluctuations (see Sec. 3.8). To achieve an accurate charge quantization, the quantum charge fluctuations should occur at small time scales. According to Girvin [22], the characteristic time scale for the charge fluctuations is inversely proportional to the relevant excitation gap. In disordered zigzag ribbons there is a small gap between the occupied and unoccupied midgap states in the weak-disorder regime, and this gap induces very small timescales for quantum fluctuations [1]. The Zak phase of a polyacetylene specimen can be $\pi/2$ mod 2π (provided that an appropriate unit cell is chosen). So one might conclude that its end charge is $e^-/2$. However, there is as yet no conclusive experimental evidence for it (but spin–charge separation was observed). In our opinion, this is because quantum charge fluctuations between quasi-degenerate states destroy an accurate charge quantization.

Now let us examine charge fluctuations induced by disorder in the energy interval $[E - \delta E, E + \delta E]$. Some states in this interval are more localized on the left zigzag edge while some other states are more localized on the right zigzag edge. Numerical calculations indeed show that, although their average boundary charge on one zigzag edge is $e^-/2$, its variance is significant [1]. However, near the mid gap energy $|E| \approx 0$, these fluctuations are small.

> **Exercise 18.3.** Discuss whether a fractional charge can exist outside of zigzag ribbons. Is it analogous to the color confinement, i.e., that color charged particles (such as quarks and gluons) cannot be isolated? Read also Ref. [22].

The $e^-/2$ charges are a result of the subtle interplay between topology of the underlying lattice, electron correlations and disorder. The Su–Schrieffer–Heeger effect is a part of it: half of soliton's spectral weight originates from the conduction band and the other half from the valence band (this is a consequence of particle–hole symmetry). In each disorder, realization particle–hole symmetry (chiral symmetry) is broken, but after disorder, averaging the symmetry is partially restored. This implies that the **average** edge charge is $e^-/2$ at each energy $|E| < \Delta/2$, but with a significant charge variance. However, the charge variance near zero energy is negligible in the **weak-disorder limit**, see Fig. 18.14. In addition, a singular disorder potential gives rise to a soft gap in the tunneling DOS [10,20], and such a gap protects fractional charges from quantum fluctuations. What is the physical origin of a soft gap? The essential physics is the **electron correlation**: it is difficult for the tunneling electron to avoid other electrons since it takes long time for interacting electrons to diffuse away from each other (see Ref. [17, pp. 290 and 645]).

Bibliography

[1] Y. H. Jeong, S.-R. Eric Yang, and M. C. Cha, Soliton fractional charge of disordered graphene nanoribbon, *J. Phys.: Condens. Matter* **31**, 265601 (2019).

[2] S.-R. Eric Yang. M. C. Cha, H. J. Lee, and Y. H. Kim, Topologically ordered zigzag nanoribbon: $e/2$ fractional edge charge, spin–charge separation, and ground-state degeneracy, *Phys. Rev. Res.* **2**, 033109 (2020).

[3] V. Dobrosavljevic, "Typical-medium theory of Mott–Anderson localization," in *50 Years of Anderson Localization*, edited by E. Abrahams (World Scientific, Singapore, 2010).

[4] E. Abrahams, *50 Years of Anderson Localization* (World Scientific, Singapore, 2010).

[5] M. V. Berry, Singular limits, *Phys. Today* **55**, 5, 10 (2002). •

[6] L. R. F. Lima, F. A. Pinheiro, R. B. Capaz, C. H. Lewenkopf, and E. R. Mucciolo, Effects of disorder range and electronic energy on the perfect transmission in graphene nanoribbons, *Phys. Rev. B* **86**, 205111 (2012).

[7] T. Ando, Theory of electronic states and transport in carbon nanotubes, *J. Phys. Soc. Jpn.* **74** 777 (2005).

[8] S. C. Kim and S.-R. Eric Yang, Coulomb impurity problem of graphene in magnetic fields, *Ann. Phys* **347** 21 (2014); S.-R. Eric Yang and Hyun C. Lee, X-ray edge problem of graphene, *Phys. Rev. B* **76**, 245411 (2007).

[9] T. Stauber, P. Parida, M. Trushin, M.V. Ulybyshev, D.L. Boyda, and J. Schliemann, Interacting electrons in graphene: Fermi velocity renormalization and optical response, *Phys. Rev. Lett.* **118**, 266801 (2017).

[10] S.-R. Eric Yang and A. H. MacDonald, Coulomb gaps in a strong magnetic field, *Phys. Rev. Lett.* **70**, 4110 (1993).

[11] M. Fujita, K. Wakabayashi, K. Nakada, and K. Kusakabe, Peculiar localized state at zigzag graphite edge, *J. Phys. Soc. Jpn.* **65**, 1920 (1996).

[12] L. Yang, C.-H. Park, Y.-W. Son, M. L. Cohen, and S. G. Louie, Quasiparticle energies and band gaps in graphene nanoribbons, *Phys. Rev. Lett.* **99**, 186801 (2007).

[13] J. Hinze, MC-SCF. I. The multi-configuration self-consistent-field method, *J. Chem. Phys.* **59**, 6424 (1973).

[14] S. R. White, Density-matrix algorithms for quantum renormalization groups, *Phys. Rev. B* **48**, 10345 (1993).

[15] Y. H. Kim, H. J. Lee, Hyun-Yong Lee, and S.-R. Eric Yang, New disordered anyon phase of doped graphene zigzag nanoribbon, *Sci. Rep.* **12**, 14551 (2022).

[16] A. H. Castro Neto, F. Guinea, N. M. R. Peres, K. S. Novoselov, and A. K. Geim, The electronic properties of graphene, *Rev. Mod. Phys.* **81**, 109 (2009). •

[17] S. M. Girvin and K. Yang, *Modern Condensed Matter Physics* (Cambridge University Press, Cambridge, 2019). •

[18] Y. H. Jeong and S.-R. Eric Yang, Topological end states and Zak phase of rectangular armchair ribbon, *Ann. Phys.* **385**, 688 (2017).

[19] C. Kittel, *Quantum Theory of Solids* (Wiley, Hoboken, 1987). •

[20] S.-R. Eric Yang, A. H. MacDonald, and B. Huckenstein, Interactions, Localization, and the integer quantum Hall effect, *Phys. Rev. Lett.* **74**, 3229 (1995).

[21] X.-G. Wen, Colloquium: Zoo of quantum-topological phases of matter, *Rev. Mod. Phys.* **89**, 041004 (2017); X.-G. Wen, Topological order: From long-range entangled quantum matter to a unified origin of light and electrons, ISRN *Condens. Matter Phys.* **2013**, 198710 (2013).

[22] S. M. Girvin, "The quantum Hall effect: Novel excitations and broken symmetries," in *Les Houches Lecture Notes: Topological Aspects of Low Dimensional Systems*, edited by A. Comtet, T. Joliceur, S. Ouvry, and F. David (Springer-Verlag, Berlin and Les Editions de Physique, Les Ulis, 2000). •

Chapter 19

Anyons and Spin–Charge Separation in Zigzag Ribbon

"When he finished his talk, I wondered what kind of impression his approximate derivation of the expected result had made on the distinguished theoreticians in the front row. The first person to comment was not a theoretician at all, however, but a little man with a three day's growth of beard who looked as if he had just crawled out of the basement of MIT. He said, "Hey, da spin ain't one. It's t'ree. Dey measured it!" Suddenly, I understood the main function of the theoretician: not to impress the professors in the front row but to agree with observation."

Murray Gell–Mann [1]

"It doesn't matter how beautiful your theory is, it doesn't matter how smart you are. If it doesn't agree with experiment, it's wrong."

Richard P. Feynman

In this chapter, we first argue that $e^-/2$ fractional charges are anyons with a statistical phase factor $e^{i\pi/2}$. We discuss measurable experimental implications of $e^-/2$ fractional charges. At half-filling, the fractional charges give rise to a linear tunneling DOS at a certain critical value of disorder strength. Upon doping, the soft gap of the tunneling density of states is replaced by a sharp peak at the midgap energy. But, upon further doping, this peak disappears. The detection of this peak will provide strong evidence for the presence of $e^-/2$ fractional charges. In addition, we argue that the ground state displays spin–charge separation, which may be observable in transport measurements.

19.1. Anyons and Tunneling Density of States

The following argument suggests that $e^-/2$ fractional charges are anyons. Consider two Hartree–Fock mixed chiral states that display $e^-/2$ fractional charges, see Fig. 19.1. If we exchange these two electrons, the total many-body wave function of N electrons acquires a statistical phase $e^{i\pi} = -1$ (see the Slater determinant wave functions given in Eq. (18.7)). This exchange is equivalent to the exchanges of two $e^-/2$ charges on the left zigzag edge and two others on the right zigzag edge. Each of these exchanges must generate the statistical phase $e^{i\pi/2}$ to yield the final phase of $e^{i\pi} = -1$. Upon exchange of two fractional charges, the total wave function thus changes as $\psi(\vec{r}_i, \vec{r}_j) \to e^{i\pi/2}\psi(\vec{r}_j, \vec{r}_i)$. An anyon with the Abelian statistical phase $e^{i\pi/2}$ is called a semion.

Assume that a tunneling electron fractionalizes into $m = 1/\nu$ fractionally charged quasiparticles (see Fig. 19.2)

$$e^- \to e^-/m + \cdots + e^-/m. \tag{19.1}$$

The chiral Luttinger liquid theory of fractional quantum Hall edges, [2, 3] described in Chapter 12, predicts the following DOS

$$\rho(E) \propto E^{1/\nu-1}, \tag{19.2}$$

where ν is the filling factor and the energy E is measured from the Fermi energy. (Its derivation is given in Sec. 12.3.) In this theory, electrons tunnel

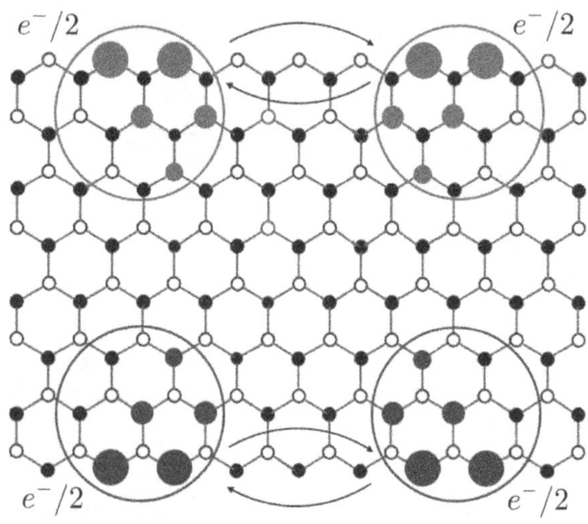

Fig. 19.1. Two Hartree–Fock mixed chiral states are shown. Exchanges of two $e^-/2$ charges on the left zigzag edge and two others on the right zigzag edge are displayed.

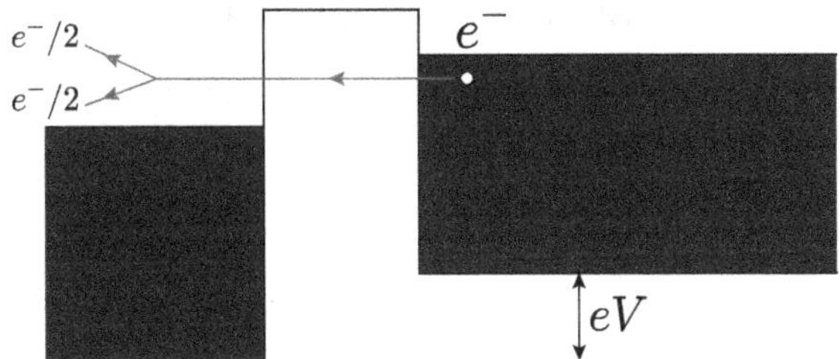

Fig. 19.2.　When a tunneling electron enters a disordered zigzag ribbon, it fractionalizes into two $e^-/2$ charges. Here the electron tunnels along the z axis and the ribbon is in the xy-plane. Reprinted with permission from Ref. [4].

into the edges that support gapless excitations and fractionalize into anyons, see Sec. 12.2. This result leads to the following I–V curve

$$I \propto V^{1/\nu}. \tag{19.3}$$

This predicted $I-V$ curve has been experimentally confirmed [5] (the DOS of the lead is assumed to be a constant near the Fermi energy). It should be noted that Laughlin quasiparticles have an odd denominator fractional charge $e\nu$, and an even denominator fractional charge $e^-/2$ is not found in fractional quantum Hall systems. In the case of semions with $e^-/2$ charges, the expected $I-V$ curve is given by [6]

$$I \propto \int d\epsilon_1 \int d\epsilon_2 \theta(eV - \epsilon_1 - \epsilon_2) \propto V^2, \tag{19.4}$$

where θ and $\epsilon_{1,2}$ are the step function and quasiparticle energies, respectively. This $I-V$ curve is equivalent to a **linear** tunneling DOS.

The following result was used in the deriving the $I-V$ current between two materials of the tunneling junction

$$I \propto \int_{\infty}^{\infty} \rho_{\text{zigzag}}(E)\rho_{\text{lead}}(E + eV) \left[f(E) - f(E + eV)\right] dE, \tag{19.5}$$

where $\rho_{\text{zigzag}}(E)$ and $\rho_{\text{lead}}(E)$ are the DOS of the zigzag ribbon and the lead, respectively (see Fig. 19.2). When electrons are tunneling **out** of the ribbon the current is given by

$$I \propto \int_{\infty}^{\infty} \rho_{\text{lead}}(E)\rho_{\text{zigzag}}(E + eV)[f(E) - f(E + eV)]dE. \tag{19.6}$$

Exercise 19.1. Show Eq. (19.4).

19.2. Linear Density of States of Undoped System

An excellent opportunity to observe fractional boundary charges of zigzag ribbons has recently arisen, as rapid progress has been made in the fabrication of atomically precise graphene nanoribbons [7]. Here we propose an experiment that may provide evidence of the presence of $e^-/2$ fractional charges in interacting disordered zigzag graphene nanoribbons.

The gap-edge states are all localized along the ribbon direction, in contrast to the fractional quantum Hall edge states. Moreover, significant disorder-induced charge fluctuations may occur in a given energy interval: Some of the states in the interval are more localized on the left or right zigzag edges. This tendency increases as the electron energy deviates from $\pm\Delta_s/2$. For these reasons, the chiral Luttinger liquid theory does not apply to zigzag graphene nanoribbons. Despite this, we can apply the above heuristic argument to a zigzag graphene nanoribbon with $m = 2$.

To check the $I-V$ result given above, we compute the DOS of an interacting disordered zigzag graphene nanoribbon and find that, for the critical disorder strength where the zigzag graphene nanoribbon supports gapless excitations (i.e., where Δ_s vanishes), our computed Hartree–Fock DOS is linear near the Fermi energy. This finding is in agreement with the heuristic argument given above. We performed finite-size calculations and computed the DOS given by

$$\rho(E) = \frac{D_{\delta E}(E)}{L N_D \delta E}, \tag{19.7}$$

where $D_{\delta E}(E)$ is the total number of states in the energy histogram interval δE. We defined the critical point Γ_c as the value where Δ_s is zero, i.e., where the gap closes. The DOS results for $\Gamma = 0.03t < \Gamma_c$ and $\Gamma = 0.18t \gtrsim \Gamma_c$ are plotted in Figs. 18.7 and 19.3, respectively. Below $\Gamma < \Gamma_c$, an exponentially small gap develops in the DOS, and even a weak potential can form a soft gap in the limit of infinitely long ribbon. The DOS near $\Gamma \approx \Gamma_c$ is **linear**, as shown in Fig. 19.3. Our numerical results show that the energy range where the DOS is linear increases as the ribbon length grows and that fluctuations in the DOS also decreases. Note also that Γ_c decreases as the ribbon length increases (this is a finite-size effect). Γ_c does **not** represent a metal-insulator transition point since the gap-edge states are all localized in the interacting disordered zigzag graphene nanoribbon. (It is interesting to note that the shape of a Coulomb gap of the integer quantum Hall state is also linear, see Fig. 10.14.)

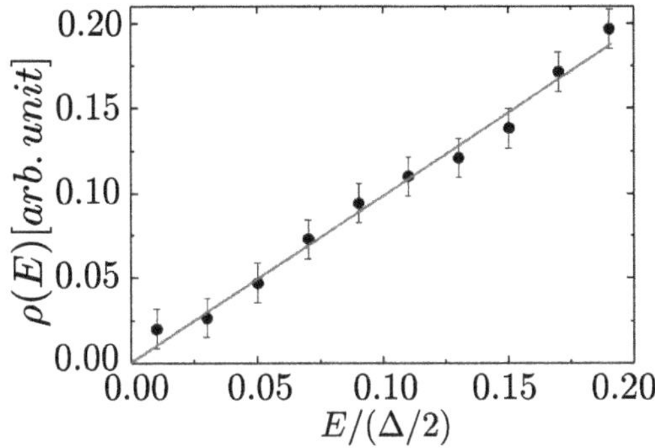

Fig. 19.3. Plot of DOS $\rho(E)$ for disorder potential strength $\Gamma = 0.18t$. Here, the disorder realization number $N_D = 4409$, ribbon width $w = 7.1$ Å, and ribbon length $L = 740.46$ Å. The histogram interval is $\delta E = 0.02\frac{\Delta}{2}$. The solid line represents a linear fit to $\rho(E)$. Reprinted with permission from Ref. [4].

Further reading. Fractional quantum states may be simulated on a Haldane honeycomb lattice model. Utilizing the quantized Hall response of such a system, one can demonstrate statistics of the quasiparticles of the non-Abelian Moore-Read state [8]. It is not clear how to perform a similar calculation in a Mott–Anderson insulator of disordered zigzag nanoribbon with Abelian quasiparticles, where electron localization is crucial.

19.3. Spin–Charge Separation

We saw that polyacetylene, superconductors (Sec. 8.4), and integer quantum Hall (Sec. 12.5) states exhibit spin–charge separation. Do interacting disordered zigzag graphene nanoribbons also exhibit spin–charge separation?

Figure 18.5 displays the net site spin values of a zigzag ribbon. An interacting disordered zigzag graphene nanoribbon displays antiferromagnetism that is perturbed in comparison to the one in the absence of disorder. Away from the boundary sites, inside the ribbon are mostly antiferromagnetically coupled. Magnetization is mostly ferromagnetic on each of the two boundary zigzag edges but the two boundary zigzag edges are approximately antiferromagnetically coupled. On the left boundary zigzag edge, the site spin direction flips at two places, where magnetic domain walls are formed. Also note that on the right boundary zigzag edge, the site spin values are nearly zero in two regions. We will show below that this effect is associated with spin–charge separation.

Let us try to understand how the singular disorder potential disrupts the symmetry protected topological phase. Suppose that the site occupation numbers of the disorder-free left edge are $n_{i\uparrow} = 0.7$ and $n_{i\downarrow} = 0.3$. Then, those of the right edge are $n_{i\uparrow} = 0.3$ and $n_{i\downarrow} = 0.7$, respectively. Assume that disorder generates one spin-up and one spin-down occupied soliton state near the gap-edge displaying charge fractionalization. In other words, a spin-up electron on the left zigzag edge of the interacting zigzag graphene nanoribbon is replaced by two $e^-/2$ fractional charges, one of which resides on the left zigzag edge while the other resides on the right zigzag edge (Fig. 19.4). However, such a configuration is energetically costly at half-filling since the mean site occupation number becomes greater than

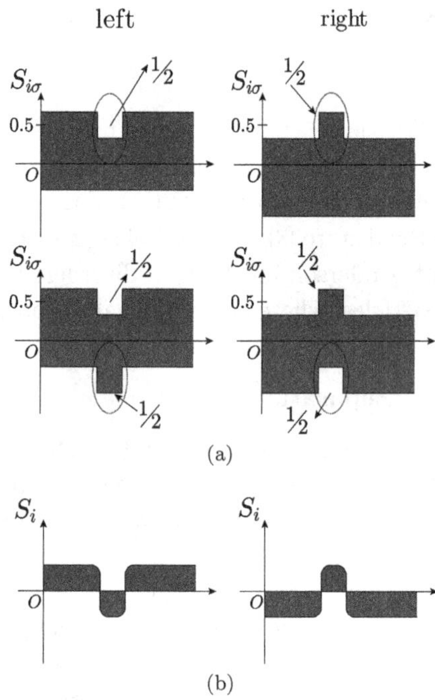

Fig. 19.4. (a) Suppose there is a spin-up electron inside the dashed curve shown in the figure. This spin-up electron on the left edge fractionalizes and, as a result, one $e^-/2$ resides on the left edge and another $e^-/2$ resides on the right edge. **Lower**: (a) Site-spin z-components, $s_{i\uparrow} = n_{i\uparrow}$ and $s_{i\downarrow} = n_{i\downarrow}$, are plotted along edges for replacement of left-edge spin-up electron by two fractional charges. Here, $n_{i\sigma}$ is the site occupation number. The left and right figures correspond to the left and right zigzag edges, respectively. The number 1/2 indicates a removed or added electron occupation number. An additional spin-down electron of the right edge is replaced by two fractional charges. (b) Plot of total z-component of ground state site spin s_i. Reprinted with permission from Ref. [4].

one. If, in addition, a spin-down electron on the right edge fractionalizes, one $e^-/2$ charge resides on the left and the other $e^-/2$ on the right zigzag edge, see Fig. 19.4. Now the total occupation number of each site is now close to one (i.e., the zigzag graphene nanoribbon is half-filled) and double occupancy is avoided. But the total z-component of the site spin on the zigzag edges changes sign along the edge direction, as shown in Fig. 19.4. Note that the disorder potential creates an **even number of solitons** to minimize the energy cost of double occupancy of a site (a soliton consists of a pair of fractional charges). In other words, two fractional charges of opposite spins are involved in creating a magnetic domain with a **zero** spin. (The same mechanism is responsible for spin–charge separation in polyacetylene, see Figs. 4.11 and 4.12). Its charge is e^- and not $e^-/2$. Thus spin–charge separation is present.

In addition, the magnetic zigzag edge reconstruction can also lead to a **spin–charge separation** [6, 9]. (We encountered this phenomenon in polyacetylene (Sec. 4.4) and integer quantum Hall edges (Sec. 12.5).) Figures 19.5(a) and 19.5(b) show how a charge fractionalization process

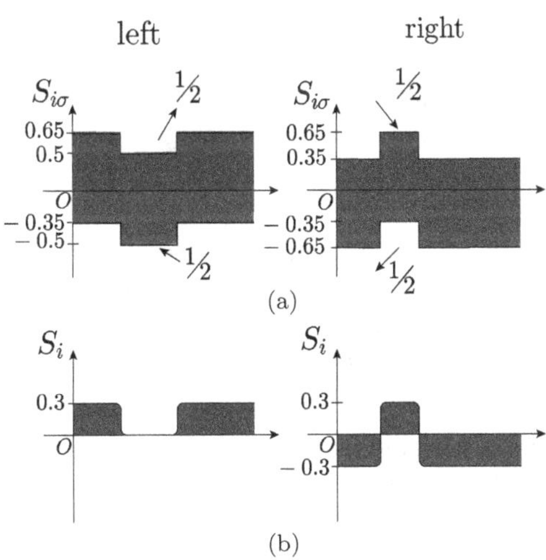

Fig. 19.5. (a) Another rearrangement of the edge structure is shown. Site-spin z-components, $s_{i\uparrow} = n_{i\uparrow}$ and $s_{i\downarrow} = n_{i\downarrow}$, are schematically plotted along edges for replacement of left-edge spin-up electron by two fractional charges. Here, $n_{i\sigma}$ is the site occupation number. The left and right figures correspond to the left and right zigzag edges, respectively. The number 1/2 indicates a removed or added electron occupation number. An additional spin-down electron of the right edge is replaced by two fractional charges. (b) Plot of total z-component of ground state site spin s_i. The resulting spin–charge separation is displayed. Reprinted with permission from Ref. [4].

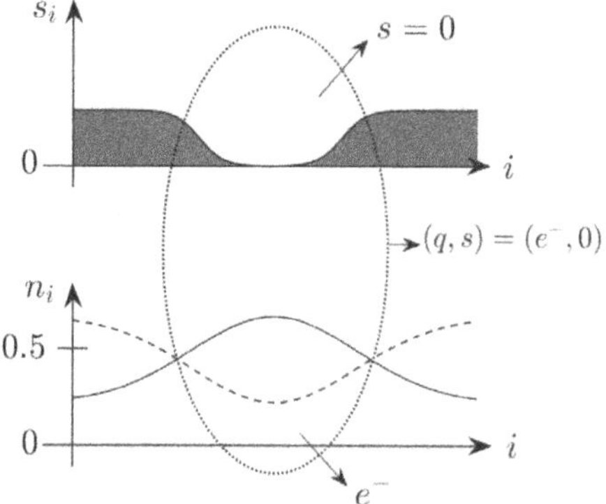

Fig. 19.6. Illustration of a particle displaying spin–charge separation $(q, s) = (e^-, 0)$. Upper: spin occupation number is shown as a function of site index i. Lower: The dashed (straight) line displays the occupation number of spin-up (-down) electrons.

results in an object $(-e, 0)$ that display spin–charge separation, see Fig. 19.6. Here $-e$ denotes an electron charge located on the left edge and number 0 means no spin. When such an object moves along the zigzag edge, it will carry charge but **no** spin. It is similar to spin–charge separation occurring in polyacetylene, see Fig. 4.11. However, in superconductors spin–charge occurs differently: charge is zero but spin is $1/2$, see Fig. 8.7. When the spin–charge separation takes place, the site occupation number also gets modified, see Fig. 19.6. In the integer quantum Hall state, both spin and charge of an electron are finite but they move independently of each other, see Sec. 12.5.

Magnetic domain walls and spin-charge separation may also lead to a rearrangement of site spins located away from the ribbon boundary, see Fig. 19.5. We will see later in Chapter 20 that this feature may be utilized in the calculation of topological entanglement entropy.

Spin–charge separation in polyacetylene was observed in transport and spin susceptibility measurements of a doped system, see Sec. 4.3. It would also be interesting to perform a similar experiment in slightly doped disordered graphene zigzag nanoribbons.

19.4. Tunneling DOS of Doped Disordered Zigzag Ribbon

How does the tunneling DOS of a disordered ribbon look like in the limit of small **doping**? A tunneling electron enters into a soliton state and divides into two fractionally charged quasiparticles.

$$e^- \rightarrow e^-/2 + e^-/2. \tag{19.8}$$

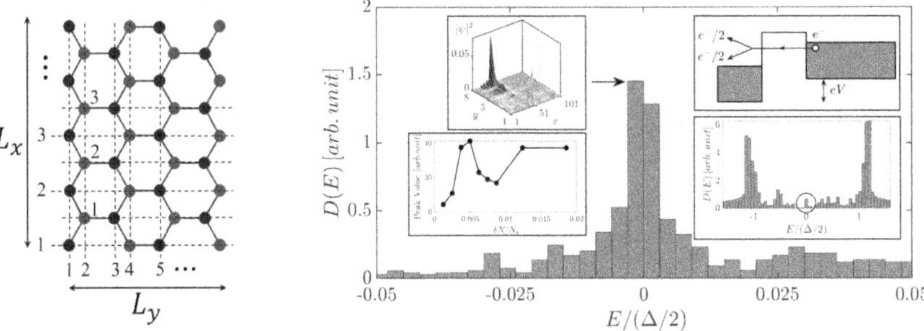

Fig. 19.7. The right figure shows the HF DOS of a slightly doped ribbon away from half-filling in the weak disorder region. The results are for $N_e = N_s + 3(\delta N = N_s = 0.0037)$, $\Gamma = 0.01t$, $L_x = 101$, $L_y = 8$, $n_{\text{imp}} = 0.1$, and $U = t(g = 0.0032)$. The units of L_x and L_y are shown in the left figure (L_x and L_y are equal to the length and width of the ribbon, respectively). Sharp peak is present inside the soft gap at the midgap energy $E = 0$ (the magnitude of this peak is rather *small* in comparison to the peaks at $E = \pm\Delta/2$ shown in Fig. 1.4). Since there are excess electrons, the Fermi energy $E_F/(\Delta/2) = 0.14$ is *above* the mid gap energy. The DOS of $L = 300$ in a larger energy interval $E < \Delta$ is shown in the lower right inset. Upper left inset displays a HF eigenstate exhibiting charge fractionalization. Note that energy is measured in units of $\Delta = 2$. The number of disorder realization is $N_D \approx 400$. Upper right inset shows fractionalization of a tunneling electron. Differential I–V measurement gives the DOS. Lower left inset displays the dependence of the midgap peak on doping concentration for $L = 100$. Reprinted with permission from Ref. [11].

(Ref. [10] gives a good account of this process.) If fractionalized charges move away from each other their mutual correlation energy may be neglected, i.e., they move independently of each other (however, they may still be entangled). The minimum energy cost to create an $e^-/2$ fractional charge from the ground state is $\Delta_s/2$, which corresponds to the midgap energy $E = 0$ in the excitation spectrum.[a]

Adding a few extra electrons to the half-filled ribbon in the weak disorder region creates a sharp peak inside an exponentially decaying small soft gap near the midgap energy $E = 0$, as shown in Fig. 19.7. There are

---

[a]These fractional charges are analogous to quasiparticles of the $1/m$ Laughlin state of fractional quantum Hall effect (m is an odd integer and the charge of a quasiparticle is e^-/m). Fractional quasiparticles of quantum Hall states are stable for the same reason as explained above for a more detailed explanation, see Ref. [10]. With extra quasiparticles, the total energy of the quantum Hall state is thus $E_- = E_m + \delta N m \Delta_-$, where E_m is the ground state energy and Δ_- is the quasiparticle excitation energy [10]. Although these particles form quasi-degenerate states their fractional charges are well-defined. The electron localization helps to stabilize charge fractionalization, see Ref. [10] for an explanation.

also two side peaks, one on each side of the sharp peak at the midgap energy. The width of the central peak is $\sim 0.01t$. The shape of the tunneling DOS in the low doping limit is qualitatively different from that of the half-filled undoped state. In the low doping limit, when an entering electron has a non-zero energy $E \neq 0$, it has a significant chance to not split into $e^-/2$ charges because fractionalization is only approximate at non-zero energies [4]. Note that a soliton state is described by a non-local wave function, as shown in the upper left inset of Fig. 19.7. The lower left inset of Fig. 19.7 shows the highly non-linear dependence of the peak value at $E = 0$ on doping density $\delta N/N_s$ (δN is the number of doped electrons and N_s is the number of carbon atoms in the ribbon). The zero energy peak in the DOS disappears for $\delta N/N_s > 0.01$. Such a non-linear behavior is rather unusual and provides a strong evidence for fractional charges.

The midgap peak consists of quasi-degenerate states. Are fractional charges of these states stable against quantum fluctuations? Each added electron fractionalizes into two quasiparticles that are localized by disorder. Localization effects suppress quantum charge fluctuations [10] and states nearby in energy tend to be separated spatially by large distances. (Localization means that quantum tunneling is shut off, unlike in the double well case; see Sec. 3.8.) In addition, in the presence of an excitation gap, quantum charge fluctuations are suppressed further and fractional charges are expected to be well-defined.

Edge site occupation numbers and site spins are displayed in Fig. 19.8. Note also that most of the added fractional charges go to sites that have occupation numbers $n_i \leq 1$, as shown in Fig. 19.8 (the occupation numbers of the sites that were occupied in the undoped state are nearly unaltered). The nature of the weakly doped states will be elucidated in Sec. 21.7 by performing density matrix renormalization calculations.

Other experiments may be used to detect fractional charges [12, 13]. Investigation of tunneling between zigzag edges, as in fractional quantum Hall bar systems [14], may be fruitful. Quantum shot noise may directly measure [15] the tunneling fractional charge of a zigzag graphene nanoribbon. Resonant tunneling measurement through a quantum dot structure made of a rectangular zigzag graphene nanoribbon may also be explored [16]. Finally, it would be interesting to investigate other zigzag nanoribbon systems that exhibit antiferromagnetism, e.g., silecene and boron nitride nanoribbons [17, 18]. Disorder can couple the left and right zigzag edges and lead to charge fractionalization.

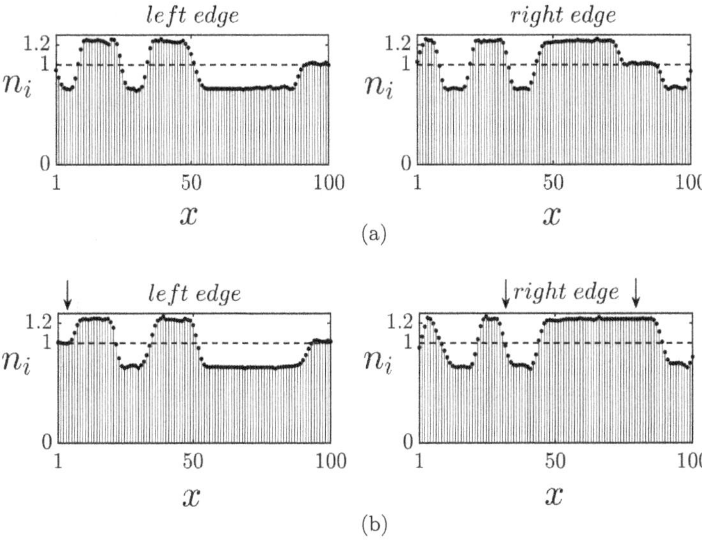

Fig. 19.8. (a) The Hartree-Fock occupation numbers for the undoped ribbon. The parameters are $\Gamma = 0.01t$, $L_x = 101$, $L_y = 8$, $n_{\text{imp}} = 0.1$, and $U = t$. (b) Same as in (a) but with extra electrons added $N_e = N_s + 3$ ($\delta N = 3$ and $\delta N/N_s = 0.0037$). Arrows indicate sites where added particles reside. The background occupation numbers are given in (a) and they remain almost unchanged.

Bibliography

[1] M. Gell-Mann, *The Quark and the Jaguar: Adventures in the Simple and the Complex* (St. Martin's Griffin, New York, 1995). •

[2] X.-G. Wen, Colloquium: Zoo of quantum-topological phases of matter, *Rev. Mod. Phys.* **89**, 041004 (2017); X.-G. Wen, Topological order: From long-range entangled quantum matter to a unified origin of light and electrons, *ISRN Condensed Matter Physics* **2013**, 198710 (2013).

[3] C. de C. Chamon and E. Fradkin, Distinct universal conductances in tunneling to quantum Hall states: The role of contacts, *Phys. Rev. B* **56**, 2012 (1997).

[4] S.-R. Eric Yang, M. C. Cha, H. J. Lee, and Y. H. Kim, Topologically ordered zigzag nanoribbon: $e/2$ fractional edge charge, spin–charge separation and ground-state degeneracy, *Phys. Rev. Res.* **2**, 033109 (2020).

[5] M. Grayson, D. C. Tsui, L. N. Pfeiffer, K. W. West, and A. M. Chang, Continuum of chiral Luttinger liquids at the fractional quantum Hall edge, *Phys. Rev. Lett.* **80**, 1062 (1998).

[6] S. M. Girvin and K. Yang, *Modern Condensed Matter Physics* (Cambridge University Press, Cambridge, 2019).

[7] P. Ruffieux, S. Wang, B. Yang, C. Sanchez-Sanchez, J. Liu, T. Dienel, L. Talirz, P. Shinde, C. A. Pignedoli, and D. Passerone, On-surface synthesis of

graphene nanoribbons with zigzag edge topology, *Nature* **531**, 489 (2016).

[8] W. Zhu, S. S. Gong, F. D. M. Haldane, and D. N. Sheng, Topological characterization of the non-abelian moore-read state using density-matrix renormalization group, *Phys. Rev. B* **92**, 16510 (2015).

[9] W. P. Su, J. R. Schrieffer, and A. J. Heeger, Solitons in Polyacetylene, *Phys. Rev. Lett.* **42**, 1698 (1979); A. J. Heeger, S. Kivelson, J. R. Schrieffer, and W. P. Su, Solitons in conducting polymers, *Rev. Mod. Phys.* **60**, 781 (1988).

[10] S. M. Girvin, "The Quantum Hall Effect: Novel Excitations and Broken Symmetries," in *Les Houches Lecture Notes: Topological Aspects of Low Dimensional Systems*, edited by A. Comtet, T. Joliceur, S. Ouvry, and F. David (Springer Verlag, Berlin and Les Editions de Physique, Les Ulis, 2000). •

[11] Y. H. Kim, H. J. Lee, Hyun-Yong Lee, and S.-R. Eric Yang, New disordered anyon phase of doped graphene zigzag nanoribbon, *Sci. Rep.* **12**, 14551 (2022).

[12] J. Nakamura, S. Liang, G. C. Gardner, and M. J. Manfra, Direct observation of anyonic braiding statistics, *Nature Phys*, **16**, 931 (2020). 18

[13] H. Bartolomei, M. Kumar, R. Bisognin, A. Marguerite, J.-M. Berroir, E. Bocquillon, B. Plaçcais, A. Cavanna, Q. Dong, U. Gennser, Y. Jin, and G. Féve, Fractional statistics in anyon collisions, *Science* **368**, 173 (2020).

[14] W. Kang, H. L. Stormer, L. N. Pfeiffer, K. W. Baldwin, and K. W. West, Tunnelling between the edges of two lateral quantum Hall systems, *Nature* **403**, 59 (2000).; I. Yang, W. Kang, K. W. Baldwin, L. N. Pfeiffer, and K. W. West, Cascade of quantum phase transitions in tunnel-coupled edge states, *Phys. Rev. Lett.* **92**, 056802 (2004).

[15] R. de-Picciotto, M. Reznikov, M. Heiblum, V. Umansky, G. Bunin, and D. Mahalu, Direct observation of a fractional charge, *Nature* **389**, 162 (1997).; L. Saminadayar, D. C. Glattli, Y. Jin, and B. Etienne, Observation of the $e/3$ fractionally charged Laughlin quasiparticle, *Phys. Rev. Lett.* **79**, 2526 (1997).

[16] V. J. Goldman and B. Su, Resonant Tunneling in the quantum Hall regime: Measurement of fractional charge, *Science* **267**, 1010 (1995).

[17] Y. Yao, A. Liu, J. Bai, X. Zhang, and R. Wang, Electronic structures of silicene nanoribbons: Two-edge-chemistry modification and first-principles study, *Nanoscale Res. Lett.* **11**, 371 (2016).

[18] V. Barone and J. E. Peralta, Magnetic boron nitride nanoribbons with tunable electronic properties, *Nano Lett.* **8**, 8, 2210 (2008).

Chapter 20

Topological Order and Entanglement Spectrum of Zigzag Ribbon

"... so we are obliged either to stick to a pragmatic approach or strict instrumentalist interpretation, or else to accept the existence of a strange non-locality that seems hard to reconcile with our normal concepts of spatial separation between independent entities."

C. J. Isham

"... different measurements produce different realities. Not just different results, but different realities– and what's more, ones that are not necessarily compatible with one another."

Philip Ball

"...does that mean the Moon is not there when I am not looking at it."

Albert Einstein

A one-dimensional system with a hard gap cannot be topologically ordered, i.e., no long-range entangled phase can exist [1]. However, our previous results of ground state degeneracy, fractional charges, and spin–charge separation suggest that interacting disordered zigzag nanoribbons may be a topologically ordered insulator. It is unclear whether zigzag ribbons really have non-zero topological entanglement entropy (TEE) [2] since the presence of an exponentially small soft gap may be compatible with zero TEE.

To prove that interacting disordered zigzag nanoribbons are indeed topologically ordered, one must show that TEE is non-zero. TEE may be represented by a Wilson loop, which may be thought of as a non-local[a] topological order parameter. We numerically compute the TEE and show that it is finite and universal, independent of the strength of both electron–electron interaction and disorder. Our results also show that disorder profoundly affects the symmetry protected phase and can generate topological order (this is a topological phase transition). Disorder makes a locally entangled state even more entangled and generates non-local topological entanglement (the relevant entanglement is between non-local fractional charges and not between spins). However, there is a competition between disorder and on-site repulsion. When disorder is strong, edge magnetism vanishes and when on-site repulsion is large, no fractional edge charges are generated. Note that the phase boundary is not sharp and there is a significant crossover region where the TEE displays non-universal values.

> **Exercise 20.1.** Study non-local entanglement and the Bell inequalities. They show that if quantum mechanics is complete then a non-local interaction must exist. Hint: A nice presentation of this can be found in pp. 171–175 and 211–218 of Ref. [3].

20.1. Quantum Entanglement and Reduced Density Matrix

• Entropy and Density Matrix

Entropy may be computed from the density matrix, which we already encountered in Sec. 5.2 (students should review Appendix C before reading further). The entropy of a system can be defined in terms of its density matrix ρ

$$S = -\operatorname{Tr} \rho \log \rho, \tag{20.1}$$

where $\operatorname{Tr} = \sum_i \langle i | \ldots | i \rangle$ with a complete set of states $\{|i\rangle\}$. The entropy of the system is thus given by the eigenvalues of ρ

$$S = -\sum_i \lambda_i \ln \lambda_i. \tag{20.2}$$

---

[a]Quantum nonlocality means that the measured results of a quantum system cannot be understood in terms of a realistic theory in which an object is influenced directly only by its immediate surroundings.

The binary entropy is

$$S = -\sum_i \left[\lambda_i \ln \lambda_i + (1 - \lambda_i) \ln(1 - \lambda_i) \right]. \tag{20.3}$$

Let us explain the result Eq. (20.2) by giving a simple heuristic argument. Suppose there are M macroscopic states and that each macroscopic state is equally likely to occur (a macroscopic state is a state that can be easily distinguished experimentally). Then the entropy is $S = \ln M = -\langle \ln p_i \rangle = -\sum_i p_i \ln p_i$ since $p_i = 1/M$ (we have set the Boltzmann constant $k_B = 1$). This result is identical to that of Eq. (20.2) since the eigenvalue ρ_i may be identified with the probability p_i to find the macroscopic state i in a measurement.

For example, if a particle is in rooms L **and** R at the same time, its quantum state is

$$|\Psi\rangle = \frac{1}{\sqrt{2}}(|L\rangle + |R\rangle). \tag{20.4}$$

The density matrix of the particle is

$$\rho = |\Psi\rangle\langle\Psi| = \begin{pmatrix} 1/2 & 1/2 \\ 1/2 & 1/2 \end{pmatrix}. \tag{20.5}$$

The off-diagonal matrix elements represent quantum interference terms. The eigenvalues are $\lambda_1 = 0$ and 1 and the entropy is $S = 0$. This value of entropy is consistent with the fact that a pure quantum state has no entropy.

• **Reduced Density Matrix**

Let us first discuss the meaning of quantum entanglement. A quantum system can be partitioned into two disjoint regions A and B (an example is given in Fig. 20.1). If the state of the total system can be written as a direct product between a state of A and another state of B

$$|\psi\rangle = |\psi_A\rangle \otimes |\psi_B\rangle, \tag{20.6}$$

then A and B are not entangled. Otherwise A and B are **entangled.** Entanglement is correlations that cannot be explained by classical physics.

In order to quantify entanglement entropy, we first need to define the reduced density matrix of a subsystem. The state of the composite system

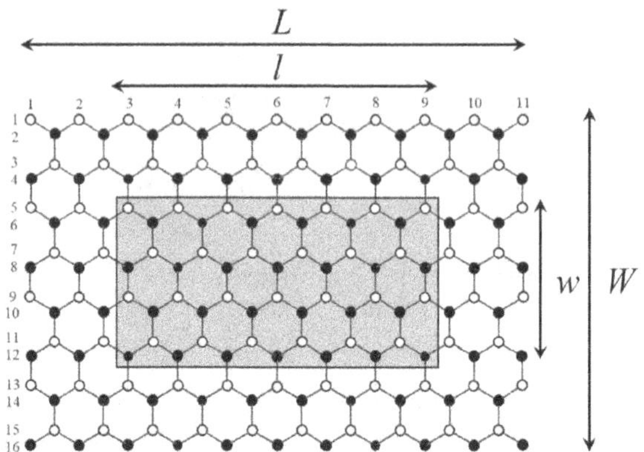

Fig. 20.1. Zigzag ribbon is divided into two regions A and B.

$A + B$ can be written as a direct product of states of A and B

$$
|\psi\rangle = \sum_{i,j} c_{ij} |a_i\rangle \otimes |b_j\rangle,
$$
$$
\langle\psi| = \sum_{l,k} c_{lk}^* \langle a_l| \otimes \langle b_k|.
$$

(20.7)

Here $\{|a_i\rangle\}$ ($\{|b_i\rangle\}$) are a complete set of states of part A (B). One can show that $|\psi\rangle$ is entangled if and only if the determinant of c_{ij} is non-zero. All the properties of the composite system can be calculated from its density matrix

$$
\rho_{AB} = |\psi\rangle\langle\psi| = \sum_{ijkl} c_{ij} c_{lk}^* |a_i\rangle\langle a_l| \otimes |b_j\rangle\langle b_k|.
$$

(20.8)

The entanglement entropy of A may be computed from the eigenvalues of its reduced density matrix, see Eq. (20.2). The reduced density matrix of a subsystem $A(B)$ (see Fig. 20.1) is obtained by performing partial trace over the states of $B(A)$

$$
\rho_A = \mathrm{Tr}_B \rho_{AB} = \sum_{ijklp} c_{ij} c_{lk}^* |a_i\rangle\langle a_l| \langle b_p|b_j\rangle\langle b_k|b_p\rangle = \sum_{ilp} c_{ip} c_{lp}^* |a_i\rangle\langle a_l|,
$$
$$
\rho_B = \mathrm{Tr}_A \rho_{AB} = \sum_{ijklp} c_{ij}^* c_{lk} \langle a_p|a_i\rangle\langle a_l|a_p\rangle |b_j\rangle\langle b_k| = \sum_{jkp} c_{pj}^* c_{pk} |b_j\rangle\langle b_k|.
$$

(20.9)

Note that in the first (second) equation, the density matrix is given as a matrix product $\mathbf{CC}^\dagger$ ($\mathbf{C}^\dagger\mathbf{C}$). All the properties of system A can be calculated from its density matrix. The **entanglement entropy** [4,5] may be computed using Eq. (20.2).

To understand better the meaning of these results, let us give an example of a reduced density matrix. Consider an entangled state of two bosons with zero total spin far away from each other, one in Seoul and the other in Copenhagen. Their quantum mechanical state is pure and is given by[b]

$$|\psi\rangle = \frac{1}{\sqrt{2}}(|\uparrow_A\rangle|\downarrow_B\rangle + |\downarrow_A\rangle|\uparrow_B\rangle). \tag{20.11}$$

(Sometimes I will use the notation $|ij\rangle$ or $|i\rangle|j\rangle$ as a shorthand for $|i\rangle \otimes |j\rangle$. Note that $|i\rangle \otimes |j\rangle$ is different from $|j\rangle \otimes |i\rangle$.) This **entangled** state represents a **superposition** of two states, one in which particle A has spin up and B spin down and the other with the opposite configuration. Its entropy is zero since it is a pure state. However, its reduced density matrix is

$$\rho_A = \mathrm{Tr}_B\rho_{AB} = \langle\uparrow_B |\psi\rangle\langle\psi| \uparrow_B\rangle + \langle\downarrow_B |\psi\rangle\langle\psi| \downarrow_B\rangle$$

$$= \frac{1}{2}(|\uparrow_A\rangle\langle\uparrow_A | + |\downarrow_A\rangle\langle\downarrow_A |), \tag{20.12}$$

which is a density matrix of a **mixed** state,[c] see Eq. (20.13). Decoherence gives rise to a mixed state with a finite entropy. Here tracing out the B-part corresponds to the effect of decoherence (B-particle is considered to be the environment that provides the source of decoherence). Thus when **decoherence** is present, the quantum superposition disappears [6]. The density matrix of A can be written as

$$\rho = \begin{pmatrix} 1/2 & 0 \\ 0 & 1/2 \end{pmatrix}. \tag{20.13}$$

---

[b]If the spin measurements are made along the z-axis then the relevant state with zero total spin is

$$|\psi\rangle = \frac{1}{\sqrt{2}}(|\rightarrow_A\rangle|\leftarrow_B\rangle + |\leftarrow_A\rangle|\rightarrow_B\rangle). \tag{20.10}$$

This can be verified using $|\rightarrow\rangle = \frac{1}{\sqrt{2}}(|\uparrow\rangle + |\downarrow\rangle)$ and $|\leftarrow\rangle = \frac{1}{\sqrt{2}}(|\uparrow\rangle - |\downarrow\rangle)$.

[c]A mixed state is not in the Hilbert space of the system. It describes the probabilities to find the system in different quantum states. This randomness may be present, for example, due to noise in sample preparation or to decoherence.

The diagonal matrix elements represent the probabilities to find A in spin-up and -down states, respectively. This density matrix gives the maximum entanglement entropy of $\ln 2$ (see Eq. (20.2)), meaning that system A is in the spin-up or -down state with equal probability. In other words, although the total system has zero entropy, a subsystem has **finite entropy**. This is an example of the **paradoxical** nature of quantum mechanics.

> **Exercise 20.2.** Show that the entanglement entropy of the pure state given in Eq. (20.11) is zero.

> **Exercise 20.3.** Compute the entanglement entropy using the density matrix given in Eq. (20.12).

- Hartree–Fock Reduced Density Matrix

The Hartree–Fock approximation provides an excellent description of zigzag nanoribbons. We will use it to investigate the reduced density matrix of zigzag nanoribbons. The Hartree–Fock density matrix (see Sec. 5.2) is

$$\rho = \sum_k n_k |k\rangle\langle k|, \tag{20.14}$$

where n_k is the occupation number of the single-particle Hartree–Fock states $|k\rangle$. This gives

$$\rho_{ij} = \sum_k n_k \langle i|k\rangle\langle k|j\rangle = \sum_k n_k \mathbf{B}_{ik}\mathbf{B}_{kj}. \tag{20.15}$$

We divide the zigzag ribbon into two parts A and B. We restrict the indices $i, j \in A$ and obtain the reduced density matrix [7, 8].

The spin-up **or** -down destruction operator of a Hartree–Fock single particle state $|k\rangle$ is

$$a_k = \sum_i \mathbf{A}_{ki} c_i, \tag{20.16}$$

where c_i is the destruction the operator of site i. Inverting this relation, we find

$$c_i = \sum_k (\mathbf{A}^{-1})_{ik} a_k = \sum_k \mathbf{B}_{ik} a_k. \tag{20.17}$$

Let us define the correlation function

$$\mathbf{C}_{ij} = \langle\Psi|c_i^\dagger c_j|\Psi\rangle, \tag{20.18}$$

where $|\Psi\rangle$ is the Hartree–Fock ground state of $A + B$. (Here only spin-up electrons are considered. Spin-down electrons give the same result.) It can be written as

$$\mathbf{C}_{ij} = \langle\Psi| \left(\sum_k \mathbf{B}_{ik} a_k\right)^\dagger \left(\sum_{k'} \mathbf{B}_{jk'} a_{k'}\right) |\Psi\rangle$$

$$= \langle\Psi| \left(\sum_k a_k^\dagger \mathbf{B}_{ki}^*\right) \left(\sum_{k'} \mathbf{B}_{jk'} a_{k'}\right) |\Psi\rangle$$

$$= \sum_k \mathbf{B}_{ki}^* \mathbf{B}_{jk} n_k = \sum_k n_k \mathbf{B}_{ik} \mathbf{B}_{kj}^*, \tag{20.19}$$

where the sum is over the **occupied** Hartree–Fock single particle k states. Here we have used

$$\langle\Psi| a_k^\dagger a_{k'} |\Psi\rangle = \delta_{kk'} n_k. \tag{20.20}$$

The correlation function is identical to the density matrix, see Exercise 20.4. To compute the entanglement entropy we need to find the eigenvalues ϵ_α of the matrix $\mathbf{C}$.

Exercise 20.4. Show that Eq. (20.15) is identical to Eq. (20.19). Hint: $\mathbf{B}_{ik}$ is a real matrix, i.e., the Hartree–Fock eigenvectors can all be chosen to be real because our Hartree–Fock Hamiltonian is real and symmetric. Let A be a real and symmetric matrix with eigenvectors and eigenvalues v and λ. Then $Av = \lambda v$ implies $A\bar{v} = \lambda\bar{v}$ ($\bar{v}$ is the complex conjugate of v). Use this to show that the eigenvectors can all be chosen to be real. Mathematica and MATLAB automatically generate real eigenvectors.

20.2. Schmidt Decomposition

What is the physical meaning of the eigenvalues of the reduced density matrix? The eigenvalue spectrum may be analyzed using the Schmidt decomposition. The Schmidt decomposition may be proved using the singular value decomposition (SVD) [9, 10].[d]

---

[d]The singular value decomposition is also an essential ingredient in the density matrix renormalization group approach, which we will study in Chapter 21.

- Singular Value Decomposition

Consider an arbitrary $m \times n$ matrix $\mathbf{C}$. According to SVD, any matrix can be written as

$$\mathbf{C} = \mathbf{U}S\mathbf{V}^{\dagger}. \tag{20.21}$$

Here $\mathbf{U}$ is an $m \times p$ matrix with orthonormal columns ($p = \min(m, n)$) while $\mathbf{V}^{\dagger}$ is a $p \times n$ matrix with orthonormal rows. S is a $p \times p$ diagonal matrix with real elements $\sigma_k > 0$ (see Fig. 20.2). If $m = n$ then U and V are unitary. From the result (20.21) it follows that

$$\mathbf{C} = \sum_{k=1}^{p} \sum_{ij} U_{ik} S_{kk} (V^{\dagger})_{kj} = \sum_{k=1}^{p} \sigma_k \mathbf{u}_k \mathbf{v}_k^{\dagger}, \tag{20.22}$$

where $\{\mathbf{u}_i\}$ and $\{\mathbf{v}_i\}$ are, respectively, the column vectors of the matrices $\mathbf{U}$ and $\mathbf{V}$. This means $\mathbf{C}$ can be decomposed into a weighted, ordered sum of separable **matrices** $\sigma_i \mathbf{u}_i \mathbf{v}_i^{\dagger}$.

In real-world applications often $m \gg n$. The column vectors of $\mathbf{C}$ may represent, for example, data for facial expressions of n persons. Data for time-dependent turbulent flow may also be stored in $\mathbf{C}$. Matrices with small values of σ_i may be neglected in Eq. (20.22).

> **Exercise 20.5.** Show $\mathbf{C}\mathbf{v}_i = \sigma_i \mathbf{u}_i$ and $\mathbf{C}^{\dagger}\mathbf{u}_i = \sigma_i \mathbf{v}_i$. Hint: Use $\mathbf{CV} = \mathbf{US}$ and $\mathbf{C}^{\dagger}\mathbf{U} = \mathbf{VS}^{\mathbf{T}}$ (when $m = n$ then $\mathbf{S}^{\mathbf{T}} = \mathbf{S}$).

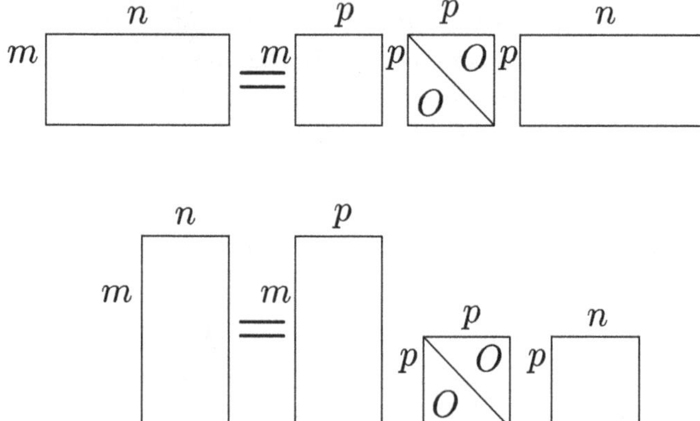

Fig. 20.2. Matrix representation of the SVD: $\mathbf{C} = \mathbf{U}S\mathbf{V}^{\dagger}$. Each box represents a matrix. The box with a diagonal represents the diagonal matrix S. **Upper:** $m < n$. **Lower:** $m > n$.

• Derivation of Schmidt Decomposition

Let $\{|a_i\rangle\}$ and $\{|b_j\rangle\}$ be any orthonormal basis sets of A and B, respectively. Let A and B have Hilbert spaces of dimensions N_A and N_B, respectively. Then any pure state of $A + B$ can be written as

$$|\psi\rangle = \sum_{i,j} C_{ij}|a_i\rangle \otimes |b_j\rangle. \tag{20.23}$$

Let us apply the singular value decomposition to the matrix C_{ij}

$$\mathbf{C} = \sum_k \sigma'_k |\psi'^k_A\rangle \otimes \langle\psi'^k_B|, \tag{20.24}$$

or equivalently its matrix elements are

$$C_{ij} = \sum_k \sigma'_k \langle a_i|\psi'^k_A\rangle \otimes \langle\psi'^k_B|b_j\rangle. \tag{20.25}$$

Plugging this into Eq. (20.23) we find

$$|\psi\rangle = \sum_k \sigma'_k \left(\sum_i \langle a_i|\psi'^k_A\rangle\langle|a_i\rangle\right) \otimes \left(\sum_j \langle\psi'^k_B|b_j\rangle|b_j\rangle\right)$$

$$= \sum_k \sigma'_k |\psi'^k_A\rangle \otimes |\psi'^{k*}_B\rangle = \sum_k |\sigma'_k|(e^{i\phi_k}|\psi'^k_A\rangle) \otimes (|\psi'^{k*}_B\rangle). \tag{20.26}$$

This can be written as the Schmidt decomposition

$$|\psi\rangle = \frac{1}{\sqrt{M}} \sum_k^M \sigma_k |\psi^k_A\rangle \otimes |\psi^k_B\rangle \tag{20.27}$$

with the identification of $|\sigma'_k|$ with σ_k, $e^{i\phi_k}|\psi'^k_A\rangle$ with $|\psi^k_A\rangle$, and $|\psi'^{k*}_B\rangle$ with $|\psi^k_B\rangle$, respectively. Here the rank (the Schmidt number) is $M = \min(N_A, N_B)$ and $\sum_i \sigma_i^2 = 1$ with $\sigma_i \geq 0$.[e] If all $\sigma_i = 0$ except for one, the resulting state is a product state with no entanglement. A diagrammatic representation is shown in Fig. 20.3. Note that $|\psi^k_A\rangle$ and $|\psi^k_B\rangle$ are orthonormal sets.

---

[e]The strictly positive values σ_i in the Schmidt decomposition of ψ are its Schmidt coefficients. The number of Schmidt coefficients, counted with multiplicity, is called its rank.

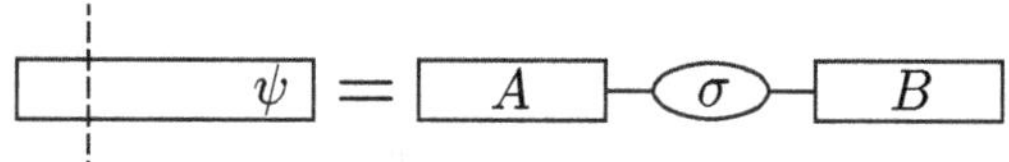

Fig. 20.3. A diagrammatic representation of the Schmidt decomposition. $|\psi_A^i\rangle(|\psi_B^i\rangle)$ are states of $A(B)$ and σ is a diagonal matrix.

From the definition $\rho_A = \sum_j^M \langle\psi_B^j|\psi\rangle\langle\psi|\psi_B^j\rangle$ it follows

$$\rho_A = \sum_{i=1}^{M} |\sigma_i|^2 |\psi_A^i\rangle\langle\psi_A^i|. \tag{20.28}$$

Similarly

$$\rho_B = \sum_{j=1}^{M} |\sigma_j|^2 |\psi_B^j\rangle\langle\psi_B^j|. \tag{20.29}$$

Thus ρ_A and ρ_B are both diagonal in the basis sets $\{|\psi_A^i\rangle\}$ and $\{|\psi_B^i\rangle\}$ used in the Schmidt decomposition. Their number M is equal to the number of non-zero eigenvalues of ρ_A and ρ_B. Thus

$$S = -\mathrm{Tr}\,\rho_A \log\rho_A = -\mathrm{Tr}\,\rho_B \log\rho_B. \tag{20.30}$$

20.3. Topological Entanglement Entropy of Disordered Zigzag Ribbon

Consider a region A of a gapped two-dimensional system **without** topological order. In such a system, only the boundary region within the finite correlation length contributes to the entanglement entropy of A. In other words, the entanglement between two parts of the system A and B should be proportional to the length L of the boundary between them. It obeys the following area-law scaling of entanglement entropy:

$$S = \alpha L. \tag{20.31}$$

Here α is a non-universal constant. In topologically ordered insulators, there is a correction term to the entanglement entropy called topological entanglement entropy. In this section, we will investigate this new term for interacting disordered zigzag graphene nanoribbons.

Suppose a system is topologically ordered. How can one know that a system has a topologically order? One way is to compute the so-called TEE. It is defined as the TEE value of a Wilson loop [11], which we already encountered in connection with the non-Abelian Berry phase (Eq. (2.85)). Both Abelian and non-Abelian statistics may be associated with a Wilson loop [2]. A **simply connected** region D of a topologically ordered gapful system with one boundary has an extra contribution $-\beta$ to the entanglement entropy [2, 12, 13]

$$S_D = \alpha L - \beta. \tag{20.32}$$

The first term is due to local correlations (see Eq. (20.31)). The second term represents the sub-dominant and universal TEE [11, 14] due to non-local correlations. The entanglement entropy $S_D = -\mathrm{Tr}[\rho_D \ln \rho_D]$ is given by the reduced density matrix ρ_D of region D. The value of β of a spin liquid can be numerically computed, see, for example, Jiang *et al.* [15]. The TEE of fractional quantum Hall states can be also computed, see Ref. [16].

- Topological Entanglement Entropy of Toric Code

Let us analytically compute the entanglement entropy of the toric code (see also Ref. [17]). This will give us some intuitive understanding of what TEE is. The total system is $S = A + \bar{A}$, where $\bar{A}$ is defined to be the outside area of A, see Fig. 20.4. Since the ground state $|\Psi\rangle$ is given by equal-weight superposition of closed loop states we find [1]

$$|\Psi\rangle = \sum_{Q \text{ even}} |\Psi_Q\rangle |\bar{\Psi}_Q\rangle, \tag{20.33}$$

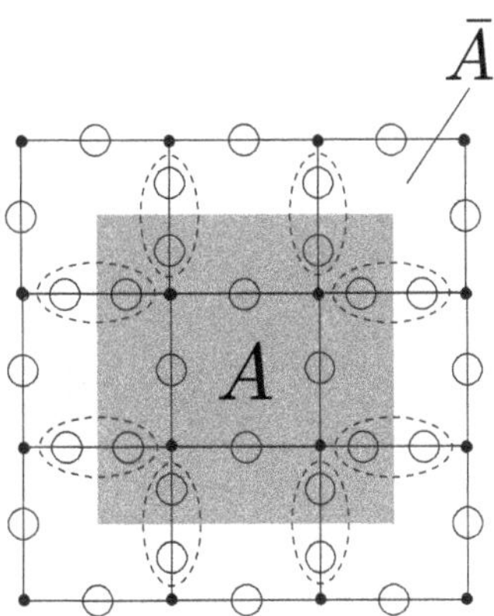

Fig. 20.4. Regions A and $\bar{A}$ share the same links, i.e., the same qubits. Each of these qubits is drawn twice in a dashed circle.

where $|\Psi_Q\rangle$ is a state of A and $|\bar{\Psi}_Q\rangle$ of $\bar{A}$. The boundary between A and $\bar{A}$ is depicted in Fig. 20.4. Here the quantity Q labels a total spin value of the boundary consisting of M links (legs)

$$Q = q_1 + \cdots + q_M, \qquad (20.34)$$

where q_i represents spin value at a link i attached to $\bar{A}$ (0 and 1 for spin-up and -down). From Eq. (20.33) we infer that the eigenvalues of the reduced density matrix is $\frac{1}{N}$, where N denotes the number of components in the Schmidt decomposition (this follows from Eqs. (20.27), (20.28), and (20.29)). The entanglement entropy of A is thus $S = -\frac{1}{N} \log \frac{1}{N} = \log N$. What is the value of N? Suppose M links (legs) are connected to region A. But effectively there are only $M - 1$ independent links because the total value of Q must be even, see Eq. (20.34) (this is illustrated in Fig. 20.5). We find $N = 2^{M-1}$. Thus the entanglement entropy is

$$S = (M - 1)\ln 2. \qquad (20.35)$$

The term $M \ln 2$ is the usual extensive entanglement entropy. The other term $-\ln 2$ is a constant and represents the TEE. This means the toric code is a topologically ordered insulator.

- Topological Entanglement Entropy of Zigzag Ribbon

Using the Hartree–Fock approximation, we investigate the entanglement in many-body topological insulators of interacting disordered zigzag graphene nanoribbons. First we try to compute S/L of interacting zigzag graphene nanoribbons as a function of $1/L$ and extract the TEE from the result.

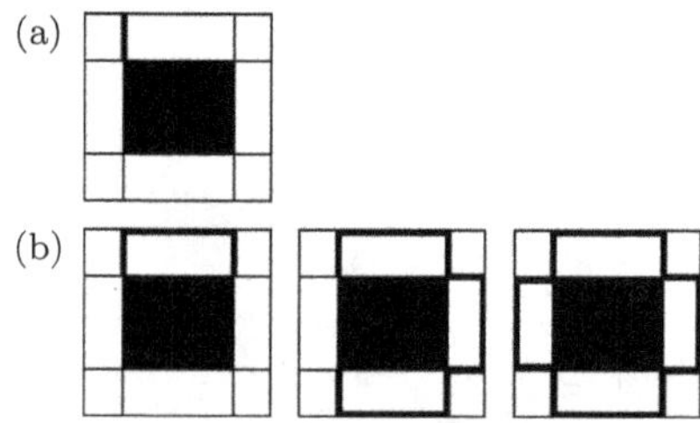

Fig. 20.5. (a) When an odd number of spin-down links are attached to the black region A, a closed loop cannot be formed. (b) Only an even number of spin-down links attached to the black region gives rise to a closed loop. In (a) the value of $Q = 1$ and a closed loop state is not formed. Some possible states are displayed in (b). Their Q values are $Q = 2, 6, 8$ and closed loop states are formed.

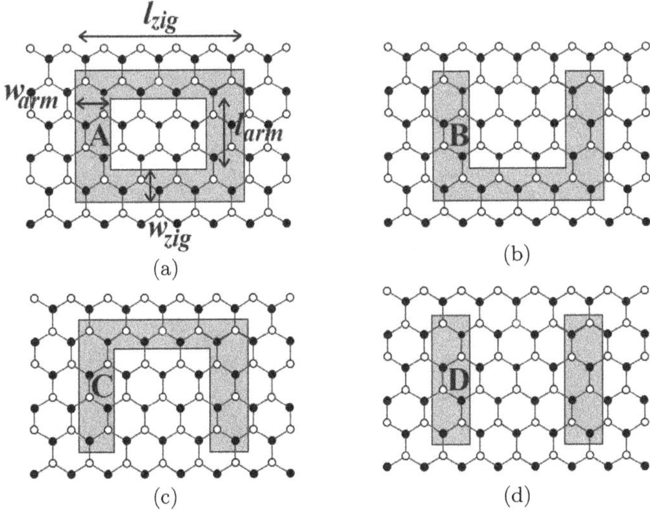

Fig. 20.6. Regions of the ribbon used in Eq. (20.36) to compute the TEE. Vertical lengths are measured in number of horizontal carbon lines and horizontal lengths in number of vertical carbon lines, see Fig. 18.5. Reprinted with permission from Ref. [18].

For an interacting but disorder-free system, we find $\beta \approx 0$. Note that, in this case, a hard gap exists in the DOS and *no* zero-energy zigzag edge states exist, as shown in Fig. 1.3. However, in the presence of disorder, this method of computing the TEE from Eq. (20.32) is not accurate: the data we obtained display significant disorder fluctuations that depend on the size of the system.

To eliminate non-topological contributions from the shape of the entangled area we adopt the following method given in Ref. [11]. The ribbon region R is partitioned using four different ways $(A, \bar{A})$, $(B, \bar{B})$, $(C, \bar{C})$, and $(D, \bar{D})$ (a bar means the complementary region of the ribbon). The regions should be as in Fig. 20.6 or a smooth deformation thereof without changing how the regions border on each other. One of the regions consists of a rectangular ring A (the Wilson loop) that lies well inside the ribbon. (A is **not** simply connected.)[f] The TEE of such a

---

[f]Kitaev and Preskill [14] uses a different partition. The entire system is first divided into two regions A and $\bar{A}$. (A is chosen to be a simply connected region inside the system.) Then A is divided into three regions. The two approaches of Refs. [11, 14] give the same result. However, in some systems one approach may give more accurate numerical result than the other.

ring [2, 11] can be written as

$$S_{\text{top}} = (S_A - S_B) - (S_C - S_D) = -2\beta < 0. \tag{20.36}$$

In this equation, the non-universal contributions from the corners cancel each other. Since the entire ribbon is topologically ordered these regions need not include a part of the zigzag edges. S_A is the entanglement entropy of the region consisting of the sites in A, see Fig. 20.6(a). Other entanglement entropies are defined similarly. Suppose that the entanglement entropies of these regions have only non-universal contributions. The areas $A - B$ and $C - D$ represent the same area, see Fig. 20.6. Then the entanglement entropy $S_A - S_B$ has the same non-universal contribution as that of $S_C - S_D$ and they will cancel each other. But now assume that the entanglement entropies of these regions also have universal contributions. If they all have the same universal contribution then S_{top} would be zero. However, the multiply-simply connected loop area A has an extra contribution. The region A is multiply connected with **two boundaries**, one inner and one outer. Its topological entanglement entropy is thus -2β. This is because, in a sense, the TEE is a **boundary contribution**. (This point can be understood by analyzing how the TEE was computed in the toric code, see Sec. 20.3.) The TEE is

$$S_{\text{top}} = -(-2\beta + \beta) - (-\beta + 2\beta) = -2\beta. \tag{20.37}$$

Except near a *critical point*, the value of the TEE is universal and is independent of system parameters. Simple dimensional analysis suggests that the dimensionless TEE β depends on the dimensionless quantity Γ/U, given by a scaling function $\beta = f(\Gamma/U)$.

The bulk TEE of zigzag ribbons is computed as follows. A careful analysis is required to reduce finite-size effects, see Jiang *et al.* [15]. To compute β one must use a thick and large rectangular ring so that long range entanglement is properly included: the width of the ring $w = w_{\text{zig}} = w_{\text{arm}}$ should be larger than the correlation length ξ. (In zigzag ribbons, the correlation length $\xi \sim \hbar v_F/\Delta$ is of the order of the lattice constant, estimated from the size of disorder-free gap $\sim 2\text{eV}$ [19] and the Fermi velocity v_F.) When the width increases, the size of ribbon and the ring must be increased simultaneously, see Fig. 20.7. At the same time, the distance from the zigzag edges to the ring is also increased beyond ξ. As a check on our method, we computed the TEE of a gapful armchair nanoribbon.

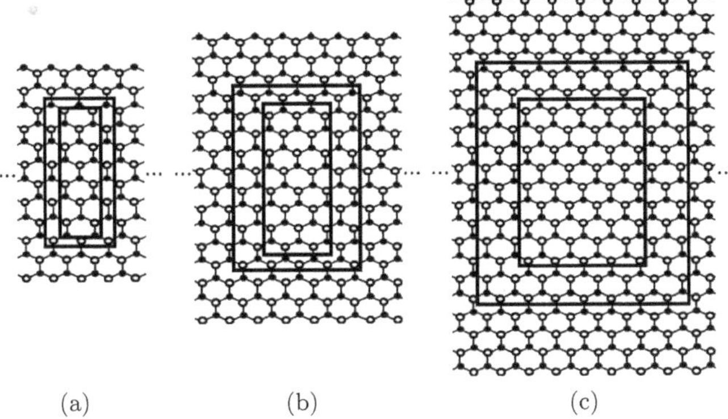

(a)   (b)   (c)

Fig. 20.7. As the ring width increases sizes of ring and ribbon are proportionally increased.

We found the expected value of zero both in the absence and presence of disorder.

> **Exercise 20.6.** Can spin–spin and charge–charge correlation lengths be different in interacting zigzag graphene nanoribbons? The correlation length is estimated to be of the order of few lattice constant. Which correlation length is it? Hint: What kind of correlation is computed in Eq. (20.18)?

The TEE result of interacting disordered zigzag graphene nanoribbons is displayed in Fig. 20.8 for two values of the ring width. We see that, as the ring width increases, the numerical uncertainties decrease. We find that the numerical uncertainty of the computed TEE is small when the ring width is larger than $\gtrsim 7$ (the uncertainty is estimated by varying the ring width at the fixed values of the ribbon length and width $L \sim 150$ and $W \sim 100$ and the rectangular ring sizes $l_{\text{zig}} \sim 125$ and $l_{\text{arm}} \sim 50$; vertical lengths are measured in number of horizontal carbon lines and horizontal lengths in number of vertical carbon line, see Fig. 18.5).

Except for small values of the ratio between the disorder strength and on-site repulsion, Γ/U, the results are independent of different values of L, W, N_{imp}, l_{arm}, and l_{zig}. From the data collapse, we infer that $\beta \approx 0.016 \pm 0.003$ and that it is, within numerical uncertainty, independent

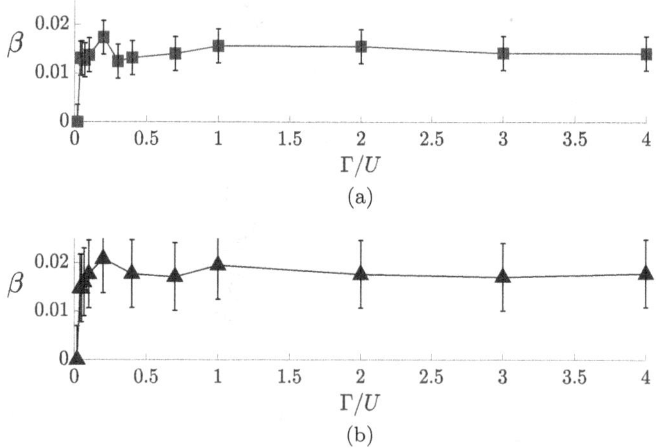

Fig. 20.8. TEE β of an interacting disordered zigzag nanoribbon is computed using a rectangular ring with different ring widths w. Each point represents a different value of (U, Γ). In (a) the ring width, ribbon length, ribbon width are, respectively, $w = 7$, $L = 130$, $W = 112$. The side lengths of the ring are $l_{\text{zig}} = 111$ and $l_{\text{arm}} = 46$ (units of length are explained in Fig. 20.1). In (b) they are $w = 4$, $L = 100$, $W = 64$, $l_{\text{zig}} = 81$, $l_{\text{arm}} = 28$. The number of disorder realizations N_D are, respectively, 10 and 20 in (a) and (b). Reprinted with permission from Ref. [18].

of the strength of the interaction and disorder. It thus takes a **universal** value. We consider this small value of β to be related to the presence of an exponentially small soft gap in interacting disordered ribbons. We see from the data that, as the critical point is approached, $\Gamma/U \to 0$, the value of the TEE varies rather abruptly to zero[g] in a non-universal manner [1].

Can one use a different partition the ribbon? One possibility is shown in Fig. 20.9. The TEE of the ring can be written as

$$S_{\text{top}} = (S_{ABC} - S_{BC}) - (S_{AB} - S_B). \tag{20.38}$$

However, a numerical work showed that this partition displays significant non-universal contributions [18] (see Exercise 20.7).

---

[g]In the absence of disorder ($\Gamma = 0$) translational symmetry is restored. In this case it is not straightforward to compute the TEE when U/t is small. This is because the gap is small and the correlation length is large. One needs prohibitively large ribbons to compute the TEE accurately. The presence of disorder reduces the correlation length, which may be evaluated by investigating the spatial dependence of the reduced density matrix.

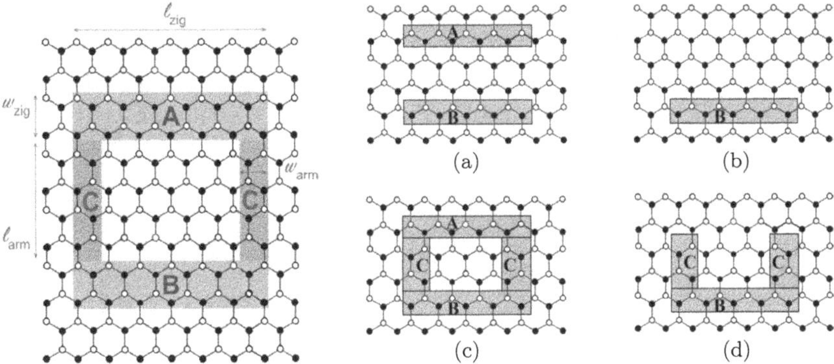

Fig. 20.9. **Left**: Another rectangular ring region investigated to compute the TEE. The ring consists of two rectangular ribbons (A and B) consisting of zigzag lines. The other two rectangular ribbons (both denoted C) consist of armchair lines. The length and width of the rectangular zigzag (armchair) ribbons are ℓ_{zig} (ℓ_{arm}) and w_{zig} (w_{arm}), respectively. **Right**: Entanglement entropy is computed for different regions shown in (a), (b), (c), and (d).

> **Exercise 20.7.** Consider the partition shown in Fig. 20.9. Why can one use, as a measure of non-universal contributions, the difference of two entanglement entropies $S_A - (S_{AB} - S_B)$ or $S_A - (S_{ABC} - S_{BC})$? Hint: In the absence of non-universal contributions the difference $S_A - (S_{AB} - S_B)$ should go to zero in thermodynamic limit. If the TEE is zero then $S_A - (S_{ABC} - S_{BC})$ should also decrease to zero.

Are there other condensed matter systems with a pseudogap that have the same value of the topological entanglement entropy? In order to define a universality class, one must compute topological entanglement entropies for different systems. This may be an interesting research to pursue.

20.4. Entanglement Spectrum of Disordered Zigzag Ribbon

It is believed that the ground state entanglement spectrum of a topological insulator resembles the corresponding edge spectrum of the system [20–22]. In other words, the entanglement spectrum of an interior region of a system contains information about the edge modes. For example, the entanglement spectrum of the Affleck–Kennedy–Lieb–Tasaki state consists of two degenerate non-zero eigenvalues, which mimic the doubly degenerate edge energy spectrum of the system with a physical boundary.

- Entanglement Spectrum of Hartree–Fock Approximation⋆

For a single-particle Hamiltonian the reduced density matrix for region A (see Fig. 20.1) can be written as [7]

$$\rho_A = Ke^{-\tilde{h}}, \tag{20.39}$$

where $\tilde{h}$ must be a **bilinear** "Hamiltonian"

$$\tilde{h} = \sum_{ij} h_{ij} c_i^\dagger c_j \tag{20.40}$$

with a normalization constant K. The "Hamiltonian" must be of this form because the non-interacting version of Wick's theorem must apply (see Exercise 20.4). Note that this particular density matrix describes a Fermi gas at **temperature** $k_B T = 1$. When this Hamiltonian is diagonalized we get the following Hamiltonian matrix:

$$\tilde{h}_{ij} = \sum_k \psi_k^*(i)\psi_k(j)\tilde{\epsilon}_k, \tag{20.41}$$

where $\tilde{\epsilon}_k$ are **eigenvalues** of the "Hamiltonian" $\tilde{h}$. The diagonalized Hamiltonian in operator form is

$$\tilde{h} = \sum_k \tilde{\epsilon}_k a_k^\dagger a_k. \tag{20.42}$$

Since the system is a Fermi gas at $k_B T = 1$ the reduced density matrix is equal to

$$\rho_{A,ij} = \mathrm{Tr}(\rho c_i^\dagger c_j) = \sum_k \psi_k^*(i)\psi_k(j)\frac{1}{e^{\tilde{\epsilon}_k}+1}, \tag{20.43}$$

where $c_i = \sum_k \psi_k(i) a_k$. The distribution of the eigenvalues $\tilde{\epsilon}_i$ of $\tilde{h}$ is called the **entanglement spectrum.**

We saw how to compute the reduced density matrix within the single-particle Hartree–Fock Hamiltonian. The eigenvalues of a density matrix are (see Eq. (20.29))

$$\lambda_k = |\sigma_k|^2 = \frac{1}{e^{\tilde{\epsilon}_k} + 1}. \tag{20.44}$$

(Note that $\tilde{\epsilon}$ are not the eigenenergies of the Hartree–Fock Hamiltonian.) One can thus relate the eigenvalues λ_i of ρ_A to the eigenvalues of $\tilde{\epsilon}_i$ of $\tilde{h}$ through this equation. The values $\tilde{\epsilon}_i \approx 0$, corresponding to $\lambda_k \approx 1/2$, dominate entanglement [23].

Using this method, we have computed the entanglement spectrum of zigzag nanoribbons. In symmetry protected topological insulators, the entanglement spectrum should mimic [21,22] the energy spectrum of gapless edge modes. Indeed the DOS and entanglement spectrum of a symmetry protected zigzag ribbon resemble each other: there are nearly flat band states near the zone boundaries (see Fig. 17.3) which resemble the nearly zero eigenvalues. The similarity is also present in the case of topologically ordered phase of disordered zigzag ribbons, as can be seen by comparing the DOS and the entanglement spectrum (see Figs. 18.7 and 20.11).

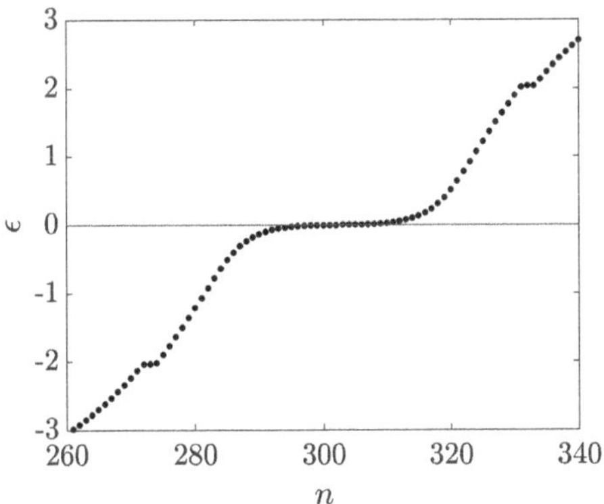

Fig. 20.10. Entanglement spectrum in the absence of disorder. The zigzag edges are not included in region A (see Fig. 20.1). On-site repulsion is $U = 0.5t$ and the width and length of the ribbon are $W = 16$, and $L = 300$. For the region A the length and width are $l = 151$ and $w = 8$. For units of the widths and lengths see Fig. 20.1(a).

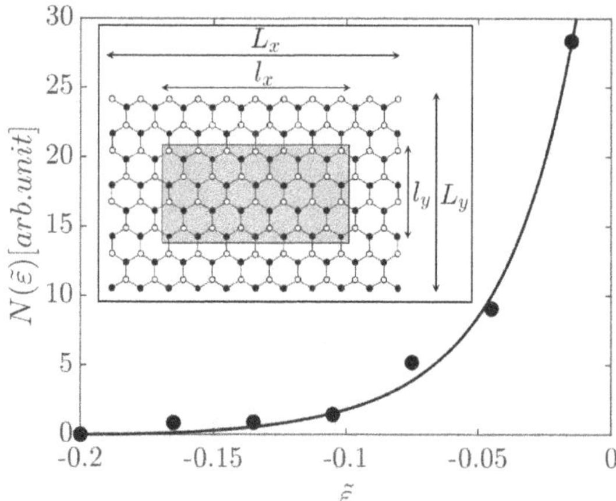

Fig. 20.11. Distribution of eigenvalues $\tilde{\epsilon}_k$ of a reduced density matrix is plotted (there are also positive values of $\tilde{\epsilon}_k$, but the distribution is identical). It is for a rectangular region inside the ribbon with length and width $l_x = 76$ and $l_y = 36$. The ribbon itself has length and width $L_x = 150$ and $L_y = 72$ (L_x and L_y are equal to the length and width of the ribbon, respectively). The distribution follows the exponential curve $B[e^{(\tilde{\epsilon}-\tilde{\epsilon}_0)^2/\delta^2} - 1]$ (black solid line). Parameters are $\Gamma = 0.3t$, $n_{\mathrm{imp}} = 0.1$, and $U = 0.5t$. The number of disorder realization is $N_D \sim 50$. Reprinted with permission from Ref. [24].

Exercise 20.8. Let us illustrate Wick's theorem at finite temperature $k_B T = 1$ for a free scalar field. (a) Show that the statistical average is

$$\langle x_i x_j \rangle = \frac{\int_{-\infty}^{\infty} \cdots \int_{-\infty}^{\infty} dx_1 \ldots dx_M e^{-\frac{1}{2}\mathbf{x}^T \cdot A \cdot \mathbf{x}} x_i x_j}{\int_{-\infty}^{\infty} \cdots \int_{-\infty}^{\infty} dx_1 \ldots dx_M e^{-\frac{1}{2}\mathbf{x}^T \cdot A \cdot \mathbf{x}}} = A_{ij}^{-1},$$

(20.45)

where $\mathbf{x} = (x_1, \ldots, x_M)$ and A is a matrix. Hint: Differentiate Eq. (G.11) two times with respect to f_i and f_j and then set $f = 0$. (b) Show

$$\langle x_i x_j x_k x_l \rangle = \frac{\int_{-\infty}^{\infty} \cdots \int_{-\infty}^{\infty} dx_1 \ldots dx_M e^{-\frac{1}{2}\mathbf{x}^T \cdot A \cdot \mathbf{x}} x_i x_j x_k x_l}{\int_{-\infty}^{\infty} \cdots \int_{-\infty}^{\infty} dx_1 \ldots dx_M e^{-\frac{1}{2}\mathbf{x}^T \cdot A \cdot \mathbf{x}}}$$

$$= A_{ij}^{-1} A_{kl}^{-1} + A_{il}^{-1} A_{jk}^{-1} + A_{ik}^{-1} A_{jl}^{-1}.$$

(20.46)

Hint: Differentiate Eq. (G.11) four times with respect to f_i, f_j, f_k, and f_l and then set $f = 0$.

Similar results may be obtained for fermions by representing creation and destruction operators using anticommuting Grassmann variables θ

$$\prod_i \int d\theta_i^* d\theta_i \theta_m \theta_l^* e^{\theta_j A_{jk} \theta_k^*} = A_{ml}^{-1} \det A. \qquad (20.47)$$

For commuting boson variables $1/\det A$ appears instead of $\det A$. The determinant can be absorbed in the definition of the expectation value $\langle \dots \rangle$. Grassmann variables are an advanced subject and we do not pursue it here.

20.5. Nature of Topological Ordered State: Non-Local Wave Functions

As mentioned before, in interacting disordered zigzag ribbons, spins of the opposite edges are mostly correlated antiferromagnetically in the weak disorder regime. However, long-range topological entanglement should not be confused with the conventional long-range spin correlations.

To elucidate the nature of disordered interacting zigzag ribbons in the weak disorder regime at half-filling, it is useful to compare ground state properties for various phases appearing for different values of Γ and U. We do this by comparing different results of (q_A, E), as shown in Fig. 20.12. In a non-interacting ribbon, bonding and antibonding states with $q_A = 1/2$ (see Figs. 20.13(a) and (b)) are destroyed by a disorder potential. Disorder-free interacting ribbons are a symmetry protected topological insulator, but fractional charges do not appear, only integer charges appear in the presence of on-site U, see Fig. 20.12(c). In contrast, in the presence of disorder, electron interactions generate mixed chiral solitons in the exponentially decaying **soft gap**, see Fig. 20.12(d). Each of these solitons supports two $e^-/2$ fractional charges with A and B chirality, respectively.

Let us remind ourselves of the salient properties of a **mixed chiral edge state** of an electron. Its wave function has two components

$$\psi_\sigma(i) = (\phi_\sigma(i), \chi_\sigma(i))^T, \qquad (20.48)$$

where i is site index and $\sigma = \uparrow, \downarrow$ is the z-component of spin. Here we are using the pseudospin representation (see Sec. 14.5): the first (second) component represents the probability amplitude to find the electron at site

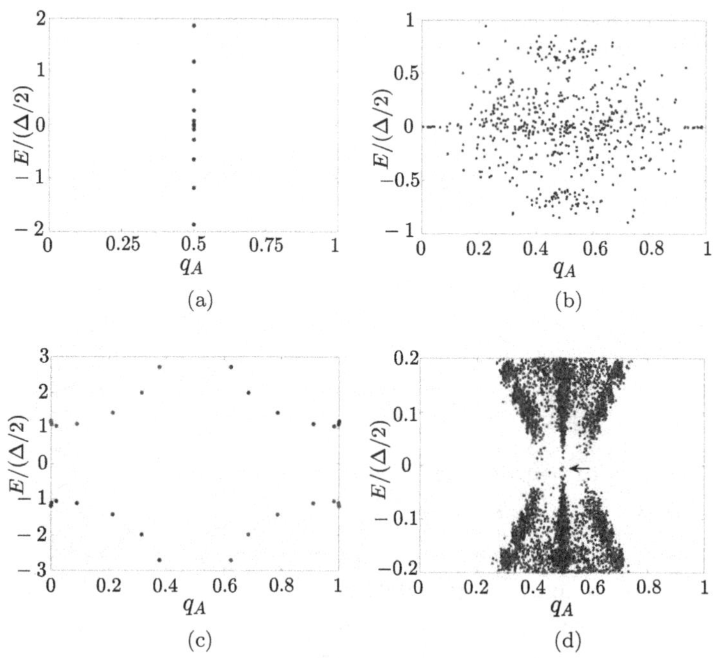

Fig. 20.12. (a) (q_A, E) plots for the following zigzag nanoribbons: disorder-free non-interacting, (b) disordered non-interacting (note that in this case the magnetic ground state is ill-defined because antiferromagnetic and paramagnetic states are degenerate), (c) disorder-free interacting, and (d) disordered interacting. A state indicated by an arrow is a soliton state with a pair of well defined $e^-/2$ fractional charges. Dots represent both spin up and down states. Note that spin-splitting is only present in (d). Reprinted with permission from Ref. [25].

$i \in A(B)$ sublattice. $\phi_\sigma(i)$ is localized near the left zigzag edge while $\chi_\sigma(i)$ is localized near the right zigzag edge, see Fig. 20.13(b). Two fractional charges $e^-/2$ of a mixed chiral edge state are located, respectively, on the left (A-type) and right (B-type) zigzag edges and have different chirality. Note that the locations of these two $e^-/2$ fractional charges on the left and right zigzag edges are **nearly identical**, see Fig. 18.15. The fractional charges of a mixed chiral state thus exhibit long-range correlations. There are also numerous Hartree–Fock single particle states with energy not only in the gap Δ but also outside it (see the plot of (q_A, E) in Ref. [26]). These bulk states decay more slowly from the zigzag edges. These mixed chiral soliton states also distort antiferromagnetism in the interior of zigzag ribbons. Note also the **small** magnetic domains and spin-charge separated objects inside the ribbon (see Fig. 18.15) that are similar to those of the

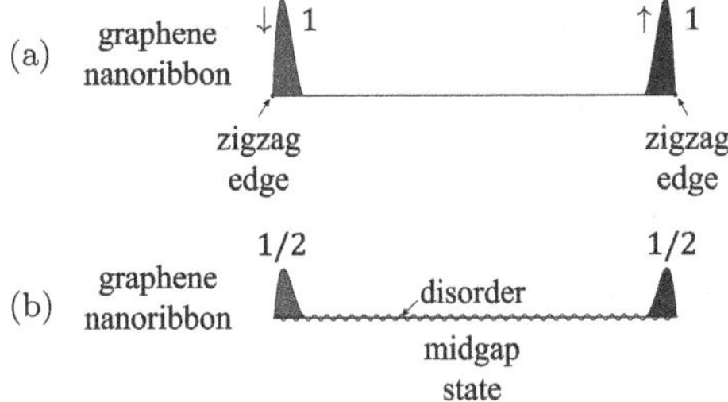

Fig. 20.13. (a) Illustrates a **chiral** edge state in a clean and gapfull zigzag graphene nanoribbon. Such an edge state displays an integer boundary charge of one. The shaded areas in the figure indicate the probability density of an edge state. Note that the opposite zigzag edges have different chiralities. (b) A **mixed** chiral state with fractional edge charges on the opposite zigzag edges in the presence of disorder. The wavy line indicates a disorder potential. Reprinted with permission from Ref. [27].

zigzag edges. The locations of these occupation number modulations in a rectangular region, displayed in Fig. 20.14, are approximately identical along the ribbon direction [28]. Many such rectangular regions are present. We believe that these interior modulations play a key role in generating a finite TEE for a region **deep inside** the zigzag ribbon, see Fig. 18.5.

The Hartree–Fock ground state of an interacting disordered zigzag nanoribbon contains all these mixed chiral edge states

$$|\Psi\rangle \sim \prod_{m\in\text{occ}} a^\dagger_{m\uparrow} \prod_{m'\in\text{occ}} a^\dagger_{m'\downarrow}, \tag{20.49}$$

where $a^\dagger_{m\sigma}$ creates a mixed chiral (non-local) state given by Eq. (20.48) with energy either in the gap or outside the gap. A mixed chiral state can be written as a linear combination of a left and right chiral states

$$a^\dagger_{m\sigma} = u_{m\sigma}L^\dagger_{m\sigma} + v_{m\sigma}R^\dagger_{m\sigma}, \tag{20.50}$$

where $L^\dagger_{m\sigma}$ ($R^\dagger_{m\sigma}$) creates a chiral state on the left (right) line of carbon atoms. The other important ingredient in the state in Eq. (20.49) is the presence of another state $m' \downarrow$ which has nearly identical probability density profile as that of the state $m \uparrow$ but with opposite spin. This combination of spin-up and -down states will keep the electron occupation numbers as uniform as possible which will minimize the repulsive electron–electron interactions (this **correlation effect** is graphically illustrated in Fig. 20.4). We believe that these mixed states are the origin of the long-range topological entanglement. In terms of the chiral states,

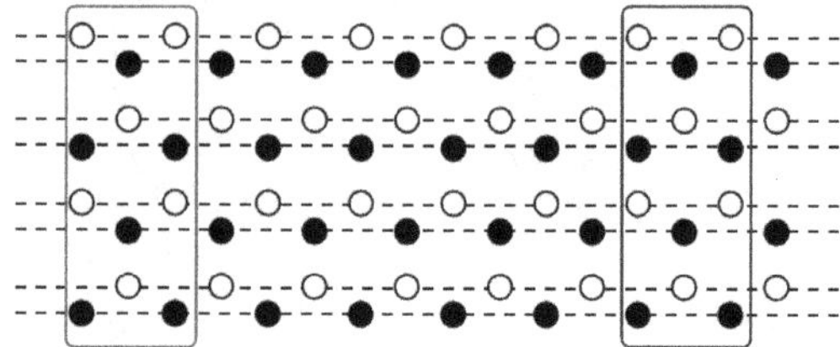

Fig. 20.14. The modulation positions along the ribbon direction of the electron occupation number in a rectangle are correlated with those of all other carbon atoms in the same box. Outside these rectangles they are uncorrelated. The positions and number of these rectangles are random and their size may not be the same.

the Hartree–Fock ground state can be written as

$$|\Psi\rangle \sim \prod_{m\in\mathrm{occ}} (u_{m\uparrow}L^{\dagger}_{m\uparrow} + v_{m\uparrow}R^{\dagger}_{m\uparrow}) \times \prod_{m'\in\mathrm{occ}} (u_{m'\downarrow}L^{\dagger}_{m'\downarrow} + v_{m'\downarrow}R^{\dagger}_{m'\downarrow})|0\rangle.$$

$$(20.51)$$

When the products are multiplied out, numerous terms containing states $m\uparrow$ and $m'\downarrow$ are present in the linear combination of states, suggesting that the Hartree–Fock ground state of interacting disordered zigzag nanoribbon is much more **entangled** than that of the disorder-free ground state.

Further reading. D. H. Lee and J. D. Joannopoulos suggested that the Anderson resonating valence bond state might constitute another example of fractional quantum Hall behavior. It is unclear how similar the ground state of the interacting disordered zigzag nanoribbon is to the $p = 1/2$ Laughlin state, whose excitations are also semions. Read about anyon superconductivity [29], which is a theory of superconductivity based on the idea that the relevant quasi-particles in the superconductors are semions.

A short-range entangled state may be changed into a product state by a generalized stochastic local transformation [1], while a long-range entangled state cannot be changed in this way.

Students should also read an excellent introduction to topological order by Sachdev in Scientific American [30].

Bibliography

[1] B. Zeng, X. Chen, D.-L. Zhou, and X.-G. Wen, *Quantum Information Meets Quantum Matter* (Springer, 2019).

[2] J. K. Pachos, *Introduction to Topological Quantum Computation* (Cambridge University Press, Cambridge, 2012).

[3] C. J. Isham, *Lectures on Quantum Theory* (Imperial College Press, London, 1995). •

[4] L. Amico, R. Fazio, A. Osterloh, and V. Vedral, Entanglement in many-body systems, *Rev. Mod. Phys.* **80**, 517 (2008).

[5] S. M. Girvin and K. Yang, *Modern Condensed Matter Physics* (Cambridge University Press, 2019). •

[6] W. H. Zurek, Decoherence and the transition from quantum to classical — revisited, *Los Alamos Sci.* **27**, 2 (2002). •

[7] I. Peschel, Calculation of reduced density matrices from correlation functions, *J. Phys. A Math. Gen.* **36**, L205 (2003).

[8] J. I. Latorre and A. Riera, A short review on entanglement in quantum spin systems, *J. Phys. A: Math. Theor.* **42**, 504002 (2009).

[9] K. F. Riley, M. P. Hobson, and S.J. Bence, *Mathematical Methods for Physics and Engineering* (Cambridge University Press, Cambridge, 2006). •

[10] S. L. Brunton and J. N. Nathan, *Data-Driven Science and Engineering: Machine Learning, Dynamical Systems, and Control* (Cambridge University Press, Cambridge, 2019).

[11] M. Levin and X.-G. Wen, Detecting topological order in a ground state wave function, *Phys. Rev. Lett.* **96**, 110405 (2006).

[12] J. Eisert, M. Cramer, and M.B. Plenio, Colloquium: Area laws for the entanglement entropy, *Rev. Mod. Phys.* **82**, 277 (2010).

[13] X.-G. Wen, Zoo of quantum-topological phases of matter, *Rev. Mod. Phys.* **89**, 041004 (2017).

[14] A. Kitaev and J. Preskill, Topological entanglement entropy, *Phys. Rev. Lett.* **96**, 110404 (2006).

[15] H.-C. Jiang, Z. Wang, and L. Balents, Identifying topological order by entanglement entropy, *Nat. Phys.* **8**, 902 (2012).

[16] M. Haque, O. Zozulya, and K. Schoutens, Entanglement entropy in fermionic Laughlin states, *Phys. Rev. Lett.* **98**, 060401 (2007).

[17] H. Yao and X.-L. Qi, Entanglement entropy and entanglement spectrum of the Kitaev model, *Phys. Rev. Lett.* **105**, 080501 (2010).

[18] Y. H. Kim, H. J. Lee, and S.-R. Eric Yang, Topological entanglement entropy of interacting disordered zigzag graphene ribbons, *Phys. Rev. B* **103**, 115151 (2021).

[19] L. Yang, C.-H. Park, Y.-W. Son, M. L. Cohen, and S. G. Louie, Quasiparticle energies and band gaps in graphene nanoribbons, *Phys. Rev. Lett.* **99**, 186801 (2007).

[20] H. Li and F. D. M. Haldane, Entanglement spectrum as a generalization of entanglement entropy: Identification of topological order in non-Abelian fractional quantum Hall effect states, *Phys. Rev. Lett.* **101**, 010504 (2008).

[21] L. Fidkowski, Entanglement spectrum of topological insulators and superconductors, *Phys. Rev. Lett.* **104**, 130502 (2010).

[22] A. M. Turner, Y. Zhang, A. Vishwanath, Entanglement and inversion symmetry in topological insulators, *Phys. Rev. B* **82**, 241102R (2010).

[23] M. Pouranvari and Kun Yang, Entanglement spectrum and entangled modes of random XY spin chains, *Phys. Rev. B* **88**, 075123 (2013).

[24] Y. H. Kim, H. J. Lee, Hyun-Yong Lee, and S.-R. Eric Yang, New disordered anyon phase of doped graphene zigzag nanoribbon, *Sci. Rep.* **12**, 14551 (2022).

[25] S.-R. Eric Yang. M. C. Cha, H. J. Lee, and Y. H. Kim, Topologically ordered zigzag nanoribbon: $e/2$ fractional edge charge, spin-charge separation, and ground-state degeneracy, *Phys. Rev. Res.* **2**, 033109 (2020).

[26] Y. H. Jeong, S.-R. Eric Yang, and M. C. Cha, Soliton fractional charge of disordered graphene nanoribbon, *J. Phys.: Condens. Matter* **31**, 265601 (2019).

[27] S.-R. Eric Yang, Soliton fractional charges in graphene nanoribbon and polyacetylene: Similarities and differences, *Nanomaterials* **9**, 885 (2019). •

[28] Lê Hoàng Anh, Young Heon Kim, In Hwan Lee, and S.-R. Eric Yang, work in progress.

[29] V. Kalmeyer and R. B. Laughlin, Equivalence of the resonating-valence-bond and fractional quantum hall states. *Phys. Rev. Lett.* **59**, 2095 (1987).

[30] S. Sachdev, Strange and Stringy, *Sci. American* **308**, 1, 44 (2013). •

Chapter 21

Matrix Product States and Disordered Anyon Phase⋆

"Physics is mathematical, not because we know so much about the physical world, but because we know so little: it is only its mathematical properties that we can discover."

Bertrand Russell

"This idea of computational irreducibility that says that even though you may know the rules by which something operates, that does not mean that you can readily... be smarter than it and jump ahead and figure out what it's going to do."

Stephen Wolfram

"The problem of computational irreducibility doesn't keep physicists awake at night, since there are approximate methods available that give eminently good insight into the problem."

Per Bak

In undoped ribbons, the Hartree–Fock approximation gives several nearly degenerate local minimum states. The topologically ordered state with fractional charges is one of these local minimum states. Is it the true ground state? The importance of quantum correlations [1] beyond the HF approximation is unclear and the properties of the true ground state are not well-known.

This is especially so in doped ribbons because the density of states (DOS) is non-zero and quantum correlations are expected to be larger.

Upon significant doping, such a system is strictly speaking, not expected to be a topologically ordered insulator as no hard gap exists. However, the system at low doping is still an insulator with localized edge states near the Fermi energy, displaying doubly degenerate ground states. In addition, the ground state of a doped disorder-free ribbon displays an edge spin density wave, in contrast to the uniform spin density of undoped ribbons. The effect of localization and charge quantization on the nature of the ground state is not understood. Is the system still a topologically ordered insulator?

To investigate the properties of low-doped zigzag ribbons, we apply the density matrix renormalization group (DMRG) approach in the matrix product states (MPSs) representation [2–6]. The MPS representation is a powerful tool for solving eigenvalue problems of quantum many-body systems with strong correlations. It can be regarded as an optimization algorithm based on the Schmidt decomposition of many-body quantum state which can be efficiently represented as the so-called MPS. Matrix product and tensor network states are actively used to investigate topological phases of matter. (A recent review can be found in Ref. [7].) This method showed that all gapped topological quantum states in one dimension are in symmetry protected phases. In addition, they can be used to classify symmetry protected topological insulators [8]. In some systems, the DMRG approach may be useful in classifying the topological order. Although the DMRG generally fails in large two-dimensional systems, it may be applied to narrow quasi-one-dimensional zigzag ribbons.

We have applied the DMRG to disordered interacting zigzag ribbon. Such a calculation shows that quantum fluctuations mix nearly degenerate ground states of the Hartree–Fock approximation and that the true ground state displays fractional charges. In this chapter, we will show that, in doped ribbons, the fractional charges form a new disordered anyon phase.[a] The new phase is still topologically ordered in the low doping limit. This phase is ideally suited for the observation of fractional charges because the DOS displays an unusual dependence on doping density. As doping level increases, a topological phase transition into a topologically trivial state is expected. Doped zigzag ribbons may display unusual transport, magnetic, and inter-edge tunneling properties.

---

[a]This is largely a technical chapter and students who are not interested in numerical computations may skip it at first reading.

21.1. Density Matrix Renormalization Group

Before studying the MPS, let us briefly explain the core ideas of the density matrix renormalization group method [3]. It uses the renormalization group theory of K. G. Wilson. Wilson's approach is based on a discrete lattice model. Everytime some degrees of freedom are "integrated" out, the Hamiltonian renormalizes, i.e., its coupling constants change (this corresponds to step 3 below). The Hamiltonian is expected to move closer to a fixed point. A good and pedagogical introduction to the renormalization group theory is given by its originator K. G. Wilson in Ref. [2] (a student should read this paper before further reading). The algorithm is as follows:

1. Find the ground state of the full Hamiltonian of a superblock shown in Fig. 21.1(b) using exact diagonalization method. The superblock has to be small enough so that it can be exactly diagonalized.
2. Divide the superblock into the left and right blocks A and B (see Fig. 21.1(c)). Write the ground state as $|\Psi\rangle = \sum_{ij} \psi_{i,j} |i\rangle \otimes |j\rangle$, where $\{|i\rangle\}$ and $\{|j\rangle\}$ are the basis states of the left and right blocks, respectively.
3. Calculate the reduced density matrix of the left block $\rho_{ii'} = \sum_{j \in B} \psi_{i,j} \psi_{i',j}^*$. Diagonalize the density matrix $\rho_{ii'}$. This step explains why the method is called **density matrix** renormalization group. Keep the m eigenvectors with the largest eigenvalues. This is done using the SVD, explained in Exercise 20.25.
4. Project the Hamiltonian using these m eigenvectors. In this way, the new Hamiltonian for the block is formed. See Sec. 21.2.
5. Add a new site to each of the left and right blocks A and B, as shown in Fig. 21.1(a). The Hamiltonians are changed to include these additional sites (see Appendix 21.2). Combining these left and right Hamiltonians, the new Hamiltonian for the superblock is formed. Go to step 1.

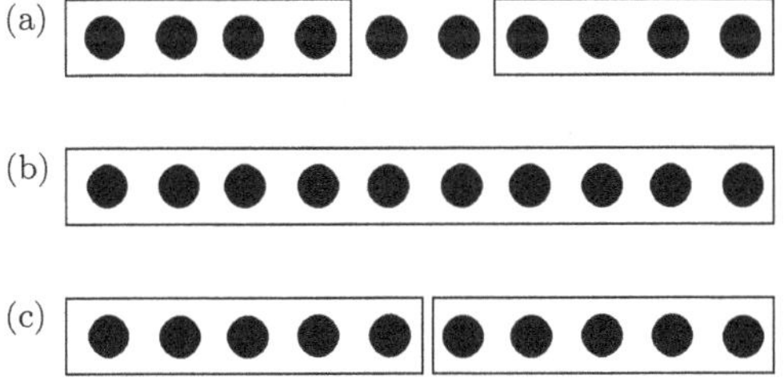

Fig. 21.1. Basic procedure of the infinite system density matrix renormalization group method is shown schematically. Dots represent spin sites.

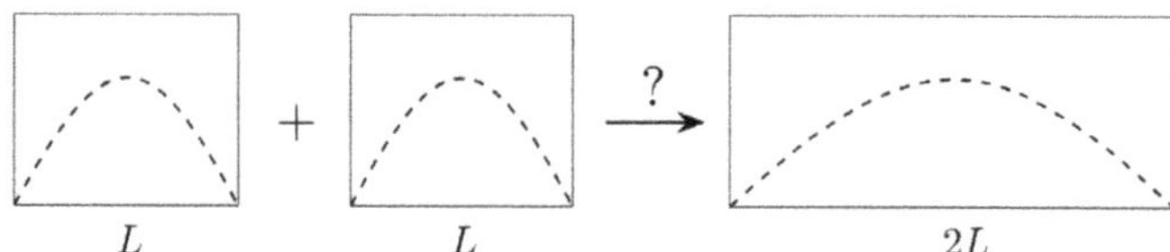

Fig. 21.2. The wave functions of the lowest-energy states of A, B, and AB are shown. If only the lowest-energy states of A and B are retained it is not possible to reproduce the lowest-energy state of AB. Many states of A and B are needed to get accurate results.

Many more sites may be added by repeating the whole process numerous times. We explain some of these steps more in detail in the next Section.

Exercise 21.1. Suppose the basis states of A and B are, respectively, $\{|i\rangle\}$ and $\{|j\rangle\}$. One combines A and B and forms a superblock. Then any state of the superblock AB may be written as

$$|k\rangle = \sum_{ij} \psi_{ij} |i\rangle |j\rangle. \tag{21.1}$$

The total Hamiltonian of AB, H_{AB}, may be represented using these $\{|k\rangle\}$ states. The m lowest energy eigenstates of H_{AB} and use them as the new basis set. One can similarly form another superblock and repeat the whole process. However, this approach does not give accurate results. Discuss the reason for its failure. Hint: See Fig. 21.2.

21.2. Effective Hamiltonian Matrices

The following questions arise in carrying out the renormalization group process described in the previous section. How does one project a Hamiltonian onto a Hilbert subspace? The other question is how does one incorporate the effects of adding extra sites into a Hamiltonian matrix?

Sometimes it is useful to project the Hamiltonian on the vector space spanned by some m orthogonal vectors $\{\vec{u}_i\}$ with $i = 1, \ldots, m$. In this way, an effective Hamiltonian can be generated. This is done by performing the following unitary transformation

$$H' = \tilde{U}^\dagger H \tilde{U}, \tag{21.2}$$

where the transformation matrix $\tilde{U}$ consists of m eigenvectors

$$\tilde{U} = \begin{pmatrix} u_{11} & \cdots & u_{1m} \\ \vdots & \ddots & \vdots \\ u_{n1} & \cdots & u_{nm} \end{pmatrix}, \tag{21.3}$$

where n is the number of components in the vectors. Each column stands for ith vector $\vec{u}_i = (u_{1i}, \ldots, u_{ni})^T$. Note that $\tilde{U}$ is an $n \times m$ matrix.

It is also useful to know how to build the combined Hamiltonian matrix of two subsystems described by the Hamiltonian matrices H_A and H_B. The total Hamiltonian matrix is

$$H = H_A \otimes I + I \otimes H_B, \tag{21.4}$$

where I is a unit matrix. Equivalently this can be written as

$$H = H_A \otimes H_B. \tag{21.5}$$

Now if systems A and B are connected, there will be a Hamiltonian that describes the boundary terms connecting the two systems. For example, consider the Ising spin-interaction term connecting two boundary sites $i \in A$ and $i + 1 \in B$ (see Fig. 21.3)

$$\sigma_{z,i}\sigma_{z,i+1}. \tag{21.6}$$

Representing the operators σ_i and σ_{i+1} as matrices, we find

$$[\sigma_{z,i}]_{k_1,k_1'} \otimes [\sigma_{z,i+1}]_{k_2,k_2'}. \tag{21.7}$$

Here k_1 and k_1' label many-body states of A and k_2 and k_2' label those of B. The basis states of the combined system AB are

$$|k_1 k_2\rangle = |k_1\rangle|k_2\rangle. \tag{21.8}$$

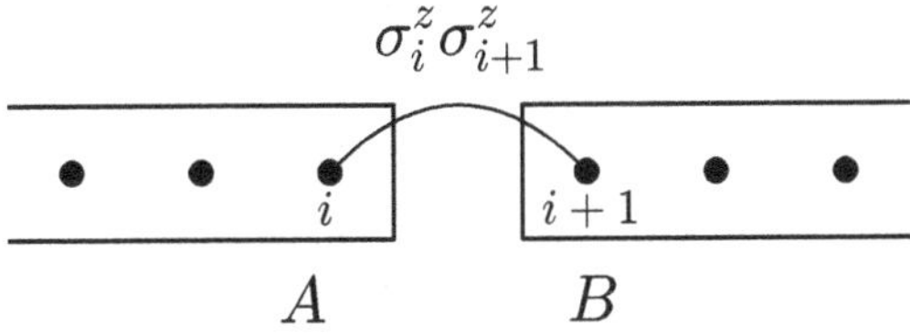

Fig. 21.3. Heisenberg spin-interaction term connecting two boundary sites $i \in A$ and $i + 1 \in B$.

Note that site spin operators may be expressed in terms of site creation and destruction operators, see Eq. (6.93). These results may be applied in renormalization group approaches.

21.3. Matrix Product States

Here we give a glimpse of the basic concepts of MPS (see also Ref. [4]). An introduction to the MPS may be found in Refs. [5,6]. Let us write the ground state of a spin model or the Hubbard model as

$$|\Psi\rangle = \sum_{s_1\ldots s_N} \psi_{s_1\ldots s_N}|s_1\ldots s_N\rangle, \tag{21.9}$$

where s_i denotes a possible quantum basis state of site i. (For a one-dimensional Hubbard model, s represents basis states $|0\rangle$, $|\uparrow\rangle$, $|\downarrow\rangle$ and $|\uparrow\downarrow\rangle$.) A matrix product state representation of $|\Psi\rangle$ has the following form

$$|\Psi\rangle = \sum_{s_1\ldots s_N} \mathrm{Tr}(A^{s_1}\ldots A^{s_N})|s_1\ldots s_N\rangle, \tag{21.10}$$

where A^{s_i} are matrices of dimension m (it is called the inner/bond dimension). Exercise 21.2 will make clear the meaning of the notations used in this equation. One of the key points of this approach is that for some quantum systems, notably one-dimensional gapful systems, the matrix dimension of A^{s_i} is not huge. Matrix product states thus provide a more concise representation of the ground state than that of exact diagonalization method.

Exercise 21.2. Consider the Affleck–Kennedy–Lieb–Tasaki (AKLT) model of spin 1 chains. Here the site basis states are $|m_z\rangle = |-1\rangle$, $|0\rangle$, and $|1\rangle$ and the inner dimension is $m = 2$. The **exact** ground state may be written as a matrix product state. Show that the A matrices are (except for a normalization factor)

$$A_{-1} = -(\sigma_x - i\sigma_y)/\sqrt{2},$$
$$A_0 = -\sigma_z, \tag{21.11}$$
$$A_1 = (\sigma_x + i\sigma_y)/\sqrt{2}.$$

These matrices are site-independent. Hint: Read about the AKLT model in Ref. [9]. The ALKT state is constructed by having each spin 1/2 form a singlet pair with one of the spin 1/2 on the neighboring sites.

• Matrix Product States and Singular Value Decomposition

Where does the MPS of Eq. (21.10) come from? Consider a chain consisting of N sites. Let us divide this chain into 1 site and $N-1$ sites; in other words, the chain is divided into A and B regions. Assume that each site has d quantum states. Applying the Schmidt decomposition (see Sec. 20.4) to any state we find

$$|\Psi^1\rangle = \sum_{\alpha_1} s_{\alpha_1}|\alpha_1\rangle_A|\alpha_1\rangle_B,\qquad(21.12)$$

where s_{α_1} are the eigenvalues of the relevant diagonal matrix of the SVD, $|\alpha_1\rangle_A$ and $|\alpha_1\rangle_B$ are **eigenstates of the reduced density matrices** for A and B, respectively. Next, divide the $N-1$ sites into 1 site and $N-2$ sites. When this process is repeated we find that any state of $N-i$ sites can be written as

$$|\Psi^i\rangle = \sum_{\alpha_i} s_{\alpha_i}|\alpha_i\rangle_A|\alpha_i\rangle_B.\qquad(21.13)$$

Here $|\alpha_i\rangle_A$ represent eigenstates of the reduced density matrix of ith sites while $|\alpha_i\rangle_B$ represent those corresponding to the rest of the $N-i$ sites. Note the orthogonality between the eigenstates of each region

$$\begin{aligned} {}_A\langle\alpha_i|\alpha_i'\rangle_A &= \delta_{\alpha_i,\alpha_i'}\\ {}_B\langle\alpha_i|\alpha_i'\rangle_B &= \delta_{\alpha_i,\alpha_i'}. \end{aligned}\qquad(21.14)$$

The processes may be described in a different but equivalent way, which will explain how the matrix products appear. The expansion coefficients $\psi_{s_1\ldots s_N}$ of the ground state given in Eq. (21.9) are converted into a matrix with dimension $d\times d^{N-1}$. We then apply an SVD to this matrix (see Sec. 20.4) and find that the amplitudes may be rewritten as a product

$$\psi_{s_1,(s_2\ldots s_N)} = \sum_{\alpha_1} U_{s_1,\alpha_1}\overbrace{[S_{\alpha_1}V^\dagger_{\alpha_1,(s_2 s_3\ldots s_N)}]}^{\psi_{(\alpha_1 s_2),(s_3\ldots s_N)}} = \sum_{\alpha_1} U_{s_1,\alpha_1}\psi_{(\alpha_1 s_2),(s_3\ldots s_N)},$$

$$(21.15)$$

where U, S, and $V^\dagger$ originate from the SVD (see Figs. 20.2 and 21.4). Note that the set of numbers in $(\ldots)$ represents a matrix index. Two matrix indices are separated by a comma. After this procedure, the chain consisting of $N-1$ sites is divided into 1 site and $N-2$ sites, followed by an SVD,

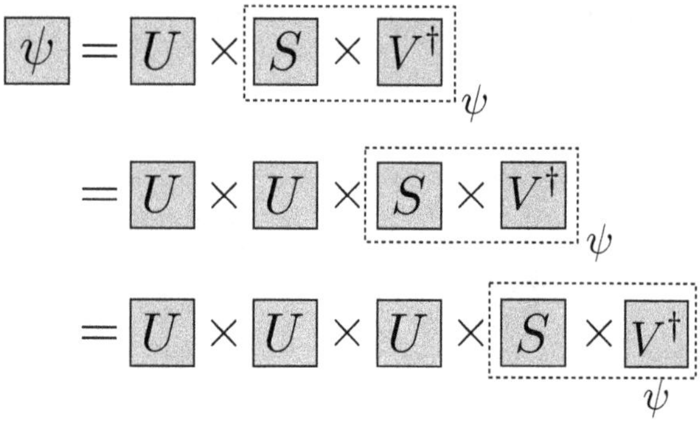

$$\boxed{\psi} = \boxed{U} \times \boxed{S} \times \boxed{V^\dagger}$$

$$= \boxed{U} \times \boxed{U} \times \boxed{S} \times \boxed{V^\dagger}$$

$$= \boxed{U} \times \boxed{U} \times \boxed{U} \times \boxed{S} \times \boxed{V^\dagger}$$

Fig. 21.4. The SVD is repeatedly applied. In each step, one finds the most important d states of the composite system consisting of the previous spin block and the added spin site. The quantities in the dashed circles represent the new matrices ψ.

i.e., $\psi_{(\alpha_1 s_2),(s_3...s_N)}$ is converted into a product (see Fig. 21.4)

$$\psi_{(\alpha_1 s_2),(s_3...s_N)} = \sum_{\alpha_2} \overbrace{U_{(\alpha_1 s_2)} \delta_{\alpha_2 \alpha_2}}^{U_{(\alpha_1 s_2),\alpha_2}} \overbrace{S_{\alpha_2} V^\dagger_{\alpha_2(s_3...s_N)}}^{\psi_{(\alpha_2 s_3),(s_4...s_N)}}$$

$$= \sum_{\alpha_2} U_{(\alpha_1 s_2),\alpha_2} \psi_{(\alpha_2 s_3),(s_4...s_N)}. \tag{21.16}$$

This is repeated until the last site in the chain is reached. The final result after the $N-1$ steps is

$$|\Psi\rangle = \sum_{s_1...s_N} \sum_{\alpha_1,...,\alpha_{N-1}} U_{s_1,\alpha_1} U_{(\alpha_1 s_2),\alpha_2} \cdots U_{(\alpha_{N-2}s_{N-1}),\alpha_{N-1}}$$

$$\psi_{\alpha_{N-1},s_N} |s_1 \ldots s_N\rangle. \tag{21.17}$$

No approximation is made so far but this expression is not in the form of a matrix product state.

Let us rewrite this result as a **matrix product state** by rewriting U and ψ as

$$U_{s_1,\alpha_1} = A_{\alpha_1}^{s_1} \quad \text{for } i = 1,$$

$$U_{(\alpha_{i-1}s_i),\alpha_i} = A_{\alpha_{i-1},\alpha_i}^{s_i} \quad \text{for } i < 1 < N, \tag{21.18}$$

$$\psi_{\alpha_{N-1},s_N} = A_{\alpha_{N-1}}^{s_N}.$$

Here s_i labels each matrix and α_{i-1} and α_i serve as matrix indices. For $i = 1,\ldots,N-1$ the dimensions of A^{s_i}'s vary, respectively, from

$(1 \times d), (d \times d^2), \ldots, (d^{N/2-1} \times d^{N/2}), (d^{N/2} \times d^{N/2-1}), \ldots, (d^2 \times d)$ to $(d \times 1)$. The resulting state may be expressed as

$$|\Psi\rangle = \sum_{\{\alpha_1,\ldots,\alpha_{N-1}\}\{s_1,\ldots,s_N\}} A_{\alpha_1}^{s_1} A_{\alpha_1,\alpha_2}^{s_2} \cdots A_{\alpha_{N-1}}^{s_N} |s_1 \ldots s_N\rangle. \qquad (21.19)$$

The construction of this state is illustrated graphically in Fig. 21.5. For later use it is convenient to define a left- and right canonical MPS. A left-canonical MPS of k sites is given by

$$|\alpha_k\rangle = \sum_{\{\alpha_1,\ldots,\alpha_{k-1}\}\{s_1,\ldots,s_k\}} A_{\alpha_1}^{s_1} A_{\alpha_1,\alpha_2}^{s_2} \cdots A_{\alpha_{k-1}}^{s_k} |s_1 \ldots s_k\rangle. \qquad (21.20)$$

Similarly a right-canonical MPS is constructed and displayed in Fig. 21.6. These states are diagrammatically represented in Figs. 21.5 and 21.6. The diagrammatic rules are explained in Fig. 21.7. Under periodic boundary conditions $\alpha_{N+1} = \alpha_1$, see Fig. 21.8.

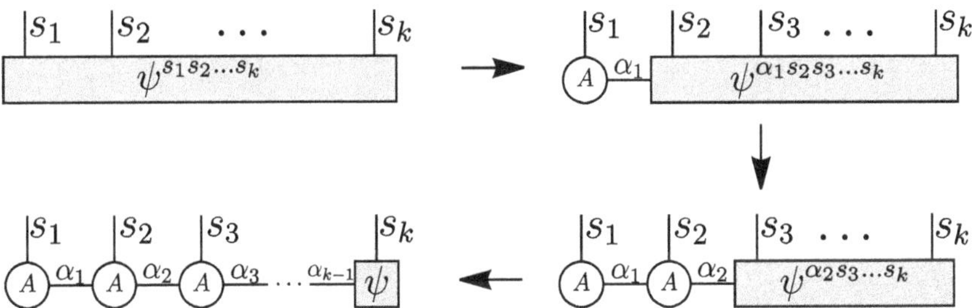

Fig. 21.5. Graphic illustration of the construction of a so-called left-canonical MPS. One starts from the left end site of the chain. The final state is the left-canonical MPS. Note that A^{s_1} of the starting first site is a vector and not a matrix.

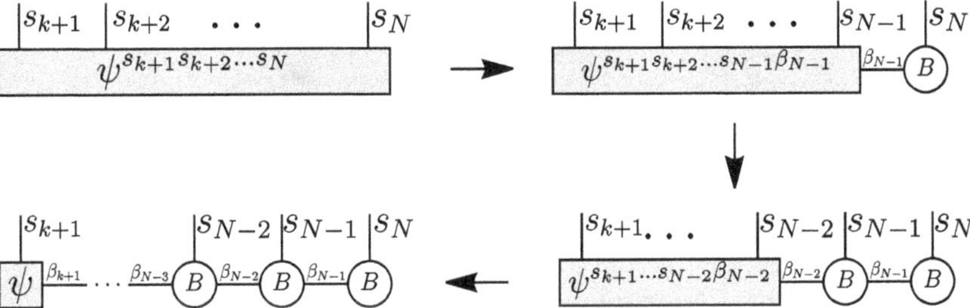

Fig. 21.6. Graphic illustration of the way to construct a right-canonical MPS. One starts from the right end site of the chain. Note that B^{s_L} of the starting last site is a vector and not a matrix. $\beta_{k+1}, \ldots, \beta_{N-1}$ are summed over.

$$A^s_{\alpha,\beta} \quad = \quad$$

$$A^s_\alpha \quad = \quad$$

$$\sum_{\alpha_0} A^{s_1}_{\alpha_0,\alpha_1} A^{s_2}_{\alpha_1,\alpha_2} =$$

Fig. 21.7. Basic diagrammatic rules. Each matrix $A^s_{\alpha,\beta}$ is associated with a vertex with three legs, i.e., three indices. The row and column indices of $A^s_{\alpha,\beta}$ are also called bond indices. A vertical line is associated with the index s_i, which denotes a local state at site i, e.g., $|s_i\rangle = |\uparrow\rangle, |\downarrow\rangle$ for spin-half sites. When two vertices are joined together the inner line is contracted, i.e., α_1 is summed over.

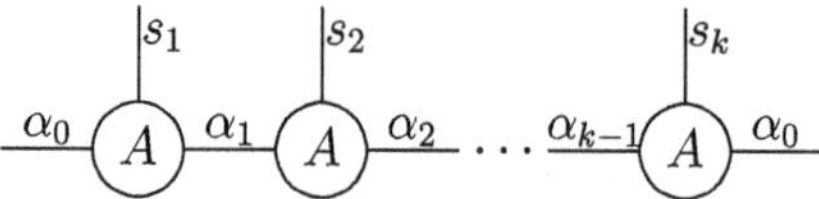

Fig. 21.8. Diagrammatic representation of a matrix product state under periodic boundary conditions.

In a mixed representation, using the left- and right-canonical states of Fig. 21.5 and 21.6, we can write the many-body state as

$$|\Psi\rangle = \sum_{\alpha_k,\beta_k} \langle \alpha_k \beta_{k+1} | \Psi \rangle | \alpha_k \beta_{k+1} \rangle$$

$$= \sum_{\{s\}} \sum_{\alpha_1,\dots,\alpha_k} \sum_{\beta_k,\dots,\beta_N} A^{s_1}_{\alpha_1} A^{s_2}_{\alpha_1,\alpha_2} \cdots A^{s_k}_{\alpha_{k-1},\alpha_k} \langle \alpha_k \beta_{k+1} | \Psi \rangle$$

$$\times B^{s_{k+1}}_{\beta_k,\beta_{k+1}} \cdots B^{s_{N-1}}_{\beta_{N-1},\beta_N} B^{s_N}_{\beta_N} |s_1 \dots s_N\rangle. \tag{21.21}$$

(See the diagrammatic representation of this in Fig. 21.9.) The quantity $\langle \alpha_l \beta_{l+1} | \Psi \rangle$ may be absorbed in one of the matrices near it.

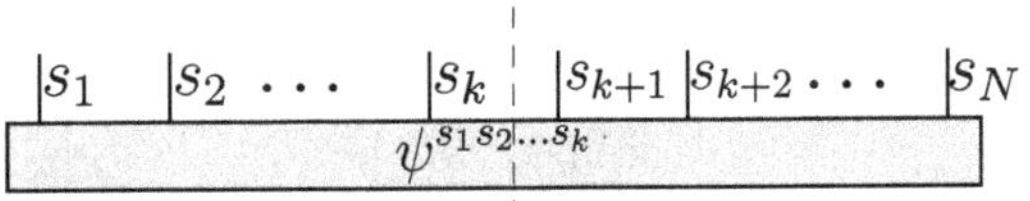

$$|\alpha_k\rangle = \sum_{\{\alpha\},\{s\}} A^{s_1}_{\alpha_1} A^{s_2}_{\alpha_1,\alpha_2} \cdots A^{s_k}_{\alpha_{k-1}} |s_1 s_2 \cdots s_k\rangle$$

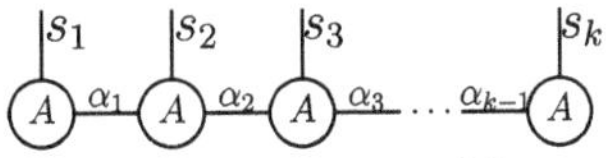

(b)

$$|\beta_{k+1}\rangle = \sum_{\{\beta\},\{s\}} B^{s_{k+1}}_{\beta_{k+1}} B^{s_{k+2}}_{\beta_{k+1},\beta_{k+2}} \cdots B^{s_N}_{\beta_{N-1}} |s_{k+1} s_{k+2} \cdots s_N\rangle$$

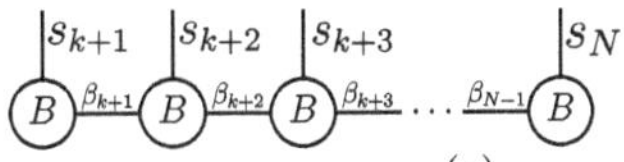

(c)

Fig. 21.9. (a) Division of a state into the left- and right-canonical MPSs. (b) Graphic illustration of the construction of the left-canonical MPS. (c) Graphic illustration of the construction of the right-canonical MPS.

21.4. Matrix Product Representation of Operators

• Inner Product

Let us compute the inner product between two matrix product states $\langle\Psi|\Psi'\rangle$ using the left-canonical form (Fig. 21.5), where

$$\begin{aligned}
|\Psi\rangle &= \sum_{s_1\ldots s_N} A^{s_1}\ldots A^{s_N}|s_1\ldots s_N\rangle, \\
|\Psi'\rangle &= \sum_{s_1\ldots s_N} B^{s_1}\ldots B^{s_N}|s_1\ldots s_N\rangle.
\end{aligned} \tag{21.22}$$

(Although we use B^{s_i} in the state $|\Psi'\rangle$ it should not be confused with the right-canonical form.) We find

$$\begin{aligned}
\langle\Psi|\Psi'\rangle &= \sum_{s_1\ldots s_N} (A^{s_1}\ldots A^{s_N})^{\dagger} B^{s_1}\ldots B^{s_N} \\
&= \sum_{s_1\ldots s_N} A^{s_N\dagger}(\ldots A^{s_i\dagger}(\ldots (A^{s_1\dagger}B^{s_1})\ldots)B^{s_i}\ldots)B^{s_N}. \tag{21.23}
\end{aligned}$$

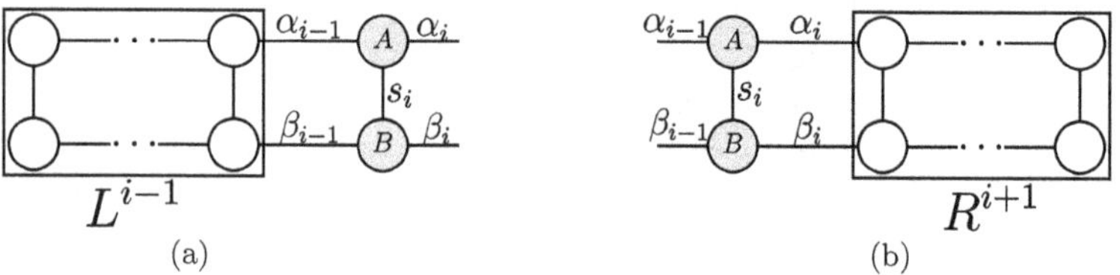

Fig. 21.10.   (a) Matrix $L^{i-1}_{\alpha_{i-1}\beta_{i-1}}$ represents an overlap between the states ψ and ψ' of the left side of i. Vertical lines representing s_i are all contracted. (b) Matrix $R^{i+1}_{\alpha_i\beta_i}$ represents an overlap between the states ψ and ψ' of the right side of i.

In this equation, we have suppressed vector and matrix indices. Note that $A^{s_1\dagger}$, B^{s_1}, $A^{s_N\dagger}$, and B^{s_N} are vectors while other $A^{s_i\dagger}$ and B^{s_i} are matrices. The overlap may be conveniently computed by introducing the quantity $L^i_{\alpha_i\beta_i}$ (Fig. 21.10(a)) satisfying the following recursive relation

$$
\begin{aligned}
L^1_{\alpha_1\beta_1} &= \sum_{s_1} A^{*s_1}_{\alpha_1} B^{s_1}_{\beta_1} \\
L^i_{\alpha_i\beta_i} &= \sum_{s_i,\alpha_{i-1},\beta_{i-1}} A^{s_i*}_{\alpha_{i-1}\alpha_i,} L^{i-1}_{\alpha_{i-1}\beta_{i-1}} B^{s_i}_{\beta_{i-1},\beta_i}.
\end{aligned}
\tag{21.24}
$$

Note that $A^{s_i\dagger}_{\alpha_i\alpha_{i-1}} = A^{s_i*}_{\alpha_{i-1}\alpha_i}$. We find

$$
\langle\Psi|\Psi'\rangle = \sum_{s_i,\alpha_{i-1},\beta_{i-1}} A^{*s_i}_{\alpha_{i-1}s_i} L^{i-1}_{\alpha_{i-1}\beta_{i-1}} B^{s_i}_{\beta_{i-1}s_i}.
\tag{21.25}
$$

Similarly we define R_{i+1} for the right canonical form, see Fig. 21.10(b). For the right canonical form (see Fig. 21.6) we have

$$
\langle\Psi|\Psi'\rangle = \sum_{s_i,\alpha_i,\beta_i} A^{*s_i}_{\alpha_i s_i} R^{i+1}_{\alpha_i\beta_i} B^{s_i}_{\beta_i s_i}.
\tag{21.26}
$$

In the mixed representation (21.21) we get

$$
\langle\Psi|\Psi'\rangle = \sum_{s_i,\alpha_{i-1},\beta_{i-1}} A^{*s_i}_{\alpha_{i-1}\alpha_i} L^{i-1}_{\alpha_{i-1}\beta_{i-1}} R^{i+1}_{\alpha_i\beta_i} B^{s_i}_{\beta_{i-1}\beta_i},
\tag{21.27}
$$

which is easier to remember in the diagrammatic representation, see Fig. 21.11.

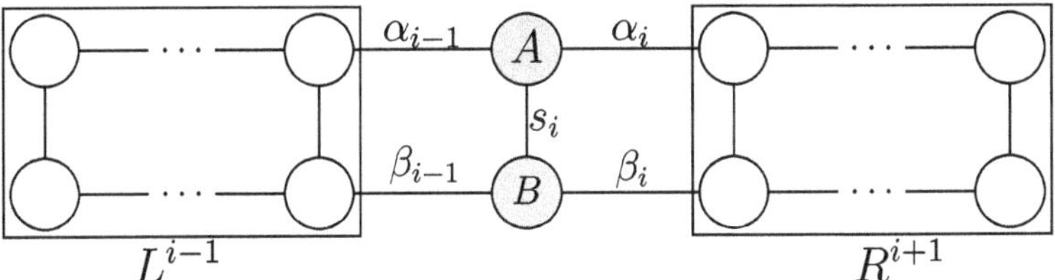

Fig. 21.11.   The inner product $\langle\Psi|\Psi'\rangle$. The horizontal and vertical lines stand for contractions.

- Expectation Value of an Operator

Using the basis set $\{|s_1\cdots s_k\rangle\}$ we may represent the matrix product state expectation value of an operator $\langle\Psi|O|\Psi\rangle$. An operator can be written as

$$O = \sum_{\{s\},\{s'\}} O^{s_1\cdots s_k}_{s'_1\cdots s'_k}|s_1\cdots s_k\rangle\langle s'_1\cdots s'_k|. \qquad (21.28)$$

To evaluate the expectation value $\langle\Psi|O|\Psi\rangle$ we first need to compute the matrix elements of O with respect to the basis set $\{|s_1\cdots s_k\rangle\}$

$$O^{s_1\cdots s_k}_{s'_1\cdots s'_k} = \langle s_1\ldots s_k|O|s'_1\ldots s'_k\rangle = W[1]^{s_1 s'_1}\ldots W[k]^{s_k s'_k}, \qquad (21.29)$$

where

$$W[i]^{s_i s'_i} = W[i]^{s_i,s'_i}_{\alpha'_{i-1},\alpha_i}. \qquad (21.30)$$

Let us use the Hamiltonian operator to explain in more detail the meaning of these notations. The Hamiltonian operator can be written as

$$H = \sum_{\{s\},\{s'\}} H^{s_1\cdots s_k}_{s'_1\cdots s'_k}|s_1\cdots s_k\rangle\langle s'_1\cdots s'_k|, \qquad (21.31)$$

where the matrix elements are

$$H^{s_1\cdots s_k}_{s'_1\cdots s'_k} = \langle s_1\cdots s_k|H|s'_1\cdots s'_k\rangle = h[1]^{s_1 s'_1}\cdots h[i]^{s_i s'_i}\cdots h[k]^{s_k s'_k}$$

$$(21.32)$$

and

$$h[i]^{s_i s'_i} = W[i]^{s_i,s'_i} \rightarrow h[i]^{s_i s'_i}_{l_i r_i} \qquad (21.33)$$

are **matrices**. These are called the matrix product operator representation, see Figs. 21.12 and 21.13. For $s_i \neq s'_i$, the off-diagonal matrix elements may

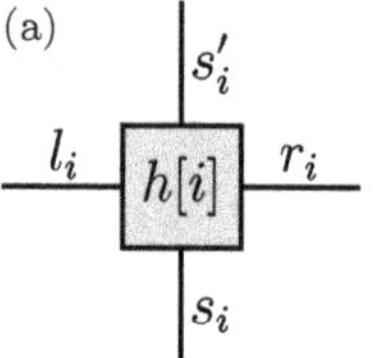

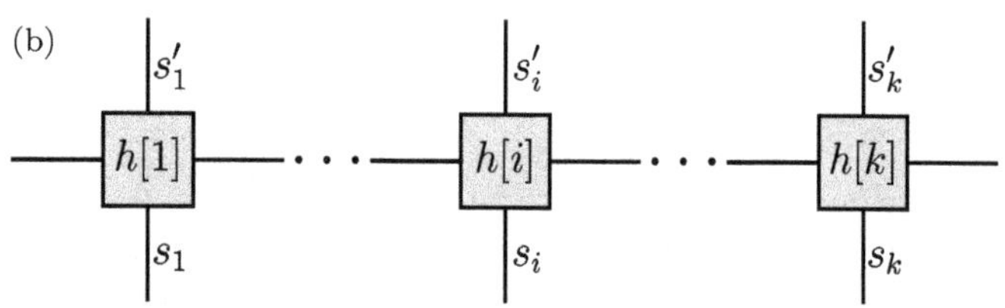

Fig. 21.12. Graphic illustration of (a) $h[i]^{s_i s_i}$ and (b) the matrix product operator $\langle s_1 \cdots s_k | H | s_1' \cdots s_k' \rangle = h[1]^{s_1 s_1'} \cdots h[i]^{s_i s_i'} \cdots h[k]^{s_k s_k'}$.

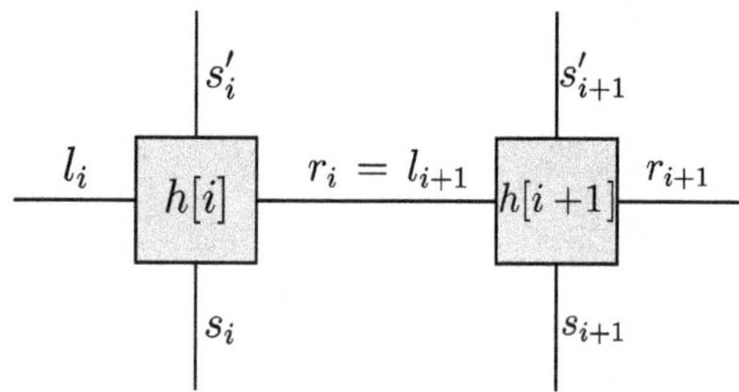

Fig. 21.13. Graphic illustration of the matrix product operator of two matrices of type $h[i]^{s_i s_i}$.

be non-zero. For a one-dimensional Hubbard model, we find

$$
h_{lr}^{ss'} =
\begin{pmatrix}
(I_i)_{ss'} & 0 & 0 & 0 & 0 & 0 \\
(c_{i\uparrow}^{\dagger})_{ss'} & 0 & 0 & 0 & 0 & 0 \\
(c_{i\downarrow}^{\dagger})_{ss'} & 0 & 0 & 0 & 0 & 0 \\
(c_{i\uparrow})_{ss'} & 0 & 0 & 0 & 0 & 0 \\
(c_{i\downarrow})_{ss'} & 0 & 0 & 0 & 0 & 0 \\
\begin{matrix}(Un_{i\uparrow}n_{i\downarrow}\\+v_i(n_{i\uparrow}+n_{i\downarrow}))_{ss'}\end{matrix} & t(c_{i\uparrow})_{ss'} & t(c_{i\downarrow})_{ss'} & -t(c_{i\uparrow}^{\dagger})_{ss'} & -t(c_{i\downarrow}^{\dagger})_{ss'} & (I_i)_{ss'}
\end{pmatrix}_{lr},
$$

$$
(21.34)
$$

where I_i stands for the trivial operator acting on site i, $c_{i\sigma}^{(\dagger)}$ for the creation operator, v_i for the random potential strength, and U for the Hubbard interaction. One can easily generalize the above W-tensor and matrix product operators into the proper quantities of the model of the quasi-one-dimensional honeycomb lattice.

Exercise 21.3. Using Eq. (21.34) show the following result:

$$h_{l_i r_i}^{ss'} \delta_{r_i l_{i+1}} h_{l_{i+1} r_{i+1}}^{ss'} = -tc_{i+1,\uparrow}^{\dagger} c_{i,\uparrow} - tc_{i+1,\downarrow}^{\dagger} c_{i,\downarrow}$$

$$-tc_{i,\uparrow}^{\dagger} c_{i+1,\uparrow} - tc_{i,\downarrow}^{\dagger} c_{i+1,\downarrow} + U n_{i,\uparrow} n_{i,\downarrow}.$$

$$(21.35)$$

Here the subscripts ss' on the operators are suppressed.

The Hamiltonian at each site consists of several terms one- and two-particle operators. One can rewrite the Hamiltonian matrix elements of Eq. (21.32) in a more compact form. We write the matrix elements as a sum of contributions from different operators

$$H_{s_1' \cdots s_k'}^{s_1 \cdots s_k} = \langle s_1, \ldots, s_k | H | s_1', \ldots, s_k' \rangle = \sum_I W[1]^{I, s_1 s_1'} \ldots W[k]^{I, s_k s_k'},$$

$$(21.36)$$

where index I labels a term in the Hamiltonian. For simplicity, we consider only electronic Hamiltonians where $W[i]^{s_i s_i'}$ does **not** depend on the indices α_{i-1}' and α_i, i.e., they are scalars: $W[i]^{s_i s_i'}$ is $\langle s_i | c_{i\sigma}^{\dagger} | s_i' \rangle$, $\langle s_i | c_{i\sigma}^{\dagger} c_{i\sigma} | s_i' \rangle$, etc. This holds also for two-particle operators. The expectation value of H, i.e., the total energy, is then given by a sum over I (see Ref. [6])

$$\langle \Psi | H | \Psi \rangle = \sum_I \sum_{s_i, s_i'} \sum_{\alpha_{i-1}', \alpha_i} \left(\sum_{\alpha_{i-1}} A_{\alpha_{i-1}\alpha_i}^{*s_i} L_{\alpha_{i-1}\alpha_{i-1}'}^{I, i-1} \right) W[i]^{I, s_i, s_i'}$$

$$\times \left(\sum_{\alpha_i'} R_{\alpha_i \alpha_i'}^{I, i+1} A_{\alpha_{i-1}'\alpha_i'}^{s_i'} \right).$$

$$(21.37)$$

This result is easier to remember in the diagrammatic representation, see Fig. 21.14.

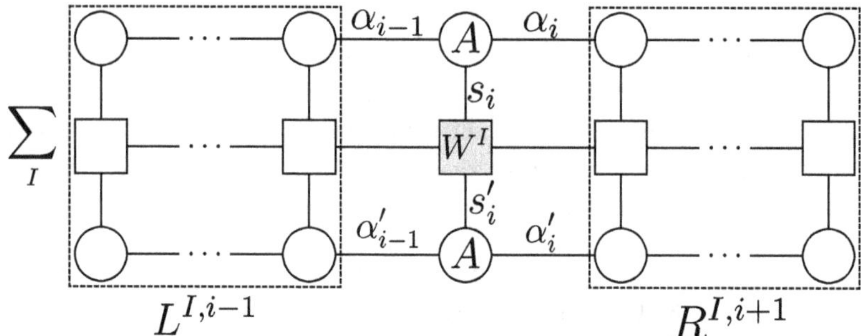

Fig. 21.14. Diagrammatic representation of the energy expectation value $\langle\Psi|H|\Psi\rangle$. As explained in the text, the horizontal legs attached to the squares are not needed for the operators considered here, see Eq. (21.36).

21.5. How to Determine Matrix Product States Numerically?

The next obvious question is: how does one actually determine these matrices? One may apply the SVD. But this may generate a bond dimension that is too large [5]. Instead one may use a variational technique. One starts from an Ansatz MPS of the desired bond dimension and variationally optimize the matrices. Here we briefly explain this variational method.

In the numerical work, for simplicity, each matrix is assumed to have the same dimension. (In this case, the number of matrices is considerably reduced in a periodic system with unit cells.) One starts with a MPS and then variationally changes the matrix elements by minimizing the total energy

$$E_{\text{tot}} = \frac{\langle\Psi|H|\Psi\rangle}{\langle\Psi|\Psi\rangle} = \frac{\sum_{\{s_i\},\{s_i'\}}(\Psi_{\cdots s_i'\cdots})^* H_{\cdots s_i\cdots}^{\cdots s_i'\cdots}(\Psi_{\cdots s_i\cdots})}{\sum_{\{s_i\}}(\Psi_{\cdots s_i\cdots})^*(\Psi_{\cdots s_i\cdots})}. \tag{21.38}$$

It is useful to employ the Lagrange multiplier method with the constraint $\langle\Psi|\Psi\rangle = 1$. The functional to be minimized is

$$L = \langle\Psi|H|\Psi\rangle - \lambda(\langle\Psi|\Psi\rangle - 1), \tag{21.39}$$

where λ is a Lagrange multiplier.

This result may be converted into a form that is more suitable for numerical computation. One picks a site i and varies the matrix elements of A^i while keeping all other sites unchanged. (Initially the matrix A^i may be chosen randomly.) The functional L must be stationary with respect to this variation. Differentiating it with respect to $A^{*s_i}_{\alpha_{i-1}\alpha_i}$ we find the matrix equation

$$\sum_I \sum_{s_i'} \sum_{\alpha_{i-1}'\alpha_i'} (L^{I,i-1}_{\alpha_{i-1}\alpha_{i-1}'} W[i]^{I s_i s_i'} R^{I,i+1}_{\alpha_i \alpha_i'}) A^{s_i'}_{\alpha_{i-1}'\alpha_i'}$$

$$= \lambda \sum_{s_i',\alpha_{i-1}',\alpha_i'} L^{I,i-1}_{\alpha_{i-1}\alpha_{i-1}'} R^{I,i+1}_{\alpha_i\alpha_i'} A^{s_i}_{\alpha_{i-1}'\alpha_i'}$$

$$= \lambda \sum_{s_i',\alpha_{i-1}',\alpha_i'} L^{I,i-1}_{\alpha_{i-1}\alpha_{i-1}'} \delta_{s_i s_i'} R^{I,i+1}_{\alpha_i\alpha_i'} A^{s_i'}_{\alpha_{i-1}'\alpha_i'} \qquad (21.40)$$

Defining the quantity in the bracket as H, this may be rewritten in a more familiar matrix eigenvalue equation

$$\sum_{s_i'\alpha_{i-1}'\alpha_i'} H_{s_i,\alpha_{i-1}\alpha_i;s_i',\alpha_{i-1}'\alpha_i'} A^{s_i'}_{\alpha_{i-1}'\alpha_i'} = \lambda A^{s_i}_{\alpha_{i-1}\alpha_i}. \qquad (21.41)$$

(From the overlap $\langle\Psi|\Psi\rangle = 1$, given in Eq. (21.27), one can derive the right-hand side.) The right-hand side of this equation is graphically depicted in Fig. 21.15. One solves this matrix equation and then updates the values of site i with the new values. After this, one picks another site and optimizes again. One sweeps through all sites serval times and updates the relevant matrices at each site until the energy converges. The whole process is somewhat similar to the Hartree–Fock process. (The Hartree–Fock Hamiltonian matrix is also variationally determined until self-consistency is achieved.)

Further reading. Gapped one-dimensional **boson** systems can be described by matrix product states, and one can show that they do not have topological order. A one-dimensional fermion system can be mapped into a one-dimensional boson system via the Jordan-Wigner transformation. One-dimensional systems with a hard gap thus do not have topological order (for a review on this subject matter, see Ref. [7].)

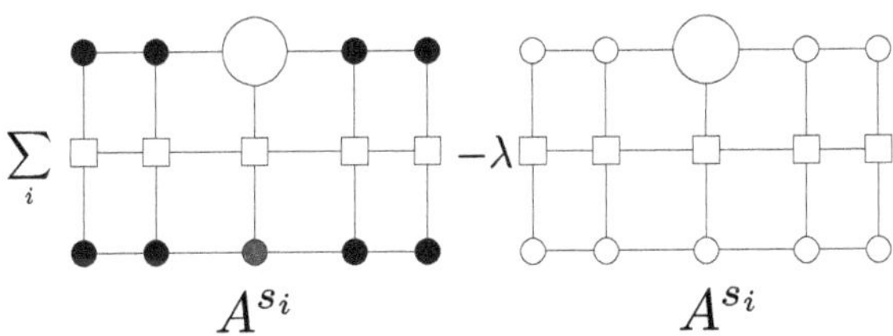

Fig. 21.15. Graphic illustration of the right-hand side of Eq. (21.41). Compare this figure with Fig. 21.14.

21.6. Comparison with Hartree–Fock Result

In graphene systems, mean-field approximations are widely used because they usually give accurate results. This means that quantum fluctuations not included in the Hartree–Fock approximation are small. Thus the bond dimension needs not be large. Required bond dimension depends strongly on the entanglement and correlation of the ground state. In fact, one keeps track of the truncation errors and accuracy of the relevant quantities, such as the variance of the norm of wave function, energy variance and maximal singular value in the truncation and concludes that the largest bond dimension needs not be large to capture the physics. Also, it is widely known that the Hubbard model of the graphene nanoribbon without disorder exhibits the zigzag magnetic order that is not largely entangled, and thus the bond dimension needs not be large. The obtained results thus agree with the previous theoretical works. Regarding the gapless or gapped nature of the ground state, a scaling analysis of the correlation length as a function of the bond dimension clearly shows that it does not diverge but converge.

There are several nearly degenerate HF ground states. One of them is the soliton state displaying fractional charges. Whether this state is the true ground state needs to be examined. One can perform a matrix product version of the DMRG to test the validity of the ground state properties of zigzag ribbons obtained within the HF approximation [10]. One uses the snake pattern of the matrix product state defined on the zigzag nanoribbon lattice, as shown in Fig. 21.16. (The actual numerical implementation of the matrix product version of the DMRG is rather sophisticated. For technical details of this calculation, see Refs. [5, 6, 10].) In such a calculation

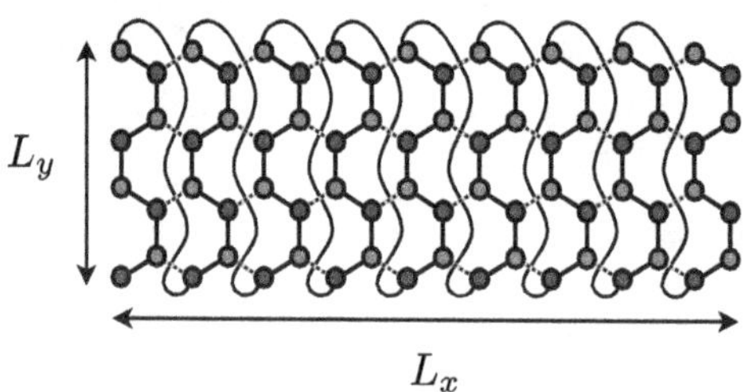

Fig. 21.16.　Schematic figure of the snake pattern of the matrix product state defined on the zigzag nanoribbon lattice (L_x and L_y are equal to the length and width of the ribbon, respectively). Curved lines connecting carbon atoms do not indicate a bond between carbon atoms. Instead they show ordering of MPS sites ($s_1, \ldots, s_k$), see Eq. (21.20). Reprinted with permission from Ref. [10].

for a ribbon, first, one tests the results for the disorder-free zigzag ribbon, which is in the symmetry protected phase. The two approaches give consistent results. The results for a disordered interacting zigzag nanoribbon half-filling are also in agreement: site spin values computed in the two approaches agree qualitatively, see Figs. 21.17(b) and 21.17(c). Note that the HF ground state of a doped disorder-free ribbon displays an **edge spin density wave** [11] (see Fig. 21.18), in contrast to the uniform spin density of undoped ribbons (this result is obtained using the paramagnetic initial state with a small spin splitting, see Eq. (18.3)). The DMRG gives a similar result, see Ref. [10].

21.7. New Disordered Anyon Phase

Let us add electrons to the top of the undoped interacting disordered state of a zigzag ribbon. In the low doped regime, the added electrons divide into fractional charges. As more electrons are added, the fractional charges form a new disordered anyon phase [10].[b]

In the new disordered anyon phase, the sharp peak in the DOS at the midgap energy, present in the weak disorder region (see Fig. 19.7), *disappears* with increasing doping level, but the two side peaks near $E \sim \pm 0.05\Delta/2$ persist, as shown in Fig. 21.19(d). Simultaneously, the edge occupation number and site spin profiles become highly nonuniform, as shown in Figs. 21.19(a) and 21.19(b). They look rather different in comparison to those with less electrons, see Figs. 19.8. In addition, we observe that local magnetic moments extended over several sites with nonzero values of s_{iz} are present. Many of the added electrons with energy near the midgap have q_A values $\sim 1/2$ (q_A is defined in Sec. 18.3). However, not all of them are fractionalized, as the probability density of a gap-edge state shown in the left inset of Fig. 21.19(d) demonstrates. This means that charge fractionalization is *not* exact.

The nature of the disordered ground state of the doped system is as follows. First, as we already mentioned, the ground state of doped disorder-free zigzag ribbons is different from that of undoped ribbons: the density matrix renormalization group (see the last chapter) displays **spin density** type modulations on the zigzag edges, see Ref. [10] (the magnitude

---

[b]When the density of quasiparticles reaches a certain value in a Laughlin state the quasiparticles also form a correlated state. The resulting state is a hierarchical fractional quantum Hall state.

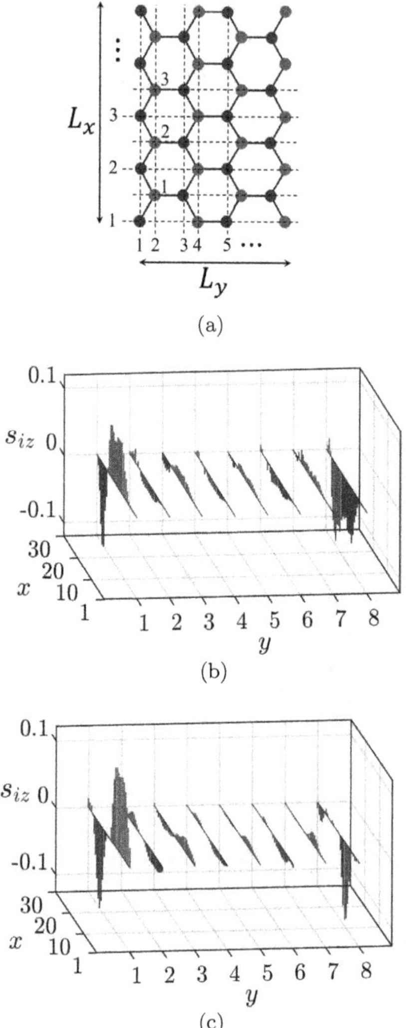

(a)

(b)

(c)

Fig. 21.17. (a) Vertical and horizontal lines of carbon atoms are numbered. All lengths and widths in this paper are measured in the number of these lines. (b) DMRG result of the ground state site spins s_{iz} at half-filling for $U = t$, $n_{\mathrm{imp}} = 1$, and $\Gamma = 0.5t$ ($g = \frac{\Gamma\sqrt{n_{\mathrm{imp}}}}{U} = 0.5$). We found that other spin components s_{ix} and s_{iy} are very small. (c) Hartree–Fock site spin values at half-filling are shown. Here, $U = t$, $n_{\mathrm{imp}} = 0.1$, and $\Gamma = 0.5t$ ($g = 0.16$ is smaller compared with the value used in (b)). Reprinted with permission from Ref. [10].

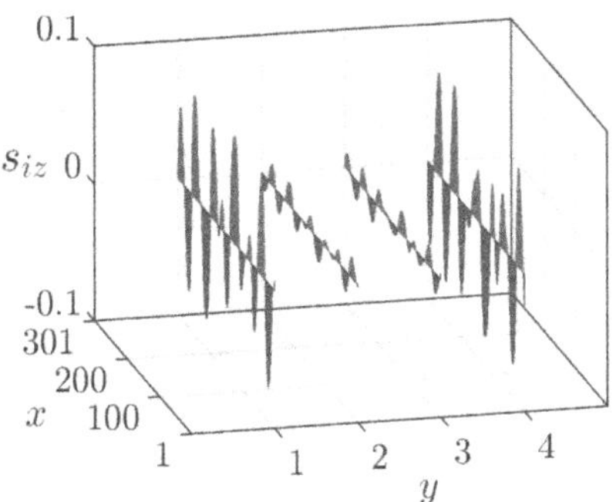

Fig. 21.18. Site spin values s_{iz} of a disorder-free zigzag ribbon. The HF results for $N_e = N_s + 20(\delta N/N_s = 0.017)$, $L_x = 301$, $L_y = 4$, and $U = t$. Reprinted with permission from Ref. [10].

of the modulation increases with increasing doping; the opposite edges are still antiferromagnetically coupled). In the presence of disorder, the ground state changes again and the periodic spin density is destroyed. Instead modulated edge ferromagnetism transforms into a highly distorted spin density wave. This is because a disorder potential is a singular perturbation, see Sec. 18.1. The DMRG results, which include quantum fluctuations, are in qualitative agreement with those of the solitonic Hartree–Fock state (note several local minimum Hartree–Fock states).

As the vertical dashed lines in Figs. 21.19(b) and (c) show values of s_{iz} for each zigzag edge change sign where n_i abruptly changes. Moreover, sites i, where the values of $n_{i\sigma}$ and s_{iz} abruptly change, have almost identical values for the x-coordinate on the left and right zigzag edges, see the vertical dashed lines (some peaks may have different x-coordinates, see Fig. 18.15). Thus the left and right edges are non-locally connected. These objects proliferate compared with the undoped ribbon. There are also objects extended over several sites with rather small values of $s_i = \frac{1}{2}(n_{i\uparrow} - n_{i\downarrow}) \approx 0$. In such an object, spin–charge separation would take place. Doped zigzag ribbons with fractional charges and spin-charge separated objects may display unusual transport, magnetic, and inter-edge tunneling properties [12, 13].

21.8. Phase Diagram of Interacting Disordered Zigzag Ribbons

What is the nature of the ground state in the presence of strong disorder [14] and/or large doping [10]? A systematic investigation of this problem

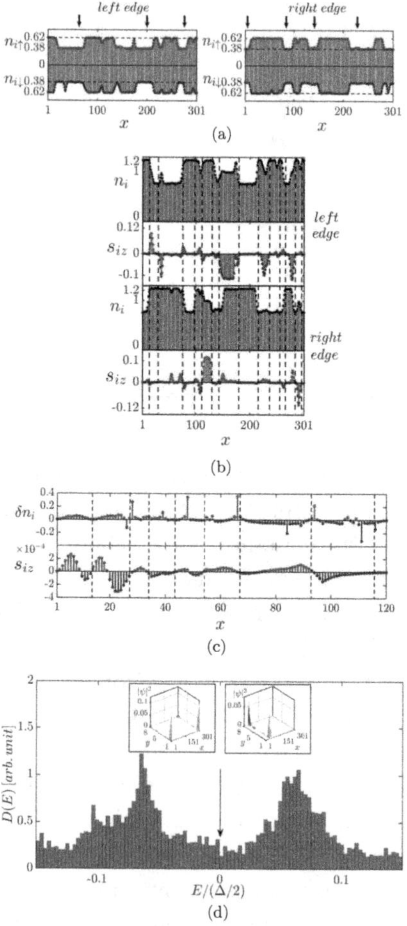

Fig. 21.19. (a) Site occupation numbers $n_{i\sigma}$ of a disordered ribbon. The HF results for $N_e = N_s + 20$ ($\delta N/N_s = 0.0083$), $\Gamma = 0.06t$, $L_x = 301$, $L_y = 8$, $n_{\mathrm{imp}} = 0.1$, and $U = t$ ($g = 0.019$). Arrows give some examples of sites where spin–charge separation is present. (b) Vertical lines indicate sites where the Hartree–Fock values of n_i or s_{iz} abruptly change. Numerous localized magnetic moments are present on the zigzag edges. (c) DMRG results for $\delta n_i \equiv n_i - n_i^{\mathrm{clean}}$ plotted as a function of x and y. Here, n_i^{clean} is the site occupation number for $\Gamma = 0$. Site spins s_{iz} are also shown. The parameters are as follows: $N_e = N_s + 12$ ($\delta N/N_s = 0.025$), $\Gamma = t$, $L_x = 120$, $L_y = 4$, $n_{\mathrm{imp}} = 0.2$, and $U = t$. Note that the length of this ribbon is considerably shorter than the one used in (a) and (b). These results are for a more strongly disordered ribbon with $g = 0.45$, and the overall magnetization is considerably weakened. (d) Hartree–Fock DOS with two side peaks. The profile of these two peaks becomes symmetric in the limit of large ribbon length or, equivalently, in the limit of zero doping. Since there are excess electrons, the Fermi energy $E_F/(\Delta/2) = 0.46$ (half-filling corresponds to $E = 0$). The number of disorder realization is $N_D \sim 200$. Insets show probability densities of two states with adjacent energies. Reprinted with permission from Ref. [10].

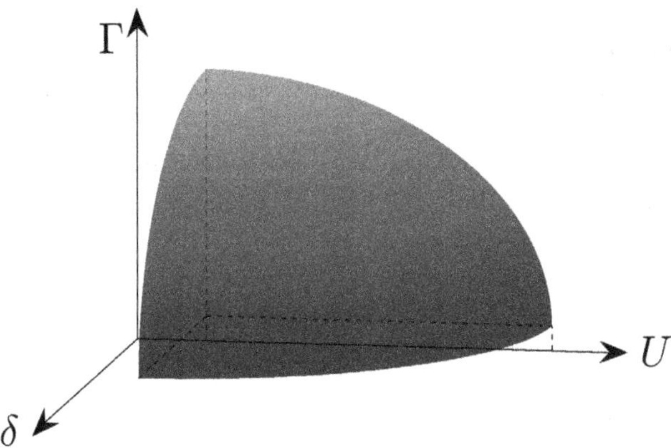

Fig. 21.20. Schematic phase diagram obtained using the Hartree–Fock approximation. Shaded region indicates the topologically ordered region. Note the manifold lies above the $(\Gamma = 0, U, \delta)$ plane.

may provide more insight into the phase change from the modulated ferro-magnetic edges at zero doping to distorted spin-wave edges at finite doping. Edge antiferromagnetism is expected to vanish and thereby also topological order. But this may happen in a non-trivial way. The topological order may not be immediately destroyed upon doping because electron localization partially suppresses quantum fluctuations. The system may thus still be an insulator with fractional charges. The following questions need to be addressed for large Γ and/or δ where the topological order is expected to disappear (δ is doping level): what are the properties of the TEE and what are the physical properties of the ground state?

We have computed the phase diagram of topologically ordered zigzag nanoribbons in the parameter space of (Γ, U, δ), see Fig. 21.20. Our numerical calculation in the strong disorder and/or large doping regimes suggests that the value of the topological entanglement entropy is non-universal. Note that the phase boundary is likely to be not sharp because the crossover region is expected to be substantial. Our results suggest that, away from the low doping and weak disorder regimes, the ground state may not display magnetic domain walls and spin–charge separation (these effects both require correlation between two fractionalized electrons with opposite spins from different zigzag edges, as we explained in Sec. 19.3). Instead single $e^-/2$ fractional charges may exist on the zigzag edges. We will define such a phase to exhibit *quasi-topological order*. Note that the system is not in the topologically ordered phase in the two planes $(\Gamma = 0, U, \delta)$ and $(\Gamma, U = 0, \delta)$. In addition, for small U and large Γ it is difficult to compute accurately the TEE. Strong disorder affects site occupation numbers substantially and they display large disorder-induced fluctuations. Thus the computed TEE shows large variances. Very large ribbons are needed to compute the TEE accurately (finite-size scaling analysis may be applied). A detailed account of the above findings will be reported elsewhere [16].

Further reading. What is the fractional spin value of an anyon? The spin-statistics relation suggests a value of $s = \theta/2\pi$, where θ is the statistical angle. This result is correct on a sphere. However, there is no consensus on whether it also applies on the plane [17, 18]. Moreover, the role of the intrinsic spin is unclear.

Bibliography

[1] S. M. Girvin, "The Quantum Hall Effect: Novel Excitations and Broken Symmetries," in *Les Houches Lecture Notes: Topological Aspects of Low Dimensional Systems*, edited by A. Comtet, T. Joliceur, S. Ouvry, and F. David (Springer Verlag, Berlin and Les Editions de Physique, Les Ulis, 2000). •

[2] K. G. Wilson, Problems in physics with many scales of length, *Sci. American* **241**, 2, 158 (1979). •

[3] S. R. White, Density-matrix algorithms for quantum renormalization groups, *Phys. Rev. B* **48**, 10345 (1993).

[4] S. Rommer and S. Östlund, Class of Ansatz wave functions for one-dimensional spin systems and their relation to the density matrix renormalization group, *Phys. Rev. B* **55**, 2164 (1997).

[5] U. Schollwöck, The density-matrix renormalization group: a short introduction, *Phil. Trans. R. Soc. A* **369**, 2643 (2011).

[6] N. Nakatani, "Matrix Product States and Density Matrix Renormalization Group Algorithm," in *Reference Module in Chemistry, Molecular Sciences and Chemical Engineering* (Elsevier, Oxford, 2018).

[7] B. Zeng, X. Chen, D.-L. Zhou, and X.-G. Wen, *Quantum Information Meets Quantum Matter: From Quantum Entanglement to Topological Phases of Many-Body Systems* (Springer, New York, 2019).

[8] X. Chen, Z.-C. Gu, Z.-X. Liu, and X.-G. Wen, Symmetry protected topological orders and the group cohomology of their symmetry group, *Phys. Rev. B* **87**, 155114 (2013).

[9] S. M. Girvin and K. Yang, *Modern Condensed Matter Physics* (Cambridge University Press, Cambridge, 2019). •

[10] Y. H. Kim, H. J. Lee, Hyun-Yong Lee, and S.-R. Eric Yang, New disordered anyon phase of doped graphene zigzag nanoribbon, *Sci. Rep.* **12**, 14551 (2022).

[11] S. Brown and G. Grüner, Charge and Spin Density Waves, *Sci. American* **270**, 4, 50 (1994). •

[12] T.-C. Chung, F. Moraes, J. D. Flood, and A. J. Heeger, Topologically ordered zigzag nanoribbon: Solitons at high density in trans– $(CH)_x$: Collective transport by mobile, spinless charged solitons, *Phys. Rev. B* **29**, 2341 (1984). •

[13] W. Kang, H. L. Stormer, L. N. Pfeiffer, K. W. Baldwin, and K. W. West, Tunnelling between the edges of two lateral quantum Hall systems, *Nature* **403**, 59 (2000).; I. Yang, W. Kang, K. W. Baldwin, L. N. Pfeiffer, and K. W. West, Cascade of Quantum Phase Transitions in Tunnel-Coupled Edge States, *Phys. Rev. Lett.* **92**, 056802 (2004).

[14] Y. H. Jeong, S.-R. Eric Yang, and M. C. Cha, Soliton fractional charge of disordered graphene nanoribbon, *J. Phys.: Condens. Matter* **31**, 265601 (2019).

[15] K. Wakabayashi, K.-I. Sasaki, T. Nakanishi, and T. Enoki, Electronic states of graphene nanoribbons and analytical solutions, *Sci. Technol. Adv. Mater.* **11**, 054504 (2010).

[16] Lê Hoàng Anh, Young Heon Kim, In Hwan Lee, and S.-R. Eric Yang, work in progress.

[17] T. Einarsson, S. L. Sondhi, S. M. Girvin and D. P. Arovas, Fractional spin for quantum Hall effect quasiparticles, *Nucl. Phys. B* **441**, 515 (1995).

[18] D. H. Lee and X.-G. Wen, Orbital spins of the collective excitations in Hall liquids, *Phys. Rev. B* **49**, 11066 (1994).

Appendix A

Envelope Wave Functions

"How often have I said to you that when you have eliminated the impossible, whatever remains, however improbable, must be the truth?"

Sherlock Holmes

Let us consider an electron moving in an external potential in addition to the periodic potential of a crystalline solid. Under certain conditions, one can derive a simple effective Hamiltonian that describes a renormalized electron moving in the external potential but in the absence of the periodic potential. The renormalized electron resembles the bare electron except that it has a new effective mass and not old bare electron mass.

Let us explain the basis idea of the envelop wave function of the renormalized electron. The electron Hamiltonian is in the presence of an external potential $V_{\text{ext}}(\vec{r})$ is

$$H = -\frac{-\hbar^2}{2m}\nabla^2 + V(\vec{r}) + V_{\text{ext}}(\vec{r}), \tag{A.1}$$

where $V(\vec{r})$ is a periodic potential and m is the electron bare mass. Consider a band that has an **extremum** at a wave vector $\vec{K}$. We are seeking approximate solutions of type

$$\psi(\vec{r}) = F(\vec{r})\frac{1}{\sqrt{N}}\sum_n e^{i\vec{K}\cdot\vec{R}}a(\vec{r}-\vec{R}_n) = F(\vec{r})\psi_{\vec{K}}(\vec{r}), \tag{A.2}$$

where $\psi_{\vec{K}}(\vec{r})$ is the Bloch wave function at a reciprocal wave vector $\vec{K}$ and N is the number of atoms. The atomic positions are labeled by $\vec{R}_n$.

How can we find the envelope wave function $F(\vec{r})$? We will show that, for a slowly varying $V_{\text{ext}}(\vec{r})$, this wave function satisfies a relatively simple equation. In the **absence** of an external potential the envelope wave function $F(\vec{r})$ satisfies

$$-\frac{-\hbar^2}{2m^*}\nabla^2 F_{\vec{q}}(\vec{r}) = [E - E_{\vec{K}}]F_{\vec{q}}(\vec{r}), \tag{A.3}$$

where $E_{\vec{K}}$ is the band energy at $\vec{K}$. Here it is assumed that the band structure near $\vec{K}$ is parabolic and the bare electron mass is replaced by the effective electron mass m^*.[a] The solution is $F(\vec{r}) = e^{i\vec{q}\cdot\vec{r}}$ with the eigenenergy

$$E = E_{\vec{K}} + \frac{\hbar^2 q^2}{2m^*}. \tag{A.4}$$

The solution can be written as the Bloch wave function $\psi_{\vec{K}}(\vec{r})$ multiply by this envelope function

$$\psi_{\vec{K}+\vec{q}}(\vec{r}) = e^{i\vec{q}\cdot\vec{r}}\sum_n e^{i\vec{K}\cdot\vec{R}}a(\vec{r} - \vec{R}_n) = e^{i\vec{q}\cdot\vec{r}}\psi_{\vec{K}}(\vec{r}). \tag{A.5}$$

Note that $\vec{k} = \vec{q} + \vec{K}$ so $\vec{q}$ denotes how much the wave vector deviates from $\vec{K}$.

In the **presence** of an external potential, $F(\vec{r})$ satisfies

$$-\frac{-\hbar^2}{2m^*}\nabla^2 F_{\vec{q}}(\vec{r}) + V_{\text{ext}}(\vec{r})F(\vec{r}) = [E - E_{\vec{K}}]F_{\vec{q}}(\vec{r}). \tag{A.6}$$

If $V_{\text{ext}}(\vec{r})$ is small and slowly varying then $F(\vec{r})$ will be also slowly varying but somewhat perturbed from $e^{i\vec{q}\cdot\vec{r}}$. Solutions of Eq. (A.1) have the form

$$\psi(\vec{r}) \approx F(\vec{r})\psi_{\vec{K}}(\vec{r}). \tag{A.7}$$

---

[a]The effect of the periodic potential is thus absorbed in m^*; for an explanation of this approach, look up a textbook on solid state physics.

Appendix B

Matrix Diagonalization

"..., whenever I mentioned the Lagrangian, a Nobel certified condensed matter theorist would pretend that he had never heard of it. He would ask rhetorically, "What do I need the Lagrangian for? Give me the Hamiltonian and I will try to find you its eigenvalues and eigenfunctions." Bad attitude."

A. Zee

Consider a unitary transformation that diagonalizes a Hamiltonian matrix H

$$H' = U^\dagger H U. \tag{B.1}$$

Let $\{\vec{e}_\alpha\}$ be a basis set before the transformation

$$\vec{e}_\alpha = \begin{bmatrix} 0 \\ \vdots \\ 1 \\ \vdots \\ 0 \end{bmatrix}, \tag{B.2}$$

where only the αth component is 1. Let the transformed basis vectors, i.e., eigenvectors of H, be

$$\vec{v}_\alpha = U\vec{e}_\alpha. \tag{B.3}$$

According to linear algebra, the transformation matrix is

$$U = (\vec{v}_1, \dots, \vec{v}_n) = \begin{pmatrix} v_{11} & \cdots & v_{n1} \\ \vdots & \ddots & \vdots \\ v_{1n} & \cdots & v_{nn} \end{pmatrix}. \tag{B.4}$$

Here we have used the notation

$$\vec{v}_\alpha = \begin{bmatrix} v_{\alpha 1} \\ \vdots \\ v_{\alpha n} \end{bmatrix}. \tag{B.5}$$

Note $U^\dagger$ stores $\{\vec{v}_\alpha^{\,\dagger}\}$ in rowwise

$$U^\dagger = \begin{pmatrix} v_{11}^* & \cdots & v_{1n}^* \\ \vdots & \vdots & \vdots \\ v_{n1}^* & \vdots & v_{nn}^* \end{pmatrix}. \tag{B.6}$$

($\vec{v}_\alpha$ is a column vector and $\vec{v}_\alpha^{\,\dagger}$ is a row vector.) Note that the unitarity condition of U is equivalent to orthonormality of the eigenvectors

$$(U^\dagger U)_{\alpha\beta} = \vec{v}_\alpha^{\,\dagger} \cdot \vec{v}_\beta = \delta_{\alpha\beta}. \tag{B.7}$$

The eigenvectors $\{\vec{v}_\alpha\}$ constitute the transformed basis set. It is related to the old basis set as follows:

$$\vec{v}_\alpha = U \vec{e}_\alpha. \tag{B.8}$$

The vector components are

$$v_{\alpha i} = \sum_j U_{ij} e_{\alpha j}. \tag{B.9}$$

Appendix C

Density Matrix

"They think that intelligence is about noticing things that are relevant (detecting patterns); in a complex world, intelligence consists in ignoring things that are irrelevant (avoiding false patterns)."

Nassim Nicholas Taleb

Density matrix is a widely used in physics. One reason is that a physical system is often entangled with its environment and this means that its quantum-mechanical state is no longer pure. If we do not know which quantum state the system is in, how can we describe the system? The expectation value of a physical quantity of such a system can be conveniently expressed by the reduced density matrix. It can, for example, be used to find exact ground states of one-dimensional interacting systems. Even if the total system is in a pure state, the various subsystems that make it up will typically be in mixed states, see Sec. 20.1. The density matrix of a subsystem is a very useful concept in describing topological insulators because it can be used to quantify the topological entanglement entropy (in Sec.20.3). Moreover, the eigenvalue spectrum of the reduced density matrix can also be used to discern topologically ordered phases.

A well-known example of the use of density matrix is from statistical mechanics. In statistical mechanics, the density matrix of a system of particles is given by the Boltzmann factor

$$\rho = \sum_k w_k |k\rangle\langle k| = \sum_k \frac{e^{-\beta E_k}}{Z} |k\rangle\langle k|, \tag{C.1}$$

where Z is the partition function, $\{|k\rangle\}$ is a complete set of quantum states, E_k is the energy of $|k\rangle$, and $\beta = 1/k_B T$ is the inverse temperature (k_B is

517

the Boltzmann constant). In this density matrix, a fraction particles with relative population w_1 is described by $|k_1\rangle$, some other fraction with relative population w_2, by $|k_2\rangle$, and etc.

The general density matrix of a mixed state is

$$\rho = \sum_k f_k |k\rangle\langle k|, \tag{C.2}$$

where f_k is the probability to find the system in a state $|k\rangle$. States can be either many-body or single particle states (including Hartree–Fock single particle states). For a pure state, the density matrix reduces to

$$\rho = |i\rangle\langle i|. \tag{C.3}$$

The one-particle density matrix in site representation is

$$\rho(r, r') = \langle r|i\rangle\langle i|r'\rangle = \phi_i(r)\phi_i^*(r'). \tag{C.4}$$

One can show that a maximally mixed state has the lowest possible value of purity $\mathrm{tr}\rho^2$ while a pure state has $\mathrm{tr}\rho^2 = 1$.

The expectation value of a physical quantity of a system is

$$\langle O\rangle = \mathrm{Tr}\{\rho O\} = \sum_i \langle i|\rho O|i\rangle = \sum_{ij} \langle i|\rho|j\rangle\langle j|O|i\rangle. \tag{C.5}$$

For a pure state

$$\langle O\rangle = \mathrm{Tr}\{\rho O\} = \sum_i \langle i|k\rangle\langle k|O|i\rangle = \langle k|O|k\rangle, \tag{C.6}$$

and for a mixed state

$$\langle O\rangle = \sum_k f_k \langle k|O|k\rangle. \tag{C.7}$$

Time evolution of the density matrix satisfies the von Neumann equation

$$\frac{\partial\rho}{\partial t} = \frac{\partial}{\partial t}(|\psi\rangle\langle\psi|) = \left(\frac{\partial}{\partial t}|\psi\rangle\right)\langle\psi| + |\psi\rangle\frac{\partial}{\partial t}\langle\psi|$$

$$= -\frac{i}{\hbar}H|\psi\rangle\langle\psi| + \frac{i}{\hbar}|\psi\rangle\langle\psi|H = -\frac{i}{\hbar}[H, \rho]. \tag{C.8}$$

Note the minus sign in front of the commutator. A density operator is a positive semi-definite, **Hermitian** operator of trace 1.

It is instructive to consider how a mixed state may appear from a pure state when the interaction with the environment is taken into account. Consider the following spin state

$$|\psi\rangle = a|\uparrow\rangle + b|\downarrow\rangle. \tag{C.9}$$

Its density operator is

$$\rho = |\psi\rangle\langle\psi| = aa^*|\uparrow\rangle\langle\uparrow| + ba^*|\downarrow\rangle\langle\uparrow| + ab^*|\uparrow\rangle\langle\downarrow| + bb^*|\downarrow\rangle\langle\downarrow|. \tag{C.10}$$

It can be written as a matrix of a pure state

$$\rho = \begin{pmatrix} aa^* & ab^* \\ ba^* & bb^* \end{pmatrix}. \tag{C.11}$$

The off-diagonal elements are the interference terms. Inserting $a = |a|e^{i\phi_1}$ and $b = |b|e^{i\phi_2}$, we can write this density matrix as

$$\rho = \begin{pmatrix} |a|^2 & |a||b|e^{i\phi} \\ |a||b|e^{-i\phi} & |b|^2 \end{pmatrix}, \tag{C.12}$$

where $\phi = \phi_1 - \phi_2$. The phases ϕ of spins of an ensemble will fluctuate due to the interaction with the environment, i.e., decoherence, and the average value of the phase factor will average out to zero, $\langle e^{i\phi}\rangle = 0$. The resulting density matrix is in a mixed form

$$\rho = \begin{pmatrix} |a|^2 & 0 \\ 0 & |b|^2 \end{pmatrix}. \tag{C.13}$$

Further reading. For a good introduction of density matrix formalism, see R. P. Feynman, *Statistical Mechanics: A Set of Lectures* (W.A. Benjamin, San Francisco, 1972). Another good book is by K. Blum, *Density Matrix Theory and Applications* (Springer, Berlin, 2012).

Some physicists believe that decoherence processes solve the quantum measurement problem. However, there is no universal agreement on this. Decoherence will not completely eliminate the interference terms. Even a very small interference term gives rise to conceptual problems: "Little pregnant is still pregnant". Read more about the quantum measurement problem. A good introduction is *"Beyond Weird: Why Everything You Thought You Knew About Quantum Physics is Different"* by P. Ball (The University of Chicago Press, Chicago, 2018).

Appendix D

Unitary Transformation of Operators

"To climb hard routes, it's better to try every conceivable option... It'll save you time in the long run."

Ethan Pringle

Let us consider the expectation value an operator X

$$\langle A|X|A\rangle = \langle A|U^\dagger U X U^\dagger U|A\rangle = \langle A'|X'|A'\rangle \tag{D.1}$$

where

$$|A'\rangle = U|A\rangle \tag{D.2}$$

and

$$X' = UXU^\dagger \tag{D.3}$$

The last equation shows how an operator X transforms under a unitary transformation U.

This transformation should not be confused with how a matrix transforms under a unitary transformation. Matrix elements transform differently. Consider a state $|\alpha\rangle$ under a unitary transformation

$$|\alpha\rangle \to U|\alpha\rangle. \tag{D.4}$$

Matrix element transform as

$$\langle\alpha|X|\beta\rangle \to \langle U\alpha|X|U\beta\rangle = \langle\alpha|U^\dagger XU|\beta\rangle \tag{D.5}$$

Thus a matrix transforms as $U^\dagger XU$. See Fig. D.1 for a more detailed explanation.

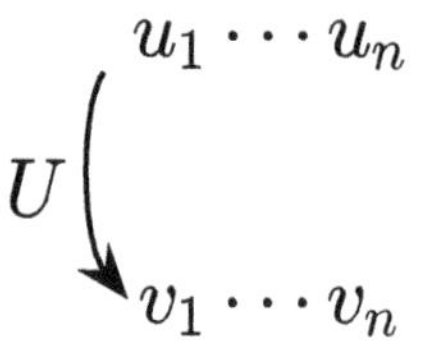

$$u_1 \cdots u_n$$

$$U \Big\downarrow$$

$$v_1 \cdots v_n$$

$$X'(\{\vec{v}_\alpha\}) = U^\dagger X(\{\vec{u}_\alpha\})U$$

Fig. D.1. The old (new) basis states are $\{\vec{u}_\alpha\}$ ($\{\vec{v}_\alpha\}$). The new transformed matrix is related to the previous matrix as follows $X'(\{\vec{v}_\alpha\}) = U^\dagger X(\{\vec{u}_\alpha\})U$.

Let us first explain an active transformation. Suppose you rotate a two-dimensional vector v through an angle θ about z-axis in **counterclockwise** direction while the coordinate system remains fixed. The new coordinates of the vector is $R_z(\theta)v$, where the rotation matrix is

$$R_z(\theta) = \begin{pmatrix} \cos\theta & -\sin\theta \\ \sin\theta & \cos\theta \end{pmatrix}. \tag{D.6}$$

In quantum mechanics, such a rotation is facilitated by the operator $R = e^{-\frac{i}{\hbar}\vec{\theta}\cdot\vec{L}}$, where $\vec{L}$ is the angular momentum operator and $\vec{\theta}$ is the rotation angle vector. Translation of a system is facilitated by the operator

$$T = e^{-\frac{i\vec{r}\cdot\vec{p}}{\hbar}}. \tag{D.7}$$

Note the minus sign in the exponent.

What is a passive transformation? If the coordinate system is rotated counterclockwise, the relevant matrix is $R_z(-\theta)$ and not $R_z(\theta)$. Suppose you rotate the coordinate system an angle θ in **clockwise** direction while v remains fixed. Then the coordinates of the vector in the new coordinate system is $R_z(\theta)v$. So $R_z(\theta)$ describes both active and passive transformations, one in counterclockwise and the other in clockwise directions. Translation of a coordinate system is facilitated by the operator

$$T = e^{+\frac{i\vec{r}\cdot\vec{p}}{\hbar}}. \tag{D.8}$$

and rotation is facilitated by $R = e^{+\frac{i}{\hbar}\vec{\theta}\cdot\vec{L}}$. Note the plus sign in the exponent.

Appendix E

Commutation Relations

"Physics made me sick the whole time I learned it. What I couldn't stand was this shrinking everything into ... hideous, cramped, scorpion–lettered formulas."

Sylvia Plath[b]

- General Commutation Relations

The following commutation relations hold both for boson and fermion operators. Commutation relations of products of operators are

$$[AB, CD] = A[B, CD] + [A, CD]B \qquad (E.1)$$

and

$$[A, BC] = [A, B]_+ C - B[A, C]_+. \qquad (E.2)$$

The Hausdorff formulas are

$$e^{\lambda A} B e^{-\lambda A} = B + \lambda[A, B] + \frac{\lambda^2}{2}[A, [A, B]] + \cdots \qquad (E.3)$$

and

$$e^A e^B e^{-A} = e^{e^A B e^{-A}}. \qquad (E.4)$$

If A and B do not commute

$$e^{\epsilon A} e^{\epsilon B} \approx e^{\epsilon(A+B) + \frac{\epsilon^2}{2}[A,B]}, \qquad (E.5)$$

---

[b]Sylvia Plath was a tragic American poet.

where ϵ is very small. If two operators A and B commute with their commutator, i.e., if $[A, [A, B]] = [B, [A, B]] = 0$, then

$$e^A e^B = e^{[A,B]} e^B e^A = e^{A+B} e^{\frac{1}{2}[A,B]},$$

$$[A, f(B)] = [A, B] f'(B). \tag{E.6}$$

• Useful Results for Boson Operators

When b and $b^\dagger$ are boson operators and α, β, and γ are constants one can show

$$[b, e^{\alpha b^\dagger}] = \alpha e^{\alpha b^+},$$

$$[b, (b^\dagger)^n] = n(b^\dagger)^{n-1}. \tag{E.7}$$

Bose coherent states $|\alpha\rangle$ satisfy

$$b e^{\alpha b^\dagger} |0\rangle = b|\alpha\rangle = [b, e^{\alpha b^\dagger}]|0\rangle = \alpha|\alpha\rangle. \tag{E.8}$$

The Baker–Campbell–Hausdorff formula is

$$e^{\alpha b^\dagger + \gamma b} = e^{\alpha b^\dagger} e^{\gamma b + \alpha\gamma/2} = e^{\gamma b} e^{\alpha b^\dagger - \alpha\gamma/2} \tag{E.9}$$

and

$$\langle e^{\alpha b^\dagger + \gamma b} \rangle = e^{\alpha\gamma(\langle b^\dagger b\rangle + 1/2)}. \tag{E.10}$$

If A and B are linear combinations of b and b^+ then

$$\langle e^A \rangle = e^{\langle A^2\rangle/2},$$

$$\langle e^A e^B \rangle = e^{\langle (A^2+B^2)/2 + AB\rangle}. \tag{E.11}$$

Further reading. Proofs of some of the relations given in this section can be found in Y. Kuramoto, *Quantum Many-Body Physics: A Perspective on Strong Correlations* (Springer, Tokyo, 2020). See also Z. F. Ezawa, *Quantum Hall Effects: Recent Theoretical and Experimental Developments* (World Scientific, Singapore, 2013), and Wikipedia under Baker–Campbell–Hausdorff formula.

Appendix F

Conductivity and Correlation Function

"Wheeler's First Moral Principle: Never make a calculation until you know the answer. Make an estimate before every calculation, try a simple physical argument (symmetry! invariance! conservation!) before every derivation, guess the answer to every paradox and puzzle. Courage: No one else needs to know what the guess is. Therefore make it quickly, by instinct. A right guess reinforces this instinct. A wrong guess brings the refreshment of surprise. In either case life as a spacetime expert, however long, is more fun!"

John Wheeler

- Diagonal Conductivity

Consider a time-dependent electric field along the x-axis $E_x(x,t) = E_0 e^{i(qx-\omega t)}$. The expectation value of the current density is

$$\langle j_x \rangle = \sum_{kk'} (j_x)_{kk'} \rho_{k'k}, \tag{F.1}$$

where $\rho_{k'k}$ is the density matrix (see Sec. 5.2). The x-component of the current density operator is

$$j_x(x) = \frac{-e}{V} \frac{p_x}{m}, \tag{F.2}$$

where p_x is the x-component momentum operator, m is the particle mass, and V is the system volume. Its Fourier transform is

$$j_x(q) = \frac{-e}{V} e^{-iqx} \frac{p_x}{m}. \tag{F.3}$$

525

The matrix element is

$$(j_x(q))_{kk'} = -\frac{e}{mV}\langle k|e^{-iqx}p_x|k'\rangle. \tag{F.4}$$

Since the electric field is related to the vector potential via $E_x(x,t) = -\frac{1}{c}\frac{\partial A_x(x,t)}{\partial t}$ the perturbing term in the Hamiltonian is

$$\frac{e}{mc}A_x(x,t)p_x \rightarrow V(x,\omega) = -\frac{ie}{m\omega}E_x(x,t)p_x = -\frac{ie}{m\omega}E_x(t)e^{iqx}p_x. \tag{F.5}$$

This result leads to

$$\langle k'|V|k\rangle = -E_0\frac{ie}{m\omega}\langle k'|e^{iqx}p_x|k\rangle. \tag{F.6}$$

The current is related to the density matrix as

$$\langle j\rangle = \sum_{kk'}j_{kk'}\rho_{k'k}(\omega). \tag{F.7}$$

We then use the result for the density matrix given in Eq. (6.144) and the expression for the current given in Eq.(F.4). This leads to the following result for the diagonal conductivity

$$\sigma(q,\omega) = \frac{ie^2}{m^2V\omega}\sum_{kk'}\left[\frac{f(\epsilon_{k'}) - f(\epsilon_k)}{\epsilon_{k'} - \epsilon_k - \hbar(\omega + i\eta)}\right]|\langle k'|e^{iqx}p_x|k\rangle|^2. \tag{F.8}$$

(We used the current density to be proportional to the conductivity, $j = \sigma E$.)

- Off-Diagonal Conductivity and Kubo formula

The off-diagonal current–current correlation function $\langle J_x(0)J_y(t)\rangle$ can be related to the off-diagonal conductivity. In the presence of an electric field, the perturbation in the Hamiltonian can be written in terms of the current density $j_y(t)$ and vector potential $A_y(t)$

$$V(t) = -\frac{1}{c}j_y(t)A_y(t), \tag{F.9}$$

where the vector potential is related to the electric field

$$A_y = \frac{cE_y}{i\omega}e^{-i\omega t} \tag{F.10}$$

with the amplitude of the electric field E_y. Note that, in this way, translation invariance can be preserved. Here the wavelength of photon is much larger than the system size, i.e., we are in the limit where the wave vector $q \to 0$. In the interaction picture, time evolution of a state vector is

$$|\psi(t)\rangle = U(t, t_0)|\psi(t_0)\rangle, \tag{F.11}$$

where the time evolution operator is

$$U(t, t_0) = T e^{-\frac{i}{\hbar} \int_{t_0}^{t} V(t')dt'} \approx 1 - \frac{i}{\hbar} \int_{t_0}^{t} V(t')dt'. \tag{F.12}$$

The expectation value of the x-component of the current density is

$$\langle j_x(t)\rangle = \langle \psi_0(t)|j_x(t)|\psi_0(t)\rangle = \langle \psi_0|U^{-1}(t, t_0)j_x(t)U(t, t_0)|\psi_0\rangle$$

$$= \langle \psi_0| \left[1 + \frac{i}{\hbar} \int_{t_0}^{t} dt' V(t') \right] j_x(t) \left[1 - \frac{i}{\hbar} \int_{t_0}^{t} dt' V(t') \right] |\psi_0\rangle, \tag{F.13}$$

where ψ_0 is the ground state. Since the perturbation $V \sim j_y$ (from Eq.(F.9)) we can rewrite this as

$$\langle j_x(t)\rangle \approx \langle \psi_0|j_x(t)|\psi_0\rangle + \frac{E_y}{\hbar \omega} \int_{t_0}^{t} dt' e^{-i\omega t'} \langle \psi_0|[j_y(t'), j_x(t)]|\psi_0\rangle, \tag{F.14}$$

We let $t_0 = -\infty$, $E = \hbar \omega$, and $t'' = t - t'$, which gives

$$\langle j_x\rangle = \frac{E_y}{\omega} \left[\int_0^{\infty} dt'' e^{i\omega t''} \langle \psi_0|[j_x(0), j_y(t'')]|\psi_0\rangle \right]. \tag{F.15}$$

Since $j_x = \sigma_{xy} E_y$ we find that the energy-dependent zero temperature conductivity is

$$\sigma_{xy}(\omega) = \frac{1}{\omega} \int_0^{\infty} dt e^{i\omega t} \langle \psi_0|[j_y(0), j_x(t)]|\psi_0\rangle. \tag{F.16}$$

Time dependence of the current density operator is (we are using the interaction picture)

$$j(t) = e^{iH_0 t/\hbar} j(0) e^{-iH_0 t/\hbar}, \tag{F.17}$$

where H_0 is the non-interacting Hamiltonian. Plugging this into the expression for the conductivity we find

$$\sigma_{xy}(\omega) = \frac{1}{\omega} \int_0^{\infty} dt e^{i\omega t} \sum_n [\langle \psi_0|j_y|n\rangle \langle \psi_n|j_x|\psi_0\rangle e^{i(E_n - E_0)t/\hbar}$$

$$- \langle \psi_0|j_x|n\rangle \langle \psi_n|j_y|\psi_0\rangle e^{i(E_0 - E_n)t/\hbar}], \tag{F.18}$$

where E_n is the energy of ψ_n. Integration leads to

$$\sigma_{xy}(\omega) = \frac{i}{\omega} \sum_{n \neq 0} \left[\frac{\langle\psi_0|j_x|\psi_n\rangle\langle\psi_n|j_y|\psi_0\rangle}{\hbar\omega + i\eta + [E_0 - E_n]} - \frac{\langle\psi_0|j_y|\psi_n\rangle\langle\psi_n|j_y|\psi_0\rangle}{\hbar\omega + i\eta + [E_n - E_0]} \right].$$

$$(\text{F}.19)$$

Here we used $\hbar\omega \to \hbar\omega + i\eta$ ($\eta \to 0^+$) to make the integration over t convergent.

- Correlation Function

According to fluctuation–dissipation, theorem dissipation is related to fluctuations. The theorem connects the conductivity to the current–current correlation function. This result is a general result in linear response theory. The correlation function is defined as

$$\chi(t) = \theta(t)\langle\psi_0|[A(t), B(0)]|\psi_0\rangle, \qquad (\text{F}.20)$$

where $|\psi_0\rangle$ is the ground state and A and B are operators representing some physical quantities (in the following we will use $A(0) = A$ and $B(0) = B$).

Consider the time evolution operator

$$U(t, t') = Te^{-\frac{i}{\hbar}\int_{t'}^{t} dt H(t')} \qquad (\text{F}.21)$$

where the Hamiltonian contains a perturbation $-\delta(t)B$. Acting the time evolution operator on the initial state $|\psi_0\rangle$, we find that the perturbed wave function in the limit $t = 0^+$

$$|\psi(0^+)\rangle \approx |\psi_0\rangle + \frac{i}{\hbar} \sum_{m \neq 0} |\psi_m\rangle\langle\psi_m|B|\psi_0\rangle. \qquad (\text{F}.22)$$

This gives the following time-dependent wave function

$$|\psi(t)\rangle = e^{-i\epsilon_0 t/\hbar}|\psi_0\rangle + \frac{i}{\hbar} \sum_{m \neq 0} e^{-i\epsilon_m t/\hbar}|\psi_m\rangle\langle\psi_m|B|\psi_0\rangle, \qquad (\text{F}.23)$$

where $|\psi_m\rangle$ and ϵ_m are the eigenstates and eigenenergies of the system. Using this wave function we find

$$\theta(t)\langle\psi(t)|A|\psi(t)\rangle = \frac{i}{\hbar}\theta(t) \sum_{m \neq 0} [e^{-i\omega_{0m}t}\langle\psi_0|A|\psi_m\rangle\langle\psi_m|B|\psi_0\rangle$$

$$-e^{i\omega_{0m}t}\langle\psi_0|B(0)|\psi_m\rangle\langle\psi_m|A|\psi_0\rangle]. \qquad (\text{F}.24)$$

Here we used the following the identity:

$$I = \sum_m |\psi_m\rangle\langle\psi_m|. \tag{F.25}$$

It is easy to show that the result of Eq. (F.24) is equal to the correlation function

$$\chi(t) = \frac{i}{\hbar}\theta(t)\left[\langle\psi_0|A(t)B(0)|\psi_0\rangle - \langle\psi_0|B(0)A(t)|\psi_0\rangle\right]$$

$$= \frac{i}{\hbar}\theta(t)\langle\psi_0|[A(t),B(0)]|\psi_0\rangle. \tag{F.26}$$

Using the identity

$$\int_{-\infty}^{\infty} dt\, e^{i\omega}\theta(t) = P\frac{1}{\omega} - \pi\delta(\omega) = \frac{1}{\omega + i\eta}, \tag{F.27}$$

we compute the Fourier transform of the correlation function

$$\chi(\epsilon) = -i\int_{-\infty}^{\infty} dt\, e^{i\epsilon t/\hbar}\theta(t)\langle\psi_0|[B(t),A(0)]|\psi_0\rangle. \tag{F.28}$$

This gives the desired result for the spectral density of the correlation function

$$\chi(\epsilon) = \sum_{n\neq 0}\left[\frac{\langle\psi_0|B|\psi_n\rangle\langle\psi_n|A|\psi_0\rangle}{\epsilon + i\eta - [\epsilon_n - \epsilon_0]} - \frac{\langle\psi_0|A|\psi_n\rangle\langle\psi_n|B|\psi_0\rangle}{\epsilon + i\eta + [\epsilon_n - \epsilon_0]}\right]. \tag{F.29}$$

Appendix G

Functional Derivatives and Integrals

"...every discovery contains an irrational element,
or a creative intuition."

Karl R. Popper

"...there are no objective standards
by which to establish truth. Anything goes."

Paul Feyerabend

- **Functional Derivatives**

In condensed matter physics, one often has to minimize the Ginzburg–Landau free-energy functional or action of a field theory. The resulting equation of motion describes the elementary modes of the system. In this regard, functional differentiation serves as a useful tool. The functional differentiation is defined as

$$\frac{\delta F}{\delta \eta(\vec{r})} = \lim_{\epsilon \to 0} \frac{F[\eta(\vec{r}\,') + \epsilon \delta(\vec{r}\,' - \vec{r}\,)] - F[\eta(\vec{r}\,')]}{\epsilon}. \tag{G.1}$$

One of the best examples of the use of functional differentiation is the Euler–Lagrange equation of classical mechanics. The action is

$$F[\eta] = \int d\vec{r}\,' f(\eta(\vec{r}\,'), \nabla \eta(\vec{r}\,')). \tag{G.2}$$

Setting its functional differentiation to zero gives the Euler–Lagrange equation

$$\frac{\delta F}{\partial \eta(\vec{r}\,)} = \frac{\partial f}{\partial \eta(\vec{r}\,)} - \nabla \cdot \frac{\partial f}{\partial \nabla \eta(\vec{r}\,)} = 0. \tag{G.3}$$

531

The following rules may be easily derived:

$$\frac{\delta}{\delta\eta(\vec{r}\,)}\eta(\vec{r}\,') = \delta(\vec{r} - \vec{r}\,'), \tag{G.4}$$

$$\frac{\delta}{\delta\eta(\vec{r}\,)}\int d\vec{r}\,'\eta(\vec{r}\,') = 1, \tag{G.5}$$

$$\frac{\delta}{\delta\eta(\vec{r}\,)}\int d\vec{r}\,'\frac{1}{2}(\nabla\eta(\vec{r}\,'))^2 = -\nabla^2\eta(\vec{r}\,), \tag{G.6}$$

$$\frac{\delta f(g(\eta(\vec{r}\,')))}{\delta\eta(\vec{r}\,')} = f'g'\delta(\vec{r} - \vec{r}\,'), \tag{G.7}$$

$$\frac{\delta e^{\int d\vec{r}\,'\eta(\vec{r}\,')g(\vec{r}\,')}}{\delta\eta(\vec{r}\,)} = g(\vec{r})e^{\int d\vec{r}\,'\eta(\vec{r}\,')g(\vec{r}\,')}. \tag{G.8}$$

> **Exercise G.1.** Derive the result of Eq. (G.4) using the definition of the functional derivative (G.1). Answer:
>
> $$\frac{\delta F}{\delta\eta(\vec{r}\,)} = \frac{\eta(\vec{r}\,') + \epsilon\delta(\vec{r} - \vec{r}\,') - \eta(\vec{r}\,')}{\epsilon} = \delta(\vec{r} - \vec{r}\,'). \tag{G.9}$$

> **Exercise G.2.** Derive the result of Eq. (G.3) using the standard method of classical mechanics. Then derive it using functional derivative. Hint: Use
>
> $$\frac{\delta F}{\partial\eta(\vec{r}\,)} = \int d\vec{r}\,'\left(\frac{\delta f}{\partial\eta(\vec{r}\,')}\delta(\vec{r} - \vec{r}\,') + \frac{\delta f}{\partial\nabla\eta(\vec{r}\,')}\cdot\nabla\delta(\vec{r} - \vec{r}\,')\right). \tag{G.10}$$
>
> Then integrate the second term by partial integration.

> **Exercise G.3.** Derive the result of Eq. (G.6). Hint: Use $\frac{\delta(\nabla\eta(\vec{r}\,'))^2}{\delta\eta(\vec{r}\,)} = 2\nabla\eta(\vec{r}\,')\frac{\delta\nabla\eta(\vec{r}\,')}{\delta\eta(\vec{r}\,)}$ and $\frac{\delta\nabla\eta(\vec{r}\,')}{\delta\eta(\vec{r}\,)} = \frac{\nabla\delta\eta(\vec{r}\,')}{\delta\eta(\vec{r}\,)}$ followed by integration by parts.

- Functional Integrals

In condensed matter physics, it is often useful to integrate out certain degrees of freedom that are not directly relevant to the problem at hand. The resulting Hamiltonian can look different from the original Hamiltonian.

The new Hamiltonian is called the low-energy effective Hamiltonian. The functional integration technique may be used in this procedure.

The following result of functional integration is very useful

$$\int D[A]e^{-\frac{1}{2}A^T K A + f^T A} = \left(\frac{(2\pi)^N}{\det K}\right)^{1/2} e^{\frac{1}{2}f^T K^{-1}f}. \tag{G.11}$$

A corollary of this result is

$$\int D[A]e^{\frac{i}{2}A^T K A + if^T A} = \left(\frac{(2\pi i)^N}{\det K}\right)^{1/2} e^{-\frac{i}{2}f^T K^{-1}f}. \tag{G.12}$$

Here K is a real symmetric matrix, f and A are column vectors, f^T and A^T are row vectors, and $D[A] = dA_1 \cdots dA_N$. The dimension of the vectors is N.

To derive these results, we first prove a simpler result (then it should be easy to prove the main result). Consider the following one-dimensional integral over A:

$$I = \int_{-\infty}^{\infty} dA e^{-\frac{1}{2}AKA + fA}. \tag{G.13}$$

We rewrite the exponent and find

$$I = \int_{-\infty}^{\infty} dA e^{-\frac{K}{2}(A - f/K)^2 + \frac{f^2}{2K}}. \tag{G.14}$$

Changing the integration variable to $B = A - f/K$ gives

$$e^{\frac{f^2}{2K}} \int_{-\infty}^{\infty} dB e^{-\frac{KB^2}{2}} = \sqrt{\frac{2\pi}{K}} e^{\frac{1}{2}f\frac{1}{K}f}. \tag{G.15}$$

Exercise G.4. Prove the functional integral given in Eq. (G.11). Hint: Use the following results. (1) The matrix K may be diagonalized using an orthogonal transformation O: $K = O^{-1}DO$, where D is the diagonal matrix. Note that $O^T O = I$ and $(Of)^T = f^T O^T = f^T O^{-1}$. (2) Transform $B = OA$, i.e., $A = O^{-1}B$. (3) $\int dA_1 \cdots dA_N = \int dB_1 \cdots dB_N$ since the Jacobian is one. (4) Use the result given in Eq. (G.15).

Further reading. For derivations of functional differentiation and integrals, see, for example, N. Goldenfeld, *Lectures on Phase Transitions and the Renormalization Group* (CRC Press, Boca Raton, 1992).

Appendix H

Noether's Theorem

"Guessing before proving! Need I remind you that it is so that all important discoveries have been made?"

Henri Poincaré

Let us show Noether's theorem which states that every continuous symmetry of a system is to be associated with a conserved quantity. Consider the variation of the field $\Psi \to \Psi + \delta\Psi$ near the extremum field. The resulting change in the action is zero

$$
\delta S = \int dt dr \left(\frac{\delta L}{\delta \Psi} \delta\Psi + \sum_i \delta\partial_i\Psi \frac{\delta L}{\delta\partial_i\Psi} \right)
$$

$$
= \int dt dr \left(\frac{\delta L}{\delta\Psi} - \sum_i \partial_i \left[\frac{\delta L}{\delta\partial_i\Psi} \right] \right) \delta\Psi
$$

$$
+ \int dt dr \left(\sum_i \partial_i \left[\frac{\delta L}{\delta\partial_i\Psi} \delta\Psi \right] \right) = 0, \tag{H.1}
$$

where the L is Ginzburg–Landau functional density ($S = \int dt dr L$) and the summation index i is t or r. The first term vanishes (it is the equation of motion). From the second term, we have

$$
\sum_i \partial_i \frac{\delta L}{\delta\partial_i\Psi} \delta\Psi = 0. \tag{H.2}
$$

There is also variation with respect to $\delta\Psi^*$, which gives a similar equation

$$\sum_i \partial_i \frac{\delta L}{\delta \partial_i \Psi^*} \delta\Psi^* = 0. \tag{H.3}$$

So

$$\sum_i \partial_i \frac{\delta L}{\delta \partial_i \Psi} \delta\Psi + \sum_i \partial_i \frac{\delta L}{\delta \partial_i \Psi^*} \delta\Psi^* = 0. \tag{H.4}$$

This is equivalent to the continuity equation

$$\sum_i \partial_i j_i = 0, \tag{H.5}$$

where the generalized current is defined as

$$j_i = \frac{\delta L}{\delta \partial_i \Psi} \delta\Psi + \frac{\delta L}{\delta \partial_i \Psi^*} \delta\Psi^* = \partial^i \Psi^* \delta\Psi + \partial^i \Psi \delta\Psi^*. \tag{H.6}$$

Let us apply the continuity equation and find out what quantities are conserved under global gauge transformation. Since the field transforms as $\Psi' = e^{-i\alpha}\Psi$ under global U(1) transformation, the change in the field for small values of α is

$$\delta\Psi = -i\alpha\Psi \quad \text{and} \quad \delta\Psi^* = i\alpha\Psi^*. \tag{H.7}$$

From this we see that the generalized current is

$$j_i = i\alpha[\Psi^* \partial_i \Psi - \Psi \partial_i \Psi^*]. \tag{H.8}$$

So the total probability is conserved. (Here $\partial_t = \frac{\partial}{\partial t}$, $\partial_x = \frac{\partial}{\partial x}$, etc.)

Let us apply Noether's theorem to the following action based on the time-dependent Ginzberg–Landau energy functional

$$S = \int dt dr \left[\left| \frac{\partial \Psi(r)}{\partial t} \right|^2 - \alpha |\Psi(r)|^2 - \beta |\Psi(r)|^4 - \gamma^2 |\nabla\Psi(r)|^2 \right]. \tag{H.9}$$

The current density is

$$j_i = \frac{\delta L}{\delta \partial_i \Psi} \delta\Psi + \frac{\delta L}{\delta \partial_i \Psi^*} \delta\Psi^*. \tag{H.10}$$

Choosing the constant α judiciously, we find that the density and current density are

$$j_0 = \rho = i \left[\frac{\partial \Psi(r)^*}{\partial t} \Psi - \Psi^* \frac{\partial \Psi(r)}{\partial t} \right],$$

$$\vec{j} = i\gamma^2 [\nabla \Psi^* \Psi - \Psi^* \nabla \Psi]. \tag{H.11}$$

This is the conservation law for superfluid density.

Further reading. For a pedagogical presentation of Noether's theorem, see J. Schwichtenberg, *Physics from Symmetry* (Springer, New York, 2015).

Index

CPSIA information can be obtained
at www.ICGtesting.com
Printed in the USA
JSHW010938190423
40468JS00002B/5